Rice: Germplasm, Genetics and Improvement

Rice: Germplasm, Genetics and Improvement

Editor:

Valeria Severino, Ph.D.

Delve Publishing
www.delvepublishing.com

Rice: Germplasm, Genetics and Improvement

Editor: Valeria Severino, Ph.D.

708 3rd Avenue, 6th Floor
New York, NY 10017
United States of America
www.delvepublishing.com

ISBN: 978-1-68095-765-5

Library of Congress Control Number: 2016951521

About the Editor

Valeria Severino, Ph.D.

Valeria Severino obtained her PhD in Biological Processes and Biomolecules from the Second University of Naples in 2009. Her scientific interests concern the structural and functional characterization of proteins and peptides by using biochemical approaches. Fields of expertise comprises food chemistry, cellular biology, natural products and related applications. She published more than 40 papers on international peer-reviewed journals.

Contents

	List of Contributors	*xi*
	Preface	*xxiii*
	Introduction	*xxv*
Chapter 1	**Genome-Wide Association Mapping in Rice**	**1**
1.	Results	2
2.	Discussion	8
3.	Methods	9
	Acknowledgments	12
	References	12
Chapter 2	**Genome-Wide Resequencing of Krice_core For Breeding**	**15**
1.	Background	15
2.	Results	16
3.	Discussion	20
4.	Methods	21
	Conclusion	22
	Acknowledgements	22
	Authors' Contributions	22
	Reference	22
Chapter 3	**Genome-Wide Association Study for Traits Related to Plant and Grain Morphology, and Root Architecture in Temperate Rice Accessions**	**25**
1.	Introduction	25
2.	Materials and Methods	26
3.	Results and Discussion	29
	Conclusions	35
	Supporting Information	35
	Acknowledgments	39
	Author Contributions	39
	References	39
Chapter 4	**Variability Assessment of Aromatic and Fine Rice Germplasm**	**45**
1.	Introduction	45
2.	Materials and Methods	46
3.	Result	48
4.	Discussion	52
	Conclusion	53
	Acknowledgments	53
	References	53
Chapter 5	**Variation in Cooking and Eating Quality Traits in Japanese Rice Germplasm Accessions**	**57**
1.	Introduction	57
2.	Materials and Methods	58
3.	Results	60
4.	Discussion	63
	Acknowledgments	65

References 65

Chapter 6 **Association of SNP Haplotypes of Hkt Family Genes With Salt Tolerance in Indian Wild Rice Germplasm 69**

1. Background 69
2. Results 70

Conclusions 75

Methods 76

Acknowledgements 77

Authors' Contributions 77

References 77

Chapter 7 **Indica Rice Genome Assembly, Annotation and Mining of Blast Disease Resistance Genes 81**

1. Background 81
2. Methods 81
3. Results and Discussion 83

Conclusions 87

Acknowledgements 88

Authors' Contributions 88

References 88

Chapter 8 **Genome-Wide Association Study of Grain Appearance and Milling Quality in a Worldwide Collection of Indica Rice Germplasm 91**

1. Introduction 91
2. Materials and Methods 92
3. Results 94
4. Discussion 99

Conclusions 101

Supporting Information 102

Acknowledgments 104

Author Contributions 105

References 105

Chapter 9 **Complete Chloroplast Genome Sequence and Annotation of the Tropical Japonica Group of Asian Cultivated Rice (Oryza Sativa L.) 109**

1. Genome Announcement 109

References 110

Chapter 10 **Genome-Wide Dissection of the Microrna Expression Profile in Rice Embryo During Early Stages of Seed Germination 111**

1. Introduction 111
2. Materials and Methods 112
3. Results 113
4. Discussion 118

Acknowledgments 120

Author Contributions 120

References 120

Chapter 11 **Identification of a Candidate Gene For Panicle Length in Rice (Oryza Sativa L.) Via Association and Linkage Analysis 123**

1. Introduction 123
2. Materials and Methods 124

3. Results ... 125
4. Discussion ... 129
Author Contributions ... 130
Conflict of Interest Statement ... 130
References ... 130

Chapter 12 Population Structure, Diversity and Trait Association Analysis in Rice (Oryza Sativa L.) Germplasm for Early Seedling Vigor (ESV) Using Trait Linked SSR Markers ... 135
1. Introduction ... 135
2. Materials and Methods ... 136
3. Results ... 138
4. Discussion ... 142
Conclusion ... 144
Author Contributions ... 144
References ... 145

Chapter 13 The Rice Minute-Like1 (RML1) Gene, Encoding A Ribosomal Large Subunit Protein L3b, Regulates Leaf Morphology and Plant Architecture in Rice ... 147
1. Introduction ... 147
2. Materials and Methods ... 148
3. Results ... 149
4. Discussion ... 153
Acknowledgments ... 155
References ... 155

Chapter 14 The Rice TCM5 Gene Encoding a Novel Deg Protease Protein is Essential For Chloroplast Development Under High Temperatures ... 159
1. Background ... 159
2. Results ... 160
3. Discussion ... 163
Conclusions ... 164
Methods ... 164
Acknowledgments ... 166
Authors' Contributions ... 166
References ... 166

Chapter 15 Metabolic Prediction of Important Agronomic Traits In Hybrid Rice (Oryza Sativa L.) ... 169
1. Introduction ... 169
2. Results ... 170
3. Discussion ... 172
4. Methods ... 173
Acknowledgments ... 174
References ... 175

Chapter 16 SLG Controls Grain Size and Leaf Angle by Modulating Brassinosteroid Homeostasis in Rice ... 177
1. Introduction ... 177
2. Materials and Methods ... 178
3. Results ... 179
4. Discussion ... 185
Acknowledgements ... 186
References ... 186

Chapter 17 **Overcoming Inter-Subspecific Hybrid Sterility in Rice by Developing Indica-Compatible Japonica Lines** **189**

1. Introduction 189
2. Results 190
3. Discussion 192
4. Methods 193

References 193

Index **197**

List of Contributors

Susan R. McCouch
School of Integrative Plant Sciences, Plant Breeding and Genetics section, Cornell University, Ithaca, 14850 New York, USA
Department of Biological Statistics and Computational Biology, Cornell University, Ithaca, 14850 New York, USA

Mark H. Wright
School of Integrative Plant Sciences, Plant Breeding and Genetics section, Cornell University, Ithaca, 14850 New York, USA

Chih-Wei Tung
School of Integrative Plant Sciences, Plant Breeding and Genetics section, Cornell University, Ithaca, 14850 New York, USA

Lyza G. Maron
School of Integrative Plant Sciences, Plant Breeding and Genetics section, Cornell University, Ithaca, 14850 New York, USA

Kenneth L. McNally
International Rice Research Institute, 1301 Metro Manila, Philippines

Melissa Fitzgerald
International Rice Research Institute, 1301 Metro Manila, Philippines
School of Food Science, University of Queensland, St Lucia, 4072 Queensland, Australia

Namrata Singh
School of Integrative Plant Sciences, Plant Breeding and Genetics section, Cornell University, Ithaca, 14850 New York, USA

Genevieve DeClerck
School of Integrative Plant Sciences, Plant Breeding and Genetics section, Cornell University, Ithaca, 14850 New York, USA

Francisco Agosto-Perez
School of Integrative Plant Sciences, Plant Breeding and Genetics section, Cornell University, Ithaca, 14850 New York, USA
Department of Biological Statistics and Computational Biology, Cornell University, Ithaca, 14850 New York, USA

Pavel Korniliev
School of Integrative Plant Sciences, Plant Breeding and Genetics section, Cornell University, Ithaca, 14850 New York, USA
Department of Biological Statistics and Computational Biology, Cornell University, Ithaca, 14850 New York, USA

Anthony J. Greenberg
School of Integrative Plant Sciences, Plant Breeding and Genetics section, Cornell University, Ithaca, 14850 New York, USA
Department of Biological Statistics and Computational Biology, Cornell University, Ithaca, 14850 New York, USA

Ma. Elizabeth B. Naredo
International Rice Research Institute, 1301 Metro Manila, Philippines

Sheila Mae Q. Mercado
International Rice Research Institute, 1301 Metro Manila, Philippines

Sandra E. Harrington
School of Integrative Plant Sciences, Plant Breeding and Genetics section, Cornell University, Ithaca, 14850 New York, USA

Yuxin Shi
School of Integrative Plant Sciences, Plant Breeding and Genetics section, Cornell University, Ithaca, 14850 New York, USA

Darcy A. Branchini
Transnational Learning Center, Cornell University, Ithaca, 14850 New York, USA

Paula R. Kuser-Falcão
School of Integrative Plant Sciences, Plant Breeding and Genetics section, Cornell University, Ithaca, 14850 New York, USA

Hei Leung
International Rice Research Institute, 1301 Metro Manila, Philippines

Kowaru Ebana
National Institute of Agrobiological Sciences, 2-1-2 Kannondai, Tsukuba, 305-8602 Ibaraki, Japan

Masahiro Yano
National Institute of Agrobiological Sciences, 2-1-2 Kannondai, Tsukuba, 305-8602 Ibaraki, Japan

Georgia Eizenga
USDA–ARS Dale Bumpers National Rice Research Center, 2890 Hwy. 130 E., Stuttgart, Arkansas 72160, USA

Anna McClung
USDA–ARS Dale Bumpers National Rice Research Center, 2890 Hwy. 130 E., Stuttgart, Arkansas 72160, USA

Jason Mezey
Department of Biological Statistics and Computational Biology, Cornell University, Ithaca, 14850 New York, USA

Tae-Sung Kim
Department of Plant Resources, College of Industrial Sciences, Kongju National University, Yesan 340-702, Republic of Korea
Legume Bio-Resource Center of Green Manure (LBRCGM), Kongju National University, Yesan 340-702, Republic of Korea

Qiang He
Department of Plant Resources, College of Industrial Sciences, Kongju National University, Yesan 340-702, Republic of Korea

Kyu-Won Kim
Department of Plant Resources, College of Industrial Sciences, Kongju National University, Yesan 340-702, Republic of Korea
Legume Bio-Resource Center of Green Manure (LBRCGM), Kongju National University, Yesan 340-702, Republic of Korea

Min-Young Yoon
Department of Plant Resources, College of Industrial Sciences, Kongju National University, Yesan 340-702, Republic of Korea

Won-Hee Ra
Department of Plant Resources, College of Industrial Sciences, Kongju National University, Yesan 340-702, Republic of Korea

Feng Peng Li
Department of Plant Resources, College of Industrial Sciences, Kongju National University, Yesan 340-702, Republic of Korea

Wei Tong
Department of Plant Resources, College of Industrial Sciences, Kongju National University, Yesan 340-702, Republic of Korea

Jie Yu
Department of Plant Resources, College of Industrial Sciences, Kongju National University, Yesan 340-702, Republic of Korea

Win Htet Oo
Department of Plant Resources, College of Industrial Sciences, Kongju National University, Yesan 340-702, Republic of Korea

Buung Choi
Department of Plant Resources, College of Industrial Sciences, Kongju National University, Yesan 340-702, Republic of Korea

Eun-Beom Heo
Department of Plant Resources, College of Industrial Sciences, Kongju National University, Yesan 340-702, Republic of Korea.

Byoung-Kook Yun
Department of Industrial & Systems Engineering, Kongju National University, Cheonan 330-717, Republic of Korea

Soon-Jae Kwon
Korea Atomic Energy Research Institute, Advanced Radiation Technology Institute, 29 Geumgu-gil, Jeongeup-si, Jeollabuk-do 580-185, Korea

University of Science and Technology, Radiation Biotechnology and Applied Radioisotope Science, 217 Gajungro Yuseong-gu, Daejeon 305-350, Korea

Soon-Wook Kwon
Department of Plant Bioscience, College of Natural Resources and Life Science, Pusan National University, Milyang 627-706, Republic of Korea

Yoo-Hyun Cho
Seedpia, 85, Maesil-ro, Kwonsun-ku, Suwon 441-882, Republic of Korea

Chang-Yong Lee
Department of Industrial & Systems Engineering, Kongju National University, Cheonan 330-717, Republic of Korea

Beom-Seok Park
The Agricultural Genome Center, National Acedemy of Agricultural Science, Rural Development Adiministration, Wanju 565-851, Republic of Korea

Yong-Jin Park
Department of Plant Resources, College of Industrial Sciences, Kongju National University, Yesan 340-702, Republic of Korea
Legume Bio-Resource Center of Green Manure (LBRCGM), Kongju National University, Yesan 340-702, Republic of Korea

Filippo Biscarini
PTP Science Park, 26900 Lodi, Italy

Paolo Cozzi
IBBA-CNR, Lodi, Italy

Laura Casella
PTP Science Park, 26900 Lodi, Italy

Paolo Riccardi
PTP Science Park, 26900 Lodi, Italy

Alessandra Vattari
PTP Science Park, 26900 Lodi, Italy

Gabriele Orasen
CREA-Council for Agricultural Research and Economics, Rice Research Unit, 13100 Vercelli, Italy

Rosaria Perrini
CREA-Council for Agricultural Research and Economics, Rice Research Unit, 13100 Vercelli, Italy

Gianni Tacconi
CREA-Council for Agricultural Research and Economics, Genomics Research Centre, 29017 Fiorenzuola d'Arda (Piacenza), Italy

Alessandro Tondelli
CREA-Council for Agricultural Research and Economics, Genomics Research Centre, 29017 Fiorenzuola d'Arda (Piacenza), Italy

Chiara Biselli
CREA-Council for Agricultural Research and Economics, Rice Research Unit, 13100 Vercelli, Italy

Luigi Cattivelli
CREA-Council for Agricultural Research and Economics, Genomics Research Centre, 29017 Fiorenzuola d'Arda (Piacenza), Italy

Jennifer Spindel
Department of Plant Breeding & Genetics, Cornell University, Ithaca, NY, United States of America

Susan McCouch
Department of Plant Breeding & Genetics, Cornell University, Ithaca, NY, United States of America

Pamela Abbruscato
PTP Science Park, 26900 Lodi, Italy

Giampiero Valé
CREA-Council for Agricultural Research and Economics, Rice Research Unit, 13100 Vercelli, Italy
CREA-Council for Agricultural Research and Economics, Genomics Research Centre, 29017 Fiorenzuola d'Arda (Piacenza), Italy

Pietro Piffanelli
PTP Science Park, 26900 Lodi, Italy

Raffaella Greco
PTP Science Park, 26900 Lodi, Italy

M. Z. Islam,
Genetic Resources and Seed Division, Bangladesh Rice Research Institute (BRRI), Gazipur 1701, Bangladesh

M. Khalequzzaman
Genetic Resources and Seed Division, Bangladesh Rice Research Institute (BRRI), Gazipur 1701, Bangladesh

M. K. Bashar
CIAT, HarvestPlus, Banani, Dhaka 1213, Bangladesh

N. A. Ivy
Bangabandhu Sheikh Mujibur Rahman Agricultural University (BSMRAU), Gazipur 1706, Bangladesh

M. M. Haque
Bangabandhu Sheikh Mujibur Rahman Agricultural University (BSMRAU), Gazipur 1706, Bangladesh

M. A. K. Mian
Bangabandhu Sheikh Mujibur Rahman Agricultural University (BSMRAU), Gazipur 1706, Bangladesh

Kiyosumi Hori
National Institute of Agrobiological Sciences

Keitaro Suzuki
NARO Institute of Crop Science

Ken Iijima
National Institute of Agrobiological Sciences

Kaworu Ebana
National Institute of Agrobiological Sciences

Mishra S
Jacob School of Biotechnology and Bioengineering, Sam Higginbottom Institute of Agriculture, Technology and Sciences, Allahabad, 211007, India
National Research Centre on Plant Biotechnology, Pusa Campus, New Delhi, 110012, India

Singh B
National Research Centre on Plant Biotechnology, Pusa Campus, New Delhi, 110012, India

Panda K
National Research Centre on Plant Biotechnology, Pusa Campus, New Delhi, 110012, India

Singh BP
National Research Centre on Plant Biotechnology, Pusa Campus, New Delhi, 110012, India

Singh N

National Research Centre on Plant Biotechnology, Pusa Campus, New Delhi, 110012, India

Misra P
Jacob School of Biotechnology and Bioengineering, Sam Higginbottom Institute of Agriculture, Technology and Sciences, Allahabad, 211007, India

Rai V
National Research Centre on Plant Biotechnology, Pusa Campus, New Delhi, 110012, India

Singh NK
National Research Centre on Plant Biotechnology, Pusa Campus, New Delhi, 110012, India

H. B. Mahesh
Marker Assisted Selection Laboratory, Department of Genetics and Plant Breeding, University of Agricultural Sciences, Bengaluru 560065, India
Department of Plant Pathology, College of Food, Agricultural and Environmental Sciences, Ohio State University, Columbus 43210, USA
Genomics Laboratory, Centre for Cellular and Molecular Platforms (C-CAMP), National Centre for Biological Sciences (NCBS), Bengaluru 560065, India

Meghana Deepak Shirke
Manipal University, Manipal 576104, India
Genomics Laboratory, Centre for Cellular and Molecular Platforms (C-CAMP), National Centre for Biological Sciences (NCBS), Bengaluru 560065, India

Siddarth Singh
Pacific Biosciences, Boon Lay Way, Singapore 609964, Singapore

Anantharamanan Rajamani
Genomics Laboratory, Centre for Cellular and Molecular Platforms (C-CAMP), National Centre for Biological Sciences (NCBS), Bengaluru 560065, India

Shailaja Hittalmani
Marker Assisted Selection Laboratory, Department of Genetics and Plant Breeding, University of Agricultural Sciences, Bengaluru 560065, India

Guo-Liang Wang
Department of Plant Pathology, College of Food, Agricultural and Environmental Sciences, Ohio State University, Columbus 43210, USA

Malali Gowda
Genomics Laboratory, Centre for Cellular and Molecular Platforms (C-CAMP), National Centre for Biological Sciences (NCBS), Bengaluru 560065, India
Genomics Discovery Program, School of Conservation, Life Science and Health Sciences, TransDisciplinary University, Foundation of Revitalization of Local Health Traditions, Bengaluru 560064, India

Xianjin Qiu
Engineering Research Center of Ecology and Agricultural Use of Wetland, Ministry of Education/College of Agriculture, Yangtze University, Jingzhou 434025, China

Yunlong Pang
Institute of Crop Science/National Key Facility for Crop Gene Resources and Genetic Improvement, Chinese Academy of Agricultural Sciences, Beijing 100081, China

Zhihua Yuan
Engineering Research Center of Ecology and Agricultural Use of Wetland, Ministry of Education/College of Agriculture, Yangtze University, Jingzhou 434025, China

Danying Xing
Engineering Research Center of Ecology and Agricultural Use of Wetland, Ministry of Education/College of Agriculture, Yangtze University, Jingzhou 434025, China

Jianlong Xu
Institute of Crop Science/National Key Facility for Crop Gene Resources and Genetic Improvement, Chinese Academy of Agricultural Sciences, Beijing 100081, China
Shenzhen Institute of Breeding & Innovation, Chinese Academy of Agricultural Sciences, Shenzhen 518120, China

Michael Dingkuhn
International Rice Research Institute, Metro Manila, Philippines
CIRAD, UMR AGAP, F-34398 Montpellier, France

Zhikang Li
Institute of Crop Science/National Key Facility for Crop Gene Resources and Genetic Improvement, Chinese Academy of Agricultural Sciences, Beijing 100081, China

Guoyou Ye
International Rice Research Institute, Metro Manila, Philippines

Shuo Wang
Faculty of Environmental Science and Engineering, Kunming University of Science and Technology, Kunming, China
Faculty of Life Science and Technology, Kunming University of Science and Technology, Kunming, China

Li-zhi Gao
Faculty of Environmental Science and Engineering, Kunming University of Science and Technology, Kunming, China
Faculty of Life Science and Technology, Kunming University of Science and Technology, Kunming, China

Dongli He
Key Laboratory of Plant Germplasm Enhancement and Speciality Agriculture, Wuhan Botanical Garden, Chinese Academy of Sciences, Wuhan 430074, China

Qiong Wang
University of Chinese Academy of Sciences, Beijing 100049, China
Key Laboratory of Plant Germplasm Enhancement and Speciality Agriculture, Wuhan Botanical Garden, Chinese Academy of Sciences, Wuhan 430074, China

Kun Wang
College of life science, Wuhan University, Wuhan 430072, China

Pingfang Yang
Key Laboratory of Plant Germplasm Enhancement and Speciality Agriculture, Wuhan Botanical Garden, Chinese Academy of Sciences, Wuhan 430074, China

Erbao Liu
State Key Laboratory of Crop Genetics and Germplasm Enhancement, Nanjing Agricultural University, Nanjing, China

Yang Liu
State Key Laboratory of Crop Genetics and Germplasm Enhancement, Nanjing Agricultural University, Nanjing, China

Guocan Wu
State Key Laboratory of Crop Genetics and Germplasm Enhancement, Nanjing Agricultural University, Nanjing, China

Siyuan Zeng
State Key Laboratory of Crop Genetics and Germplasm Enhancement, Nanjing Agricultural University, Nanjing, China

Thu G. Tran Thi
State Key Laboratory of Crop Genetics and Germplasm Enhancement, Nanjing Agricultural University, Nanjing, China
College of Agronomy, Hue University of Agriculture and Forestry, Hue University, Hue, Vietnam

Lijun Liang
State Key Laboratory of Crop Genetics and Germplasm Enhancement, Nanjing Agricultural University, Nanjing, China

Yinfeng Liang
State Key Laboratory of Crop Genetics and Germplasm Enhancement, Nanjing Agricultural University, Nanjing, China

Zhiyao Dong
State Key Laboratory of Crop Genetics and Germplasm Enhancement, Nanjing Agricultural University, Nanjing, China

Dong She
State Key Laboratory of Crop Genetics and Germplasm Enhancement, Nanjing Agricultural University, Nanjing, China

Hui Wang
State Key Laboratory of Crop Genetics and Germplasm Enhancement, Nanjing Agricultural University, Nanjing, China

Imdad U. Zaid
State Key Laboratory of Crop Genetics and Germplasm Enhancement, Nanjing Agricultural University, Nanjing, China

Delin Hong
State Key Laboratory of Crop Genetics and Germplasm Enhancement, Nanjing Agricultural University, Nanjing, China

Annamalai Anandan
Division of Crop Improvement, National Rice Research Institute, Cuttack, Odisha, India

Mahender Anumalla
Division of Crop Improvement, National Rice Research Institute, Cuttack, Odisha, India

Sharat Kumar Pradhan
Division of Crop Improvement, National Rice Research Institute, Cuttack, Odisha, India

Jauhar Ali
Plant Breeding, Genetics and Biotechnology Division, International Rice Research Institute, Los Baños, Philippines

Ming Zheng
State Key Laboratory for Crop Genetics and Germplasm Enhancement, Nanjing Agricultural University, Nanjing 210095, P.R. China

Yihua Wang
State Key Laboratory for Crop Genetics and Germplasm Enhancement, Nanjing Agricultural University, Nanjing 210095, P.R. China

Xi Liu
State Key Laboratory for Crop Genetics and Germplasm Enhancement, Nanjing Agricultural University, Nanjing 210095, P.R. China

Juan Sun
State Key Laboratory for Crop Genetics and Germplasm Enhancement, Nanjing Agricultural University, Nanjing 210095, P.R. China

Yunlong Wang
State Key Laboratory for Crop Genetics and Germplasm Enhancement, Nanjing Agricultural University, Nanjing 210095, P.R. China

Yang Xu
State Key Laboratory for Crop Genetics and Germplasm Enhancement, Nanjing Agricultural University, Nanjing 210095, P.R. China

Jia Lv
State Key Laboratory for Crop Genetics and Germplasm Enhancement, Nanjing Agricultural University, Nanjing 210095, P.R. China

Wuhua Long
State Key Laboratory for Crop Genetics and Germplasm Enhancement, Nanjing Agricultural University, Nanjing 210095, P.R. China

Xiaopin Zhu
State Key Laboratory for Crop Genetics and Germplasm Enhancement, Nanjing Agricultural University, Nanjing 210095, P.R. China

Xiuping Guo
National Key Facility for Crop Gene Resources and Genetic Improvement, Institute of Crop Science, Chinese Academy of Agricultural Sciences, Beijing 100081, P.R. China

Ling Jiang
State Key Laboratory for Crop Genetics and Germplasm Enhancement, Nanjing Agricultural University, Nanjing 210095, P.R. China

Chunming Wang
State Key Laboratory for Crop Genetics and Germplasm Enhancement, Nanjing Agricultural University, Nanjing 210095, P.R. China

Jianmin Wan
State Key Laboratory for Crop Genetics and Germplasm Enhancement, Nanjing Agricultural University, Nanjing 210095, P.R. China
National Key Facility for Crop Gene Resources and Genetic Improvement, Institute of Crop Science, Chinese Academy of Agricultural Sciences, Beijing 100081, P.R. China

Kailun Zheng
Development Center of Plant Germplasm Resources, College of Life and Environmental Sciences, Shanghai Normal University, Shanghai 200234, China

Jian Zhao
Shanghai Institute for Biological Sciences, Chinese Academy of Sciences, Shanghai 200032, China
Development Center of Plant Germplasm Resources, College of Life and Environmental Sciences, Shanghai Normal University, Shanghai 200234, China

Dongzhi Lin
Development Center of Plant Germplasm Resources, College of Life and Environmental Sciences, Shanghai Normal University, Shanghai 200234, China

Jiaying Chen
Shanghai Institute for Biological Sciences, Chinese Academy of Sciences, Shanghai 200032, China
Development Center of Plant Germplasm Resources, College of Life and Environmental Sciences, Shanghai Normal University, Shanghai 200234, China

Hua Zhou
Agricultural Faculty, Hokkaido University, Sappro 060-0817, Japan
Development Center of Plant Germplasm Resources, College of Life and Environmental Sciences, Shanghai Normal University, Shanghai 200234, China

Sheng Teng
Shanghai Institute for Biological Sciences, Chinese Academy of Sciences, Shanghai 200032, China

Yanjun Dong
Development Center of Plant Germplasm Resources, College of Life and Environmental Sciences, Shanghai Normal University, Shanghai 200234, China

Zhiwu Dan
State Key Laboratory of Hybrid Rice, College of Life Sciences, Wuhan University, Wuhan 430072, China
Engineering Research Center for Plant Biotechnology and Germplasm Utilization, Ministry of Education, Wuhan University, Wuhan 430072, China

Jun Hu
State Key Laboratory of Hybrid Rice, College of Life Sciences, Wuhan University, Wuhan 430072, China
Engineering Research Center for Plant Biotechnology and Germplasm Utilization, Ministry of Education, Wuhan University, Wuhan 430072, China

Wei Zhou
State Key Laboratory of Hybrid Rice, College of Life Sciences, Wuhan University, Wuhan 430072, China
Engineering Research Center for Plant Biotechnology and Germplasm Utilization, Ministry of Education, Wuhan University, Wuhan 430072, China

Guoxin Yao
State Key Laboratory of Hybrid Rice, College of Life Sciences, Wuhan University, Wuhan 430072, China
Engineering Research Center for Plant Biotechnology and Germplasm Utilization, Ministry of Education, Wuhan University, Wuhan 430072, China

Renshan Zhu
State Key Laboratory of Hybrid Rice, College of Life Sciences, Wuhan University, Wuhan 430072, China
Engineering Research Center for Plant Biotechnology and Germplasm Utilization, Ministry of Education, Wuhan University, Wuhan 430072, China

Yingguo Zhu
State Key Laboratory of Hybrid Rice, College of Life Sciences, Wuhan University, Wuhan 430072, China
Engineering Research Center for Plant Biotechnology and Germplasm Utilization, Ministry of Education, Wuhan University, Wuhan 430072, China

Wenchao Huang
State Key Laboratory of Hybrid Rice, College of Life Sciences, Wuhan University, Wuhan 430072, China
Engineering Research Center for Plant Biotechnology and Germplasm Utilization, Ministry of Education, Wuhan University, Wuhan 430072, China

Zhiming Feng
State Key Laboratory for Crop Genetics and Germplasm Enhancement, Jiangsu Plant Gene Engineering Research Center, Nanjing Agricultural University, Nanjing 210095, China

Chuanyin Wu
National Key Facility for Crop Resources and Genetic Improvement, Institute of Crop Science, Chinese Academy of Agricultural Sciences, Beijing 100081, China

Jeehee Roh
Department of Life Science, Chung-Ang University, Seoul 156-756, Korea

Long Zhang
State Key Laboratory for Crop Genetics and Germplasm Enhancement, Jiangsu Plant Gene Engineering Research Center, Nanjing Agricultural University, Nanjing 210095, China

Jun Chen
National Key Facility for Crop Resources and Genetic Improvement, Institute of Crop Science, Chinese Academy of Agricultural Sciences, Beijing 100081, China

Shengzhong Zhang
State Key Laboratory for Crop Genetics and Germplasm Enhancement, Jiangsu Plant Gene Engineering Research Center, Nanjing Agricultural University, Nanjing 210095, China

Huan Zhang
State Key Laboratory for Crop Genetics and Germplasm Enhancement, Jiangsu Plant Gene Engineering Research Center, Nanjing Agricultural University, Nanjing 210095, China

Chunyan Yang
State Key Laboratory for Crop Genetics and Germplasm Enhancement, Jiangsu Plant Gene Engineering Research Center, Nanjing Agricultural University, Nanjing 210095, China

Jinlong Hu
State Key Laboratory for Crop Genetics and Germplasm Enhancement, Jiangsu Plant Gene Engineering Research Center, Nanjing Agricultural University, Nanjing 210095, China

Xiaoman You
State Key Laboratory for Crop Genetics and Germplasm Enhancement, Jiangsu Plant Gene Engineering Research Center, Nanjing Agricultural University, Nanjing 210095, China

Xi Liu
State Key Laboratory for Crop Genetics and Germplasm Enhancement, Jiangsu Plant Gene Engineering Research Center, Nanjing Agricultural University, Nanjing 210095, China

Xiaoming Yang
State Key Laboratory for Crop Genetics and Germplasm Enhancement, Jiangsu Plant Gene Engineering Research Center, Nanjing Agricultural University, Nanjing 210095, China

Xiuping Guo
National Key Facility for Crop Resources and Genetic Improvement, Institute of Crop Science, Chinese Academy of Agricultural Sciences, Beijing 100081, China

Xin Zhang
National Key Facility for Crop Resources and Genetic Improvement, Institute of Crop Science, Chinese Academy of Agricultural Sciences, Beijing 100081, China

Fuqing Wu
National Key Facility for Crop Resources and Genetic Improvement, Institute of Crop Science, Chinese Academy of Agricultural Sciences, Beijing 100081, China

William Terzaghi
Department of Biology, Wilkes University, Wilkes-Barre, PA 18766, USA

Seong-Ki Kim
Department of Life Science, Chung-Ang University, Seoul 156-756, Korea

Guo J
State Key Laboratory for Conservation and Utilization of Subtropical Agro-Bioresources, South China Agricultural University, Guangzhou 510642, China

Xu X
State Key Laboratory for Conservation and Utilization of Subtropical Agro-Bioresources, South China Agricultural University, Guangzhou 510642, China

Li W
State Key Laboratory for Conservation and Utilization of Subtropical Agro-Bioresources, South China Agricultural University, Guangzhou 510642, China

Zhu W
State Key Laboratory for Conservation and Utilization of Subtropical Agro-Bioresources, South China Agricultural University, Guangzhou 510642, China

Zhu H
State Key Laboratory for Conservation and Utilization of Subtropical Agro-Bioresources, South China Agricultural University, Guangzhou 510642, China

Liu Z
State Key Laboratory for Conservation and Utilization of Subtropical Agro-Bioresources, South China Agricultural University, Guangzhou 510642, China

Luan X
State Key Laboratory for Conservation and Utilization of Subtropical Agro-Bioresources, South China Agricultural University, Guangzhou 510642, China

Dai Z
State Key Laboratory for Conservation and Utilization of Subtropical Agro-Bioresources, South China Agricultural University, Guangzhou 510642, China

Liu G
State Key Laboratory for Conservation and Utilization of Subtropical Agro-Bioresources, South China Agricultural University, Guangzhou 510642, China.

Zhang Z
State Key Laboratory for Conservation and Utilization of Subtropical Agro-Bioresources, South China Agricultural University, Guangzhou 510642, China

Zeng R
State Key Laboratory for Conservation and Utilization of Subtropical Agro-Bioresources, South China Agricultural University, Guangzhou 510642, China

Tang G
State Key Laboratory for Conservation and Utilization of Subtropical Agro-Bioresources, South China Agricultural University, Guangzhou 510642, China

Fu X
State Key Laboratory for Conservation and Utilization of Subtropical Agro-Bioresources, South China Agricultural University, Guangzhou 510642, China

Wang S
State Key Laboratory for Conservation and Utilization of Subtropical Agro-Bioresources, South China Agricultural University, Guangzhou 510642, China

Zhang G
State Key Laboratory for Conservation and Utilization of Subtropical Agro-Bioresources, South China Agricultural University, Guangzhou 510642, China

Preface

Rice is a cereal grain and represents the most consumed staple food for the major part of human population, especially in Asia. Rice is the seed of the grass specie Oryza sativa with two main subspecies (indica that is prevalent in tropical regions, and japonica, prevalent in the subtropical and temperate regions of Asia). In Africa, the most commonly cultivated species is Oryza glaberrima. Rice production has been improved during years, especially with genetic studies. In fact, to meet the growing necessity of food, more resistant and strong rice varieties with higher yield potential are needed. Moreover, geographic specific characterization and improvement for the different cultivars of rice represent a key point. In this book, latest papers about rice germplasm, genetic and improvement are compiled. The state-of-the-art of genomic studies on rice cultivars is reported. First, open-access resources for genome-wide association mapping in rice are proposed. Then, germplasm analyses of rice accessions from India, Europe, Bangladesh, and Japan are discussed, along with some specific properties. Detailed genetic studies about chloroplast genome sequencing, microRNA expression profile, SSP marker analysis and characterization of resistance-related genes are also presented. Finally, a new promising approach to overcome the inter-subspecific hybrid sterility is presented. Traditional rice varieties encompass a huge range of potentially valuable genes. These can be used to develop superior varieties for farmers to take part in the uphill battle of feeding an ever-increasing world population. The genes linked to valuable traits can help breeders create new rice varieties that have improved yield potential, higher nutritional quality, better ability to grow in problem soils, and improved tolerance of pests, diseases, and the stresses, such as flood and drought, that will be inevitable with future climate change. The high quality genome sequence of rice has provided a rich resource to mine information about diversity of genes and alleles which can contribute to improvement of useful agronomic traits.

Editor
Valeria Severino, Ph.D.

INTRODUCTION

Rice is the world's most important food crop and a staple food for more than half of the world's population. More than 90% of the world's rice is produced and consumed in Asia, where 60% of the people live. In India, rice is grown on about 44.5 Mha and provides food for more than 70% of the population and serves as the principal energy source for most of the people. In the last six decades, rice production has steadily kept in pace with the population growth rate, mainly due to the gains from the technologies of green revolution era such as semi-dwarf, fertilizer responsive high yielding varieties and other associated managerial technologies. Of late, rice yield growth has reached a plateau and no significant increase is being realized in productivity levels. Keeping in view the annual average population growth rate of ~1.5% and estimated per capita consumption of about 250 g of rice per day, the demand for rice is expected to increase to 40% by 2025. This projected demand can be met only if there is a steady increase in productivity and production. Further, the increases in production have to be achieved under conditions of declining and deteriorating land, soil and water resources and at the same time preserving the environmental quality. It will be indeed difficult to meet these daunting challenges only with the application of conventional plant breeding techniques and tools, as no headway is being made through these approaches. Modern tools of biotechnology can be helpful to meet the rice production and productivity targets as the novel variability can be created and harnessed across the species and kingdoms.

The major countermeasure against genetic erosion is the conservation of seeds of land races and their wild relatives in gene banks. Such genetic conservation is being carried on at international, national, and institutional levels. The Inter-national Board for Plant Genetic Resources (IBPGR) was established in Rome in 1973, for the purpose of establishing liaison between these conservation programs and to support their activities. The International Rice Research Institute (IRRI) maintains the international gene bank for rice in its International Rice Germplasm Center, which has about 83,000 accessions. Large collections of rice germplasm are also conserved in national gene banks in India, Thailand, China, Indonesia, Japan, and several other countries. While the gene pools existing in nature are dynamic and evolving, those conserved in the seed banks are static. The conservation programs consist of collection from the field, multiplication, cold storage of dried seeds, documentation, evaluation, record-keeping, and use of selected stocks for breeding. The evolutionary forces in operation in nature are thus curtailed and man's work in this direction is much limited in range and activity. Moreover, the conservation programs are always endangered by personnel changes, changes in budgetary policy, and other human factors. The system of management of core collections is even more important than the total number of accessions kept in a gene bank.

The unit of operation in a rice conservation program is an accession, which is handled usually as a pure line. This is almost unavoidable when a large number of accessions are to be conserved. However, seed samples collected from a field population cannot be represented by a single accession. The number of lines necessary for preserving 95% of genes present in a population with a 5% frequency is about 60. The number of lines necessary for representing a population are 20 to 100 depending upon the genetic diversity and other conditions (Oka 1983). Of the large collection kept at IRRI, some 9,700 or about 12% of the total were found to have useful genes for tolerance for drought, cold, deep-water, and adverse soils, and resistance to diseases, insects, and nematodes. When a particular resistance is sought for, however, the probability of finding a useful gene is higher in wild germplasm than among ordinary cultivars. This indicates the importance of wild species, particularly the wild relatives of cultigens, in conservation programs.

Knowledge of higher plant genetics is essential for advancing the frontiers of molecular genetics and for application of techniques of molecular and cellular genetics to crop improvement. Until recently knowledge of basic genetics of rice lagged far behind that of other major food crops such as maize, wheat, barley, and tomato. During the last decade, however, rapid advances in higher plant and molecular genetics have made rice a model crop for research in molecular genetics. The karyotype of rice has been studied at the somatic prometaphase as well as at the pachytene stage of meiosis, and individual members of the chromosome complement have been identified. A wealth of marker genes affecting various morphological and physiological traits has been assembled and many of these marker genes have been assigned to respective linkage groups. All of the 12 linkage groups are known and have been associated with respective chromosomes. About 48 isozyme loci are known and 33 of them have been assigned to 9 of the 12 chromosomes.

Most competition in rice is for light. Competition begins early in the tillering stage and increases in intensity proportionate to the increase in plant size and the density of planting. Competition for nitrogen may occur at later growth stages but is readily overcome by adding fertilizer. But adding nitrogen aggravates the competition for light because it induces tall plants to grow taller. Competition for light is also increased by close spacing, the length and intensity of the rainy season, weeds, and other factors that reduce the penetration of light into populations. Competition is caused by differential growth rates and sizes of neighboring plants. Large plants invade the space occupied by smaller ones, shade the other plants, and capture a disproportionate share of tile solar radiation. The smaller plants have reduced tillering, produce weak and spindly culms, accumulate less dry matter, exhibit premature leaf senescence, and have pronounced spikelet sterility. In short, the smaller plants, when intermixed with larger ones, appear agronomical undesirable.

Irrigated rice

Review of rice production methods has shown that practices range from very primitive to highly mechanized. Tractors and two-wheeled power tillers are the most important agricultural machines used for rice production. In 1980 the number of power tillers used per 1 000 ha was from 0.1 to 26 in tropical Asia, 56 in China, 73 in Taiwan (province of China), 198 in the Republic of Korea and 1158 in Japan. In Asia, animals (buffalo and water buffalo, carabao) are still used for plugging and harrowing. Land preparation may be carried out while the soil is dry or wet, depending on the water supply. For irrigated rice, the soil is prepared wet or puddled in Asia, but puddling is not generally practiced in America, Europe and Africa. In areas without a hard pan, where animals and tractors sink in the mud, the soil is prepared with hand hoes. Regardless of whether the land is prepared wet or dry, the water is always held on the lowland fields by bunds.

Most irrigated rice is transplanted, although direct seeding is becoming more extensive. The seeds are pregerminated and grown in wet seed-beds for 9 to 14, 20 to 25 or 40 to 50 days after sowing and are then transplanted either by hand or by mechanical transplanters. The number of seedlings per hill may vary from one to eight. Direct seeding is done by broadcasting the pregerminated grain by hand in Asia or by water-seeding by airplane in the United States and Australia. The seeds may also be machine-drilled in puddled soil or drill-seeded into dry soil. Deep-water rice is commonly dryseeded, but it is occasionally transplanted or double transplanted.

Ideally, water is maintained in the rice field to suppress weed growth during the growing season. Hand weeding and mechanical or rotary weeders are popular. Herbicides are also economical and effective. Fertilization is normally practiced for increased yield, particularly with the modern, semi-dwarf or high-yielding varieties which respond well to fertilizer without lodging. Both inorganic and organic fertilizers are used, including green manures such as the leguminous shrub Sesbania spp. and the water plants Azolla and Anabaena spp. Modern rice varieties increase in grain yield by 6 kg per kg of applied fertilizer in the wet season and by 9 kg per kg of applied fertilizer in the dry season. Total fertilizer nutrients range from 10 to 100 kg/ha in tropical Asia and from 200 to 350 kg/ha in Japan, Taiwan and the Republic of Korea.

Rice ecosystems

Rain-fed lowland rice is grown on puddled soil in fields bounded by dykes that can pond water to depths of 0 to 25 cm (shallow) and 25 to 50 cm (medium), depths seldom exceeded in such areas. The irrigation water is not received from river diversions, storage reservoirs or deep wells, but from rainfall or runoff from a local catchment area. The prevailing climatic and soil conditions in shallow rain-fed rice areas are extremely variable. In deep-water (50 to 100 cm) rain-fed lowland rice, modern semi-dwarf varieties cannot be used. Fertilizer use is low, stand establishment difficult and pest control almost impossible, and yields are poor. Rain-fed lowland rice is next to irrigated rice in importance in terms of harvested area and production of rice.

Upland rice is grown in fields that are not bunded but are prepared and seeded under dry conditions and depend on rainfall for moisture. In Brazil, a major part of the rice crop is upland. In India and throughout Southeast Asia, upland cultivation is common along river banks as waters recede at the end of the rainy season. Soils are commonly heavy and residual moisture alone sustains growth. Upland rice farming ranges from shifting cultivation of forested hilly or mountainous areas that are cleared and burned to large-scale mechanized operations. Between these two extremes is farming of sloping hill regions that are subject to serious erosion and frequent drought, by hundreds of thousands of the poorest of rice farmers. The environmental damage here is very serious.

In deep-water rice, water depth is at least 1 m during a significant portion of the growing season. In large parts of Bangladesh as well as in portions of the Mekong and the Chao Praya Deltas, water depth may exceed 5 m, but it is normally between 1 and 3 m in other regions. Where water rises rapidly after the start of the monsoon rains, rice is commonly broadcast in unpuddled fields that are seldom bounded by dykes of any sort. The varieties planted are tall and leafy, with few tillers. They are photoperiod sensitive and mature only after the rainy season. They can elongate and float as the water level rises. Major dyking and flood control projects in the last two decades have upgraded many former deep-water rice areas into the rain-fed or irrigated category in Bangladesh, India, Thailand and southern Viet Nam.

Competition and Breeding Systems

The intense interplant competition in early segregating generations is one of the most critical factors that affect the choice of breeding systems. Strong competition is most pronounced in tall × dwarf crosses and occurs whenever the two parents of a cross are morphologically distinct. Most competition in rice is for light. Competition begins early in the tillering stage and increases in intensity proportionate to the increase in plant size and the density of planting. Competition for nitrogen may occur at later growth stages but is readily overcome by adding fertilizer. But adding nitrogen aggravates the competition for light because it induces tall plants to grow taller. Competition for light is also increased by close spacing, the length and intensity of the rainy season, weeds, and other factors that reduce the penetration of light into populations. Competition is caused by differential growth rates and sizes of neighboring plants. Large plants invade the space occupied by smaller ones, shade the other plants, and capture a disproportionate share of tile solar radiation. The smaller plants have reduced tillering, produce weak and spindly culms, accumulate less dry matter, exhibit premature leaf senescence, and have pronounced spikelet sterility. In short, the smaller plants, when intermixed with larger ones, appear agronomically undesirable. Plants of the smaller varieties, originally mixed in equal proportion with the larger ones, are rapidly reduced in number.

In extreme cases, all weakly competitive varieties are eliminated from mixtures after two or three cycles of competition. Similar in segregating progeny populations of crosses of distinct plant types are more directly related to breeding methodology. The monogenic inheritance of dwarfism dictates that 25% of the F_2 population of a tall × dwarf cross is dwarf. In the absence of competition and assuming equal fitness for tall and dwarf segregates, the theoretical percentage of dwarfs in subsequent generations is easily calculated. Thus, without competition, the dwarfs would increase from 25% of the population in the F_2 to 48.5% in the F_6. But in experiments on the competition of dwarfs and tall, with high nitrogen rates and close spacing, the dwarfs declined from 25% in the F 2 to about 6% in the F_6. This deviation from the number of expected dwarfs is theoretically a direct measure of competition. Such losses are similar but less drastic within populations that receive low nitrogen rates, that are spaced widely, or that are grown in a clear, dry season.

CHAPTER 1

GENOME-WIDE ASSOCIATION MAPPING IN RICE

Susan R. McCouch, Mark H. Wright, Chih-Wei Tung, Lyza G. Maron, Kenneth L. McNally, Melissa Fitzgerald, Namrata Singh, Genevieve DeClerck Francisco Agosto-Perez, Pavel Korniliev, Anthony J. Greenberg, Ma. Elizabeth B. Naredo, Sheila Mae Q. Mercado, Sandra E. Harrington, Yuxin Shi, Darcy A. Branchini, Paula R. Kuser-Falcão, Hei Leung, Kowaru Ebana, Masahiro Yano, Georgia Eizenga, Anna McClung, and Jason Mezey

Rice (*Oryza sativa*) is a staple food that feeds 3 billion people[1,2]. It has the largest, publicly available, single-species germplasm collection in the world[3], and a rich 'omics' toolkit. Rice was the first crop plant to be fully sequenced[4], has over 3,000 re-sequenced varieties[5,6,7,8], *de novo* assemblies within three subpopulations[9,10,11] and BAC-based sequence assemblies for 12 related *Oryza* species[12]. With these resources in hand, the challenge is to utilize this information to understand and predict how genotypic variation gives rise to the abundance of phenotypic variation that we see in nature and in agriculture.

Genome-wide association studies (GWAS) offer a powerful strategy for understanding the genetic basis of complex traits. This approach has been particularly productive for rice, where major associations have been detected for important traits, including grain size, grain weight, flowering time, disease resistance and abiotic stress tolerance, among others[13,14,15,16]. Hundreds of genes governing traits of interest have been cloned and new, powerful tools are currently being deployed to achieve targeted genome editing[17,18,19]. In applied breeding programs, major-effect loci discovered via quantitative trait loci (QTL) mapping and GWAS have been successfully introgressed from wild and exotic genetic resources into elite, high-yielding varieties via marker-assisted selection[20,21,22].

Rice has been particularly amenable to GWAS largely because of its evolutionary and domestication history. The wild ancestor of Asian rice, *O. rufipogon*, is a partially (35%) outcrossing species complex broadly dispersed across tropical South, South-East and Eastern Asia. It consists of locally or regionally adapted subpopulations with both perennial and annual forms[23]. The domestication of *O. sativa* began ~10,000–12,000 years ago, and human selection was accelerated by inbreeding which retained and fixed many favourable alleles of large effect. The unique combinations of locally adapted allele complexes in different subpopulations provide ideal characteristics for GWAS, including both universal and population-specific large effect alleles, extended linkage disequilibrium (LD) within some subpopulations that is advantageous for mapping with low marker coverage, rapid LD decay within other subpopulations offering improved mapping resolution and an abundance of well-partitioned genetic variation among subpopulations. These properties, in combination with a small, tractable genome (380 Mb), and the ability to produce immortal panels comprised of thousands of inbred, homozygous individuals, make it possible to evaluate the same traits in different environments simultaneously, and to evaluate a variety of different traits on the same germplasm over time. The ability to make controlled matings and to generate large populations each generation provides a powerful system for validating loci and for harnessing their value by introgressing favourable alleles into other genetic backgrounds.

Here, we report the development of an association mapping panel consisting of 1,568 diverse inbred rice varieties and a genotyping data set generated using a high-density rice array (HDRA) comprised of 700,000 single-nucleotide polymorphisms (SNPs). The HDRA represents the most densely populated genotyping array available for any organism at this time, based on the number of SNPs/kb interrogated across the genome. We explore the value of these resources for GWAS using grain length as the phenotype, and use numerous analytical techniques to identify significant associations that uncover both major and minor QTLs. Further, we demonstrate distinct advantages of cross-population mapping where we gain power by directly accounting for large effect alleles from

Citation: Susan R. McCouch, Mark H. Wright, Chih-Wei Tung, Lyza G. Maron, Kenneth L. McNally, Melissa Fitzgerald, Namrata Singh, Genevieve DeClerck Francisco Agosto-Perez, Pavel Korniliev, Anthony J. Greenberg, Ma. Elizabeth B. Naredo, Sheila Mae Q. Mercado, Sandra E. Harrington, Yuxin Shi, Darcy A. Branchini, Paula R. Kuser-Falcão, Hei Leung, Kowaru Ebana, Masahiro Yano, Georgia Eizenga, Anna McClung, and Jason Mezey Genome-wide association mapping in rice. Nat. Commun. 7:10532 doi: 10.1038/ncomms10532 (2016).

within the mapping model. We show how this may lead to the identification of population-specific haplotypes, which are of particular value to the plant breeder.

The GWAS resources and data sets generated in this project are designed to facilitate the collection of phenotypic information on a wide range of traits and characteristics (see companion paper by Crowell *et al.*[24]) and to rapidly and efficiently convert that information into focused hypotheses about causal genes and QTL locations. They also provide a bridge between genomics and applied plant breeding. With increasing accuracy over time, data sets generated for different purposes and at differing levels of resolution can be leveraged against each other via shared haplotypes, making it feasible to undertake cross-population imputation of both genotype and phenotype and to predict the phenotypic outcomes from genotypic data. GWAS results can also be incorporated as fixed variables into genomic selection models to improve the accuracy of genomic estimated breeding values[25,26,27].

These resources are part of a larger rice diversity research platform being developed by the international rice research community. All the materials, data sets and tools reported here are available as open-source products. Together these resources provide the basis for further exploration of phenotype–genotype relationships in rice and for expediting variety development in this important food crop.

RESULTS

The high-density rice array

The HDRA assays 700,000 SNPs (average call rate=93.5%), or ~1 SNP every 0.54 kb across the rice genome. It was designed to efficiently capture most of the haplotype variation observed in a discovery panel consisting of 16M SNPs (generated by sequencing 125 rice genomes at ~7 × genome coverage) and to maximize the inclusion of non-synonymous SNPs. An estimated 45% of HDRA SNPs map within genes, covering all 39,045 unique, non-TE rice gene models (http://rice.plantbiology.msu.edu/), while 55% of SNPs map to intergenic regions. Non-synonymous SNPs, which represent ~16% of all informative SNPs on the HDRA (Table 1), are found in 91% of unique, non-TE gene models. Of the genic SNPs, 55% are distributed within exons, 36% within introns, 3% within 5′-untranslated regions (UTRs) and 6% within 3′-UTRs. Of the intergenic SNPs, 50% are located in putative regulatory regions (PPRs) within 2 kb of a transcriptional start site.

Table 1: Summary of polymorphisms detected by the HDRA in the different rice subpopulations

	O. sativa	IND	AUS	TRJ	TEJ	ARO
Total accessions	1,571	498	187	360	240	38
Phenotyped accessions*	1,146	335	160	256	190	31
Polymorphic SNPs	671,906	619,988	547,928	551,746	470,318	392,943
Non-synonymous polymorphic SNPs	109,814	100,762	88,093	89,863	78,259	62,560
SNPs MAF>0.05	439,280	336,740	328,393	208,055	170,904	290,068
SNPs MAF>0.10	300,769	223,094	238,308	120,538	94,826	195,271
Private SNPs†	—	25,718	10,961	9,322	4,916	3,383

MAF, major allele frequency; SNP, single-nucleotide polymorphism.
*Accessions phenotyped for grain length.
†Private SNPs are the SNPs segregating only in the indicated subpopulation.

Genetic structure of the rice diversity panel

Using the HDRA, we assayed SNP variation in a collection of 1,568 diverse accessions of *O. sativa* and the resulting genotypic data set was analyzed for population structure using fastSTRUCTURE[28]. The optimum number of subpopulations was determined to be K=5. The five subpopulations were identified as *aus*, *indica*, *tropical japonica*, *temperate japonica* and *aromatic* (or*Kashmiri japonica*, Group V), with *aus* and *indica* comprising the *INDICA* varietal group, and *tropical japonica*, *temperate japonica* and *aromatic* comprising the *JAPONICA* varietal group[15,16,29,30]. (We use upper-case letters to indicate the two major varietal groups, *INDICA* and *JAPONICA*, and lower-case letters to indicate subpopulations.) Using a cutoff of 70% ancestry, accessions were classified into one of the five major subpopulation groups, or into one of three admixture classes: admixed *INDICA* (*aus-indica*), admixed*JAPONICA* (*tropical japonica-temperate japonica-aromatic*) or admixed *INDICA-JAPONICA*.

We applied principal component analysis[31] to summarize global genetic variation in the diversity panel (Fig. 1a). The top three principal components (PCs) explain 40% of the genetic variation, where the first PC separates the *INDICA* and *JAPONICA* varietal groups (23.58% of the variation), the second PC separates the*aus* and the *indica* subpopulations (10.44%), while the third PC separates the three *JAPONICA*subpopulations (6.06%). LD decays rapidly with distance in all populations (Fig. 1b), with *indica* exhibiting the most rapid LD decay and *temperate japonica* the most extended LD.

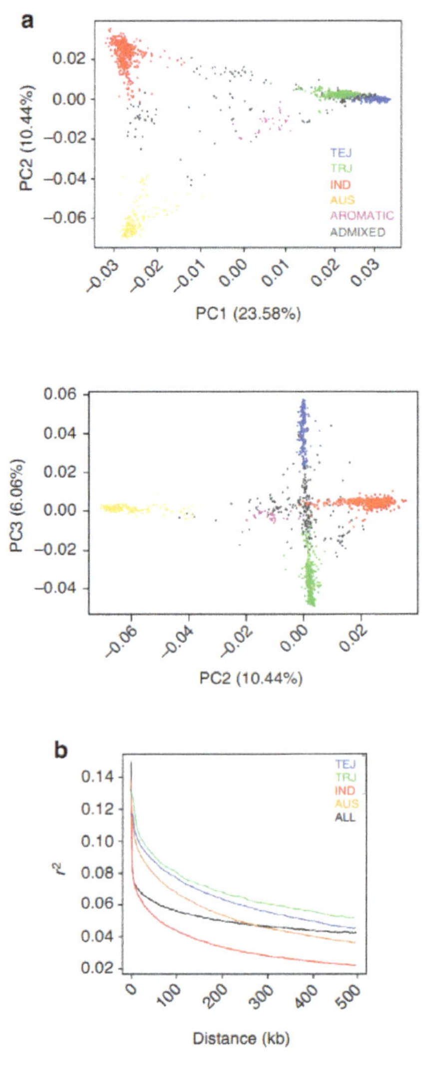

Figure 1: Genetic structure of the rice diversity panel. (a) PPCA

showing global genetic variation in the rice diversity panel. Left: first and second PCs; right: second and third PCs. (b) LD as a function of distance between SNP markers in each O. sativa subpopulation. The circles in a and lines in b are coloured based on subpopulation assignments from FastSTRUCTURE: IND—indica (red); AUS—aus (yellow); TEJ—temperate japonica (blue); TRJ—tropical japonica (green); ALL—all subpopulations combined (black). Purple circles: aromatic; Grey circles: admixed lines

The design of the HDRA provides a rich selection of shared and private SNPs distributed across the subpopulations of *O. sativa* and its closest wild relatives, *O. rufipogon/O. nivara*, with emphasis on efficiency of haplotype detection. A total of 671,906 SNPs detected polymorphism within the subpopulations of *O. sativa*, ~8% were private to individual subpopulations (Table 1), and 23,748 SNPs were private to the Asian wild progenitors. When subpopulation statistics were compared, the largest number of both private and shared SNPs was detected within *indica*, the smallest was detected in *aromatic* and the other subpopulations had intermediate numbers of informative SNPs on the array (Table 1). These figures are consistent with sequence-based estimates of relative subpopulation diversity[5,15].

There is a non-linear, decreasing relationship between the number of lines assayed and SNPs detected, indicating that the same SNPs are being assayed over and over again as the panel size increases. The diversity panel developed by the NIAS[32], by far the smallest collection of lines, is very efficient for its size in capturing a broad range of diversity, while the size of the two larger rice diversity panels, RDP1 (refs [16, 33]) and RDP2 enhance the resolution of GWAS primarily by assaying new recombinants. Given that diversity is partitioned into deeply differentiated subpopulations in *O. sativa*, we further explored the genetic relationships between pairs of subpopulations by examining 2D site frequency spectra. The majority of SNPs that are at low frequency in *indica* are also found at low frequency in *aus*, though *indica*-specific and *aus*-specific SNPs are also detected (Fig. 2a). Similarly, SNPs that are at low frequency in*tropical japonica* are also usually observed at low frequency in *temperate japonica*, with few exceptions (Fig. 2b). However, we see a very different picture when we compare the *INDICA* and *JAPONICA* varietal groups, where the sites that are at moderate-to-high frequency in one of the subpopulations are found at markedly lower frequencies in the other (Fig. 2c). Nonetheless, a common set of 291,132 SNPs (43%) segregate cross all 5 subpopulations of *O. sativa*, while 418,502 SNPs segregate in the *INDICA* varietal group and 295,984 SNPs segregate in the *JAPONICA* varietal group. This made it possible to perform GWAS using different subsets of germplasm, though many SNP frequencies vary significantly across the different populations. Interestingly, the frequency distribution of synonymous and non-synonymous SNPs is very similar across groups, with slightly more non-synonymous SNPs occurring at low frequency, but a good distribution of both across the entire site frequency spectra. The observed SNP frequencies and distributions within and between subpopulations have important implications for the way GWAS is conducted in rice.

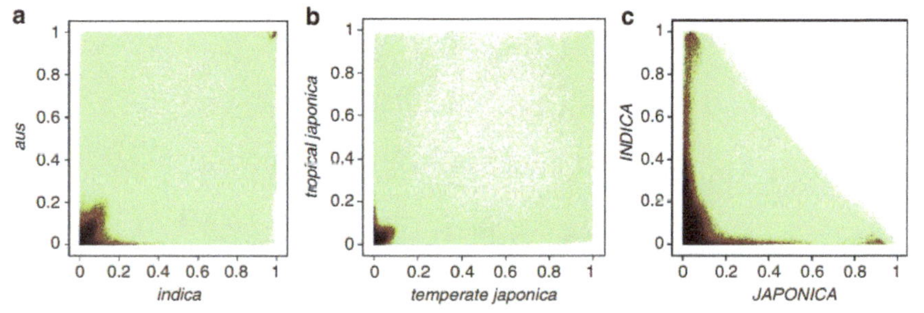

Figure 2: Allele frequencies in the different subpopulations of *O. sativa* Two-dimensional site frequency spectra (SFS) for the entire diversity panel (RDP1 and 2): (a) indica vs aus; (b) temperate japonica vs tropical japonica; (c) JAPONICA vs INDICA. The spectra were displayed by grouping frequency values for each SNP in a pair of populations into 250 bins using the hexbin package in R. Note: the joint frequency of the indica and japonica samples cannot exceed 0.5 because without an outgroup, we calculate the frequency of the minor allele in the entire RDP1 and 2 sample (that is, the frequency spectrum is folded). However, in different sub-samples (for example, within japonica) some alleles with frequencies below 0.5 in the complete sample can be more prevalent, even reaching fixation (frequency of 1.0), thus giving the appearance of an unfolded SFS.

Improved GWAS resolution using a larger number of lines

In this study we investigated the potential to resolve both major- and minor-effect GWA-QTLs using the new HDRA genotypic data set and grain length as a phenotype. Grain length is a highly heritable phenotype, with broad sense heritability of 0.976. A major QTL for grain size was previously identified on rice chromosome 3 in a number of studies across different genetic backgrounds and environments[34,35]. The gene underlying this QTL, *GRAIN SIZE 3* (*GS3*, Os03g0407400), has been cloned and characterized and the most common functional polymorphism is a C≥A transition that occurs in the second exon of the gene. Using genetic transformation, Takano-Kai *et al.*[36] demonstrated that the wild-type allele (C) confers short grain and the derived allele (A) confers long grain in natural populations. Other mutations in the same gene have also been shown to affect grain length but they occur at low frequency in *O. sativa*, making them unlikely to be detected using GWAS[37].

Rice chromosome 5 carries two closely linked genetic loci associated with grain size that have been positionally cloned and functionally characterized, namely *GRAIN WIDTH 5* (*GW5*) and *GRAIN SIZE 5*(*GS5*, LOC_Os05g06660)[38,39]. For *GW5*, the functional polymorphism is a 1,212 bp deletion in the reference genome (cv. Nipponbare) that removes the only exon of the gene, resulting in a short, wide, heavy grain[38,40]. For *GS5*, polymorphisms in the promoter region leading to differential gene expression were associated with grain size[39]. In addition, a vast number of QTL regions have been mapped, and 14 genes associated with grain shape and size have been cloned[41,42,43,44,45,46,47,48,49,50,51,52,53,54,55]. Although the function of many of these genes is still poorly understood, they provide us with clear, true-positive loci where common alleles affecting grain size can serve as the basis for assessing the mapping power and resolution of the HDRA.

Using the HDRA on the RDP1 panel alone, or the RDP1 and 2 panels together, we see two major GWAS peaks associated with grain length, one on chromosome 3 and one on chromosome 5 (Fig. 3a). Zoom-in plots of the peaks on chromosome 3 confirm that the same peaks are detected, and SNP-3.16732086 is the most significant SNP (MS-SNP) in both cases (Fig. 3b). With RDP1 alone, the MS-SNP explains 16.5% of the phenotypic variation and is detected at *P* value=4.48e−17, while when RDP1 and 2 are used, the MS-SNP is detected at *P* value=1.48e−55 and explains 15.9% of the variation (Fig. 3b). The highly significant *P*values are because of the fact that SNP-3.16732086 directly interrogates the functional polymorphism in exon 2 of *GS3*, while the second and third MS-SNPs map 11.3 kb (SNP 3.16718013) and 16.8 kb (SNP 3.16712500) upstream of the transcriptional start site of *GS3*, and explain 6% and 5.8% of the phenotypic variation, respectively. Interestingly, in addition to SNP-3.16732086, there are several HDRA SNPs that map within the *GS3* gene itself (coloured green in Fig. 3a,b) but have significance levels and allele effects similar to SNPs that are farther away. This is because of the fact that, despite being within the causal gene, they are not in perfect LD with the functional SNP that is responsible for grain length.

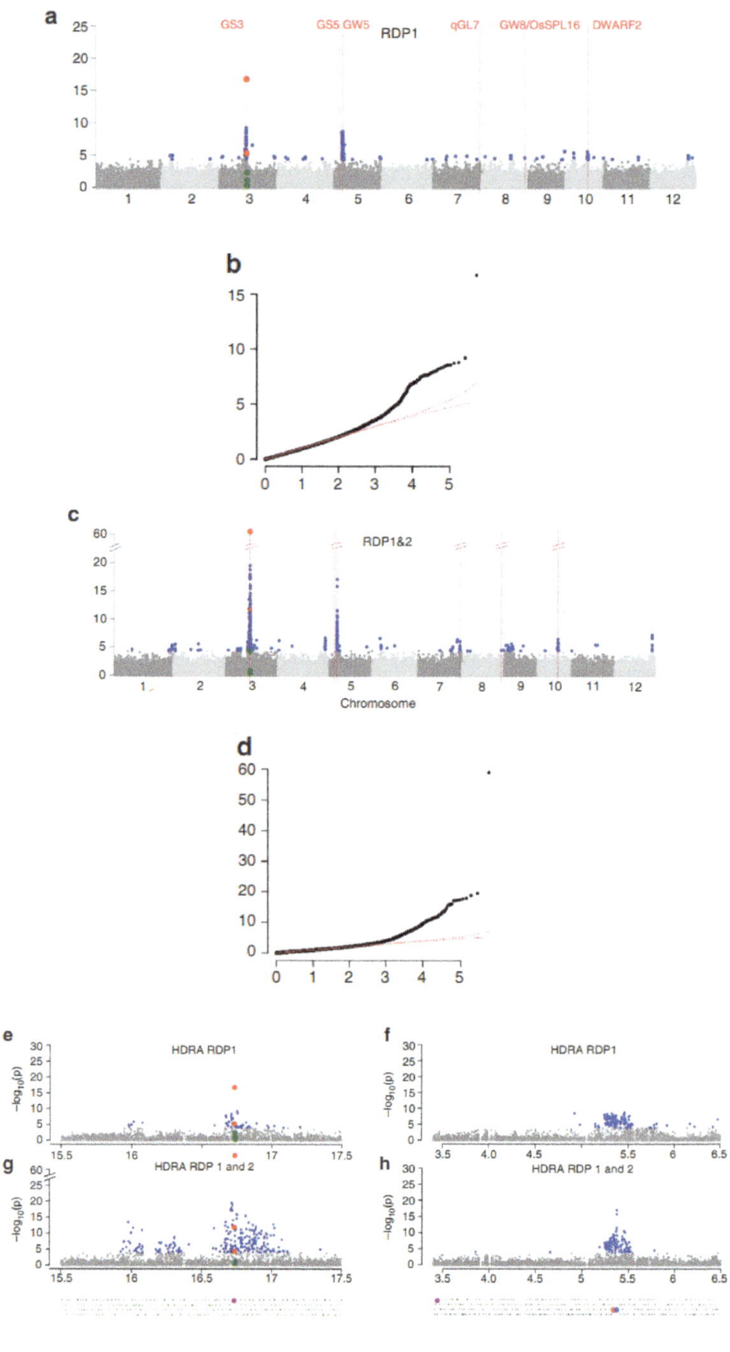

Figure 3: GWAS for rice grain length. Manhattan plots (a,c) and quantile–quantile plots (b,d) for GWAS using the HDRA on RDP1 (a,b); and

GWAS using the HDRA on RDP1 and 2 (b,d). On each Manhattan plot, the x-axis shows the SNPs along each chromosome; y-axis is the log10 (P value) for the association. Red vertical lines indicate the position of candidate genes and/or QTL regions associated with grain length. (e,g) Zoom-in on the major GWAS peaks for grain length on chromosomes 3 (15.5–17.5 Mb) using (e) the HDRA on RDP1; and (g) the HDRA on RDP1 and 2. Zoom-in of the major GWAS peaks for grain length on chromosomes 5 using (f) the HDRA on RDP1; and (h) the HDRA on RDP1 and 2. Green circles underneath each set of plots represent gene models; the gene model highlighted in magenta in e,g is GS3 (LOC_Os03g0407400); gene models highlighted in f,h are GS5 (LOC_Os05g06660, magenta) and the two gene models flanking the GW5 deletion, LOC_Os05g09510 (blue) and LOC_Os05g09520 (red). In all panels, significant SNPs (with P values at or above the 10% FDR threshold) are coloured blue. In a,c,e,g, the SNPs that fall within the coding region of GS3 are highlighted in green if they have non-significant P values and in red if they have significant P values. The SNPs highlighted are (SNP name (P value for RDP1, P value for RDP1 and 2)): SNP-3.16728637 (0.69, 0.50); SNP-3.16730230 (1.00, 0.21); SNP-3.16732086 (1.83e 17, 1.05 e 59); SNP-3.16732418 (6.25 e 06, 1.96 e 12); SNP-3.16733169 (0.01, 7.02 e 05) and SNP-3.16733346 (0.12, 0.42). NATU

Turning our attention to the peak on chromosome 5, a cursory comparison of the two plots suggests that they detect a GWA peak in roughly the same region (Fig. 3c). Using the HDRA on RDP1, a cluster of significant SNPs forms a dense plateau across a ~300 kb region, while the resolution is greatly improved with the larger panel of germplasm (RDP1 and 2).

The grain length genes *GS5* and *GW5* are located 1.9 Mb apart on rice chromosome 5. The known functional polymorphism associated with *GW5* corresponds to a 1.2 kb deletion in the reference genome that deletes the*GW5* gene[40]. The deletion is located between LOC_Os05g09510 and LOC_Os05g09520 (Fig. 4). When we verified the distance of the MS-SNPs to *GW5* and *GS5*, the MS-SNP maps only 11 kb away from the*GW5* deletion (Fig. 4) and all the MS-SNPs cluster in the vicinity of the functional deletion affecting *GW5*; no peak was observed at *GS5* (Fig. 3c). This is consistent with the fact that the functional polymorphisms at*GS5* are known to be rare in rice germplasm[39] and provides evidence that GWAS can distinguish levels of statistical support for two *a priori* candidate genes that are only 2 Mb apart.

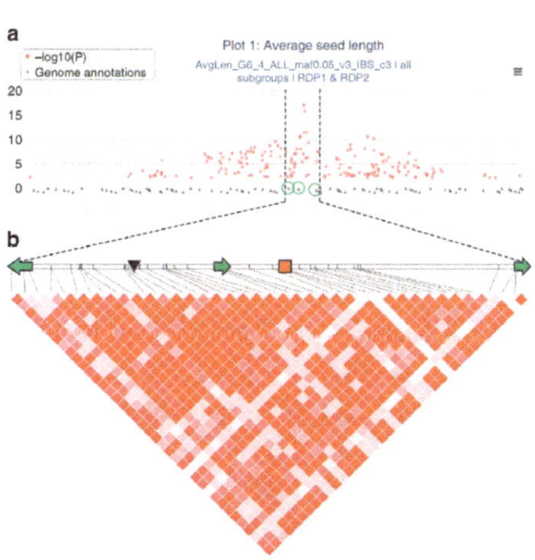

Figure 4: GWAS peak on chromosome 5 in relation to the *GW5* functional deletion. (a) screenshot from our GWAS Viewer (ricediversity.org) zoomed-in to show GWAS peak over GW5 candidate gene region in RDP1 and 2 (ALL). Three gene models under the peak are circled in green. (b) LD in the region of the GW5 functional polymorphism. The horizontal bar shows a physical map of chromosome 5 (5.35–5.39 Mb); green arrows represent the three genes circled above, from left to right: LOC_Os05g09510, LOC_Os05g09520 and LOC_Os05g09530, with direction indicating strand orientation; position of the GW5 functional deletion is shown as black triangle, and red square represents position of the MS-SNP (SNP-5.5371749, located at 5,371,772).

We previously reported a GWA study in the RDP1 using a custom-designed genotyping array with 44,100 SNPs (44 K array)[16]. Although the 44 K array also detected the same two major peaks on chromosomes 3 and 5, significant SNPs are sparse and tend to have lower *P* values than those detected by the HDRA.

GWAS models with SNP covariates

To further explore the relationships among QTLs for grain length, we re-ran GWAS analyses for the HDRA on RDP1 and 2 and included the MS-SNPs (one from each major peak on chromosomes 3 and 5) as co-variates in the GWA models. When all samples were evaluated together, additional peaks were detected on all 12 rice chromosomes, with the most significant located on chromosome 10 (Fig. 5a). The MS-SNP defining the chromosome 10 peak also had the highest per cent phenotypic variation explained (PVE), but not the largest allele effect in *ALL* with two SNP covariates.

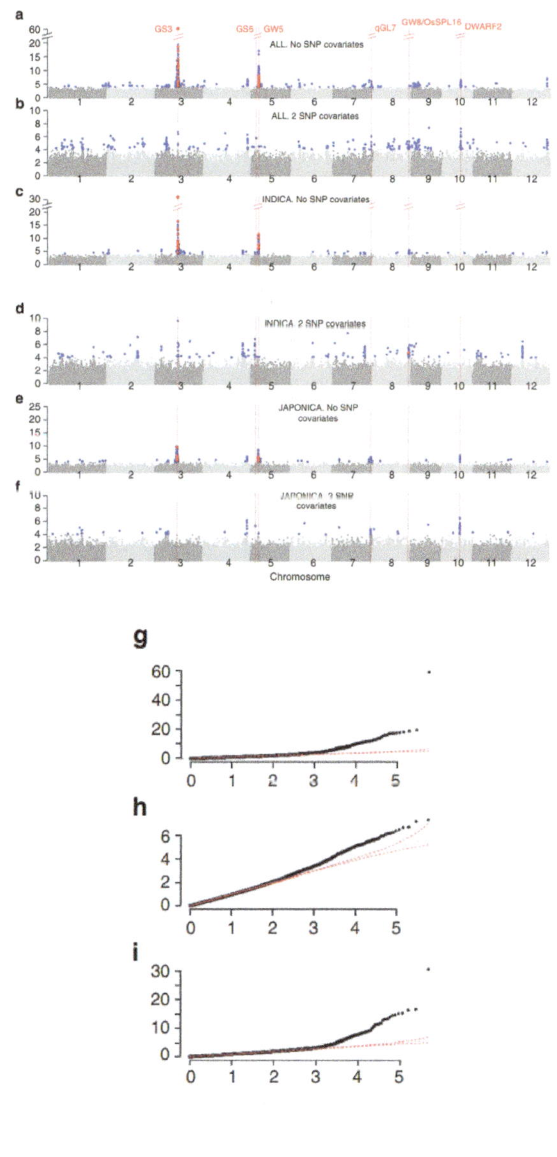

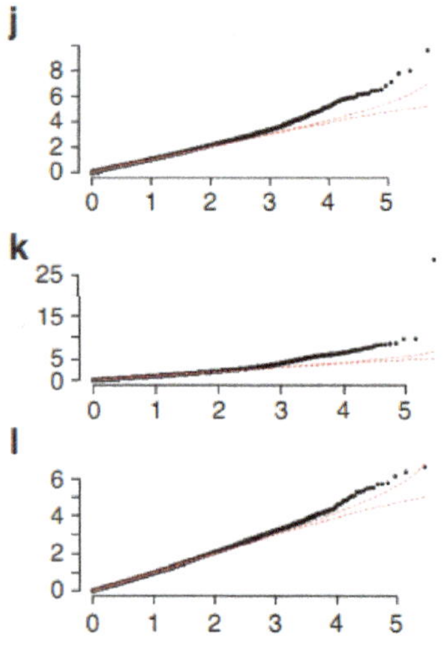

Figure 5: GWAS for grain length in the *Indica* and *Japonica* varietal groups and using SNP covariates. Manhattan plots (a–f) and quantile–quantile plots (g–l) for GWAS using the HDRA on RDP1 and 2. On each Manhattan plot, the x-axis shows the SNPs along each chromosome; y-axis is the log10 (P value) for the association. Significant SNPs (with P values at or above the 10% FDR threshold) are coloured blue; significant SNPs within a 24 kb window surrounding candidate grain length genes are coloured red. Red vertical lines indicate the position of candidate genes and/or QTL regions associated with grain length. HDRA on RDP1 and 2 ALL, with 0 (a,g) and 2 SNP covariate (b,h): SNP-3.16732086 and SNP-5.5371749; HDRA on RDP1 and 2 INDICA, with 0 (c,i) and 2 SNP covariate (d,j): SNP-3.16732086 and SNP-5.5359497; HDRA on RDP1 and 2 JAPONICA, with 0 (e,k) and 2 SNP covariate (f,l): 3.16732086 and SNP-5.5366680.

We also performed GWAS independently for the two varietal groups, *INDICA* (*aus*+*indica*+admixed *aus-indica*) and *JAPONICA* (*temperate japonica*, *tropical japonica*, *aromatic*+admixed *temperate-tropical-aromatic*; Fig. 5b,c), as well as for each subpopulation individually. From a bird's eye view, the same major GWAS peaks on chromosomes 3 and 5 were detected in the joint *INDICA* and joint*JAPONICA* analyses as described previously. Peaks were defined as regions containing at least one SNP at$P<1.0e-05$ and at least two additional SNPs ($P<1.0e-4$) located within a 100-kb window on either side of the MS-SNP. The functional SNP in the *GS3* gene (SNP 3.16732086) on chromosome 3 was consistently identified as the MS-SNP with the highest PVE and allele effect in all analyses, while on chromosome 5, different but closely linked SNPs (mapping within ~12 kb of each other) identified the same peak in different analyses (Fig. 5). When the two MS-SNPs (from chromosomes 3 and 5) were used as covariates in each analysis, significant GWAS-peaks for grain length were detected in 12 regions of the genome in *ALL*, 13 regions in the joint *INDICA* analysis, and 2 regions in the joint *JAPONICA* analysis (Fig. 5). One of the 13 peaks identified in *INDICA* (on chromosome 8) and both of the peaks detected in *JAPONICA* were also identified in *ALL*, but no peaks were shared between *INDICA*and *JAPONICA*. This can be explained by the differences in allele frequencies, allele effects and per cent variance explained by the SNPs associated with grain length in each group.

An examination of the peaks in each of the individual subpopulations revealed novel features. In the *indica* and *tropical japonica* subpopulations, the two major peaks on chromosomes 3 and 5 were clearly visible, while in *aus*, a minor peak was identified several kb away from*GS3* and no peak was observed near *GW5*. In *temperate japonica*, a very wide peak was observed 1.4 Mb to the left of *GS3* along with a minor peak at *GW5*. These results suggest the possibility that different causal genes in the *GS3* region may contribute to the observed phenotypic variation in the *aus* and *temperate japonica* subpopulations. When SNP covariates were applied to the subpopulation-specific analyses, we discovered additional minor peaks associated with grain length in each. Notably, *aus* had twice as many significant minor peaks as *indica* (along with higher average allele effect and PVE values and *indica* had twice as many as either *tropical* or *temperate japonica* . Half of the minor peaks in the subpopulation-specific analyses overlapped with those observed in *ALL* or in the joint-*INDICA* or joint-*JAPONICA* GWAS, but the other half were novel, providing interesting targets for further research and use in breeding.

In many cases, GWAS peaks were found to encompass previously described genes or QTLs associated with grain length or shape. A minor peak on chromosome 3 in the *indica* subpopulation co-localizes with *POSITIVE REGULATOR OF GRAIN LENGTH* 1 (*PGL1*); *PGL1* encodes an atypical non-DNA-binding bHLH that interacts *in vivo* with *ANTAGONIST OF PGL1* (*APG*) and increases grain length and weight in transgenic rice. *PGL1-APG* potentially represents a new grain length- and weight-controlling pathway[56,57]. A minor peak on chromosome 2 in *aus* co-localizes with *qLWR2*, a QTL for length-to-width ratio detected in a cross between an *indica* and a *japonica*[43]. A peak on chromosome 7 detected both in *ALL* and in *tropical japonica* co-localizes with a fine-mapped QTL (*qGL7*) controlling grain length, also detected in an *indica* X *japonica* cross[43]. A peak on chromosome 5 in *aus* is located in-between the candidate genes *APG*[57] and *SGL/SRS3*[44,53]. *Short Grain Length* (*SGL*) is the causal gene of a small and round seed mutant (*sgl/srs3*, obtained from a tissue culture explant of *japonica* variety, Kitaake) and encodes a protein member of the kinesin 13 family. On chromosome 8, a peak detected in *ALL* as well as in the *aus*subpopulation co-localizes with the grain size gene *GW8/OsSPL16* (ref. [42]). This gene encodes a SQUAMOSA promoter-binding protein. The allele conferring long grain at this locus was discovered in a*basmati* variety, suggesting that it may be shared between *aromatic* and *aus* varieties[34,41,58]. Finally, a broad peak was detected on chromosome 10 in *ALL*, encompassing the *DWARF2* gene ~240 kb from the MS-SNP. This gene encodes a cytochrome P450 oxidoreductase involved in brassinosteroid biosynthesis[45]. This is the causal gene of a classic rice dwarf mutant, *ebisu dwarf* (*d2*)[45]. This is the first report of natural variation associated with several of these loci, most of which were discovered using mutant analyses.

To determine whether the major genes on chromosomes 3 and 5 interact epistatically with other loci, we performed a covariate vs SNP interaction scan using SNP-3.16732086 (*GS3*) and SNP-5.5371749 (*GW5*) as covariates. We detected a significant interaction between the MS-SNP at *GW5* and several SNPs in a ~100 kb region on chromosome 4 ($P=1.86e-21$), while no significant interactions with *GS3* were identified. The chromosome 4 region showed no significant effect on grain size when tested independently, only as an interaction. The minor allele frequency (MAF) for the most significant epistatic-SNP (SNP-4.4260961) is 10% across all subpopulations; it is essentially fixed in *aus*, *indica* and *temperate japonica* and segregates only in *tropical japonica* (MAF=24%). This suggests that both epistasis and differences in allele frequency contribute to the distinctive genetic architectures associated with grain length in the different subpopulations. The SNPs with significant *P* values for the *GW5* interaction define a region on chromosome 4 containing 14 genes. Among these, the most notable is a gene encoding a C2H2 zinc-finger transcription factor. RNA-seq data indicate that this gene is expressed in inflorescences pre and post emergence, as well as the seed (embryo and endosperm;http://rice.plantbiology.msu.edu/cgi-bin/ORF_infopage.cgi?orf=LOC_Os04g08034). Further investigation on the potential impact of this gene (or region) on grain length is warranted.

Harnessing the power of subpopulation-specific GWAS

Further examination of the peaks in each subpopulation demonstrates the value of GWAS as a road map for gene discovery. Here we highlight an example where a clear peak in the *GS3* region was observed in the *aus*subpopulation. The MS-SNP in *aus* (SNP-3.16709886) is located 18 kb to the left of*GS3* (Fig. 6a), two gene models away in an intergenic region, while the previously reported functional polymorphism in exon 2 of *GS3* (SNP-3.16732086) is fixed for the wild-type allele in *aus*, and therefore not significant. The *aus* MS-SNP-3.16709886 is also significant in *ALL* and in joint *INDICA* when SNP-3.16732086 (the *GS3* FNP) is included as a covariate in the GWAS model (Fig. 6a, second panel; Fig. 5a,b). These results suggest that SNP-3.16709886 tags variation for grain length that is independent of the known*GS3* allele. The mutation in exon two of *GS3* confers long grain, while four additional mutations in *GS3* (all occurring in exon 5) confer shorter grain[34,36]. Interestingly, the derived allele at the *aus* MS-SNP is associated with longer grains (Fig. 6).

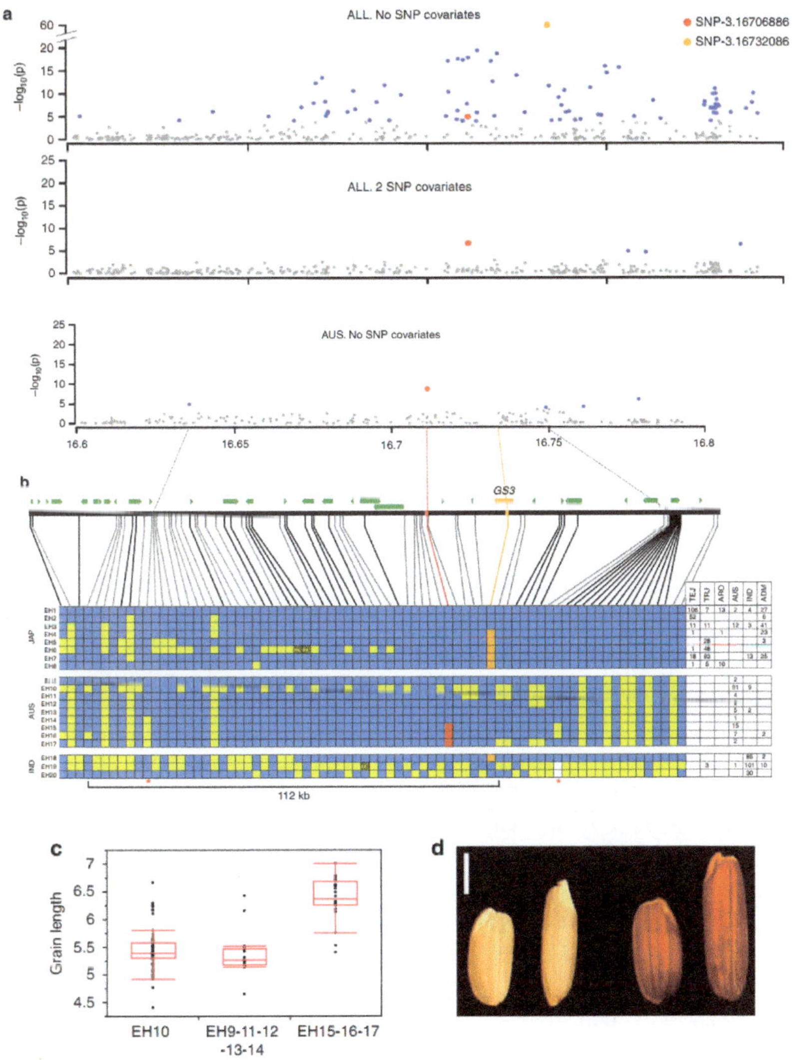

Figure 6: The *aus*-specific peak on chromosome 3. (a) Zoom-in of chromosome 3 (16.6–16.8 Mb) showing GWAS peaks over GS3 using the HDRA with RDP1 and 2 ALL (top), ALL with 2 SNP covariates (centre), and the aus subpopulation (bottom). The aus MS-SNP (SNP-3.16709886) is indicated by a red circle and the ALL MS-SNP (SNP-3.16732086) by an orange circle. Dashed lines connecting (a,b) indicate the position of the aus MS-SNP (red), the GS3 FNP (orange), and the second and third MS-SNPs in aus (grey). (b) Extended haplotype (EH) analysis across the peak region in RDP1 and 2. Top track shows the position of each SNP used in the analysis, connected by lines to the haplotype table below. Gene models are represented by green arrows; GS3 by an orange arrow. SNPs connected by a grey line are located in intergenic regions; those connected by a black line fall within genes. Rows in the haplotype table correspond to consensus haplotype groups constructed based on HDRA SNP data. Blue cells represent SNPs identical to the Nipponbare reference; yellow cells represent variant alleles; dashed yel-

low cells represent heterozygotes and white represent missing data. The aus-specific MS-SNP is highlighted in red and the GS3 FNP in orange. The second and third aus MS-SNPs are indicated by red stars under the haplotype table. The number and subpopulation identity of varieties carrying each haplotype are indicated in the table. EHs are represented as a consensus and minor variation is present within each group. For clarity, low-frequency recombinant haplotypes are not displayed (except for those aus recombinant haplotypes informative for the present analysis). (c) Quantile boxplots showing phenotypic distribution (average grain length) for varieties carrying the common aus haplotype, varieties with introgression carrying the reference allele at SNP-3.16709886 (EH9, 11, 12, 13 and 14); and with introgression carrying the derived allele at SNP-3.16709886 (EH15, 16 and 17). The ends of the each box represent the 75th and 25th quantiles (d) Representative aus accessions with short and long grain phenotypes: Left to right: Ghorbhai/NSFTV326 (EH10), Phudugey/NSFTV131 (EH17), Jamir/NSFTV328 (EH12) and PTB-30/NSFTV360 (EH15). Scale bar, 2 mm.

Our GWAS results point to a potentially novel source of variation in or near *GS3* in the *aus* subpopulation. To better characterize variation in the region we constructed extended haplotypes (EHs) based on HDRA genotypes for the 1,146 rice lines phenotyped for grain length (Fig. 6b). The EHs encompass a 190 kb region extending beyond the perimeter of the *aus*-specific peak as defined by its three MS-SNPs. Similar to what we have described previously[36], three well-differentiated major haplotype groups are observed in the region:*japonica*, *aus* and *indica* (according to the subpopulation in which each occurs at highest frequency). The most common *temperate japonica* haplotypes, EH1 and EH2 (83%) carry the wild-type allele at SNP-3.16732086, the common *tropical japonica* haplotypes, EH7, EH6 and EH5 (87%) carry the derived (long grain) *GS3* allele (indicated in orange in Fig. 6b), while the *aromatic* accessions are almost evenly split between two haplotypes carrying the wild-type and derived alleles at *GS3* (EH1 and EH8, respectively). The mutation marked by SNP-3.16732086 is known to have occurred in a *japonica* background then introgressed into *indica* (EH7 and EH18), but it is notably rare in *aus* and was not observed in any of the *aus* varieties in our panel.

The common *aus* haplotype is EH10 (62%) and the common *indica* haplotype is EH19 (40.9%). Two *aus*accessions carry the common *japonica* haplotype EH1, 12 accessions carry EH3 and two carry the recombinant EH9 haplotype, while 40 additional *aus* varieties are distributed among seven *aus*-specific haplotypes: three with the derived allele at SNP-3.16709886 (EH15/16/17) and four with the *wt* allele (EH11–14). These seven haplotypes are similar to EH3, but are differentiated by specific recombination break points within the *aus* background. Varieties carrying EH15/16/17 have significantly longer grains than those carrying the common *aus* haplotype EH10 (Fig. 6c,d), and even more so than lines carrying EH9/11/12/13/14, which are genetically similar across the region.

All 24 *aus* varieties that carry the derived MS-SNP (SNP-3.16709886) also carry a second SNP (indicated by the left red star on Fig. 6b). One short-grained *aus* variety was found to carry the second MS-SNP but not the derived allele at SNP-3.16709886 (EH14). The third MS-SNP (indicated by the right red star on Fig. 6b) is present in all 22 *aus* varieties carrying EH15/16, but not in the 2 lines carrying EH17. The recombination break points in EH16 on the left and in EH17 on the right allow us to define a 112-kb candidate region underlying this novel association with grain length (indicated by a bracket in Fig. 6b). The target region contains 13 gene models and includes *GS3* and its regulatory region. Additional random mutations across the*japonica*-like region of introgression are observed in many of the *aus* lines documented here. None of these mutations are informative in the context of this analysis, though they do suggest that the introgression is not a recent event.

Lastly, it is interesting to note that there is significant variation for grain length among the varieties carrying the common *aus* haplotype, EH10. This variation is the source of the minor GWAS-QTLs detected when SNP-3.16709886 is used as a covariate in the *aus* subpopulation-specific GWAS.

DISCUSSION

In this study, we set out to design a publicly available GWAS mapping platform that comprises genotype information on a set of accessions that will help to understand and dissect the genetic architecture of economically important traits in this crucial crop. We purified a total of 1,568 rice accessions and developed a 700,000-SNP array (HDRA) that assays one SNP approximately every 0.54 kb across the rice genome and is enriched for SNPs in and around genes.

In terms of SNP density, the HDRA offers more SNPs per kb than the highest density commercial human genotyping array (Illumina Human OMNI 5-4 BeadChips; Illumina, San Diego, CA, USA) and surpasses the largest publicly available genotyping platform in any crop species, namely a maize array with 600,000 SNPs, or 1 SNP per 4 kb (ref. [59]). While array technology has represented the gold standard in genotyping for many years, and is advantageous for generating high-quality data and efficiently processing large numbers of samples, it has inherent ascertainment bias owing to the emphasis on common SNPs in the discovery panel. While the decreasing cost and continued technical improvements in the throughput and quality of next-generation sequencing platforms is rapidly displacing the use of fixed arrays for genotyping, high-quality array data, such as that reported here, provides a valuable framework of diversity information into which new re-sequencing information can be readily integrated. In keeping with this potential, the recent re-sequencing of several thousand rice genomes[5,8] promises to enhance the utility of the resources and analysis tools described in this study. Though the rice varieties sequenced to date do not overlap significantly with the RDP1 and 2 plant materials reported here, haplotype information derived from the re-sequencing data can be readily integrated with the HDRA SNP data set to enhance the resolution of the genotypic data set, providing new opportunities to identify candidate genes and functional polymorphisms underlying many of the QTLs identified by GWAS.

The ability to identify both large- and small-effect QTL associated with a phenotype of interest is of great interest to breeders. By including the most significant SNP(s) from each experiment as a covariate(s) in the GWA model, we demonstrate our ability to identify both common and subpopulation-specific alleles. The high heritability of grain length suggests that this trait may be conditioned by only a few genes of large effect, but the range of variation for grain size detected in this collection of germplasm (4.2–7.3 mm) and the combination of both natural and artificial selection that have influenced trait expression in different subpopulations and environments, hint at greater complexity. Here, we demonstrate that, in addition to well-characterized genes of large effect, our approach allows for the detection of numerous genetic factors of small effect that contribute to the highly valued grain quality characteristics that distinguish local rice varieties, supported by the fact that several peaks co-localize with known QTL regions and candidate genes related to grain shape and/or size. In cases where well-characterized grain size genes do not co-locate with GWAS peaks, it is presumably because the functional alleles at these loci occur at low frequency in our diversity panel, or have immeasurably small allele effects in the context of other loci segregating for grain length. GWAS peaks identified in regions where no QTL for grain size has been mapped to date provide new opportunities for gene discovery.

Grain length is one of many traits for which different cultures and commercial markets show strong and often opposing preferences[58]. In Japan, Korea and northern China, short grain rice is strongly preferred. These areas grow almost exclusively *temperate japonica* varieties that have near universally short, wide grains and are adapted to cooler seasons and the long days of higher latitudes. Long, slender grains are preferred in South Asia and parts of South East Asia where *indica* varieties are grown; short, thin grains are prized in traditional *aus*-growing regions such as Bangladesh and southeastern parts of India, and very long, thin, aromatic grains are the specialty of the *basmati* and *sadri*-growing provinces of India, Pakistan, Afganistan and Iran[58,60]. Knowledge about specific genes and alleles that contribute to grain quality and yield in different genetic backgrounds will assist rice breeders in fine-tuning their selection for superior grain quality, and has implications for other crops where grain morphology is also highly valued.

In a companion paper by Crowell *et al.*[24], we explore the potential to use the rice diversity panel and accompanying HDRA SNP data set to dissect the genetics of panicle architecture on field-grown plants using a high-resolution imaging system as the basis for phenotyping. We demonstrate the potential to build 'association networks' based on GWAS analysis, and to use phenotypic covariates to dissect multi-trait GWA-peaks, with direct implications for applied plant breeding.

In addition to the SNP data set, genetic resources and analytical strategies presented here, we have also developed a suite of bioinformatics tools to facilitate the analysis of GWAS data (Fig. 7). These tools are comprised of a GWAS Viewer, Allele Finder and Genome Browser. The GWAS Viewer is a web-based interactive tool for visualization and analysis of GWAS data; the Allele Finder is a tool for querying and retrieving large genotype data sets in a user-friendly graphical interface; and our local mirror of the UCSC Genome Browser facilitates the analysis of HDRA and other SNP data sets in a genomic context. Together, these resources represent an open-source research platform that will empower the plant community to explore the genetics underlying a wide array of complex traits, providing new opportunities for geneticists and breeders to collectively understand and better utilize the wealth of natural variation that is conserved in gene banks around the world.

Figure 7: Data visualization tools. The GWAS Viewer (1) and the Allele Finder (2) were developed by the Rice Diversity Project to facilitate interpretation of GWAS results and the HDRA genotype data set. A local instance of the UCSC Genome Browser (3) was also created. All are available for public use at www.ricediversity.org. Users can select up to six plots to display in the main GWAS Viewer pane (4) by making selections in the control panel at the left (5). Points display SNP ID, position and P value when moused-over. Clicking on a data point (6) provides similar information in a pop-up window, along with a hyperlink to the Genome Browser. Users can highlight and zoom-in on a particular region by holding a click and dragging over (7,8) the region. When zoomed-in the IRGSP-1.0/MSU7 genome annotation features are displayed as grey triangles (9) and can be moused-over or clicked on for more information. Users can enter up to 100 MSU7 LOC IDs in the 'Add Candidate Genes' box (10) for display on a selected plot as orange triangles. A synchronized-zoom feature is available (11). Print or download an image by clicking at the top right of the plot (12), or click on the 'Print All Charts' button to print or download all plots displayed in the current session. Our local instance of the UCSC Genome Browser has several tracks of data (13) in addition to the IRGSP-1.0/MSU7 annotation; these include the RAP-DB genome annotation56, positions of SNPs on the Rice Diversity HDRA, the 44 K (ref. 17) and 1,536 (ref. 57) SNP chips and their functional annotations, and SNP sets from the Rice OPA breeder's mini-chips52. The Allele Finder tool allows the user to quickly find, examine and interpret specific regions of HDRA genotype data set. Enter a genome region of interest (14), select masking and display options (15), a data set and output format (16), select germplasm groups or subpopulations (17), and click the submit button. Along with the allele data the query results (18) display position information, reference and alternate allele, functional information, LOC_Os IDs, and predicted amino acid changes.

METHODS

Germplasm

The germplasm collections from which accessions used in this study were drawn are summarized. They consist of three different collections: (i) the RDP1 representing 433 accessions (423 accessions with seed for distribution) from 78 different countries, as described previously[16,33]; (ii) the RDP2 representing 1,445 accessions from 92 different countries, described here for the first time; and (iii) accessions contributed by The National

Institute of Agrobiological Sciences (NIAS)[61,62,63] consisting of 75 accessions from 19 different countries. The subpopulation distribution of lines analyzed for GWAS differed for the three diversity panels. Briefly, RDP1 consisted of 20% *indica*, 13% *aus*, 25% *temperate japonica*, 25% *tropical japonica*, 3% *aromatic* and 18% *admixed* accessions; the RDP2 consisted of 35% *indica*, 11%*aus*, 12% *temperate japonica*, 23% *tropical japonica*, 2% *aromatic* and 17% *admixed* accessions; the NIAS panel consisted of 39% *indica*, 23% *aus*, 15% *temperate japonica*, 5% *tropical japonica*, 0% *aromatic* and 18% *admixed* accessions.

Seeds were originally obtained from (i) the International Rice Genebank Collection at the International Rice Research Institute (IRRI) in the Philippines (http://irri.org/our-work/seeds), (ii) the USDA-ARS National Small Grains Collection (NSGC), Aberdeen, ID, USA, which is part of the National Plant Germplasm System (www.ars-grin.gov/npgs) and (iii) the NIAS Genebank in the National Germplasm Collection in Tsukuba, Japan (http://www.gene.affrc.go.jp). Because of the observed heterogeneity in the gene bank accessions, all materials were purified for two generations via single seed descent before genotyping. The purified lines analyzed here can be considered representative haplotypes from the original, heterogeneous landrace accessions available in the gene banks.

The collection of varieties

To acknowledge their unique genetic composition, RDP1 and RDP2 materials are registered as genetic stocks and have been assigned new accession numbers in the T.T. Chang Genetic Resources Center at IRRI, and in the Genetic Stocks-*Oryza* collection at the Dale Bumpers National Research Center in Stuttgart, AR, USA. Information about RDP1 accessions can be found in Eizenga *et al.*[33]. RDP2 accessions were selected to increase the number of accessions available for GWA, limiting the degree of potential duplication, and simultaneously to better represent the range of genetic variation in the combined RDP1 and 2 panels. In compiling the RDP2, particular emphasis was placed on the inclusion of *O. sativa* varieties adapted to parts of the world where food security is most at risk, as well as those that are or have been widely used in genetic studies and breeding at IRRI and among its national program partners.

Development of the HDRA

Methods for the development of the HDRA, including SNP discovery and selection, probe design, probe preparation and hybridization, genotype calling and quality control can be found.

SNP functional annotation

A total of 66,152 gene models were downloaded from the MSU Rice Genome Annotation Project version 7 in GFF3 format. Data were converted to the UCSC Browser's BED format and then uploaded to a local instance of the Galaxy Project bioinformatics server (http://galaxyproject.org/). All locations of exons, introns and UTR regions from the MSUv7 gene annotation were extracted using Galaxy's *Gene BED To Exon/Intron/Codon* tool. Assignment of SNPs to the extracted genomic features was determined by detecting overlaps with each feature's locus using the UCSC Browser's liftOver tool (https://genome.ucsc.edu/util.html). For those SNPs that overlapped an exon, further analysis was done using Galaxy's Amino Acid changes (aaChanges) tool in order to determine the character (for example, synonymous or non-synonymous) of each SNP's induced mutation. A script written in Python was used to merge and integrate all the results from the previous steps and assign each HDRA SNP to one or more functional categories. SNPs were classified as UTR5, UTR3, exonic, intronic based on overlap with the regions of the same name. In addition, SNPs were assigned to PPRs defined as segments within 2 kb upstream from the transcription start site. SNPs overlapping the PPRs were determined as described above. Since many SNPs were associated with more than one putative function, precedence was established for assigning SNPs to a primary category for counting purposes. The primary category was assigned using a Python script that assigned SNPs in the following priority order: exonic, intronic, UTR (5′/3′), PPR. The primary category assignment for each SNP was used as the basis for the rest of the functional analyses.

Principal components analysis

PCs were extracted from the genetic correlation matrix (using only SNPs present in all lines) by applying the svd function in the R software package[64].

Site frequency spectra

To generate site-frequency spectra, for synonymous and non-synonymous SNPs we divided the SNPs into 100 frequency intervals and counted the number of SNPs in each bin. We did this separately for synonymous and non-synonymous SNPs to compare the distributions within each category, making sure that binning frequency was the same and thus yielded comparable spectra. We made these spectra for RDP1 and 2 (*ALL*) as well as for *INDICA* and *JAPONICA* varietal groups. Since we do not have an outgroup, we plotted only minor allele frequencies, thus the frequency values do not exceed 0.5 on the plots. We also tried a range of bin numbers from 50 to 500, with the same results.

LD Decay

To assess LD among SNP markers in each of the *O. sativa* subpopulations (*aus*, *indica*, *tropical japonica*,*temperate japonica*, *aromatic*), we identified lines that belonged unambiguously to each group based on FastSTRUCTURE results (>70% ancestry). We discarded SNPs with >33% missing data and with fewer than three individuals carrying the minor allele. The samples consisted of 469 individuals (with 536,769 SNP markers) from *indica*; 170 individuals (430,077 markers) from *aus*; 356 lines (443,570 markers) from *tropical japonica*; and 229 lines (313,378 markers) from *temperate japonica*. We moved 4,000-SNP windows along each chromosome. In each window, we calculated LD between all pairs of markers, using only lines with pairwise complete data. As measures of LD, we calculated r^2 (the square of the correlation coefficient between SNP states) and the absolute value of D' (Lewontin, 1964)[65]. The latter measure is more robust to allele frequency fluctuations than r^2. As we calculated LD, we noted the distances between SNP markers. We then assigned the distances to 500 bins and calculated mean LD within each band of distances.

Phenotyping for grain length

Three replicates of the complete set of samples were grown in the 2011 wet season in IRRI fields in the Philippines, using standard crop management practices for irrigation and pest control. A subset of 129 accessions that spanned the range in grain lengths observed during 2011 was also grown during the dry season of 2012 at IRRI and grain length measured on these varieties was used to estimate broad sense heritability[66] using the following equation:

$$h_{BS}^2 = \frac{\sigma^2 g}{\sigma^2 g + \frac{\sigma^2 ge}{e} + \frac{\sigma^2 \varepsilon}{er}}$$

where σ_g^2 is the genotypic variance, σ_{ge}^2 is the genotype by environment interaction variance, ^{2} is the experimental error variance, r is replicates and e is environments. For both the full set (2011) and the subset (2012), grain was harvested at maturity and stored for 6 weeks under controlled relative humidity, to equilibrate for moisture content. A sample of paddy (10 g) was dehulled (THU35A 250V 50Hz Test Husker, Satake), and polished by placing grains into 10 ml capsules with fused aluminium oxide and then abrading gently for 1 h in a paint shaker. Grain length was measured on an average of 75 individual grains from each of three biological replicates, with a Cervitec Grain Inspector 1625 (FOSS Analytical, Hoganas, Sweden; Calingacion *et al.*[60]) at the Grain Quality and Nutrition Center at IRRI.

Genome-wide association studies

GWAS were run using a linear-mixed model in EMMAX vs beta-07Mar2010 (ref. [67]), which takes underlying population structure into account by including a kinship matrix as a covariate. The germplasm samples used for generating the HDRA SNP data set and all resulting genotyping GWA results were grown in 2011, and are subsets of the RDP collections consisting of 406 accessions from RDP1 and 1,087 from RDP2. The genotype data was limited to include only lines with phenotypes for the trait of interest (average seed length), and SNPs with MAF >0.05 in each subpopulation, as well as at least 3 minor allele count and 30% maximum missing data. These lines were included in the calculation of the kinship matrix. When the analysis was run across all available lines in the panel, SNPs present at MAF>0.05 in individual subpopulations were combined and three additional PC covariates were added to the model. No PC covariates were used in analyses within subpopulations (*indica*, *aus*, *temperate japonica* and *tropical japonica*). A significance threshold of 10% false discovery rate (R implementation of Benjamini and Hochberg[68]) was calculated. A GWAS pipeline for matching germplasm data and phenotypes, filtering and farther data processing was built in Python 2.6 using R scripts for making plots, and is available for download and use at ricediversity.org.

Allele effects, percent variation explained and epistasis scans

To calculate PVE and report this for all tests in the EMMAX results, we modified EMMAX C source code to perform this calculation using the model term variance that EMMAX already extracts to conduct the *t*-test for significance and report *P* values for the tested marker. We also modified EMMAX to calculate the total per cent variance explained of the model (including the random effects term) for every test as well as the additive genetic variance explained (without the random effects term). For total fixed effect genetic variance explained (relevant when two or more marker terms are included—see below), we extracted the residuals from an initial model fit with dummy marker terms substituted with random genotypes to produce the same degrees of freedom as each model of interest and calculated the variance. We subtracted this from the phenotypic variance to get the variance explained by the kinship matrix random effects term. In the model tests including fixed genetic effect terms, we subtracted this value from the total model variance explained to obtain the variance explained by the additive genetic effects of the markers. In the additive linear mixed model genotype variables were coded as AA=−1, AB=0, BB=1. Allele effects correspond to the regression coefficient. This is the expected absolute change in grain length for each B (non-reference) allele with twice this value equal to the expected grain length difference between the BB and AA homozygotes. This corresponds to the standard additive effects quantitative genetic model, where the dominance deviation d is constrained to be d=0 (co-dominant locus).

To perform a GWAS analysis for possible epistatic interaction effects with SNPs detected as highly significant in our single-marker scans, we modified further modified the EMMAX to allow specific markers to be specified on the command line as fixed model covariates. The modified EMMAX would then select these markers out of the input data and include the genotypes at these SNPs (more than one may be specified) as additive fixed-effect independent model terms in addition to markers being tested one-by-one in the genome-wide scan (the 'scan marker'). Allele effect size, *P* value and PVE were reported for these terms in every test as well as these values for the scan marker. As expected, fixed terms specified in this way generated similar allele effects, *P* values and PVEs as they did in their single marker test in the initial GWAS scan. Finally, we tested interactions of scan markers with these fixed covariate terms by modifying EMMAX source to calculate an epistatic interaction term as −1 for AA/AA, 0 for AA/BB and BB/AA and +1 for BB/BB; and 0 for any combination including a heterozygote genotype. This interaction term, as well as the independent effect term(s) for the specified markers, was inserted into the mixed-model fixed effect matrix (along with the scan marker). As with the independent covariate SNP terms, interaction effect size, *P* value and PVEs were output for all genome-wide tests.

Haplotype analysis across *GS3* region

EHs spanning a 190 kb region flanking *GS3* and the *aus*-specific peak in the region were constructed on the 1,146 *O. sativa* accessions from RDP1 and 2 genotyped with the HDRA and phenotyped for grain length. The HDRA carries a total of 424 total SNPs in the 190 kb region. SNPs with a MAF>0.05 and <3% missing data were initially selected. SNPs were then filtered based on a frequency test; only SNPs with a significant frequency difference between the *INDICA* and *JAPONICA* varietal groups (*P* value cutoff of 1.0e−06). The final set of SNPs used in constructing the haplotype map in Fig. 6 consisted of 76 SNPs.

ACKNOWLEDGMENTS

We gratefully acknowledge outstanding technical support from Fumio Onishi and Eileen Ceconi at Cornell, Teresa Hancock and Daniel Wood at the Dale Bumpers National Rice Research Center in Stuttgart, AR, USA, the molecular, conservation and field teams of the T.T. Chang Genetic Resources Center and the Grain Quality and Nutrition Center at IRRI, and valuable input during design of the HDRA from Elhan Ersoz. Funding was provided by the United States Department of Agriculture (USDA AFRI award # 2009-01881 to SMc); the US- National Science Foundation (NSF-PGRP awards #0606461 and #1026555 to SMc), Japan-World Bank Breeding project (DRPC2008-112 to IRRI), Bayer and Syngenta SKEP projects (DRPC2008-108 and DRPC2009-101 to IRRI).

Footnotes

Author contributions S.R.M., M.H.W. and J.M. designed the experiments; M.H.W. designed the HDRA array and methodology for calling genotypes and quality control procedures; J.M., P.K., M.H.W., A.J.G., L.G.M. and S.R.M. analyzed and/or interpreted data; M.H.W., C.-W.T., G.D.C., F.A.P., P.R.K.-F. and Y.S. performed data management; C.-W.T. and N.S. performed genotyping; K.L.M., N.S., M.E.B.N., S.M.Q.M., S.E.H., H.L., K.E., M.Y., G.E. and A.M. provided germplasm and/or germplasm management; G.E. and M.F. performed phenotyping; G.D.C., F.A.P. and D.A.B. developed database and bioinformatics tools; S.R.M. and L.G.M. wrote the manuscript.

REFERENCES

1. FAO. Increasing Crop Production Sustainably. The Perspective of Biological Processes Food and Agriculture Organization of the United Nations, Rome (2009).
2. Bouman B. A. M., Barker R., Humphreys E. & Tuong T. P. in Water for Food, Water for Life ed. Molden D. Earthscan; International Water Management Institute (2007).
3. Jackson M. T. Conservation of rice genetic resources: the role of the International Rice Genebank at IRRI. Plant Mol. Biol. 35, 61–67 (1997).
4. International Rice Genome Sequencing Project. The map-based sequence of the rice genome. Nature436, 793–800 (2005).
5. The rice genome project. The 3,000 rice genomes project. Gigascience 3, 7 (2014).
6. Huang X. *et al.* A map of rice genome variation reveals the origin of cultivated rice. Nature 490, 497–501 (2012).
7. Xu X. *et al.* Resequencing 50 accessions of cultivated and wild rice yields markers for identifying agronomically important genes. Nat. Biotechnol. 30, 105–111 (2012).
8. Duitama J. *et al.* Whole genome sequencing of elite rice cultivars as a comprehensive information resource for marker assisted selection. PLoS ONE 10, e0124617 (2015).
9. Schatz M. *et al.* Whole genome de novo assemblies of three divergent strains of rice, *Oryza sativa*, document novel gene space of *aus* and *indica*. Genome Biol. 15, 506 (2014).
10. Sakai H. *et al.* Construction of pseudomolecule sequences of the aus rice cultivar Kasalath for comparative genomics of asian cultivated rice. DNA Res. 21, 397–405 (2014).
11. Yu J. *et al.* A draft sequence of the rice genome (*Oryza sativa* L. ssp. *indica*). Science 296, 79–92 (2002).
12. Jacquemin J., Bhatia D., Singh K. & Wing R. A. The International Oryza Map Alignment Project: development of a genus-wide comparative genomics platform to help solve the 9 billion-people question. Curr. Opin. Plant Biol. 16, 147–156 (2013).
13. Famoso A. N. *et al.* Genetic architecture of aluminum tolerance in rice (*Oryza sativa*) determined through genome-wide association analysis and QTL mapping. PLoS Genet 7, e1002221 (2011).
14. Huang X. *et al.* Genome-wide association studies of 14 agronomic traits in rice landraces. Nat. Genet.42, 961–967 (2010).
15. Huang X. *et al.* Genome-wide association study of flowering time and grain yield traits in a worldwide collection of rice germplasm. Nat. Genet. 44, 32–39 (2012).
16. Zhao K. *et al.* Genome-wide association mapping reveals a rich genetic architecture of complex traits in *Oryza sativa*. Nat. Commun. 2, 467 (2011).
17. Shan Q., Wang Y., Li J. & Gao C. Genome editing in rice and wheat using the CRISPR/Cas system.Nat. Protoc. 9, 2395–2410 (2014).
18. Miao J. *et al.* Targeted mutagenesis in rice using CRISPR-Cas system. Cell Res. 23, 1233–1236 (2013).
19. Zhou H., Liu B., Weeks D. P., Spalding M. H. & Yang B. Large chromosomal deletions and heritable small genetic changes induced by CRISPR/Cas9 in rice. Nucleic Acids Res. 42, 10903–10914 (2014).
20. Collard B. C. & Mackill D. J. Marker-assisted selection: an approach for precision plant breeding in the twenty-first century. Philos. Trans. R. Soc. Lond. B Biol. Sci. 363, 557–572 (2008).
21. Septiningsih E. M. *et al.* Development of submergence-tolerant rice cultivars: the Sub1 locus and beyond. Ann. Bot. 103, 151–160 (2009).
22. Jiang G.-L. Molecular Markers and Marker-Assisted Breeding in Plants, Plant Breeding from Laboratories to Fields ed. Andersen S. B. DOI: 10.5772/52583 InTech (2013).
23. Shigesaburo T. & Takahashi N. in Developments in Crop Science Vol. 7ii, Elsevier (1984).
24. Crowell S. *et al.* Genome-wide association and high-resolution phenotyping link *Oryza sativa* panicle traits to numerous trait-specific QTL clusters. Nat. Commun. 7, 10527 (2015).
25. Zhao Y., Mette M. F., Gowda M., Longin C. F. H. & Reif J. C. Bridging the gap between marker-assisted and genomic selection of heading time and plant height in hybrid wheat. Heredity 112, 638–645 (2014).
26. Zhang Z. *et al.* Improving the accuracy of whole genome prediction for complex traits using the results of genome wide association studies. PLoS ONE 9, e93017 (2014).

27. Spindel J. E. *et al.* Genome-wide prediction models that incorporate *de novo* GWAS are a powerful new tool for tropical rice improvement. Heredity (in press).

28. Raj A., Stephens M. & Pritchard J. K. fastSTRUCTURE: variational inference of population structure in large SNP data sets. Genetics 197, 573–589 (2014).

29. Garris A. J., Tai T. H., Coburn J., Kresovich S. & McCouch S. Genetic structure and diversity in*Oryza sativa* L. Genetics 169, 1631–1638 (2005).

30. Xu Y. *et al.* Leaf-level water use efficiency determined by carbon isotope discrimination in rice seedlings: genetic variation associated with population structure and QTL mapping. Theor. Appl. Genet. 118, 1065–1081 (2009).

31. Price A. L. *et al.* Principal components analysis corrects for stratification in genome-wide association studies. Nat. Genet. 38, 904–909 (2006).

32. Kojima Y., Ebana K., Fukuoka S., Nagamine T. & Kawase M. Development of an RFLP-based rice diversity research set of germplasm. Breed Sci. 55, 431–440 (2005).

33. Eizenga G. C. *et al.* Registration of the rice diversity panel 1 for genomewide association studies. J. Plant Reg. 8, 109–116 (2014).

34. Takano-Kai N. *et al.* Evolutionary history of GS3, a gene conferring grain length in rice. Genetics 182, 1323–1334 (2009).

35. Fan C. *et al.* GS3, a major QTL for grain length and weight and minor QTL for grain width and thickness in rice, encodes a putative transmembrane protein. Theor. Appl. Genet. 112, 1164–1171 (2006).

36. Takano-Kai N. *et al.* Multiple and independent origins of short seeded alleles of *GS3* in rice. Breed. Sci. 63, 77–85 (2013).

37. Wang C., Chen S. & Yu S. Functional markers developed from multiple loci in GS3 for fine marker-assisted selection of grain length in rice. Theor. Appl. Genet. 122, 905–913 (2011).

38. Shomura A. *et al.* Deletion in a gene associated with grain size increased yields during rice domestication. Nat. Genet. 40, 1023–1028 (2008).

39. Li Y. *et al.* Natural variation in GS5 plays an important role in regulating grain size and yield in rice.Nat. Genet. 43, 1266–1269 (2011).

40. Weng J. *et al.* Isolation and initial characterization of GW5, a major QTL associated with rice grain width and weight. Cell Res. 18, 1199–1209 (2008).

41. Huang R. *et al.* Genetic bases of rice grain shape: so many genes, so little known. Trends Plant Sci.18, 218–226 (2013).

42. Wang S. *et al.* Control of grain size, shape and quality by OsSPL16 in rice. Nat. Genet. 44, 950–954 (2012).

43. Bai X. *et al.* Genetic dissection of rice grain shape using a recombinant inbred line population derived from two contrasting parents and fine mapping a pleiotropic quantitative trait locus qGL7. BMC Genet. 11, 16 (2010).

44. Kitagawa K. *et al.* A novel kinesin 13 protein regulating rice seed length. Plant Cell Physiol. 51, 1315–1329 (2010).

45. Hong Z. *et al.* A rice brassinosteroid-deficient mutant, ebisu dwarf (d2), is caused by a loss of function of a new member of cytochrome P450. Plant Cell 15, 2900–2910 (2003).

46. Ashikari M., Wu J., Yano M., Sasaki T. & Yoshimura A. Rice gibberellin-insensitive dwarf mutant gene Dwarf 1 encodes the α-subunit of GTP-binding protein. Proc. Natl Acad. Sci. 96, 10284–10289 (1999).

47. Tanabe S. *et al.* A novel cytochrome P450 is implicated in brassinosteroid biosynthesis via the characterization of a rice dwarf mutant, dwarf11, with reduced seed length. Plant Cell 17, 776–790 (2005).

48. Song X.-J., Huang W., Shi M., Zhu M.-Z. & Lin H.-X. A QTL for rice grain width and weight encodes a previously unknown RING-type E3 ubiquitin ligase. Nat. Genet. 39, 623–630 (2007).

49. Yamamuro C. *et al.* Loss of function of a rice brassinosteroid insensitive1 homolog prevents internode elongation and bending of the lamina joint. Plant Cell 12, 1591–1605 (2000).

50. Wang E. *et al.* Control of rice grain-filling and yield by a gene with a potential signature of domestication. Nat. Genet. 40, 1370–1374 (2008).

51. Abe Y. *et al.* The *SMALL AND ROUND SEED1* (*SRS1/DEP2*) gene is involved in the regulation of seed size in rice. Genes Genet. Syst. 85, 327–339 (2010).

52. Segami S. *et al.* Small and round seed 5 gene encodes alpha-tubulin regulating seed cell elongation in rice. Rice 5, 4–4 (2012).

53. Wu T. *et al.* Gene SGL, encoding a kinesin-like protein with transactivation activity, is involved in grain length and plant height in rice. Plant Cell Rep. 33, 235–244 (2014).

54. Wang Y. *et al.* Copy number variation at the GL7 locus contributes to grain size diversity in rice. Nat. Genet. 47, 944–948 (2015).

55. Wang S. *et al.* The OsSPL16-GW7 regulatory module determines grain shape and simultaneously improves rice yield and grain quality. Nat. Genet. 47, 949–954 (2015).

56. Heang D. & Sassa H. An atypical bHLH protein encoded by POSITIVE REGULATOR OF GRAIN LENGTH 2 is involved in controlling grain length and weight of rice through interaction with a typical bHLH protein APG. Breed. Sci. 62, 133–141 (2012).

57. Heang D. & Sassa H. Antagonistic actions of HLH/bHLH proteins are involved in grain length and weight in rice. Plos ONE 7, e31325 (2012).

58. Fitzgerald M. A., McCouch S. R. & Hall R. D. Not just a grain of rice: the quest for quality. Trends Plant Sci. 14, 133–139 (2009).

59. Unterseer S. *et al.* A powerful tool for genome analysis in maize: development and evaluation of the high density 600 k SNP genotyping array. BMC Genomics 15, 823 (2014).

60. Calingacion M. *et al.* Diversity of global rice markets and the science required for consumer-targeted rice

breeding. PLoS ONE 9, e85106 (2014).

61. Ebana K., Kojima Y., Fukuoka S., Nagamine T. & Kawase M. Development of mini core collection of Japanese rice landrace. Breed. Sci. 58, 281–291 (2008).

62. Yamamoto T. *et al.* Fine definition of the pedigree haplotypes of closely related rice cultivars by means of genome-wide discovery of single-nucleotide polymorphisms. BMC Genomics 11, 267 (2010).

63. Yonemaru J. *et al.* Genome-wide haplotype changes produced by artificial selection during modern rice breeding in Japan. PLoS ONE 7, e32982 (2012).

64. The R Core Team. R: A Language and Environment for Statistical Computing http://www.R-project.org (2014).

65. Lewontin R. C. The interaction of selection and linkage. I. General considerations; heterotic models.Genetics 49, 49–67 (1964).

66. Hallauer A. R. & Fo J. B. M. Quantitative Genetics in Maize Breeding 2nd edn Iowa State University (1988).

67. Zhang Z. *et al.* Mixed linear model approach adapted for genome-wide association studies. Nat. Genet. 42, 355–360 (2010).

68. Benjamini Y. & Hochberg Y. Controlling the false discovery rate–a practical and powerful approach to multiple testing. J. R. Stat. Soc. Series B Stat. Methodol. 57, 289–300 (1995).

CHAPTER 2 GENOME-WIDE RESEQUENCING OF KRICE_CORE FOR BREEDING

Tae-Sung Kim, Qiang He, Kyu-Won Kim, Min-Young Yoon, Won-Hee Ra, Feng Peng Li , Wei Tong, Jie Yu, Win Htet Oo, Buung Choi , Eun-Beom Heo, Byoung-Kook Yun, Soon-Jae Kwon, Soon-Wook Kwon, Yoo-Hyun Cho , Chang-Yong Lee , Beom-Seok Park, and Yong-Jin Park

Keywords: Whole-genome resequencing, Rice Core collection, Germplasm, SNP, INDEL, GWAS

BACKGROUND

Rice (*Oryza sativa* L.) is one of the most important staple crops in the world, providing a primary energy source for more than half of the world's population [38]. It is closely associated with economic and political stability in many developing countries, such as Asia and Africa [38]. Moreover, the amount of land suitable for agriculture is decreasing due to a variety of factors such as rapid climate changes and industrialization, while rice-eating human populations continue to grow [38]. To meet the global nutritional and socio-economic demands, dedicated efforts towards developing superior rice varieties need to be reinforced, such as accumulating and combining beneficial alleles [14, 27, 28, 37].

Rice germplasm collections continue to grow in number and size around the world [22, 35]. The International Rice Research Institute (IRRI) holds more than 11,000 accessions in the collections [22]. Since maintaining and screening such massive resources remains challenging, a significant portion of beneficial alleles in wild relatives and landraces remain under-utilized [7, 12, 22, 28, 35]. Thus, it is important to establish efficient methods to discover and exploit these unused novel alleles to maximize rice-breeding efforts [27, 28, 35].

Developing core collections has been proposed to simplify germplasm conservation and promote their effective utilization [7, 28, 31, 44]. A core collection or set refers to a subset that represents the genetic diversity of an entire genetic resource of a species [7]. A good core set minimizes redundant entries while preserving the majority of available genetic diversity of the entire collection [12], in which 10 % of the entire collection generally constitutes the core collection [7, 12]. However, if the size of the whole collection is too large, a core collection still becomes unmanageable [7, 12]. The mini core collection (about 10 % of the core) can then be subsequently developed from the core using neutral molecular markers to achieve genetic comprehensiveness [7, 12, 39].

The advent of draft genome sequences of two rice subspecies, *O. sativa* ssp. *japonica* (Nipponbare) and ssp. *indica* (93-11), along with subsequent completion of high-quality *indica, japonica*, and *aus* reference genomes has accelerated rice functional genomics research [19, 28]. In addition, these reference sequences serve as frameworks for whole-genome resequencing, which is accomplished by alignments of short sequence reads produced by the next-generation sequencing (NGS) technology [18, 20, 28, 51]. Recently, applications of genome resequencing are rapidly expanding toward various rice natural resources, providing the crop research community with unprecedented genomic resolution and scale, as well as relevant functional diversity accumulated in the rice germplasm [18, 20, 28, 51]. Under these circumstances, resequencing the germplasm core collections would be beneficial to the related community.

Here, we report the whole-genome resequencing of the 137 rice mini core collection, potentially representing 25,604 rice germplasms in the Korean genebank of the Rural Development Administration (RDA). Based on the Nipponbare reference genome, our resequencing data yielded more than 15 million (M) SNPs and 1.3 M INDELs. Phylogenetic and population analysis using 2,046,529 high-quality SNPs successfully assigned rice accessions to the relevant rice subgroups, suggesting that the SNPs capture evolutionary signatures present in rice subpopulations. We conducted genome-wide association studies (GWAS) on four agriculturally important

Citation: Kim T-S, He Q, Kim K-W, et al. Genome-wide resequencing of KRICE_CORE for breeding. BMC Genomics. 2016;17:408. doi:10.1186/s12864-016-2734-y.

traits including 'grain pericarp color', 'amylose content', 'protein content', and 'panicle number'. Among the detected association peaks, some identified previously discovered genes, indicating that KRICE_CORE can be implemented in GWAS to indentify novel alleles underlying agricultural traits. These results strongly suggest that resequencing results of KRICE_CORE are crucial for future rice breeding, as well as functional and evolutionary studies, in the post-genomic era.

RESULTS

Sequencing of the Korean heuristic rice core set

Of the 166 rice core set selected through a heuristic approach, 137 accessions that can flower in Chungcheong province, South Korea, were selected. The Korean rice core set (KRICE_CORE) included domestically adapted weedy and landrace rice and bred lines, as well as introduced lines from Africa, Europe, and America [25, 52].

We sequenced KRICE_CORE using the Illumina Hiseq 2000 or 2500 platform (Table 1). The sequencing quality of the raw reads was generally high (91.6 %). We next mapped short reads back to the IRGSP 1.0 rice genome (Methods). The mapping rate was generally high (95.48 % in average), varying from 89.5 to 98.86 %. We did not observe any significant correlations between mapping rate and sequencing depth, indicating that the poor mapping rate was not just derived from the low sequencing depth or errors. Thus, the relative differences in mapping rates may be due to sequence divergence between some accessions of KRICE_CORE and the reference sequence. Overall, effective mapping depth was 9X on average (Table 1).

Table 1: Summary of sequencing statistics for KRICE_CORE

Variation range	Sequence read	Clean read rate	Deduplication read	Deduplication rate	Mapping rate	Average depth
Max	59,334,970	95.1	54,800,574	99.4	98.9	13.9
Min	29,090,954	86.4	25,021,515	95.9	89.5	6.3
Average	39,753,733.9	91.6	36,006,200	98.8	95.5	9.0

Rice genome variation

We identified ~15 million (M) candidate SNPs from the 137 accessions (Table 2). About 13.5 % (2,046,529) of the SNPs did not contain missing data. We found that approximately 40.5 SNPs (5.5 by HQSNPs) were arrayed per kb across the KRICE_CORE genomes (Table 2 and Fig. 1a). We also found that 1.3 M INDELs existed in KRICE_CORE. Among those, 19.3 % were HQINDELs, showing an average density of INDEL per kb of 5.5 (0.7 for HQINDEL, respectively) (Table 2and Fig. 1a).

Table 2: Summary of chromosomal SNP and INDEL distribution for KRICE_CORE

Chromosome	SNP				INDEL			
	Total		High quality		Total		High quality	
	Count	Density[a]	Count	Density	Count	Density	Count	Density
1	1,492,529	34.5	245,749	5.7	157,701	3.6	34,662	0.8
2	1,194,215	33.2	219,868	6.1	123,316	3.4	28,870	0.8
3	1,117,000	30.7	227,031	6.2	116,347	3.2	30,925	0.8
4	1,482,884	41.8	151,464	4.3	118,490	3.3	19,101	0.5
5	1,136,915	37.9	190,680	6.4	98,439	3.3	23,019	0.8
6	1,279,651	41.0	175,920	5.6	117,347	3.8	22,733	0.7
7	1,268,980	42.7	152,153	5.1	112,520	3.8	20,766	0.7
8	1,323,731	46.5	148,873	5.2	113,720	4.0	18,831	0.7
9	969,987	42.2	128,492	5.6	85,643	3.7	16,566	0.7
10	1,081,145	46.6	141,153	6.1	91,306	3.9	17,727	0.8
11	1,425,491	49.1	148,897	5.1	128,449	4.4	17,902	0.6
12	1,352,277	49.1	116,249	4.2	115,237	4.2	15,040	0.5
Total	15,124,805	40.5	2,046,529	5.5	1,378,515	3.7	266,142	0.7

[a]Density = SNP/INDEL count/length of the chromosome

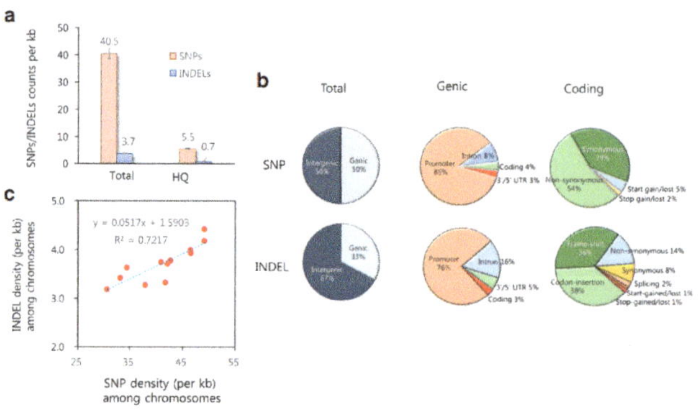

Figure 1: SNP and INDEL frequency in the KRICE_CORE population. **a** Comparison of the mean density of SNPs and Indels in KRICE_CORE. **b** SNP and INDEL distribution across various genome regions. The promoter of the genic region refers to the region 2 kb upstream of the transcription start site. **c** Correlation between SNPs and INDEL occurrences across KRICE chromosomes

About 50 % of SNPs were located in genic and intergenic regions, respectively. The majority of genic SNPs were distributed in the promoter region (85 %), followed by intronic (8 %), coding (4 %), 5′/3′UTR (3 %), and splicing (0.02 %) regions. In the coding region, we found that SNPs causing non-synonymous amino acid changes (non-synonymous SNPs) were enriched (54 %). The ratio of nonsynonymous to synonymous SNPs was 1.38 (54 % vs. 39 % of coding SNPs), which was similar to previous reports [1, 2].

A lower abundance of INDELs is located in genic regions (33 %). In addition, codon-insertion and frame-shifts are the most abundant classes of INDELs (38 and 36 %, respectively) in the coding region.

The density of SNPs and INDELs varied among rice chromosomes, ranging from 30.7 to 49.1 (4.2–6.4 with HQSNPs) and 3.2 to 4.2 (0.5–0.8 with HQINDEL), respectively. We also found that the densities of SNPs and INDELs were positively correlated in chromosomes (Fig. 1c). To determine whether this correlation also occurred across local genomic regions, we investigated the correlation between SNP and INDEL densities within 50-kb windows across rice genomes (Fig. 2a). We found that SNPs are clustered in the rice genome (Fig. 2a), suggesting that SNPs were generated non-randomly across the rice genome [2, 15, 33, 47]. Furthermore, we observed a significant positive correlation between SNPs and INDELs across the rice genome, indicaing that INDELs are also clustered (Fig. 2b). Thus, the shared SNP and INDEL clustering patterns may be affected by a genome-wide set of regions that either commonly promote or restrict the genesis of SNPs and INDELs [2, 6].

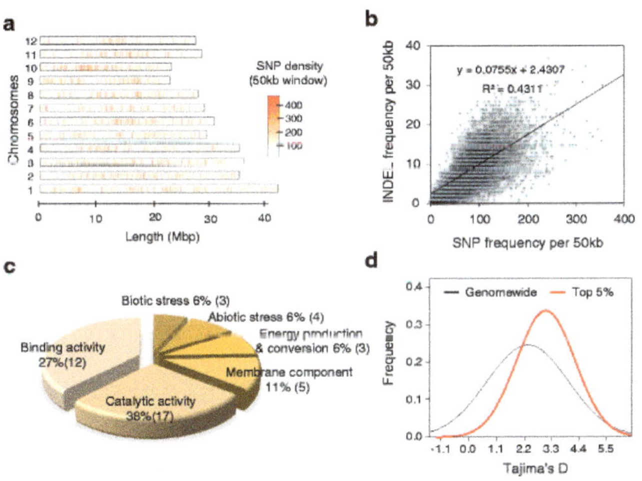

Figure 2: Genome-wide distribution of SNPs and INDELs of KRICE_CORE. **a** SNP density of the 50-kb window across the KRICE_CORE genome. **b** Correlation between SNP and INDEL occurrence across KRICE_CORE genome. **c** Functional category of genes in SNP/INDEL-enriched regions. **d** Distribution of Tajima's D values of the genome wide and top 5 % SNP enriched regions

We next investigated functional categories in the regions where SNP and INDEL frequencies were commonly high. Therefore, we performed gene enrichment tests for the top 5 % of regions. Among the genes functionally annotated, catalytic and binding activity were mostly enriched, followed by abiotic/biotic stress responses, energy production & conversion, and protein transport, which are important to maintain rice physiology. Considering that the mean Tajima's D value for the windows of the top 5 % region was significantly higher than that at the genome-wide level, the accumulated SNPs and INDELs in these genes may facilitate creation of functional diversity (Fig. 2d).

Population structure and genetic diversity of KRICE_CORE

Using 2,046,529 HQSNPs, we performed population genetic studies to explore the genetic structures of KRICE_CORE. First, we constructed a neighbor-joining tree (Fig. 3a). To gain legitimate evolutionary insights from the resultant phylogenetic subdivisions, we incorporated the previously defined rice accessions into the tree analysis (Methods) [51]. We found that our tree analysis separated into five major subgroups, which partitioned into three major groups corresponding to japonica (81 accessions, 59 %),*indica* (hereafter *IND*, 43, 31 %) and *aus* (*AUS*, 6 %), with a further subdivision of *japonica* into *temperate* (*TEJ*, 62, 45 %) and *tropical* (*TRJ*, 19, 14 %) rice groups. However, KRICE_CORE does not reflect *Group V* since only two accessions belong to the sub group.

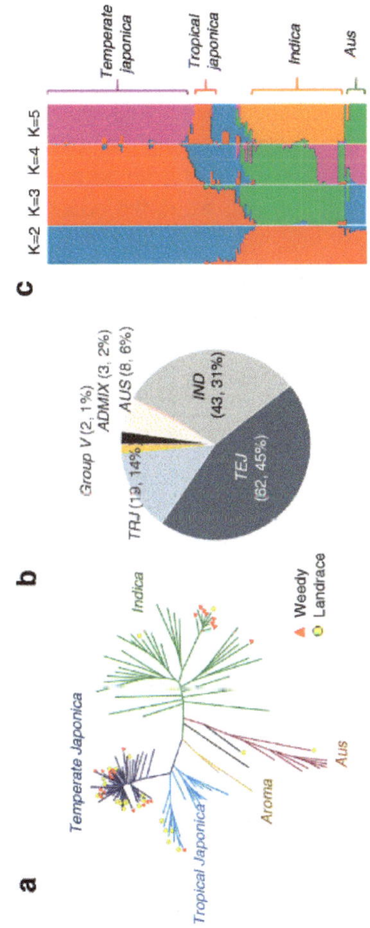

Figure 3: Population structure of KRICE_CORE. **a** Neighbor-joining analysis among KRICE_CORE using 2,046,529 HQSNPs. Landraces and weedy accessions are denoted as red triangles and yellow circles in the tree, respectively. **b** Pie graph designating the proportion of each subgroup. **c** Population structure analysis using FRAFFE. Each color represents one population. Each accession is represented by a horizontal bar, and the size of each colored segment in each horizontal bar represents the proportion contributed by ancestral population. K value represents the number of assumed clusters or populations

We found that Korean landrace and weedy accessions are mostly present in *TEJ* groups (Fig. 3a). Note that some are assigned to different subgroups, such as *IND*, *AUS* and

TRJ (Fig. 3a). This led to the hypothesis that independent domestications of *TRJ* type together with *IND* (and others) have occurred in the Korean peninsula. However, these results may be derived from the outcrossing between weedy rice and *IND and TEJ*type rice varieties (Tongil type), which are widely cultivated in Korea. Since we have limited evidence for indica landrace rice, further studies are required to support this hypothesis.

We further investigated plausible population structures using FRAPPE, which estimates individual ancestry and admixture proportions assuming that K populations exist based on a maximum likelihood method (Fig. 3c). We analyzed the data by increasing K (the number of population) from 2 to 5 (Fig. 3c). For K=2, we observed a division between *O. rufipogon*/japonica and *O. nibara*/others, including *IND* type. When K=3, *AUS* subgroups were clearly separated from the *IND* group. When, K=4, *TRJ* and *TEJ* were separated. When K=5, a new subgroup including *group V* emerged from *TRJ* (Fig. 3c). These results supported the previous hypothesis that the two cultivated rice subspecies have been domesticated independently, forming two ancient populations; one population that gave rise to *indica* and the other to *japonica*, respectively. Since we independently confirmed that *O. rufipogon* and *O. nivara* were assigned to *japonica* and *indica* subgroups, we further supported the hypothesis that one population that gave rise to *indica* may have evolved from an ancestral population closely related to *O. nivara*, while the other that gave rise to *japonica* may have originated from *O. rufipogon* [51].

We estimated the sequence diversity of KRICE_CORE. Reduction of diversity (ROD) was calculated from the sequence diversity (π) of KRICE_CORE and wild rice, *O. rufipogon,*across the entire rice genome. We found that the ROD value of KRICE_CORE was highly dynamic across the rice genome, ranging from −30 to 1. Since HQSNPs were used in the analysis, it is possible that the majority of ROD values were correctly estimated. However, some RODs may be exaggerated, especially when π of wild rice was close to zero but those of others were higher.

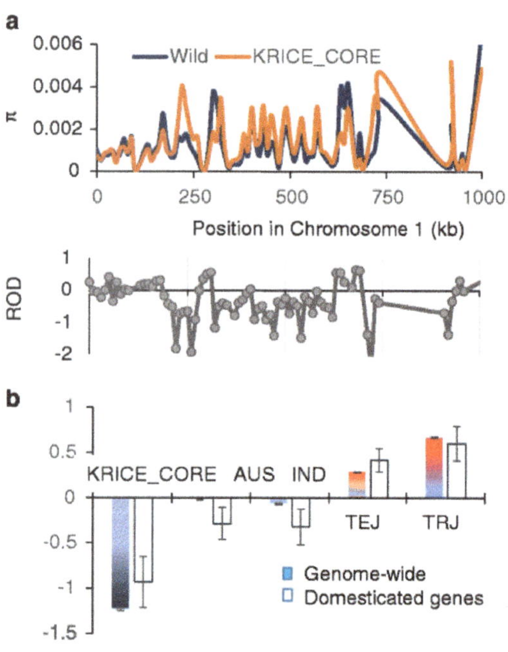

Figure 4: Genetic diversity of KRICE_CORE. **a** Nucleotide diversity (π) of wild rice (*O. rufipogan*) vs KRICE_CORE (*upper*) and the resulting ROD value (Methods). **b** Mean RODs among the subgroups in genome-wide or domesticated regions. Error bar indicates standard error of the mean (SEM). Sliding window analyses of π or ROD are shown for chromosome 1 with a 10-kb window

We found that KRICE_CORE potentially retains a high diversity level. Since *TRJ* is more closely related to *O. rufipogon* than *TEJ*, the average ROD value for *TRJ* of KRICE_CORE is much higher than that of *TEJ* (0.67 vs 0.29, $P<0.001$ [one-way ANOVA]). However, incorporation of the divergent *AUS* and *IND* to KRICE_CORE, which decreased the ROD values greatly, significantly increased the overall diversity, as measured by ROD values (Fig. 4b). We also investigated ROD for regions that contain previously reported domesticated genes. Considering that the amount of genetic diversity for the selected regions were comparable both at the genome-wide level and within each subpopulation (Fig. 4b), KRICE_CORE accommodates abundant functional diversities shaped from diverse external environments. These results will be useful for future rice breeding and fundamental studies in the post-genomic era to address current climate changes. Overall, these genetic and population structure results suggest that the size of KRIC_CORE is small but effectively retains genetic comprehensiveness.

Genome-wide association studies (GWAS)

We performed GWAS to determine whether we could exploit the natural diversity present in the KRICE_CORE to detect agriculturally useful alleles via association genetics. We chose four agriculturally important phenotypes ('pericarp color', 'amylose content', 'number of panicle per plant', and 'rice seed protein content'), which are representative of the qualitative or quantitative traits in KRICE_CORE. We observed a wide range of phenotypic variation, spanning a 1.7- to 8.6-fold difference. To determine whether the significant variabilities have a genetic basis, we employed a basic case/contro association test (for qualitative traits) using PLINK or a compressed mixed linear model (for quantitative traits) using GAPIT (Methods).

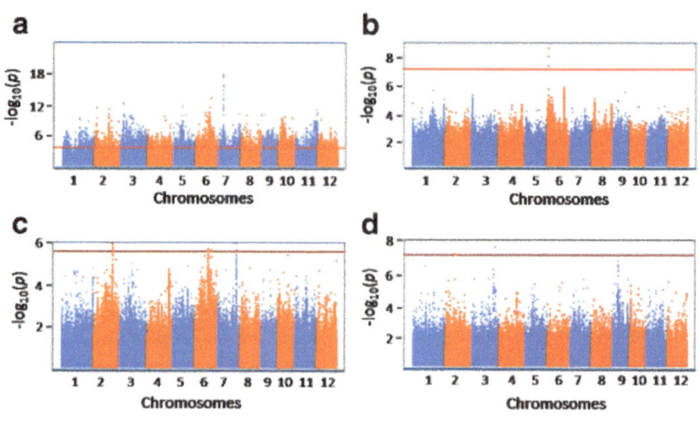

Figure 5: Genome-wide association studies of 'pericarp color' (**a**), 'amylose content' (**b**), 'rice seed protein content' (**c**), and 'number of panicles per plant' (**d**). Manhattan plots of the linear (for a) or mixed linear model (for **b–d**) are shown from negative log10-transformed *P*-values, which are plotted against the positions on each of the twelve chromosomes. A red horizontal line indicates the genome-wide significance threshold

We found that GWAS signals associated with pericarp color and grain amylose percentage pinpointed the previously reported Rc locus at chromosome (ch) 7 and Waxy locus (at ch 6), respectively. For the red pericarp color, the most significantly associated polymorphism (P=3.05e-19) that explained ~46 % of the phenotypic variance was the 14 bp INDEL at 6,068,072 bp, which is located in the seventh exon of Rc. For amylose content, one of the most significantly associated polymorphisms was located at the 5′ UTR region of the Wx locus (Os06g013300, at 1,765,761 bp of ch 6), explaining about 52 % of the phenotypic variation (Table 3).

Table 3: Genome-wide significant association signals of agronomic traits using the linear and compressed MLM

Traits	Ch	Position	Major allele (Count)	Minor allele (Count)	MAF[a]	R2	P value	Known Loci (Reference)
Pericarp color	7	6068072	A(95)	G(41)	0.30	0.465	3.05E-19	Rc(46)
	6	21304669	T(93)	C(43)	0.32	0.286	4.87E-11	
	3	3640060	G(116)	A(13)	0.12	0.268	2.48E-10	
	5	11479616	C(108)	T(26)	0.20	0.255	7.57E-10	
Amylose content	6	1765761	G(73)	T(61)	0.46	0.555	5.55E-09	Wx(21)
Protein content	7	24810573	A(77)	G(59)	0.43	0.348	1.14E-06	
	2	26791708	C(110)	T(25)	0.19	0.347	1.27E-06	
	6	18133739	G(93)	A(42)	0.31	0.339	2.15E-06	
Panicle number	3	30877559	C(128)	T(9)	0.07	0.404	3.13E-08	
	9	7760141	C(128)	T(0	0.07	0.386	1.09E-07	

[a]Minor allele frequency

We observed that the GWAS mapping precision varied between Rc and Wx regions. For pericarp color, the significantly associated SNPs were scattered around the 287-kb region of ch7 (Fig. 6a), but amylose content-related SNPs were dispersed within a 33-kb region (Fig. 6b). To determine whether the mapping accuracy was compounded by different evolutionary processes, such as specific domestication histories, we investigated the genetic diversity level and linkage disequilibrium (LD) structure across the 400-kb region around the association peaks (Fig. 6b–d; Methods). We found that the ROD score of the 400-kb region that included Rc (Rc region) was significantly higher than the Wx including the 400-kb region (Wx region) (mean ROD −0.68 (Rc region) vs-1.8 (Wx region), P<0.001; one-way ANOVA), indicative of a significant reduction in the genetic diversity in the Rc region (Fig. 6a-c). This substantial genetic reduction of the Rc region in KRICE_CORE may be derived from the strong selection in favor of the 14-bp deletion since the domestication of the white grain rice. We also investigated LD decay across Rc and Wx regions (Fig. 6a–c). We found that the average LD decay of KRICE_CORE was about 85 kb. The LD decay of the Wx region was comparable, but the local LD pattern of the Rc region was atypical, indicating that the LD decay does not neutrally occur against the physical distance (Fig. 6d). To further examine the aberrant LD decay pattern over the Rc region, we constructed LD maps of Rc/Wx genomic regions (Fig. 6e–f). We found that obvious block-like LD structures exist in the Rc region, indicating that more powerful selective pressure might have acted on the Rc region (Fig. 6e, f). In addition, this extended LD structure may cause spurious associations in the Rc region, complicating prediction of the causative gene through GWAS mapping.

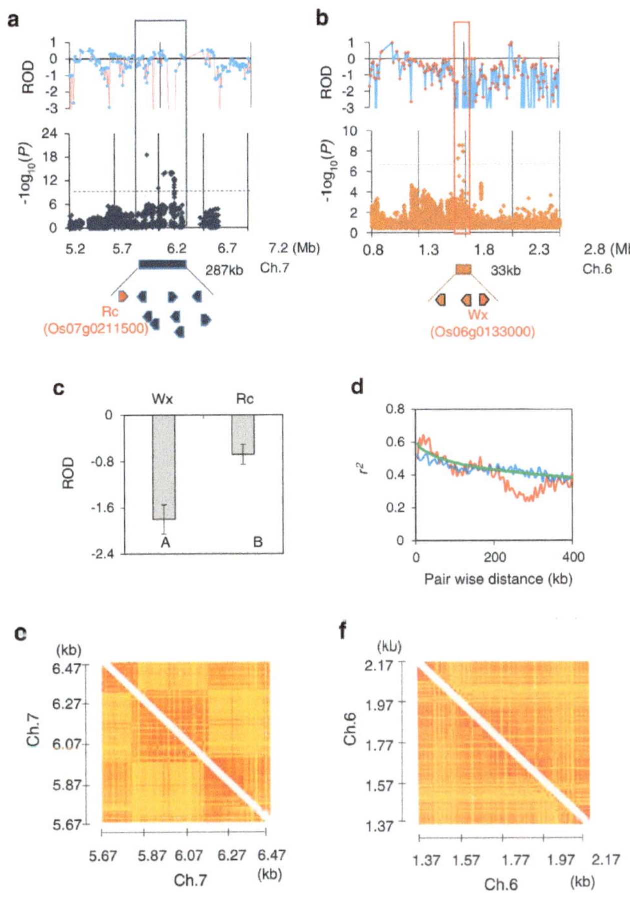

Figure 6: Regions of the associated signals near the Rc (**a**) and Wx (**b**) regions. The top of each panel shows the ROD for a 1-Mb window around the peak SNPs. Negative $\log_{10}$P-values for each SNP from the linear or compressed mixed linear model are plotted. Blue or dark orange dashed horizontal lines indicate the genome-wide significance cutoff for the Rc and Wx regions, respectively. The bottom of each panel within the blue or red box denotes the range of SNPs over the cutoff. **c** Mean ROD values for 400 kb around the Rc and Wx regions. **d** LD decay plots of the regions compared to genome-wide LD decay. **e** and **f** LD blocks of 400 kb on each side of the association peak for the Rc (**e**) and Wx (**f**) regions

Several significant association signals were found for seed protein contents and panicle number. Although we could not identify any strong candidate genes for those traits, the novel SNPs identified here can be implemented in related breeding studies.

DISCUSSION

To address upcoming challenges for food production, the efficient use of rice germplasms is critical [13, 28]. These rice germplasms, adapted from diverse ecosystems, are valuable for rice breeders/researchers, potentially providing a broad array of useful alleles that enrich gene pools of current cultivated rice varieties [17, 19, 28]. Although *ex situ* conservation efforts are important for preserving diverse rice genetic resources, identifying novel and favorable genetic variants from the vast genebank collection remains challenging, requiring extensive screening processes [12, 14, 22, 35]. Therefore, the rice core collection is a powerful approach to accelerating utilization of exotic germplasms of the entire population [7, 31, 44]. In addition, the application of whole-genome resequencing technology would establish a potent platform for rapid forward genetic studies and genome-assisted breeding, which would facilitate discovery and exploitation of useful alleles [19, 28, 35].

Genomic variations and population structure of KRICE_CORE

Here, we performed genome-wide resequencing of 137 KRICE_CORE populations originating from the 25,604 rice germplasms in the Korean genebank [52]. We identified ~15 million (M) candidate SNPs from the 137 KRICE_CORE (Table 2); 13.5 % (2,046,529) were HQSNPs at a 9x depth. Our neighbor-joining tree analysis using HQSNPs clearly showed that KRICE_CORE comprises the major rice subgroups, including *IND* together with *TEJ*,*TRJ*, and *Aus* . Regarding the 3,000 rice genome analyses, approximately 75 % of the SNPs were covered by our study, although only 4 % of rice accessions were sequenced, suggesting that KRICE-CORE represents the diversity of the whole germplasm collection with relatively few accession numbers. These data indicate that the implementation of heuristic approaches using PowerCore software effectively reduced core entry but retained maximum diversity [25].

We found that many Korean landrace and weedy accessions were assigned to distinct ecotype groups. Given that *TEJ* are dominant in the Korean peninsula, it is possible that *TRJ* and*IND* rice may have been introduced or domesticated historically. Recently, South Korea has faced gradual changes in many climate parameters, including annual temperature, rainfall amount, and precipitation [9]. The most significant climate change reported is a massive increase in the range of temperature fluctuations throughout the four seasons. This leads to heavier precipitation during the summer together with elevated temperature [24]. As a result, the Korean Peninsula is predicted to become a subtropical region [23]. To rapidly adapt to on-going climate change, these *IND* and *TRJ* weedy and landrace accessions acclimated in Korea would be valuable genetic resources. Still, many interesting evolutionary questions remain for these landrace and weedy lines, including their origins and transferring/fixation processes [42]. Addressing these questions could increase our understanding of the relevant domestication mechanisms, providing information for future improvements of Korean rice cultivars [42].

In KRICE_CORE, we observed a clear separation between *japonica* and *indica* type at $K = 2$. As the K value increased, *aus* diverged from the *indica* population. *Tropical* and *temperate* japonica types were separated from the *japonica* population (Fig. 3c). Since *O. rufipogon* and *O. nivara* are assigned to the *japonica* and *indica*at K=2, respectively, we support the hypothesis that two cultivated rice subspecies, *indica* and *japonica*, may have originated from two ancient subpopulations [43, 51]; one population that gave rise to *indica* may have evolved from an ancestral population from *O. nivara* and the other for *japonica* may have originated from *O. rufipogon* [43, 51]. However, further studies are required to determine which model explains the actual domestication processes.

Functional significance of SNPs and INDELs

It is expected that about 40 SNPs and 5.5 INDELs are arrayed per kb across the KRICE_CORE genome (Fig. 1a). However, we found that SNPs and INDELs are not evenly distributed. Rather, significant portions of SNPs and INDELs are clustered in the KRICE_CORE genome (Figs. 1c and 2a). Interestingly, the clustering patterns of those two genetic variants are strongly correlated (Fig. 2b), suggesting that SNPs and INDELs may arise non-randomly and such genesis mechanisms are highly connected [2, 6, 48]. These results are consistent with observations in mammalian and avian genomes [6, 48], indicative of the presence of mutation hot spots or cold spots across the genome [3]. The local recombination variability is reported as one of the key determinants for the mutation hot spot, driving sequence evolutions and genetic diversities [3]. Although high mutation rates may be an important driver underlying the co-clustering pattern, other evolutionary forces could create the non-randomness [2]; among these, natural selection can modulate local variability's along a chromosome from responding to the ambient environments [11, 49]; balancing selection tends to create regions of increased variability [2, 8, 50], while purifying and directional selection tend to reduce variability [40]. Thus, the effect of mutagenic hot spots in population data may be obscured by a complex interplay of these factors [3]. Varela and Amos (2010) showed that INDEL locations are conserved in the same location between humans and chimpanzees, but SNP density appears highly variable; clusters found in chimpanzees are often not found at the same site in humans, indicating that a species dependent regional context or signature may regulate the surrounding SNP density following the genesis of INDEL [16, 30, 41, 48, 50]. Further studies should explore the major determinants causing SNP/INDEL clustering and its evolutionary consequences in plant/crop species.

We found that many genes in important biological processes, including, catalytic and binding activity, abiotic/biotic stress responses, and energy production are distributed in the high SNP regions. Given that the mean Tajima's D value for the windows of the top 5 % genic SNP region is significantly higher than that at the genome-wide level (3.03 vs 2.36 in mean Tajima, $p<0.01$ by bootstrapping method), it is believed that accumulated SNPs and INDELs in these genes are more or less likely to be beneficial in creating multiple alleles, potentially resulting in functional diversity. Thus, resequencing results of KRICE_CORE would increase our understanding of the molecular basis of the functional differences, especially related to important agricultural traits [1].

Genome-wide association studies (GWAS)

GWAS has been implemented to efficiently identify candidate genes related to various useful agricultural traits in many crop species, including rice [19, 29]. Given that significant associations between genetic variations and phenotypic diversity does not require prior knowledge, GWAS with high genome coverage of SNP markers provides a genomics platform to dissect previously unknown adaptive genetic variation accumulated in plant germplasm resources over time [5]. After identifying candidate genes, GWAS allows for informed choice of parents for QTL analysis based on the haplotype information, along with suggesting targets for the following mutagenesis and transgenic studies [26]. Here, we conducted proof-of-concept GWAS analysis using four agriculturally important phenotypes (Fig. 5); the 'pericarp color' represents the qualitative trait of KRICE_CORE, while the others are for the quantitative traits.

We found that GWAS signals associated with pericarp color and grain amylose percentage pinpointed the previously reported Red pericarp (Rc) and Waxy (Wx) loci, respectively (Fig. 5). For the red pericarp color, the most significantly associated polymorphism (P=3.05e-19) is the 14 bp INDEL, which is located in the seventh exon of Rc (Fig. 6a and Table 3). Sweeney et al. (2006) reported that the recessive rc allele by the 14-bp deletion nullifies the regulatory function in the proanthocyanidin synthesis pathway, preventing development of the pigmented pericarp layer [45, 46]. We found that the majority (>94 %, 80 out of 85) of white pericarp KRICE_CORE accessions contain a 14-bp deletion in exon 7 of the gene. The exceptional five white pericarp accessions that do not have the deletion may have been generated through rc-independent mechanisms. For amylose content, one of the most significantly associated polymorphisms is located in the 5′ UTR region of the Wx locus (Fig. 6b). Very close to the association peak, we identified a splice donor site in intron 2 of Wx. Previously, it was reported that Wx^b, the GtoT point mutation at the site, is associated with the absence of amylose-characterizing glutinous rice varieties, suggesting that natural variations to the site create functional diversities for the amylose content [21]. Considering that the association peaks are found dominantly at Rc and Wx in KRICE_CORE and others [20], Rc and Wx largely underpin the red pericarp color and amylose content, respectively [20, 45, 46]. To improve the power to recover meaningful but minor associations (especially for amylose content), the sample size should be increased using diverse germplasms [4, 26, 34].

We observed that the GWAS mapping precision varies between Rc and Wx regions. For the pericarp color, the significantly associated SNPs are scattered around the 287-kb region of ch7, but amylose content-related SNPs are dispersed within the 33-kb region (Fig. 6a, b). We concluded that the much stronger LD pattern of Rc may complicate accurate prediction of the causative gene, as observed in other rice GWAS (Fig. 6e, f) [20, 45, 46]. Hence, it is believed that local LD estimations may be required when candidate genes are considered following GWAS. Besides, several significant novel associations were found for seed protein contents and panicle numbers. These results clearly demonstrate that GWAS implemented in KRICE_CORE offer a viable approach for dissecting the causal genetic variants of qualitative and quantitative phenotypes.

METHODS

Genomic DNA library construction and sequencing

Quality control (QC) for prepared genomic DNA was performed; $A_{260/280}$>1.6 and $A_{260/230}$>1.6 for UV absorbance using Nanodrop (Tecan) and ≥5 ug fluorescence concentration using the Quant-iT BR assay kit (Invitrogen). DNA was run on 0.7 % agarose gels to assess degradation.

Genomic DNA was fragmented into approximately 100–300-bp insert sizes by sonication. Each DNA fragment was end-repaired using the Paired-End DNA Sample Prep Kit (Illumina; San Diego, CA, USA), followed by the addition of a 3′-A overhang and ligation of the adapters. Size-selection for DNA fragments of 200–300 bp was performed using a gel extraction kit (Qiagen). The constructed DNA samples were quantified using Quant-iT™ dsDNA High Sensitivity Assay Kit (Invitrogen; Carlsbad, CA, USA) on an Agilent 2100 Bioanalyzer (Agilent Technologies; Santa Clara, CA, USA). After qPCR validation, the resulting libraries were subjected to paired-end sequencing with a 100-bp read length using the Illumina HiSeq 2500 platform (Illumina). Raw image files were processed by Illumina Real-Time Analysis (RTA) for image analysis and base calling.

Bioinformatics analysis

Raw sequences were first processed to obtain an average quality score (QS) per read ≤20 by trimming 3′-end of reads using SICKLE (https://github.com/najoshi/sickle). High-quality reads were aligned to the rice reference genome (IRGSP Build 5) using the Burrows-Wheeler Aligner (BWA) (version 0.7.5a) with default parameters [32]. Reads not meeting BWA quality criteria or not matching the reference genome were removed. PCR duplicate reads were removed using PICARD (version 1.88) (http://broadinstitute.github.io/picard/). Using the Genome Analysis Toolkit (GATK) (version 2.3.9 Lite) [36], regional realignment and quality score recalibration were performed, after which SNPs and InDels were identified with ≥3X read depth coverage. The resulting variants were annotated according to genomic position and overlap with current genome annotation using snpEff [10].

Plant materials

Of the 166 accessions in the rice mini core set, selected from a composite population of 4,046 accessions using a heuristic approach, 137 accessions that can flower in Chungcheong province, South Korea, were included in this study [25, 52].

Population structure inference

To confirm the subgroups of our 137 accessions, we combined our 137 sequence and the published 50 sequence data, which have clear subgroup information, and then performed a phylogenetic study using the PHYLIP 3.68 (http://evolution.genetics.washington.edu/ phylip.html) with 100 replicates for a bootstrap confidence analysis. FigTree v1.4.0 (http://tree.bio.ed.ac.uk/software/figtree/) was used to plot the constructed tree. We used the same method to construct our 137 phylogenetic tree. MEGA6 [5] was used to present the phylogenetic tree.

We further used the maximum-likelihood-based software FRAPPE to investigate the population structure of the 137 accessions. We ran 10,000 iterations and the number of clusters (*K*) was considered from 2 to 5.

Estimation of population parameters

Linkage disequilibrium parameters (r^2) for estimating the degree of LD between pair-wise SNPs (58,008 high quality SNPs with physical distance >5 kb) was calculated using the software TASSEL 5.0 with 1,000 permutations. The LD decay rate was measured as the chromosomal distance at which the average pairwise correlation coefficient (r^2) decreased to half of its maximum value. Nucleotide diversity (π) and Tajima's D were calculated by Vcftools [6] with a 50-kb slide window. The nucleotide diversity (ϖ) is defined as the average number of nucleotide differences per site between any two DNA sequences chosen randomly from the sample population:

$$\pi = \sum_{ij} x_i x_j \pi_{ij} = \sum_{i=1}^{n} \sum_{j=1}^{i} x_i x_j \pi_{ij}$$

where x_i and x_j are the respective frequencies of the *I*th and jth sequences, ϖ_{ij} is the number of nucleotide differences per nucleotide site between the *i*th and *j*th sequences, and *n* is the number of sequences in the sample. The summation is taken over all distinct pairs *i,j*, without repetition.

Tajima's D test followed the method developed by Tajima (1989). At equilibrium between genetic drift and selectively neutral mutation, the expected value of *D* is close to zero.

To measure genetic diversity, we used the reduction of diversity (ROD) as:

$$ROD = 1 - \frac{\pi_{cul}}{\pi_{wild}}$$

where ϖ_{cul} and ϖ_{wild} are the values of ϖ for the cultivated and wild varieties, respectively.

Genome-wide association analysis

For the qualitative trait (pericarp color), we performed GWAS using PLINK software. For the other three quantitative traits (amylose content, protein content and panicle number), we performed GWAS using GAPIT [7] with a mixed linear model, where the PCA matrix and relative kinship matrix were included as fixed and random effects.

CONCLUSION

This study provides strong evidence that the size of KRICE_CORE is small but contains high genetic and functional diversity across the genome. The genome information through the whole-genome resequencing technology would establish a potent platform for forward/reverse genetic studies as well as genomeassisted breeding, facilitating future discovery and exploitation of useful alleles from rice germplasms.

ACKNOWLEDGEMENTS

We thank Kyung-Ho Ma and Jong-Wook Chung, RDA, Korea, for preparing materials, technical supports and useful discussions.

AUTHORS' CONTRIBUTIONS

Conceived and designed the experiments: TK, QH, and YP. Analyzed the data: TK, QH, KK, and YP. Contributed reagents/materials/analysis tools: MY, WR, FPL, WT, JY, EH, WHO, BC, BY, S-JK, S-WK, YC, CL and BP. Wrote the paper: TK, QH, and YP. All authors have read and approved the manuscript.

REFERENCE

1. Alonso-Blanco C, Aarts MG, Bentsink L, Keurentjes JJ, Reymond M, Vreugdenhil D, et al. What has natural variation taught us about plant development, physiology, and adaptation? Plant Cell. 2009;21:1877–96.
2. Amos W. Even small SNP clusters are non-randomly distributed: is this evidence of mutational non-independence? Proc Biol Sci. 2010;277:1443–9.
3. Arbeithuber B, Betancourt AJ, Ebner T, Tiemann-Boege I. Crossovers are associated with mutation and biased gene conversion at recombination hotspots. Proc Natl Acad Sci U S A. 2015;112:2109–14.
4. Bodmer W, Bonilla C. Common and rare variants in multifactorial susceptibility to common diseases. Nat Genet. 2008;40:695–701.
5. Brachi B, Morris GP, Borevitz JO. Genome-wide association studies in plants: the missing heritability is in the field. Genome Biol. 2011;12:232.
6. Brandstrom M, Ellegren H. The genomic landscape of short insertion and deletion polymorphisms in the chicken (Gallus gallus) Genome: a high frequency of deletions in tandem duplicates. Genetics. 2007;176:1691–701.
7. Brown AHD. Core Collections: a practical approach to genetic resources management. Genome. 1984;31:818–24.
8. Bubb KL, Bovee D, Buckley D, Haugen E, Kibukawa M, Paddock M, et al. Scan of human genome reveals no new Loci under ancient balancing selection. Genetics. 2006;173:2165–77.
9. Chung Y, Yoon M, Kim H. On Climate Variations and Changes Observed in South Korea. Clim Change. 2004;66:151–61.
10. Cingolani P, Platts A, Wang IL, Coon M, Nguyen T, Wang L, et al. A program for annotating and predicting the effects of single nucleotide polymorphisms, SnpEff: SNPs in the genome of Drosophila melanogaster strain w1118; iso-2; iso-3. Fly (Austin). 2012;6:80–92.
11. Corbett-Detig RB, Hartl DL, Sackton TB. Natural selection constrains neutral diversity across a wide range of species. PLoS Biol. 2015;13:e1002112.
12. Frankel OH, Browon AHD. Current plant genetic resources-a critical appraisal. In Genetics:new frontier. 1984;4:1–11.
13. Godfray HC, Beddington JR, Crute IR, Haddad L,

Lawrence D, Muir JF, et al. Food security: the challenge of feeding 9 billion people. Science. 2010;327:812–8.

14. Govindaraj M, Vetriventhan M, Srinivasan M. Importance of genetic diversity assessment in crop plants and its recent advances: an overview of its analytical perspectives. Genet Res Int. 2015;2015:431487.
15. Hellmann I, Prufer K, Ji H, Zody MC, Paabo S, Ptak SE. Why do human diversity levels vary at a megabase scale? Genome Res. 2005;15:1222–31.
16. Hodgkinson A, Ladoukakis E, Eyre-Walker A. Cryptic variation in the human mutation rate. PLoS Biol. 2009;7:e1000027.
17. Hoisington D, Khairallah M, Reeves T, Ribaut JM, Skovmand B, Taba S, et al. Plant genetic resources: what can they contribute toward increased crop productivity? Proc Natl Acad Sci U S A. 1999;96:5937–43.
18. Huang X, Kurata N, Wei X, Wang ZX, Wang A, Zhao Q, et al. A map of rice genome variation reveals the origin of cultivated rice. Nature. 2012;490:497–501.
19. Huang X, Lu T, Han B. Resequencing rice genomes: an emerging new era of rice genomics. Trends Genet. 2013;29:225–32.
20. Huang X, Wei X, Sang T, Zhao Q, Feng Q, Zhao Y, et al. Genome-wide association studies of 14 agronomic traits in rice landraces. Nat Genet. 2010;42:961–7.
21. Isshiki M, Morino K, Nakajima M, Okagaki RJ, Wessler SR, Izawa T, et al. A naturally occurring functional allele of the rice waxy locus has a GT to TT mutation at the 5' splice site of the first intron. Plant J. 1998;15:133–8.
22. Jackson MT. Managing the world's largest collection of rice genetic resources. Proceedings of the International Symposium on Rice Germplasm Evaluation and Enhancement 1999, 22–28.
23. Jeong H. Prospect of Korea Climate change. Rural and Environmental Engineering Journal. 2010;109:22–30.
24. Kim B, Kripalani R, Oh J, Moon S. Summer monsoon rainfall patterns over South Korea and associated circulation features. Theor Appl Climatol. 2002;72:65–74.
25. Kim K, Chung H, Cho G, Ma K, Chandrabalan D, Gwag J, et al. PowerCore: a program applying the advanced M strategy with a heuristic search for establishing core sets. Bioinformatics. 2007;23.
26. Korte A, Farlow A. The advantages and limitations of trait analysis with GWAS: a review. Plant Methods. 2013;9:29.
27. Kovach MJ, McCouch SR. Leveraging natural diversity: back through the bottleneck. Curr Opin Plant Biol. 2008;11:193–200.
28. Kumar GR, Sakthivel K, Sundaram RM, Neeraja CN, Balachandran SM, Rani NS, et al. Allele mining in crops: prospects and potentials. Biotechnol Adv. 2010;28:451–61.
29. Kumar V, Singh A, Mithra SV, Krishnamurthy SL, Parida SK, Jain S, et al. Genome-wide association mapping of salinity tolerance in rice (Oryza sativa). DNA Res. 2015;22:133–45.
30. Lercher MJ, Hurst LD. Human SNP variability and mutation rate are higher in regions of high recombination. Trends Genet. 2002;18:337–40.
31. Leroy T, De BF, Legnate H, Musoli P, Kalonji A, Loor Solorzano RG, et al. Developing core collections to optimize the management and the exploitation of diversity of the coffee Coffea canephora. Genetica. 2014;142:185–99.
32. Li H, Durbin R. Fast and accurate short read alignment with Burrows-Wheeler transform. Bioinformatics. 2009;25:1754–60.
33. Lindblad-Toh K, Winchester E, Daly MJ, Wang DG, Hirschhorn JN, Laviolette JP, et al. Large-scale discovery and genotyping of single-nucleotide polymorphisms in the mouse. Nat Genet. 2000;24:381–6.
34. Manolio TA, Collins FS, Cox NJ, Goldstein DB, Hindorff LA, Hunter DJ, et al. Finding the missing heritability of complex diseases. Nature. 2009;461:747–53.
35. McCouch SR, McNally KL, Wang W, Sackville HR. Genomics of gene banks: A case study in rice. Am J Bot. 2012;99:407–23.
36. McKenna A, Hanna M, Banks E, Sivachenko A, Cibulskis K, Kernytsky A, et al. The Genome Analysis Toolkit: a MapReduce framework for analyzing next-generation DNA sequencing data. Genome Res. 2010;20:1297–303.
37. Moshelion M, Altman A. Current challenges and future perspectives of plant and agricultural biotechnology. Trends Biotechnol. 2015.
38. Muthayya S, Sugimoto JD, Montgomery S, Maberly GF. An overview of global rice production, supply, trade, and consumption. Ann N Y Acad Sci. 2014;1324:7–14.
39. Odong TL, Jansen J, van Eeuwijk FA, van Hintum TJ. Quality of core collections for effective utilisation of genetic resources review, discussion and interpretation. Theor Appl Genet. 2013;126:289 305.
40. Oleksyk TK, Zhao K, De La Vega FM, Gilbert DA, O'Brien SJ, Smith MW. Identifying selected regions from heterozygosity and divergence using a light-coverage genomic dataset from two human populations. PLoS One. 2008;3, e1712.
41. Ptak SE, Hinds DA, Koehler K, Nickel B, Patil N, Ballinger DG, et al. Fine-scale recombination patterns differ between chimpanzees and humans. Nat Genet. 2005;37:429–34.
42. Sang T, Ge S. Understanding rice domestication and implications for cultivar improvement. Curr Opin Plant Biol. 2013;16:139–46.
43. Schafleitner R, Nair RM, Rathore A, Wang YW, Lin CY, Chu SH, et al. The AVRDC - The World Vegetable Center mungbean (Vigna radiata) core and mini core collections. BMC Genomics. 2015;16:344.
44. Sweeney MT, McCouch S. The complex history of the domestication of rice. Ann Bot. 2007;100:951–7.
45. Sweeney MT, Thomson MJ, Pfeil BE, McCouch S. Caught red-handed: Rc encodes a basic helix-loop-helix protein conditioning red pericarp in rice. Plant Cell. 2006;18:283–94.
46. Tenaillon MI, Austerlitz F, Tenaillon O. Apparent

mutational hotspots and long distance linkage disequilibrium resulting from a bottleneck. J Evol Biol. 2008;21:541–50.

47. Varela MA, Amos W. Heterogeneous distribution of SNPs in the human genome: microsatellites as predictors of nucleotide diversity and divergence. Genomics. 2010;95:151–9.

48. Voight BF, Kudaravalli S, Wen X, Pritchard JK. A map of recent positive selection in the human genome. PLoS Biol. 2006;4, e72.

49. Vowles EJ, Amos W. Evidence for widespread convergent evolution around human microsatellites. PLoS Biol. 2004;2, E199.

50. Xu X, Liu X, Ge S, Jensen JD, Hu F, Li X, et al. Resequencing 50 accessions of cultivated and wild rice yields markers for identifying agronomically important genes. Nat Biotech. 2012;30.

51. Zhao W, Cho G-T, Ma K-H, Chung J-W, Gwag J-G, Park Y-J. Development of an allele-mining set in rice using a heuristic algorithm and SSR genotype data with least redundancy for the post-genomic era. Mol Breeding 2010, 26.

CHAPTER 3

GENOME-WIDE ASSOCIATION STUDY FOR TRAITS RELATED TO PLANT AND GRAIN MORPHOLOGY, AND ROOT ARCHITECTURE IN TEMPERATE RICE ACCESSIONS

Filippo Biscarini, Paolo Cozzi, Laura Casella, Paolo Riccardi, Alessandra Vattari, Gabriele Orasen, Rosaria Perrini, Gianni Tacconi , Alessandro Tondelli, Chiara Biselli, Luigi Cattivelli, Jennifer Spindel, Susan McCouch , Pamela Abbruscato, Giampiero Valé, Pietro Piffanelli , and Raffaella Greco

INTRODUCTION

Taxonomically, domesticated rice (*Oryza sativa*) can be subdivided into two subspecies -*O. sativa indica* and *O. sativa japonica*- and five major groups: indica, tropical japonica, temperate japonica, aus and aromatic [1]. Most accessions are cultivated in tropical humid areas; the recently derived group temperate japonica, adapted to temperate latitudes, however extend the area on which rice is grown, which currently amounts to over 164 MHa [2]. The temperate japonica group of *O. sativa* is mainly cultivated at temperate latitudes in the northern hemisphere (USA, southern Europe, north-western China and Japan) and accounts for ~20% of total rice production [3]. Rice cultivation in temperate areas is therefore based mostly on temperate japonica accessions, but not exclusively: also some tropical japonica accessions are fruitfully grown outside of the tropics. Accessions adapted to temperate latitudes developed specific characteristics to cope with colder climates and the four-season cycle, such as resistance to lower temperatures and reduced photoperiod sensitivity [4, 5]. Rice breeding, as is the case for most other crop species, has long focused on increasing yield (52.4 per year in 1960–2010; [6]). More recently attention is beginning to be paid also to traits related to grain quality, like shape, texture and colour (e.g. [7]). Quality of rice grains is relevant especially for niche varieties like some Italian accessions used to make "risotto" [8]. Additionally, traits related to the efficiency of cultivation have become increasingly important (e.g. [9]): for instance, root traits may be linked to nutrient acquisition from the soil and can be used in breeding to reduce water requirements in irrigated rice production [10].

Plant architecture, grain and root morphology traits largely affect plant productivity and plant response to environmental stressors, and therefore represent targets for breeding schemes designed to increase both yield and quality of the final product [11–13]. The ideal plant architecture for rice was suggested as being characterized by a relatively small number of highly productive tillers [14, 15]; grain size is a major agronomic trait, associated to grain length, width, filling and thickness [16]. Root traits are involved in rice response to drought and in the uptake of nutrients from the soil [13].

During the last 10–15 years, the increasing availability of molecular markers (first SSRs —Short Sequence Repeats, now mainly SNPs— Single Nucleotide Polymorphisms) has allowed researcher and breeders to track segments of the genome linked to specific phenotypes of interest in QTL-mapping and genome-wide association studies (see [17] for a review). SSRs or SNPs mapped to the genome have thus been (and are) extensively used for Marker (or Gene) Assisted Selection (MAS, GAS) in plant breeding programmes [18]. MAS and GAS have been successfully employed also in rice breeding (reviewed in [19]),

Citation: Biscarini F, Cozzi P, Casella L, Riccardi P, Vattari A, Orasen G, et al. (2016) Genome-Wide Association Study for Traits Related to Plant and Grain Morphology, and Root Architecture in Temperate Rice Accessions. PLoS ONE 11(5): e0155425. doi:10.1371/journal.pone.0155425.

and molecular breeding applications bear the potential to help temperate rice contribute to the worldwide need for additional rice production in the next future (estimated $116 \cdot 10^6$ tonnes by 2035; [20]). Results from QTL-mapping studies can also be used to improve the accuracy of genomic selection [21]. LD mapping based on genome-wide associations (GWAS), which exploits marker polymorphisms across all chromosomes [22], has become increasingly popular and powerful over the last few years and thanks to the emergence of more cost-effective, high-throughput genotyping platforms, has become a widely adopted approach for QTL mapping in plants [23]. Recently developed techniques based on the reduction of genome complexity, like Genotyping by sequencing (GBS) [24], are providing the marker density needed for GWAS, making the application of this procedure more feasible for different plant species.

In the present work, we report the results of a GWAS for 16 phenotypic traits related to plant morphology, grain quality (traits related to seed biometrics), and to the root apparatus, in a large collection of temperate japonica rice accessions genotyped for over 150000 SNPs. Previous GWAS works in rice have been published for a variety of traits (e.g. [25, 26]); genetic loci associated to plant architecture [14, 15], grain morphology [27] and root traits [13, 28, 29] have been identified. Most of the investigations on these agronomically relevant traits have been carried out on indica and tropical japonica groups, while very few studies are currently available in temperate japonica rice. Our study includes a wide range of shoot and root traits, employs a large set of SNPs and is focused mainly on temperate rice accessions. Detected polymorphisms may help reveal the biological processes underlying complex traits related to grain quality and other traits of agronomic importance, and could support rice breeding programmes in temperate areas.

MATERIALS AND METHODS

Plant material

The accession panel used in this study included 391 *Oryza sativa* varieties from the Rice Germplasm Collection maintained at the CREA-Rice Research Unit (Vercelli, Italy). Varieties were selected based on the results of a genetic diversity analysis with 24 independent SSR markers [30]; the aim was to include the broadest possible range of genetically and phenotypically diverse temeperate rice accessions.

The sampled collection included 9 aromatic, 11 aus, 22 indica, 96 tropical japonica and 253 temperate japonica accessions. Most of these were temperate rice developed and selected in Italy (126 accessions), or developed elsewhere and adapted to Italian agro-climatic conditions (127 accessions). This collection was representative of the worldwide rice diversity [31], and was used to evaluate the population structure of the analysed rice accession panel.

The distribution of the 391 accessions per taxonomic group can be seen in Table 1. The complete list of the accessions used in this study, with information on taxonomic group and geographic origin is reported in S1 Table. All accessions were purified through single seed descent before genotyping and phenotypic evaluation. From all 391 accessions DNA was extracted for genotyping; phenotypes were collected on a subset of temperate and tropical japonica accessions: 153 for plant morphology and grain related traits, and 93 for root traits (see Table 1)

Table 1: Taxonomic classification of accessions.

Taxonomic group	Genotypes	Shoot/grain phenotyping	Root phenotyping
Aromatic	9	2	
Aus	11		
Indica	22	5	
Temperate Japonica	253	109	68
Tropical Japonica	96	37	25
Total	391	153	93

Number of accessions falling in the different taxonomic subgroups of *O. sativa*. Distribution for the 391 genotyped accessions, and for accessions phenotyped for plant morphology and grain related traits ("shoot" traits, 153 accessions), and for root traits (93 accessions).

doi:10.1371/journal.pone.0155425.t001

DNA isolation and genotyping

Total genomic DNA was isolated from three-week old leaves using the DNeasy Plant Mini Kit (QIAGEN) with a TECAN Freedom EVO150 liquid handling robot (TECAN Group Ltd, Switzerland). For each accession, a single individual plant was used. Whole-genome genotyping was carried out using Genotyping-By-Sequencing (GBS) technology [24]; DNA digestion was performed on 100 samples in 96-well plates using ApeKI, which was shown to cut every 1 kb on average in a *in-silico* digestion of the Nipponbare reference genome. Digested DNAs were ligated to 12 of 0.6 / adapter pairs (optimised to guarantee good quality libraries in rice), and the 96-plex library constructed according to the GBS protocol. The libraries were loaded into Genome Analyzer II (Illumina, Inc., San Diego, CA) for sequencing. Raw sequence data filtering, sequence alignment to the rice reference genome (*Os-Nipponbare-Reference-IRGSP-1.0*, [32]) and SNP calling from low-coverage GBS genotyping, were carried out using the *Tassel* GBS pipeline v3.0 provided by the Buckler Lab for Maize Genetics and Diversity [33, 34]. Default memory size parameters were modified according to data size, and a minimum of 5 tag counts from sequencing data was required. Sequences were aligned to the genome with the "Burrows-Wheeler Aligner" (BWA, [35]) using default parameters for genomes smaller than 2 G. Command lines and parameters used to generate the SNP dataset are reported in S2 Table.

A total of 166418 SNPs were called from GBS genotyping. Missing SNP genotypes were then imputed using the FILLIN (Fast, Inbred Line Library ImputatioN) algorithm in the Tassel GBS pipeline, accounting for complete homozygosity of domestic rice by considering rice accessions as fully inbred (inbreeding coefficient ≥ 0.99). FILLIN is based on haplotype reconstruction around recombination break points. Haplotypes are clustered per genotype similarity using the Hamming distance function. This information is eventually used to impute the target locus in an iterative approach that attemps, through a Markov Chain MonteCarlo (MCMC) process, to maximise the likelihood of the observed SNP calls given the unobserved imputed genotypes. When the FILLIN algorithm could not find haplotypes to satisfy any of the threshold requirements, the SNP locus was not imputed. After imputation, SNPs with call-rate ≥ 90% and MAF (minor allele frequency) > 1% (109888 SNPs) were used to investigate population structure and estimate linkage disequilibrium (LD). SNPs with call-rate ≥ 90% and MAF, respectively, ≥ 5% or ≥ 10% were used in the GWAS for plant and grain traits (57179 SNP) and for root (37827 SNP) traits. The distribution of SNP per chromosome is reported in Table 2.

Table 2: Distribution of SNPs per chromosome, before and after editing for call-rate and MAF.

Chromosome	SNP tot	SNP 153	SNP 93
OSA 1	19674	6894	4005
OSA 2	15332	5031	2695
OSA 3	14697	4228	1800
OSA 4	14344	4627	3510
OSA 5	11089	3587	2873
OSA 6	14245	4071	3257
OSA 7	12918	4836	3113
OSA 8	13201	5024	3260
OSA 9	9894	3448	2028
OSA 10	12603	5478	4356
OSA 11	16794	5984	3753
OSA 12	11627	3971	3177
tot	166418	57179	37827

Number of SNPs detected through GBS in the panel of 391 rice accessions: per chromosome distribution, before and after data editing for call-rate (> 0.90) and MAF (≥ 5% for plant and grain traits; > 10% for root traits)

doi:10.1371/journal.pone.0155425.t002

Population structure

The 391 rice accessions included in this study roughly cover the entire spectrum of domesticated rice populations, belonging to the 5 major phylogenetic groups traditionally identified (*Temperate japonica*, *Tropical japonica*, *Indica*, *Aromatic* and *Aus*) -or mixtures thereof. The underlying population structure was investigated using SNP genotype data to estimate the most likely number of clusters (K) into which the accessions can be grouped, and their degree of admixture. The following likelihood model was used to estimate the value of K:

$$\mathcal{L}(Q,F|g) = \sum_i \sum_j \left\{ g_{ij} ln\left[\sum_k q_{ik} f_{kj}\right] + (2-g_{ij}) ln\left[\sum_k q_{ik}(1-f_{kj})\right]\right\} \quad (1)$$

where: g_{ij} is the observed number of copies (0 or 2, being 100% homozygous rice) of the reference allele at marker j in individual i; q_{ik} is the fraction of the genome of i contributed by population k; and f_{kj} is the reference allele frequency at marker j in population k. **Q** and **F** are the matrices of estimated q_{ik} and f_{kj}, and have dimensions $I \times K$ (n. of individuals x n. of clusters) and $J \times K$ (n. of markers x n. of clusters) respectively; $\mathcal{L}(Q,F|g)$ is the log-likelihood of the matrixes **Q** and **F** given the observed genotypes (which is equal to the their joint probability density function: $\mathcal{L}(Q,F|g) = P(g|Q,F)$).

The likelihood function in Eq 1 was maximised with a cross-validation procedure to identify the value of K that gives the best predictive ability [36].

The model was run for values of K varying from 2 to 18; a 5-fold cross validation scheme was adopted to estimate the prediction error for each value of K. Also the number of iterations needed to reach convergence (in the maximum likelihood estimation procedure) was monitored. The value of K that best fitted the data (the most likely number of clusters in the population) was determined based on the lowest prediction error and the smallest number of iterations for convergence. As described by Alexander et al. (2009, [36]), the number of iterations needed to make the solving algorithm converge rapidly increases when the data start to support poorly the tested number of clusters (K). From the matrix of contributions Q the probabilities of belonging to one of the clusters were derived, and accessions assigned accordingly.

An unweighted neighbor-joining (NJ) tree was constructed using a shared allele index based on a dissimilarity matrix estimated from the SNP dataset [37]. Based on seed availability, a subset of 153 rice accessions, predominantly temperate rice but still representative of the *Temperate*and *Tropical* japonica genetic diversity within the available rice collection, was selected and used for agronomic evaluations in field conditions, to record phenotypes related to plant morphology and grain quality. A further subset of 93 temperate and tropical japonica accessions was also evaluated for root architecture traits in controlled conditions. The list of the accessions selected for phenotyping -both plant and grain morphology and root traits- is reported in S1 Table. The genome-wide association study was therefore performed on the mentioned subsets of 153 and 93 accessions.

Phenotype recording

Two experiments were conducted to record phenotypes on the rice accessions analysed: a field experiment for plant and grain morphology traits (153 accessions), and a greenhouse experiment for root traits (93 accessions). In the field experiment, rice plants were grown at the CREA-Rice Research Unit (Vercelli, Italy). Seeds were sown directly into dry soil in a randomized complete block design with three replicates. Each replicate (plot) measured 1.7 x 0.4 m and consisted of three rows, with 0.2 m inter-row and 0.3 m inter-plot spacing. Within each row, seeds were planted at 0.15 m from each other. A standard fertilization was applied, and conventional submersion was realised at 3°—4° leaf development stage up to one month before harvesting. At the post-emergence stage, two herbicide treatments were applied. During the growing season 7 agronomic traits related to plant and grain morphology were measured on five individual plants per replicate, following the specifications of the International Union for the Protection of New Varieties of Plants (IUPOV): plant height (PH), panicle length (PL), flag leaf length (FLL), flag leaf width (FLW), seed length (SL), seed width (SW) and seed length/width ratio (SR; see Table 3 for a complete list and description).

Table 3: Recorded traits.

Trait category	Trait name	Acronym	Trait description
Plant morphology (shoot)	Plant height	PH	Height of plant from soil surface to the tip of main panicle (inflorescence) at the time of maturity (cm)
	Panicle length	PL	Length of panicle (inflorescence) from the base (panicle neck) to the tip at the time of maturity (cm)
	Flag leaf length	FLL	Length of the flag leaf measured from leaf base to leaf tip at the time of maturity (cm)
	Flag leaf width	FLW	Width of the flag leaf measured at the widest portion of flag leaf lamina at the time of maturity (cm)
	Total shoot dry weight	SDW	
Grain morphology[1]	Seed length	SL	Length of the seed with hull (palea and lemma)[2]
	Seed width	SW	Width of the seed with hull (palea and lemma)
	Seed length/width ratio	SR	Ratio of seed length/ seed width (with hull)
Root morphology[3]	Root dry weight	RDW	
	Root length	RL	
	Root surface area	RSA	
	Root volume	RV	
	Number of tips	RT	
	Total length of thick roots	RL_TK	(diameter > 0.6 mm)
	Volume of thick roots	RV_TK	(diameter > 0.6 mm)
	Volume of vertical thick roots	RV_VTK	(diameter > 0.6 mm, root angle > 57°)

[1]Grain-morphology related traits were measured on rice seeds using the WinSeedle Pro V2007 software and STD4800 scanner (Regent Instruments Inc., Quebec, Canada).

[2]Seed = Mature spikelet filled with grain and surrounded by hull (palea and lemma)

[3]The root traits considered for GWAS were measured on different root angle sectors: RL, RV, RSA and RT on angle classes 3 and 4; RL_TK and RV_TK on any root angle but with root diameter > 0.6 mm; RV_VTK from angle class 4 and diameter > 0.6 mm.

doi:10.1371/journal.pone.0155425.t003

For root phenotyping rice plants were grown under controlled conditions in plastic cylindrical mesh baskets (*Anelli s.r.l., Lodi, Italy*) placed at the top of PVC pipes (S1 Fig). PVC pipes were 60 high with a diameter of 14; mesh baskets were 5.4 cm high with a diameter of 7.2 cm and a mesh size of 2 mm. This size was large enough to allow root emergence from the basket without interference. Each PVC pipe was filled with field soil (silty clay loam, sieved to 5 mm), thoroughly mixed with 33% sand and inorganic fertilizer. The bottom of the pipe was covered with a non-woven fabric to allow free draining. PVC pipes were placed in a greenhouse based on a randomized block design with three replicates, for a total of 279 pipes. About 10 seeds for each of the 93 accessions were pre-germinated in petri-dishes for 48 h at 30°C. Afterwards, two seedlings per replicate were sown exactly in the centre of each basket at a depth of 1 cm. One week after emergence, the seedlings were thinned by selecting the healthiest ones. Plants were grown for six weeks at 28°C day / 24°C night at 75% relative humidity, under daylight conditions (16h light / 8h dark). Using a drip irrigation system, the pipes were supplied with 250 ml of water every day for the first 3 weeks and with 500 ml for the last three weeks. This amount of water —gauged in a previous test on a smaller subset of varieties (data not shown)— allowed to maintain aerobic conditions without unduly causing stress to the plants. From the third week, corresponding to the beginning of the tillering stage, 200 of inorganic fertilizer were supplied to each pipe once a week. Plant height (length from the soil to the tip of the longest leaf) was recorded on a weekly basis, as an indicator of shoot growth and plant health. At the end of the experiment (day 42), all shoots were cut at soil level, dried for three days at 80°C and weighted. Baskets were then gently extracted from the pipes and washed thoroughly to remove completely the soil from the roots. Roots were cut and classified based on their growth angle, measured by the position from where they emerged from the basket mesh. Four classes were thus defined with the following horizontal ground level angles: 0°-27° (1st class), 27°-45° (2nd class), 45°-57° (3rd class), 57°-90° (4th class). Roots were then placed on different glass trays according to the layer, taking care of spreading the smallest secondary roots, and scanned. Images were analysed with the WinRHIZO software (Regent Instruments Inc.). A total of 8 root traits were considered for the GWAS analysis: root dry weight (RDW), root length (RL), root volume (RV), root surface area (RSA), number of root tips (RT), length of thick roots (RL_TK), volume of thick roots (RV_TK) and volume of vertical thick roots (RV_VTK). RL, RV, RSA and RT were measured on roots from class 3 and 4; RL_TK and RV_TK were measured on roots of any angle but with diameter > 0.6 mm; RV_VTK on roots with diameter > 0.6 mm and angle class 4. During the root experiment, also shoot dry weight (SDW) was measured. All phenotypes were measured in 2012. Details are reported in Table 3.

Estimation of linkage disequilibrium (LD)

The pairwise linkage disequilibrium (LD) among SNPs was estimated as r^2:

$$r^2(p_A, p_B, p_{AB}) = \frac{D^2 = (p_{AB} - p_a p_B)^2}{p_A(1-p_A)p_B(1-p_B)} \quad (2)$$

where p_A, p_B and p_{AB} are the frequencies of the AB halotype and of alleles A and B at the two SNP loci, and $D^2 = (p_{AB} - p_A p_B)^2$ is the squared difference between observed and expected haplotype frequencies [38]. For LD estimation, all the 109888 SNPs with call-rate > 90% and MAF > 1% were used. Pairwise r^2 values were averaged over SNPs grouped based on stepwise increasing base-pair distance (10 kbps steps: [0–10[, [10–20[, and so forth). The average LD as a function of base-pair distance was used to estimate LD decay in the rice accessions.

Association study

Only SNPs with a call-rate ≥ 90% and minor alle frequency (MAF) ≥ 0.05 (plant and grain traits) or ≥ 0.1 (root traits) were used for the GWAS. Data filtering was applied separately to the two subsets of accessions phenotyped one for plant and grain morphology traits (153), and the other for root traits (93 accessions). This led to different numbers of SNPs used in GWAS: 57179 and 37827 for plant/grain and root traits respectively. The association between SNP genotypes and plant, grain and root phenotypes was tested fitting one SNP at a time in a mixed linear model of the following form (in matrix notation):

$$\mathbf{y} = \mathbf{1}\mu + \mathbf{X}\mathbf{b} + \mathbf{Z}\mathbf{a} + \mathbf{e} \quad (3)$$

where **y** is the vector of phenotypic observations (averages across replicates); μ is the overall mean; **b** is the vector of SNP effects, with the corresponding matrix **X** of SNP genotypes (either 0 or 1 for the two homozygous classes —AA, BB); **a** is the vector of polygenic effects with the related incidence matrix **Z**; **e** is the vector of residuals. Polygenic and residual random effects were assumed to be normally distributed $a \sim N(0, K\sigma_a^2)$ and $e \sim N(0, I\sigma_e^2)$, where **K** is the kinship matrix among all accessions based on SNP genotypes and calculated according to Astle and Balding [39], **I** is the identity matrix, and σ_a^2 and σ_e^2 are the additive genetic and residual variances.

SNP effects estimated with model 3 were thus corrected for population relatedness, thereby reducing the risk of detecting spurious phenotype-genotype associations. For every SNP and trait, the null hypothesis of no association was tested with a Student *t*-test contrasting the model including the SNP effect with a reduced model including just the polygenic effect. The issue of testing multiple hypothesis (57,179 SNPs x 8 plant and grain traits, and 37827 SNPs x 8 root traits) was addressed by monitoring the false discovery rate (FDR, [40]) of detected associations.

Software

SNP-calling, imputation of missing genotypes and the GWAS were carried out using the TASSEL pipeline and software package for genetic analysis [34, 41]. Population structure was inferred using the computer package *ADMIXTURE* [36]. Pairwise LD was estimated using the --r2 --ld-window-kb 5000 --ld-window 99999 --ld-window-r2 0 command in *Plink*[42]. Plots and data handling, preparation and processing, and NJ tree construction, were all performed with the open-source statistical programming environment *R* [43].

RESULTS AND DISCUSSION

Genotypes and population structure: selecting material

Whole-genome GBS genotyping of the 391 accessions in the European temperate rice diversity panel yielded a total of 166418 SNPs with an overall missing rate of 52%. GBS genotyping is indeed typically characterized by large rates of missing data (up to ~50% on average [33]). After partial imputation with the *Tassel* GBS pipeline, the residual missing rate was 10.8%. Only SNPs with a call-rate ≥ 90% were used for downstream biostatistics analyses. For the association study, an additional filter on MAF was applied: MAF ≥ 0.05 for plant and grain traits (153 accessions), resulting in 57179 SNP and MAF ≥ 0.10 for root traits (93 accessions), resulting in 37827 SNP (see Table 2). Considering the largest panel of SNPs available (57179 SNPs), the average distance between SNPs was found to be 6.0 Kb, with 83.2% of the SNPs at the most 10 kb away from the closest neighbouring SNP locus. 54.0% (33634) of the identified SNPs was found within the sequence of annotated genes (MSU Osa1 Rice Genome Annotation [32]); of these, 60% (20302) were located in exons, 29% (9790) in introns, and the remaining 11% in apparently non-translated regions. A relevant fraction of the identified SNPs was therefore localized in transcribed regions, higher compared to previous reports where GBS markers in coding regions were observed with lower frequency, e.g. 39.5% in soybean [44], 46% in rice [29, 45] and 20.5% in oat [46]. SNP alleles were used to assess the population structure in the panel of accessions under examination. From the admixture analysis (Eq 1), the most likely value for the number of clusters (K value) was determined based on the cross-validation error in predicting to which group each accession belonged, and on the number of iterations to convergence. The prediction error reached a minimum for $K = 5$ and plateaued afterwards, while the number of iterations needed for the model to converge sharply increased for $K > 6$. The most likely number of clusters into which the analysed panel of accessions can be subdivided was, therefore, estimated to be $K = 5$ (Fig 1), which tallies with the number of main sub-populations distinguishable in *Oryza sativa* (indica, aus, aromatic, tropical and temperate japonica; [31]). However, the value of $K = 6$ has similar support from the admixture analysis (in terms of prediction error and iterations to converge). This is not unexpected, considering that some of the accessions are likely to be crosses between populations, and that the tropical japonica subgroup has been repeatedly reported to be constituted by two main clusters [25]. Considering the admixture plot for the 391 accessions with $K = 5$, 243 were assigned to temperate japonica, 85 to tropical japonica, 22 to indica, 9 to aromatic, 11 to the aus group, with probability higher than 0.6. Twenty-one accessions had a probability of assignment lower than 0.6, and appeared to be admixed, mainly between the temperate and tropical japonica groups (S3 Table). The unweighted NJ tree constructed to illustrate the phylogenetic structure of the rice panel (Fig 2) displayed the typical bipolar structure of *O. sativa* populations, with indica and aus accessions clearly separated from aromatic and japonica accessions [31, 47]. The NJ trees for the subsets of accessions used for plant and grain (153) and root (93) phenotyping showed that temperate rice accessions were purposely selected for GWAS, with the same relative proportion between the Tropical and Temperate japonica subgroups as in the whole panel.

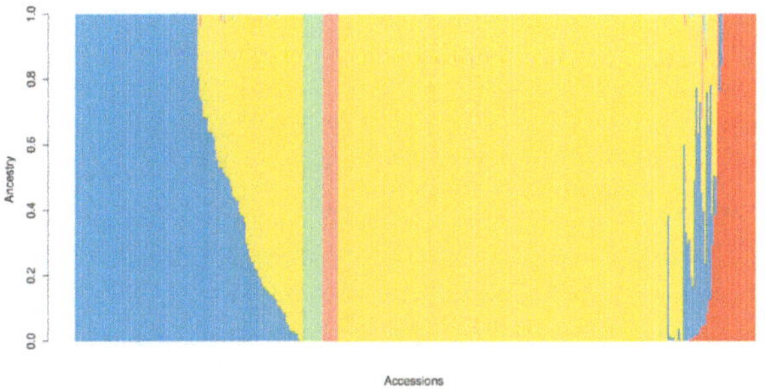

Figure 1: Stacked barplot for the ancestry of the available 391 rice accessions with K = 5.

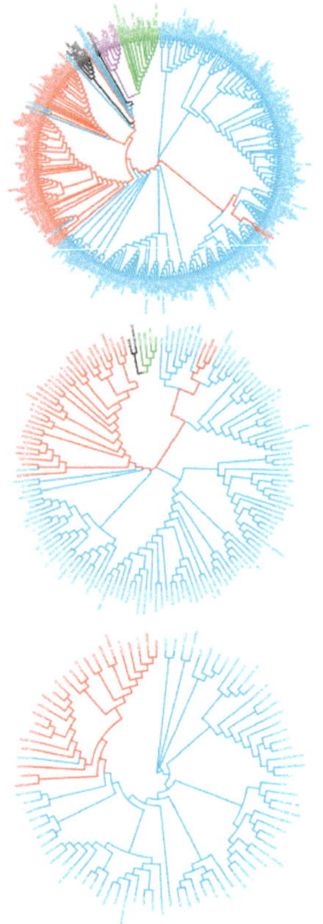

Figure 2: Neighbour-joining trees of the 391 accessions (left) and of the 153 (centre) and 93 (right) subsets used for plant & grain and root phenotyping respectively. Temperate japonica accessions in blue, tropical japonica in red, indica in green, aus in violet, and aromatic in black.

Both in the admixture and NJ tree analyses, the admixed accessions most likely originate from interspecific breeding programmes, with extensive gene exchange between specific subpopulations. Within the japonica group, the temperate and tropical clusters were clearly distinguished, and represented most of the European accessions included in the study (65% and 24% of the panel, respectively). The analysis of population structure guided the selection of the most suitable set of accessions to be phenotyped in the field and root experiments. Mostly, temperate and tropical japonica accessions were selected, with indica and aromatic accessions as outgroups.

Phenotypes and linkage disequilibrium: the basis for GWAS

On the selected set of informative temperate rice accessions, 5 plant morphology, 4 grain quality and 8 root traits were measured (Table 3). The analysis of the frequency distributions of the phenotypic classes indicates that all traits are quantitative and continuous (S2 Fig). The analysed phenotypic traits exhibited an overall broad variability, which is in principle suitable to be efficiently exploited in GWAS studies. All phenotypic traits were approximately normally distributed; a few distributions, though, were found to be slightly skewed (mainly towards lower values: e.g. SR, TRDW, RSA_4, RV_4, RL_4, RT_4), but none showed a multi-modal (e.g. bi-modal) pattern, or a clear separation in two or more classes ("discretization").

Most traits (14 out of 16) showed a larger than 2-fold difference between the minimum and the maximum values (Table 4). The coefficient of variation was on average 0.32 over all traits, and significantly larger for root traits ($\overline{c.v.} = 0.52$). Heritabilities were moderate-to-high, ranging from 0.58 for RV to 0.97 for FLW (Table 4), which is indicative of a good reproducibility of the experiments and a relevant genetic contribution to the observed variability in the measured traits. Phenotypic correlations appeared to cluster per trait group: grain quality, root morphology and, to a lesser extent, plant morphology (Fig 3). All correlations were positive, except for SL vs SW ($r_{SL,SW} = -0.48$), SW vs SL/SW ratio ($r_{SW,SR} = -0.90$) and PL vs SW ($r_{PL,SW} = -0.28$). The strongest positive correlations were $r_{RL,RSA} = 0.99$, $r_{RSA,RV} = 0.99$, $r_{RL\_TK,RV\_TK} = 0.98$, $r_{RSA,RT} = 0.98$, and $r_{RV,RT} = 0.97$. PH and PL showed a moderate positive correlation ($r_{PH,PL} = 0.65$), like PL and FLL ($r_{PL,FLL} = 0.68$). Shoot dry weight was positively correlated with all root traits; in particular, correlations with RDW, RL and RSA higher than 0.75 were observed.

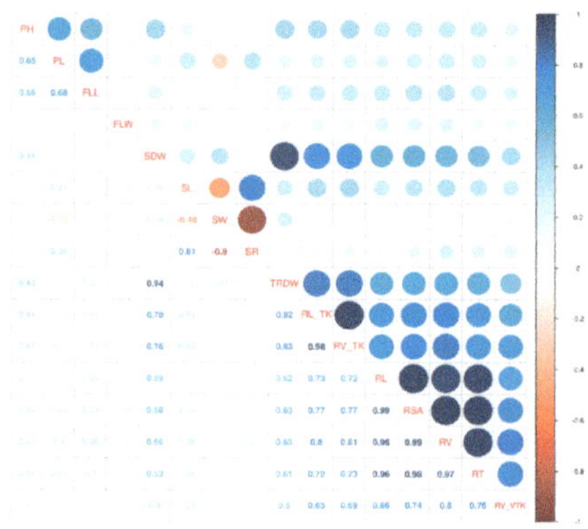

Figure 3: Phenotypic correlations among plant morphology, grain quality and root traits.

Table 4: Descriptive statistics of collected phenotypes

Trait category	Trait	Mean	SD	Max	Min	CV	h^2
Plant morphology	PH (cm)	92.36	12.21	127.40	62.90	0.13	0.79
	PL (cm)	19.48	2.87	26.40	11.60	0.15	0.82
	FLL (mm)	243.28	52.56	380.00	135.00	0.22	0.86
	FLW (mm)	10.90	1.71	16.10	6.80	0.16	0.97
	SDW (g)	0.387	0.14	0.811	0.129	0.36	0.67
Grain quality	SL (mm)	9.17	0.98	11.75	7.19	0.11	0.89
	SW (mm)	3.36	0.50	4.42	2.43	0.15	0.80
	SR (ratio)	2.82	0.66	4.30	1.89	0.23	0.91
Root traits	RDW (g)	0.131	0.05	0.287	0.052	0.38	0.66
	RL (cm)	1916	962.6	4652	440.2	0.50	0.63
	RV (cm^3)	0.805	0.416	1.933	0.206	0.52	0.58
	RSA (cm^2)	136.90	69.66	301.20	33.71	0.51	0.60
	RT (#)	4283	2181.5	9343	837	0.51	0.64
	RL_TK (cm)	260.70	93.16	497.80	69.82	0.36	0.59
	RV_TK (cm^3)	1.386	0.553	3.018	0.297	0.40	0.60
	RV_VTK (cm^3)	0.1854	0.176	0.867	0.0016	0.95	0.67

doi:10.1371/journal.pone.0155425.t004

Pairwise LD was estimated as r^2 between all SNP markers genotypes in the 391 accessions included in the study. The average r^2 as a function of inter-marker distance was used to estimate the LD decay in the population. Estimates of mean r^2 at each 10 kbps incremental steps between 0 and 5000 kbps (5 Mbps correspond to ~10 cM) indicated that average r^2starts at about 0.4 for very close markers (< 50 kbps), and decays to approximately 0.1 for SNPs as distant as 5 Mbps (Fig 4). The mean r^2 drops below 0.2 beyond 1.25 Mbps inter-marker distance. Therefore, a considerable extent of LD was observed in the analysed rice population since $r^2 = 0.2$ extends up to ~1.25 Mbps (Fig 4). This value is higher compared to previous studies in domesticated rice, where LD as high as $r^2 = 0.2$ was found from 100 kbps to maximum ~500 kbps [29, 48]. These differences are likely to be due to the different experimental settings. First, the European temperate japonica accessions in our panel are highly related to each other, and the slow LD decay suggests that few historical recombination events occurred in this population. Secondly, SNP density can explain the higher LD estimated in our study: Courtois et al used 16664 markers (both SNPs and DArTs), Mather et al only 522. LD estimates tend to be higher with denser SNP panels (e.g. [49, 50]), and LD patterns tend to emerge clearly only at higher SNP densities (e.g. [51]). Data filtering criteria may also have played a role: both studies excluded markers with a MAF < 10%; MAF is known to influence LD estimates, though mainly in terms of D', whereas r^2 estimates appear to be robust to MAF (e.g. [52]). On the other hand, smaller sample sizes (Courtois et al. used 168 accessions, mainly Tropical japonica, Mather et al. 60 accessions equally spread over the Indica, Tropical and Temperate japonica groups) are known to bias upward LD estimates (e.g. [49]).

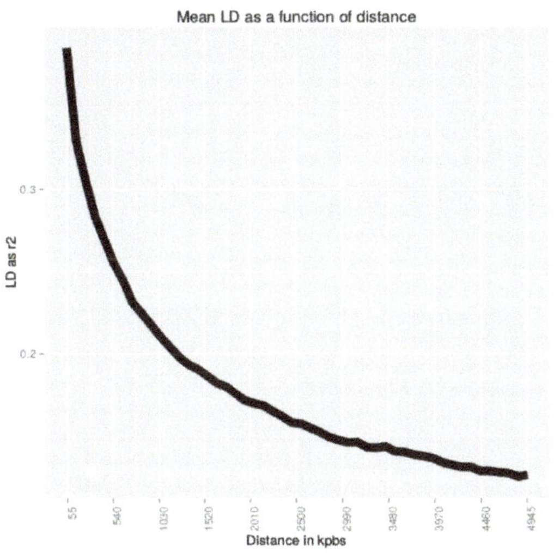

Figure 4: Average LD as a function of inter-marker distance estimated in the panel of 391 accessions.

The wide phenotypic variability recorded in the analysed plant material and the long-range estimated LD, are both preliminary building blocks for the construction of a successful model for a genome-wide association study in temperate rice.

GWAS: general aspects and specific associations

The traits included in the GWAS analysis were categorised in three classes: plant morphology, grain quality and root traits. The ample phenotypic variability and long-range LD extent, coupled with an average marker density of one SNP every 6 kbps, provided a good basis for whole-genome association mapping in temperate rice. The near normal distribution of most recorded phenotypes justified the use of a linear regression model for continuous traits in the association study (Eq 3), in which the population structure was properly accounted for through the kinship matrix.

Statistical significance of results.

The FDR was estimated to monitor the effect of multiple testing: FDR for the associations inTable 5 varied from $3 \cdot 10^{-7}$ to 0.92, with median 0.25. In GWAS studies it is however debatable whether multiple tests of a single global null hypothesis or individual tests of multiple hypothesis are concerned, and whether the threshold depends on the number of tests or on the prior probability of true SNP- phenotype associations [53]. In the latter case, the posterior odds of a true association would be given by its prior probability times the statistical power of the test ((P(reject $H_0|H_1$ is true)): probability of correctly rejecting the null hypothesis of no association when the SNP is truly associated to the phenotype), divided by estimated p-value:

$$\frac{P(\text{true association})}{1 - P(\text{true association})} = \frac{\text{prior probability} \cdot \text{power}}{\text{p} - \text{value}}$$

Table 5: Most significant associations between SNP genotypes and shoot, grain and root phenotypes in the analysed rice accessions.

	Trait	Marker	Chr	pos (kb)	p-value	FDR	Var (SNP)	interval (kb)	N. SNPs	QTL*	Candidate Gene
shoot	FLW	S1_4302318	1	4302	0.000076	0.101	0.0755		1	qFLW1	
	FLL	S1_28597986	1	28598	0.000013	0.307	0.1825	1.86	3		
	PH	S2_2826504	2	2827	0.000086	0.275	0.1113		1		LRK1 [61]
	FLL	S2_3104099	2	3104	0.000003	0.186	0.1586		1		
	FLL	S2_15780456	2	15780	0.000047	0.416	0.1061		1		
	PH	S2_29790001	2	29790	0.000051	0.241	0.1105		1	qPH-2 [58]	
	FLW	S3_12357940	3	12358	0.000035	0.065	0.0389		1		
	PL	S3_13264119	3	13264	0.000061	0.753	0.2457		1	qPLT3-2 [66]; pl3.1 [67]	
	FLW	S4_28827421	4	28827	0.000058	0.079	0.0423		1	Flw4 [54]	
	FLW	S4_31080152	4	31080	$1.37 \cdot 10^{-11}$	$3.26 \cdot 10^{-7}$	0.1914	775.7	36	qFLW4 [55]	LSCHL4 (NAL1); NarrowLeaf1)
	PL	S4_31520309	4	31520	0.000089	0.753	0.1198		1		
	FLW	S4_32193401	4	32193	0.000076	0.107	0.0383		1		
	PH	S5_17644414	5	17644	0.000009	0.0875	0.0947		1		
	FLL	S5_17644414	5	17644	0.000093	0.535	0.0554		1		
	PH	S6_22330734	6	22331	$2.04 \cdot 10^{-7}$	0.011	0.2392	278.55	11		D35 [62]
	PH	S6_23311513	6	23312	0.000100	0.275	0.1368		1		HDA702 [65]
	PH	S6_24644771	6	24645	0.000008	0.0851	0.1510	178.57	4	Qph6f [59]	
	PH	S7_16765878	7	16766	0.000021	0.148	0.0886		1		
	PL	S9_13253243	9	13253	0.000003	0.186	0.1882		1		
	FLL	S9_13253243	9	13253	0.000032	0.516	0.1487		1		
	FLL	S9_15078875	9	15079	0.000082	0.535	0.1229	50.1	2		
grain	SL	S1_42414831	1	42415	0.000066	0.766	0.0520		1		
	SL	S2_5453230	2	5453	0.000002	0.115	0.1415		1		
	SR	S2_5453230	2	5453	0.000098	0.252	0.0370		1		
	SR	S4_31243055	4	31243	0.000044	0.252	0.0299		1	gpp4	SLCHL4 (NAL1) [56]
	SW	S5_5401194	5	5401	0.000007	0.198	0.1201	254.23	13	qSW5 [77]	GW5 [71]
	SR	S5_5401194	5	5401	0.000083	0.252	0.0997	254.23	13	qSW5 [77]	GW5 [71]
	SR	S6_24916209	6	24916	0.000039	0.252	0.0201	55.57	3	qGL-6 [72]	
	SR	S7_18240854	7	18241	0.000082	0.252	0.0301		1		
	SR	S7_24575488	7	24575	0.000032	0.252	0.0962		1	grb7-2 [73]	
	SW	S7_24575488	7	24575	0.000063	0.252	0.0832		1	grb7-2 [73]	
	SR	S7_25119756	7	25120	0.000001	0.0630	0.1442	315.67	2	grb7-2 [73]	GE [70]
root	RV_TK	S1_1817023	1	1817	0.000010	0.511	0.0457	36.43	2		
	RV_TK	S2_23548832	2	23549	0.000065	0.927	0.0416		1	qRTT2-1 [78]	
	RL	S4_19970373	4	19970	0.000019	0.628	0.6834		1	qNOT4-2 [78]	
	RSA	S4_19970373	4	19970	0.000031	0.384	0.6934		1	qNOT4-2 [78]	
	RT	S4_19970373	4	19970	0.000035	0.542	0.6142		1	qNOT4-2 [78]	
	RV	S4_19970373	4	19970	0.000045	0.409	0.6959		1	qNOT4-2 [78]	
	RV_TK	S9_10237467	9	10237	0.000068	0.903	0.2890	79.87	9		
	RV	S10_15873490	10	15873	0.000082	0.173	0.0715	55.37	3	dt [79]	
	RV_TK	S10_16370558	10	16371	0.000039	0.302	0.3617	13.39	2	dt [79]	
	RV	S10_16370558	10	16371	0.000079	0.173	0.0672	13.39	2	dt [79]	

The SNP with the highest p-value is reported; FDR: false discovery rate; dt: drought tolerance;

*QTLs with no reference were obtained from the *Q-TARO* databases (http://qtaro.abr.affrc.go.jp/). The effect of the SNP is expressed in terms of proportion of the genetic variance explained by the SNP: $\sigma^2_{SNP} = \hat{a}^2 \cdot p(1-p)$, where $\hat{a}$ is the estimated effect of the SNP, and p is the minor allele frequency (Zhang et al., 2010).

doi:10.1371/journal.pone.0155425.t005

If 10 detectable signals are expected over the entire genome for any given trait, a plausible estimate of the prior odds of true association would be 10: 57179 = $1.75 \cdot 10^{-4}$ for shoot and grain traits, and 10: 37827 = $2.64 \cdot 10^{-4}$ for root traits. Assuming a power of 0.5, the posterior odds of the associations reported in Table 5 being true would range from a minimum of 0.88 ($P(x)$ = 0.46) between SNP S6_23311513 and PH on chromosome 6, to a maximum of 6387094 ($P(x)$ = 0.99) between SNP S4_31080152 and FLW on chromosome 4, with median value of 2.26 ($P(x)$ = 0.69); 13 out of 42 associations (~31%) had posterior odds > 4 ($P(x)$ > 0.8).

These results indicate the value of looking at significance of GWAS results from multiple perspectives, and point to a high credibility of the associations detected in the analysed panel of temperate rice accessions.

Association mapping.

The Manhattan plots of -log(p-values) and the Q-Q (quantile-quantile) plots of expected (under a Gaussian distribution) vs observed p-values for SNP-based genotype-phenotype associations for the three classes of plant traits under examination are reported in Figs 5, 6, 7 and 8. Additionally, the Manhattan and Q-Q plots for shoot and root dry weights are shown in Fig 9. The most significant associations ($-log(p\text{-}value) > 4$) detected in the GWAS study are listed inTable 5, where, for the largest peaks, only the most significant SNP is reported, together with the number of additional significant SNPs and the interval they span. The FDR for the associations in Table 5 ranged from $3.26 \cdot 10^{-7}$ to 0.927 (average 0.32), with seven associations having a FDR < 0.1.

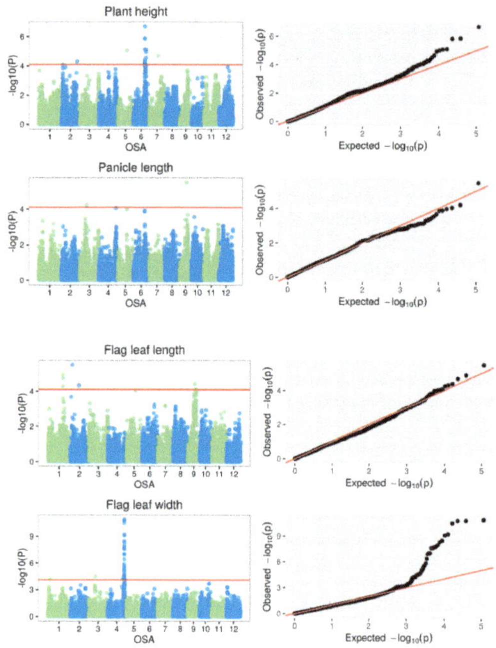

Figure 5: Manhattan and Q-Q plots of GWAS results for plant morphology traits.

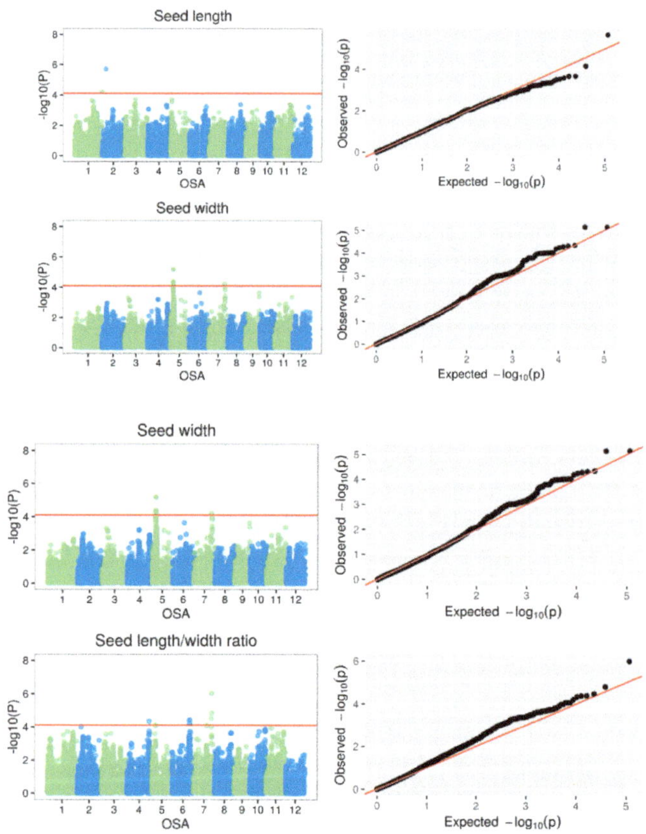

Figure 6: Manhattan and Q-Q plots of GWAS results for grain quality traits.

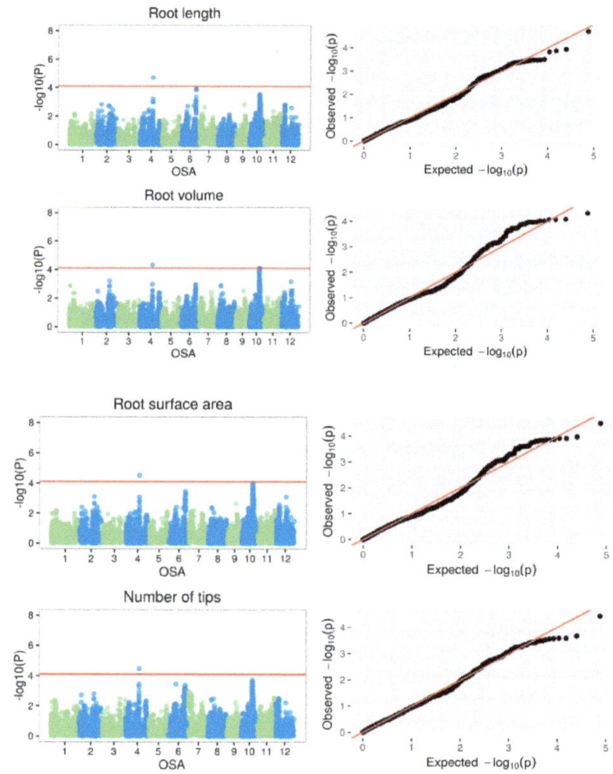

Figure 7: Manhattan and Q-Q plots of GWAS results for root traits.

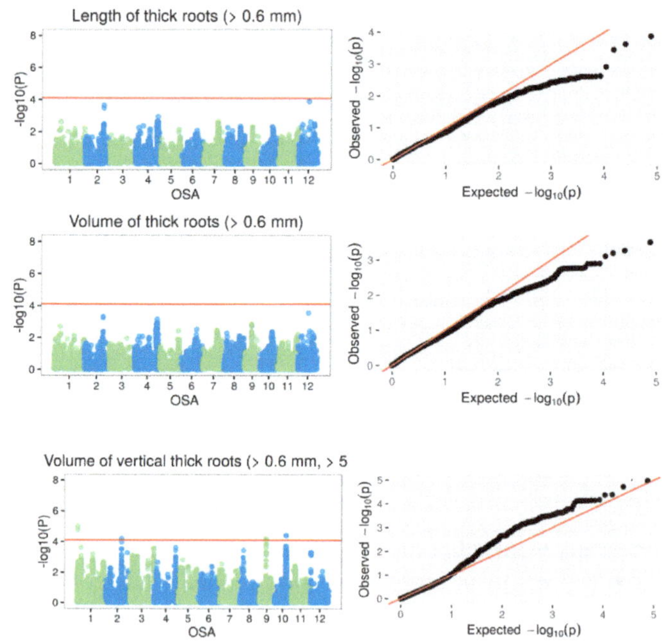

Figure 8: Manhattan and Q-Q plots of GWAS results for thick root traits.

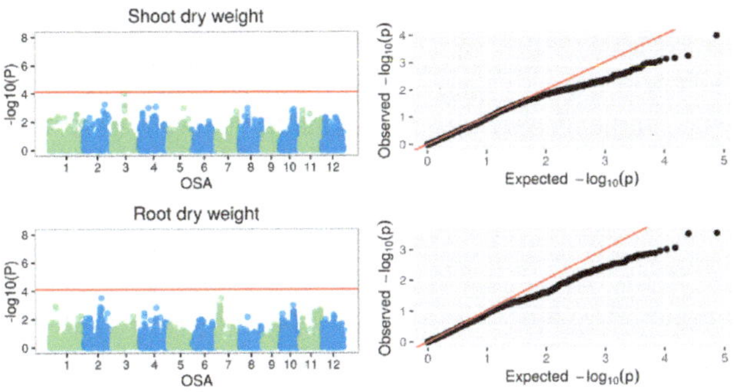

Figure 9: Manhattan and Q-Q plots of GWAS results for shoot and root dry weight.

GWAS of plant morphology traits.

A total of 21 significant associations (-log(p-value) > 4) between SNPs and the analyzed plant morphological traits were identified. The association with the lowest p-values were SNP S4_31080152 on OSA4 for FLW (p-value = $1.37 \cdot 10^{-11}$), and SNP S6_22330734 on OSA6 for PH (p-value = $2.04 \cdot 10^{-7}$). Other 4 associations for plant morphological traits had a p-value < $1 \cdot 10^{-5}$ and included two SNP for PH (S5_17644414 and S6_24644771) and one each for FLL (S2_3104099) and PL (S9_13253243). Fourteen loci were associated with a morphological trait with p-value < $1 \cdot 10^{-4}$ and one locus with p-value < $1 \cdot 10^{-3}$ (Table 5). No significant associations were detected for SDW. In five cases, multiple SNPs belonging to the same chromosome region in full LD were identified as having the same level of significance for the same trait (3 SNPs on chromosome 1 for FLL, 36 SNPs on chromosome 4 for FLW, 11 SNPs on chromosome 6 at position 22331 kb for PH, 4 SNPs on chromosome 6 at position 24645 kb for PH and 2 SNPs on chromosome 9 for FLL). The involved genomic regions ranged from 1.86 kb to 775.7 kb. Some markers were significantly associated with more than one trait:marker S5_17644415 was found to be associated with PH and FLL, marker S9_13253243 with both PL and FLL.

Among the 21 loci significantly associated with plant morphologiy traits, three were in predicted genes for proteins with characterized functions, six co-localized with QTLs known to affect the corresponding traits, and three were located proximally to QTLs controlling plant morphology (Table 5). For FLW, three positional matches were observed on chromosome 4: SNP S4_28827421 was located in correspondence to the Flw4 QTL, mapped by Lim et al. [58] through whole genome re-sequencing of 178 rice F7 RILs (Recombinant Inbred Lines); SNP S4_31080152 and S4_32193401 co-localized with the QTL Qflw4, identified by Yue et al. [54] using a recombinant inbred population obtained from a cross between indica and japonica rice cultivars. For S4_31080152 a co-localization was found with the LSCHL4 gene, tightly linked to the Narrow leafl (NAL1) gene, which was reported in a narrow leaf rice mutant [55]. The NAL1 gene affects leaf width and plant height through its effects on cell division [57]. Its protein product is a plant specific protein known to be involved in cell cycle regulation by inhibiting cell division. Jiang et al. [57] demonstrated that NAL1 affects the expression of Auxin Responsive Factors (ARFs) required for the regulation of auxin response genes expression involved in auxin signalling. Moreover, this gene may act together with the YABBY family, implicated in leaf development, to regulate the formation of mid-ribs in leaves.

Two SNPs were found to be significantly associated to PH and co-localized with two known QTLs implicated in the control of plant height in rice: SNP S2_29790001 associated to qPH2.2, which was mapped on chromosome 2 using fifteen Chromosome Segment Substitution Lines [58]; and SNP S6_24644771 associated to Qph6f on chromosome 6, which was identified by Liang and co-workers (2011, [59]) using a population of 226 RILs. qPH2.2 has been postulated to be implicated in gibberellin (GA) biosynthesis as it has been located proximal to a putative GA-encoding gene [60]. SNP S2 2826504 co-localised with the candidate gene LRK1 encoding a Leucine-rich repeat receptor-like kinase (LRR-RLK); the LRK1 gene product was demonstrated to inhibit GA biosynthesis during internode elongation by down-regulating the expression of the ent-kaurene oxidase gene OsKO2 [61]. The PH-related markers S6_22330734 and S6_23311513, close to the Qph6f QTL, were associated to two different genes known to affect PH: *D35* and *HDA702*. The D35 gene, whose mutations causes a semi-dwarf phenotype, encodes the ent-kaurene oxidase enzyme, implicated in the gibberellin (GA) biosynthetic pathway [62]. It is well known that GAs modulate plant height by promoting cell elongation [63], thus mutations in D35 result in a reduction of GA biosynthesis and plant height. HDA702 corresponds to a histone deacetylase which increases growth and affects plant architecture through an epigenetic repression of OsNAC6, encoding for a transcription factor carrying a NAC domain implicated in the control of seedling root growth [64]. Moreover, down-regulation of rice HDA702 leads to the production of narrowed leaves and stems [65].

SNP S3_13264119 associated to PL co-localised with two QTLs on chromosome 3: qPLT3-2, identified using a Double Haploid (DH) population from an indica and japonica cross grown at nine different locations across Asia [66]; and pl3.1, mapped in an advanced backcross population derived from an *O. sativa* variety crossed with the wild relative O. rufipogon [67].

The rice SD-1 semidwarfing gene is one of the most relevant genes employed in modern rice breeding, since the recessive allele results in shorter culms with improved lodging resistance and higher harvest index. The semi-dwarf phenotype was ascribed to a deficiency of active Gibberellins (GA) arising from a defective GA20-oxidase enzyme, a GA biosynthetic enzyme, which originates from a deletion of 280 bp within the coding region of Os20ox2 in indica rice, or from a substitution of a highly conserved Leu amino-acidic residue at position 266 in japonica rice ([68], a possible example of convergent selection). No SNP markers significantly associated with PH were identified in the region of sd-1 using the linear mixed model that included the kinship matrix. However, a strong association signal was detected for SNPs localized within and in proximity of the *sd-1* gene running a model that did not account for population structure (Fig 10 and S4 Table). Since the SD-1 gene is not common in upland rice varieties but it is conversely common in improved irrigated varieties, this result would imply that in the analysed panel of accessions the effects of population and polymorphisms in or close to the sd-1 gene are confounded. Apparently, in a set of temperate japonica accessions the sd-1 recessive allele linked to semi-dwarfism is still segregating, while in other accessions (e.g. the indica group) the same allele is either absent (wild type) or fixed (semi-dwarf lines).

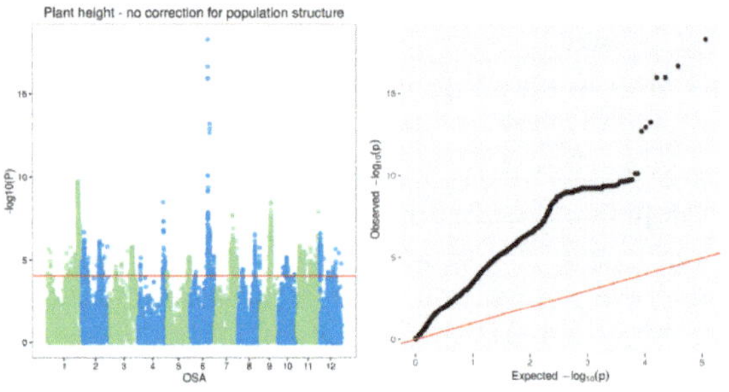

Fig 10: Manhattan and Q-Q plots of GWAS results for shoot and root dry weight.

GWAS of grain morphology traits.

The association analysis for grain traits (Table 3) was also conducted on the same set of 153 informative rice accessions (Table 1; S1 Table). A total of 11 significant associations between SNPs and grain quality traits were identified. Three associations showed a p-value $< 1 \cdot 10^{-5}$: SNP S2_5453230 with SL, SNP S5_5401194 with SW and SNP S7_25119756 with SR. Eight SNP loci were associated with grain traits with p-value $< 1 \cdot 10^{-4}$ (Table 5). In three cases, multiple SNPs from the same genomic region were in full LD and had a comparable level of significance for the same trait (13 SNPs on chromosome 5 for SW and SR, 3 SNPs on chromosome 6 for SR and 2 SNPs on chromosome 7 for SR). The size of the associated genomic regions ranged from 36.43 kb to 254.23 kb. Similarly to morphological traits, also for grain traits specific SNPSs were significantly associated with multiple traits:SNP S2_5453230 was associated with SL and SR, SNP S5_5401194 and S7_24575488 were both associated with SW and SR (Table 5), indicating either pleiotropy or genetic linkage. Among the eleven loci associated with grain morphology, two were located in predicted genes for proteins with known functions (Table 5); qSW5/GW5 is a major QTL that negatively influences grain width and weight, by regulating cell proliferation during seed development. GW5 encodes for a 144-amino-acids uncharacterized protein that interacts with ubiquitin, suggesting that GW5 might be involved in the ubiquitin pathway, which has been reported to affect seed size in plants [69]. The SR-associated SNP S7_25119756 was located close to the cytochrome p450 gene GIANT EMBRYO (GE), implicated in the determination of embryo/endosperm size in developing rice seeds [70]. In addition, three loci co-localized with QTLs affecting the corresponding traits (Table 5): the aforementioned S5_5401194 was associated to the qSW5 QTL which corresponds to the gene GW5 [71]; the SR-related S6_24916209 was located in the qGL-6 region, shown to be involved in grain length determination by a sequencing-based high resolution genotyping mapping of rice RILs [72]; SNP S7_25119756, corresponding to the GE gene, co-localized also with the grb7-2 QTL, which affects grain breadth and was mapped using 209 RILs from a cross between Basmati rice and the contrasting breeding line Pusa 1121 [73]. This QTL was also located proximally to S7_24575488, associated to both SR and SW.

Recently, Chen and co-workers [74] characterized the GE gene, which encodes for a cytochrome P450 protein belonging to the CYP78 family, and controls the proper balance between embryo and endosperm tissues. This protein family represents a plant-specific class of cytochrome P450 mono-oxygenases, with distinct roles in plant growth and organ development, and is implicated in several biosynthetic pathways: e.g. phenylpropanoids, lipids, terpenoids, alkaloids and plant hormones biosynthesis [75]. GE influences endogenous indole acetic acid production in peroxisomes of young developing seeds cells and also regulates the expression of auxin-responsive and cyclin-related genes, thus influencing embryo development, cell elongation and cell cycle.

Altogether, several QTLs and candidate genes related to plant and grain morphology were demonstrated to significantly affect rice yield when transferred to a genetic background different from that where they were identified [11, 15, 76, 77].

GWAS of root traits.

The analysis of the eight root traits (Table 3) was carried out on a set of 93 selected rice accessions representing temperate rice in the panel (Table 1; S1 Table). A total of 10 significant associations between SNPs and root traits were identified. All the associations showed p-values < 1 #x00B7; 10^{-4} (Table 5). In five cases, multiple markers belonging to the same genomic region were in full LD and showed comparable significance for the same trait: 2 SNPs on chromosome 1 for RV_TK, 9 SNPs on chromosome 9 for RV_TK, 3 SNPs on chromosome 10 for RV and 2 SNPs on chromosome 10 for RV_TK and RV. The length of the associated genomic regions ranged from 13.39 kb to 79.87 kb. Specific SNP markers were significantly associated with multiple root traits: SNP S4_19970373 on chromosome 4 was found to be associated with RL, RSA, RT and RV; SNP S10_16370558 on chromosome 10, associated with RV_TK and RV (Table 5).

Among the six SNPs significantly associated with root architecture, four co-localized with three QTLs known to influence root morphology (Table 5): S2_23548832 —associated to RV_TK— co-localised with qRTT2-1, a QTL controlling root thickness under two different developmental stages and during low moisture stress in a DH population derived from a cross between an indica and a japonica variety [78]; S4_19970373 —associated to RL, RSA, RT and RV— co-localised with qNOT4-2, a QTL that regulates the number of tillers under low moisture stress conditions in the same mapping population as qRTT2-1 [78]; S10_15873490 and S10_16370558 —associated with RV, and with RV_TK and RV, respectively— co-localised with a dt QTL (Gramene Accession Number = AQHP076), regulating root thickness during drought stress [79].

A detailed GWAS analysis of root traits in rice was recently conducted by Courtois et al [29] in a japonica rice panel. No common loci were identified between the two studies. The only positional match was observed between the q22 locus on chromosome 4 at 21386 kb in [29], associated to deep root biomass and number of roots > 30 deep, and S4_19970373 (at 19970 bps), associated to RL, RSA, RT and RV. The lack of obvious common findings is likely to be dependent upon the following: i) the germplasm panel under examination (the present panel was mainly composed of temperate japonica accessions while Courtois et al. focused their analysis on a panel tropical japonica accessions); ii) the experimental system (Courtois et al. based their phenotyping on 2D rhizoboxes, while our experiments were based on a 3D basket phenotyping); and iii) the average density of SNP markers (one marker per 22.5 kb in [29] and one marker every 11.1 kb in our GWAS analysis).

Root traits are gaining relevance in rice breeding for water-saving management systems. Currently, rice cultivation in temperate environments rely on large and consistent water supplies. Increasing concerns over climate change are urging a reduction of water for irrigation, which in turn calls for adaptability of rice varieties to reduced water availability. Root architecture plays a crucial role in adaptability of rice to such new circumstances, and introgression of QTLs and genes for root traits is demonstrated to increase rice yield under conditions of limited water availability (reviewed in [80]).

CONCLUSIONS

A genome-wide association study was successfully carried out for a set of specific traits related to plant morphology, grain quality and root architecture in a unique panel of temperate rice accessions adapted to European temperate conditions. This is one of the largest studies on temperate rice for morphological and root traits, especially where the tridimensional architecture of the roots is considered. Although much is already known about root functioning in rice, there is still need for better understanding of root traits in the context of breeding strategies for growth under water-saving management systems.

The effectiveness of the presently developed platform for high throughput phenotyping, genotyping and association studies for temperate japonica was validated by confirming the involvement of specific QTLs and known genes in the control of plant, root and seed phenotypes of relevance for high yield and adaptation to the temperate climate (e.g. like the NarrowLeaf1 gene for flag leaf width on chromosome 4, and the qSW5 QTL for seed width on chromosome 5). A number of QTLs related to the different agronomically relevant traits have been detected. These results and the developed platform bear the potential of being used in breeding programmes to improve the quality of the grain and the drought tolerance of temperate rice accessions. The European GWAS platform was developed using a panel constituted mostly of adapted temperate japonica breeding lines rather than on a diversity panel providing the opportunity to apply the obtained association results directly to ongoing and future breeding programs in Europe and other temperate rice-growing areas (e.g. Japan, northern China, USA). The identified segregating haplotypes are valuable tools to define the most suitable parents for crossing, to enhance the frequency with which these haplotypes appear in the progenies.

Marker Assisted Selection for specific favorable haplotypes will enable us to increase breeding efficiency and decrease cost by reducing the number of plants to be brought to the next generation of breeding or that need to be in-depth phenotyped.

SUPPORTING INFORMATION

S1 Figure: Root phenotyping experiment. Root phenotyping experiment. A) The various phases of the experiment. B) The four layers defined according to the root growth angle.

Trait distributions

The phenotypic distribution of the plant morphology, grain quality and root traits used for the genome-wide association study.

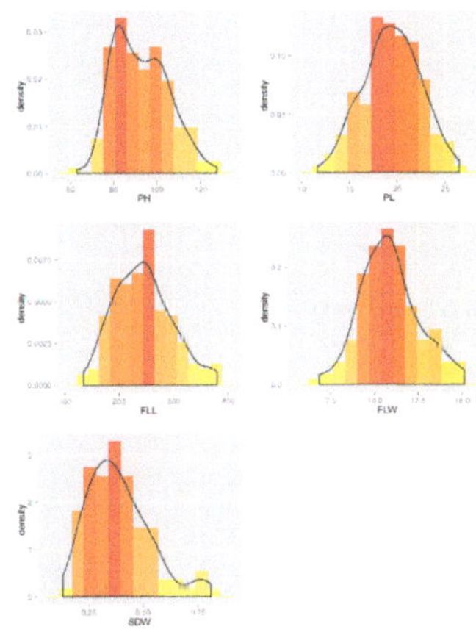

S2 Figure 1: Phenotypic distribution of plant morphology traits

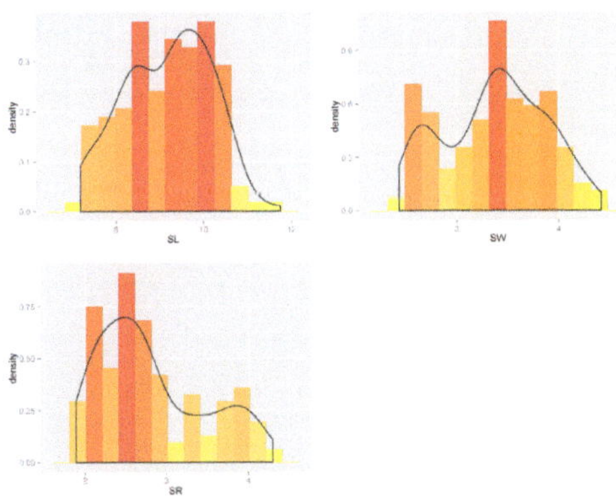

S2 Figure 2: Phenotypic distribution of grain quality traits

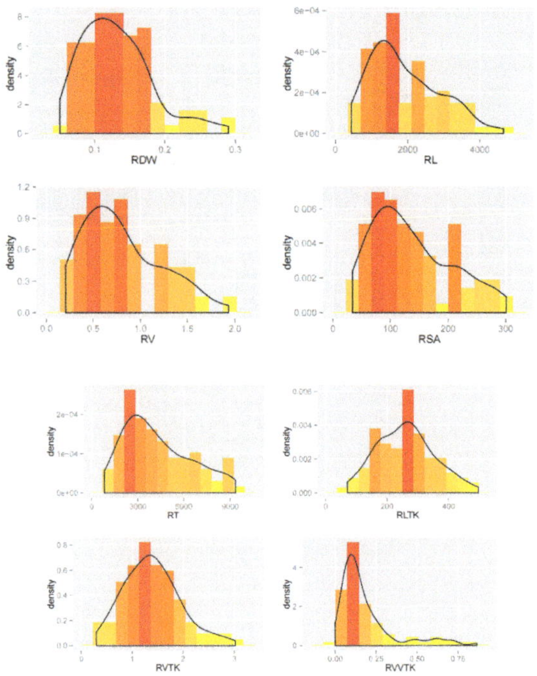

S2 Figure 3: Phenotypic distribution of root traits

S1 Table: Rice accessions. List of accessions used in the study with their geographical origin, commercial class, taxonomical group and collected pheno- types (153 accessions phenotyped for plant and grain morphology; a subset of 93 accessions phenotyped also for root traits).

Name	Origin	Class	Group	Phenotypes 153	Phenotypes 93
9311			indica		
A201	Long B	USA	tropical_japonica	Plant/grain	Root
A301	Long B	USA	tropical_japonica	Plant/grain	
ADAIR	Long B	USA	tropical_japonica		
AGOSTANO	Long A	ITALIA	temperate_japonica		
AIACE	Long A	ITALIA	tropical_japonica	Plant/grain	
AKITAKOMACHI	Round	GIAPPONE	temperate_japonica	Plant/grain	Root
ALAN	Long B	USA	tropical_japonica		
ALEXANDROS	Long B	GRECIA	tropical_japonica		
ALLORIO	Long A	ITALIA	temperate_japonica	Plant/grain	Root
ALPE	Long A	ITALIA	temperate_japonica	Plant/grain	Root
ALPHA	Long A	ITALIA	temperate_japonica		
AMBRA		ITALIA	temperate_japonica		
AMERICANO_1600	Round	ITALIA	temperate_japonica	Plant/grain	
ANTONI	Long A	BULGARIA	temperate_japonica		
APOLLO	Long B	ITALIA	tropical_japonica	Plant/grain	Root
ARBORIO	Long A	ITALIA	temperate_japonica	Plant/grain	Root
ARC_10352			aromatic		
ARGO	Medium	ITALIA	temperate_japonica	Plant/grain	Root
ARIANA		ROMANIA	indica		
Arias			tropical_japonica		
ARIETE	Long A	ITALIA	temperate_japonica	Plant/grain	Root
ARLESIENNE		FRANCIA	temperate_japonica		
ARSENAL	Long B	ITALIA	tropical_japonica		
ASIA	Long B	ITALIA	aromatic	Plant/grain	
Asse_Y_Pung			tropical_japonica		
AUGUSTO	Long A	ITALIA	temperate_japonica	Plant/grain	Root
AUZGUSTA		UNGHERIA	temperate_japonica		
Azucena		USA	tropical_japonica		
BACCO	Long A	ITALIA	temperate_japonica		
BAHIA	Medium	SPAGNA	temperate_japonica		
BAIXET	Long A	SPAGNA	temperate_japonica		
BALILLA	Round	ITALIA	temperate_japonica	Plant/grain	Root
BALILLONE_R253			temperate_japonica		
BALOCCO_G	Medium	ITALIA	temperate_japonica		
BALZARETTI	Medium	ITALIA	temperate_japonica	Plant/grain	
BARAGGIA	Round	ITALIA	temperate_japonica	Plant/grain	Root
Basmati	Long B	INDIA	aromatic		
Basmati_1	Long B	INDIA	aus		
BASMATI_C621	Long B	INDIA	aromatic		
BEIRAO	Long A	PORTOGALLO	temperate_japonica	Plant/grain	Root
BELLE_PATNA	Long B	USA	tropical_japonica		
BERTONE	Long A	ITALIA	temperate_japonica	Plant/grain	Root
BETIS			temperate_japonica		
Bico_Branco			aromatic		
Binulawan-F			indica		
BJ_1			aus		
BLACK_EGYPT			temperate_japonica		
Black_Gora			aus		
BLUE_BONNET		USA	tropical_japonica		
BOMBA	Medium	SPAGNA	temperate_japonica		
BOMBILLA	Medium	SPAGNA	temperate_japonica		
BOMBON	Medium	SPAGNA	temperate_japonica		
BOND		USA	tropical_japonica		
BRAZOS	Long A	USA	tropical_japonica		
BRIO	Round	ITALIA	temperate_japonica		
BURMA	Long A	ITALIA	tropical_japonica		
CADET	Long B	ITALIA	tropical_japonica		
CALMOCHI_101	Medium	USA	temperate_japonica	Plant/grain	Root
CAMPINO	Medium	PORTOGALLO	temperate_japonica		
CAPATAZ	Long A	SPAGNA	temperate_japonica	Plant/grain	
CARINA	Round	BULGARIA	temperate_japonica		
CARMEN	Long A	ITALIA	temperate_japonica		
CARNAROLI	Long A	ITALIA	temperate_japonica	Plant/grain	Root
CARNISE	Long A	ITALIA	temperate_japonica		
CARNISE_PRECOCE	Long A	ITALIA	temperate_japonica		
CARRICO	Round	PORTOGALLO	temperate_japonica		
CASTELMOCHI	Round	ITALIA	temperate_japonica	Plant/grain	Root
CENTAURO	Round	ITALIA	temperate_japonica	Plant/grain	Root
CENTURY_PATNA		USA	tropical_japonica		
Chau			indica		
Chiem_Chanh			indica		
Chinese			temperate_japonica		
CHIPKA	Round	BULGARIA	temperate_japonica		
CIGALON	Medium	FRANCIA	temperate_japonica	Plant/grain	
CINIA_40		CILE	temperate_japonica	Plant/grain	Root
CLOT	Medium	SPAGNA	temperate_japonica		
CNA_4081	Long B	BRASILE	indica		
COCODRIE	Long B	USA	tropical_japonica	Plant/grain	
COLINA	Round	SPAGNA	temperate_japonica	Plant/grain	
CORBETTA	Medium	ITALIA	temperate_japonica		
CRIPTO		ITALIA	temperate_japonica	Plant/grain	Root
CRLB1	Long B	ITALIA	tropical_japonica		
CT58	Long A	COLOMBIA	temperate_japonica	Plant/grain	
Dee_Geo_Woo_Gen			indica		
DELFINO	Long A	ITALIA	temperate_japonica	Plant/grain	Root
DELLROSE	Long A	USA	tropical_japonica	Plant/grain	Root
DELTA		SPAGNA	temperate_japonica		
Dhala_Shaitta			aus		
DIMITRA	Long A	GRECIA	temperate_japonica		
DIXIEBELLE	Long A	USA	tropical_japonica	Plant/grain	
Dom_Sofid			aromatic		
DRAGO	Long A	ITALIA	temperate_japonica	Plant/grain	Root
DREW	Long B	USA	tropical_japonica	Plant/grain	Root
DUCATO	Round	ITALIA	temperate_japonica		
DV85			aus		
ELIO		ITALIA	temperate_japonica		
EOLO		ITALIA	tropical_japonica		
ERCOLE	Long A	ITALIA	temperate_japonica	Plant/grain	Root
ESCARLATE	Round	PORTOGALLO	temperate_japonica		
ESTRELA_IRRADIADO	Long A	PORTOGALLO	temperate_japonica		
EUROPA	Long A	ITALIA	temperate_japonica	Plant/grain	Root
Firooz			aromatic		
FLIPPER	Long B	ITALIA	temperate_japonica		
FORTUNA	Long A	ITALIA	tropical_japonica	Plant/grain	Root
FRAGRANCE	Long B	ITALIA	aromatic	Plant/grain	
FRANCES	Medium	SPAGNA	temperate_japonica		
FULGENTE	Medium	ITALIA	temperate_japonica	Plant/grain	
GANGE	Long B	ITALIA	tropical_japonica	Plant/grain	
GARDE_SADRI	Long A	TURCHIA	temperate_japonica		
Geumbyeo			temperate_japonica		
GIADA	Long B	ITALIA	tropical_japonica	Plant/grain	Root
GIGANTE_VERCELLI	Long A	ITALIA	temperate_japonica	Plant/grain	Root
GIOVANNI_MARCHETTI	Medium	ITALIA	temperate_japonica		
GIZA_177	Medium	EGITTO	temperate_japonica	Plant/grain	
GIZA_178	Round	EGITTO	indica	Plant/grain	
GLADIO	Long B	ITALIA	tropical_japonica	Plant/grain	Root
GUADIAMAR	Medium	SPAGNA	temperate_japonica		
Guan-Yin-Tsan			indica		
GZ6296	Long A	EGITTO	indica	Plant/grain	
GZ8367		EGITTO	temperate_japonica	Plant/grain	
HANDAO_11	Round	CINA	temperate_japonica	Plant/grain	Root
HANDAO_297	Round	CINA	temperate_japonica	Plant/grain	Root
HARRA	Round	AUSTRALIA	temperate_japonica		
HONDURAS	Long A	SPAGNA	tropical_japonica	Plant/grain	
IAC32_52	Long B	BRASILE	tropical_japonica		
IBO_380-33	Long A	PORTOGALLO	temperate_japonica	Plant/grain	
IBO_400	Long A	PORTOGALLO	temperate_japonica		
ILANG_ILANG		COREA DEL SUD	tropical_japonica		
IR64	Long B	FILIPPINE	indica	Plant/grain	
ITALMOCHI	Medium	ITALIA	temperate_japonica	Plant/grain	Root
ITALPATNA_48	Long A	ITALIA	temperate_japonica	Plant/grain	Root
JACINTO	Long A	USA	tropical_japonica		
Jaya			indica		
JEFFERSON	Long A	USA	tropical_japonica	Plant/grain	Root
Jhona_349			aus		
JUBILIENI	Round	BULGARIA	temperate_japonica		
KARNAK	Long A	ITALIA	temperate_japonica	Plant/grain	
Kasalath			aus		
KATY		USA	tropical_japonica		
Khao_Gaew			aus		
KING	Long B/Long A	ITALIA	tropical_japonica		
Kitrana_508			aromatic		
KORAL	Long A	ITALIA	temperate_japonica	Plant/grain	Root
Koshihikari			temperate_japonica		
KULON	Long A	RUSSIA	temperate_japonica	Plant/grain	
KYEEMA	Long B	AUSTRALIA	tropical_japonica		
L201	Long B	USA	tropical_japonica	Plant/grain	Root
L202	Long B	USA	tropical_japonica		
L204	Long B	USA	tropical_japonica		
L205	Long B	USA	tropical_japonica		
LACASSINE	Long B	USA	tropical_japonica		
LADY_WRIGHT	Medium	USA	tropical_japonica	Plant/grain	Root
LAGRUE	Long A	USA	tropical_japonica	Plant/grain	Root
LAMONE	Long B	ITALIA	tropical_japonica	Plant/grain	Root
Lemont		USA	tropical_japonica		
LENCINO	Round	ITALIA	temperate_japonica		
LIBERO		ITALIA	tropical_japonica		
LIDO	Medium	ITALIA	temperate_japonica	Plant/grain	Root
LOMELLINO	Medium	ITALIA	temperate_japonica	Plant/grain	
LOTO	Long A	ITALIA	temperate_japonica	Plant/grain	Root
LUCERO	Round	ITALIA	temperate_japonica	Plant/grain	Root
LUNA	Medium	USA	temperate_japonica		
LUSITO_IRRADIADO	Long A	PORTOGALLO	temperate_japonica		
LUXOR	Long A	ITALIA	temperate_japonica	Plant/grain	Root

M202	Medium	USA	temperate_japonica	Plant/grain	
M203	Long A	USA	temperate_japonica		
M204	Long A	USA	temperate_japonica	Plant/grain	Root
Mansaku			temperate_japonica		
MARATELLI	Medium	ITALIA	temperate_japonica	Plant/grain	Root
MARENY	Long A	SPAGNA	temperate_japonica		
MARISMA		SPAGNA	temperate_japonica		
MARTA		ITALIA	tropical_japonica		
MARTE	Round	ITALIA	temperate_japonica	Plant/grain	Root
MAYBELLE	Long B	USA	tropical_japonica		
MERCURIO	Long B	ITALIA	tropical_japonica		
MERLE	Long B	FRANCIA	indica		
MILEV_21	Round	BULGARIA	temperate_japonica		
Miriti			tropical_japonica		
Miriti-S2			tropical_japonica		
MONTICELLI	Medium	ITALIA	temperate_japonica	Plant/grain	Root
Moroberekan			tropical_japonica		
Mudgo			indica		
MUGA	Round	PORTOGALLO	temperate_japonica		
NEMBO	Long A	ITALIA	temperate_japonica	Plant/grain	Root
NILO	Long A	ITALIA	temperate_japonica		
Nipponbare		USA	temperate_japonica		
Norin_20			temperate_japonica		
NOVARA	Medium	ITALIA	temperate_japonica	Plant/grain	
OLYMPIADA	Long B	GRECIA	indica		
OPALE	Long A	ITALIA	temperate_japonica	Plant/grain	
ORIGINARIO	Round	ITALIA	temperate_japonica		
ORIONE	Long A	ITALIA	temperate_japonica	Plant/grain	Root
OSTIGLIA	Round	ITALIA	temperate_japonica	Plant/grain	
OTA	Long A	PORTOGALLO	temperate_japonica	Plant/grain	Root
P6	Medium	ITALIA	temperate_japonica		
PADANO	Long A	ITALIA	temperate_japonica	Plant/grain	Root
PEGONIL	Medium	SPAGNA	temperate_japonica		
PELDE		AUSTRALIA	temperate_japonica	Plant/grain	
PERLA	Round	ITALIA	temperate_japonica	Plant/grain	Root
Phudugey			aus		
PIEMONTE	Long A	ITALIA	temperate_japonica	Plant/grain	Root
PLOVDIV_22	Long A	BULGARIA	temperate_japonica	Plant/grain	Root
PLOVDIV_24	Round	BULGARIA	temperate_japonica		
PLUS	Long B	ITALIA	tropical_japonica		
POLIZESTI_28		ROMANIA	temperate_japonica		
POSEIDONE	Medium	ITALIA	temperate_japonica		
Pratao			tropical_japonica		
PRECOCE_6			temperate_japonica		
PRECOCE_ROSSI			temperate_japonica		
PRECOZ_2FA		ARGENTINA	temperate_japonica		
PROMETEO	Medium	ITALIA	temperate_japonica	Plant/grain	Root
PUNTAL	Long B	SPAGNA	tropical_japonica		
R_271			temperate_japonica		
RADON		ITALIA	temperate_japonica		
RANGHINO	Round	ITALIA	temperate_japonica	Plant/grain	Root
Rathuwee			indica		
RAZZA_77	Medium	ITALIA	temperate_japonica	Plant/grain	
RB_GAMMA			temperate_japonica		
REXMONT	Long B	USA	tropical_japonica		
RIBE	Long A	ITALIA	temperate_japonica	Plant/grain	
RIBE_JAUNE			temperate_japonica		
RINALDO_BERSANI	Long A	ITALIA	temperate_japonica	Plant/grain	Root
RINGO	Long A	ITALIA	temperate_japonica	Plant/grain	
ROBBIO_SEL1	Long A	ITALIA	temperate_japonica	Plant/grain	Root
RODEO	Long A	ITALIA	temperate_japonica	Plant/grain	Root
RODINA	Round	BULGARIA	temperate_japonica	Plant/grain	
ROMA	Long A	ITALIA	temperate_japonica	Plant/grain	
ROTUNDUS	Long A	UNGHERIA	temperate_japonica		
ROXANI	Long A	GRECIA	temperate_japonica	Plant/grain	
RPC_12	Round	CINA	temperate_japonica		
RUSSO		ITALIA	temperate_japonica	Plant/grain	
S101	Medium	USA	temperate_japonica		
S102	Medium	USA	temperate_japonica		
S102_2	Medium	USA	temperate_japonica		
SAEDINENIE	Long A	BULGARIA	temperate_japonica		
SAFARI	Long A	PORTOGALLO	temperate_japonica	Plant/grain	
SAKHA_102	Medium	EGITTO	temperate_japonica	Plant/grain	Root
SAKHA_103	Round	EGITTO	temperate_japonica		
SALOIO	Long B	PORTOGALLO	temperate_japonica		
SANDORA	Long A	UNGHERIA	temperate_japonica		
SANT_ANDREA	Long A	ITALIA	temperate_japonica	Plant/grain	Root
SATURNO	Long B	ITALIA	tropical_japonica	Plant/grain	
SCUDO	Long B	ITALIA	tropical_japonica	Plant/grain	Root
SELENIO	Round	ITALIA	temperate_japonica	Plant/grain	Root
SELN_244A020	Medium	AUSTRALIA	temperate_japonica		
SENIA	Medium	SPAGNA	temperate_japonica	Plant/grain	
SEQUIAL	Medium	SPAGNA	temperate_japonica		
SESIAMOCHI	Long A/Round	ITALIA	temperate_japonica	Plant/grain	
SETANTUNO	Round	PORTOGALLO	temperate_japonica	Plant/grain	
SHANGHAI	Long A	CINA	aus		
Shinrike			temperate_japonica		
Shoemed			temperate_japonica		
SHSS_381	Long A	SPAGNA	temperate_japonica		
SHSS_53	Long A	SPAGNA	temperate_japonica		
SIRIO_CL	Long A	ITALIA	tropical_japonica		
SIS_R215	Long A	ITALIA	tropical_japonica	Plant/grain	Root
SLAVA	Medium	BULGARIA	temperate_japonica		
SMERALDO	Long A	ITALIA	temperate_japonica	Plant/grain	Root
SP55			temperate_japonica		
SPRINT	Long B	ITALIA	tropical_japonica		
SR_113	Long A	SPAGNA	temperate_japonica		
Suweon_362		KOREA	temperate_japonica		
T_1			aus		
T757		INDIA	temperate_japonica		
Taducan			indica		
TAICHUNG_65		THAILANDIA	temperate_japonica		
TEA	Medium	ITALIA	temperate_japonica		
TEJO	Long A	ITALIA	temperate_japonica	Plant/grain	Root
TEQING	Long B	CINA	indica	Plant/grain	
TEXMONT	Long A	USA	tropical_japonica		
THAIBONNET	Long B	ITALIA	tropical_japonica	Plant/grain	
THAIPERLA			temperate_japonica		
TIMICH_108		ROMANIA	temperate_japonica		
TITANIO		ITALIA	temperate_japonica	Plant/grain	Root
TITANO (ATLANTE)	Long B	ITALIA	tropical_japonica		
TOPAZIO	Medium	ITALIA	temperate_japonica		
TORIO	Long A	PORTOGALLO	temperate_japonica		
Trembese			tropical_japonica		
ULISSE	Long A	ITALIA	temperate_japonica	Plant/grain	Root
ULLAL	Round	SPAGNA	temperate_japonica		
UPLA_77	Long B	ARGENTINA	tropical_japonica	Plant/grain	Root
UPLA_79	Long B	ARGENTINA	tropical_japonica		
UPLA_91	Long B	ARGENTINA	tropical_japonica	Plant/grain	
URANO	Long B	ITALIA	tropical_japonica		
VALTEJO	Round	PORTOGALLO	temperate_japonica	Plant/grain	
VARIETA_16			temperate_japonica		
VENERE	Long B	ITALIA	temperate_japonica	Plant/grain	Root
VIALE	Long A	ITALIA	temperate_japonica		
VIALONE_190	Medium	ITALIA	temperate_japonica	Plant/grain	Root
VIALONE_NANO	Medium	ITALIA	temperate_japonica	Plant/grain	Root
VICTORIA	Round	ARGENTINA	temperate_japonica	Plant/grain	
VIRGO	Medium	ITALIA	temperate_japonica		
VOLANO	Long A	ITALIA	temperate_japonica	Plant/grain	Root
YRL_196		AUSTRALIA	temperate_japonica		
YRM_6_2	Medium	AUSTRALIA	temperate_japonica		
ZENITH	Medium	USA	tropical_japonica		
ZHENSHANG_97		CINA	indica	Plant/grain	
RUBINO	Round	ITALIA	temperate_japonica		
RONCOLO	Medium	ITALIA	temperate_japonica		
OLCENENGO	Long A	ITALIA	temperate_japonica		
SOURE	Long A	PORTOGALLO	temperate_japonica		
NIBBIO	Medium	ITALIA	temperate_japonica		
REDI	Long A	ITALIA	temperate_japonica	Plant/grain	Root
VENERIA	Long A	ITALIA	temperate_japonica		
BALDO	Long A	ITALIA	temperate_japonica	Plant/grain	Root
STRELLA	Long A	ITALIA	temperate_japonica		
CERVO		ITALIA	temperate_japonica		
ONDA		ITALIA	temperate_japonica		
M6	Long A	ITALIA	temperate_japonica		
ADELAIDE_CHIAPPELLI	Long A	ITALIA	temperate_japonica	Plant/grain	
SILLA	Long A	ITALIA	temperate_japonica	Plant/grain	Root
MOLO	Long A	ITALIA	temperate_japonica	Plant/grain	
CALENDAL	Long A	FRANCIA	temperate_japonica		
TOSCA	Long A	ITALIA	temperate_japonica		
PIERINA_MARCHETTI	Long A	ITALIA	temperate_japonica		
RIZZOTTO_51_1	Long A	ITALIA	temperate_japonica		
SESIA	Long A	ITALIA	temperate_japonica	Plant/grain	
ALICE	Long A	ITALIA	temperate_japonica	Plant/grain	Root
MANTOVA	Long A	ITALIA	temperate_japonica	Plant/grain	
FAMILIA_181	Long A	PORTOGALLO	temperate_japonica		
FAISCA			temperate_japonica		
BONNI	Long A	ITALIA	temperate_japonica		
SUPER		PORTOGALLO	temperate_japonica	Plant/grain	Root
EUROSE	Long A	ITALIA	temperate_japonica		
GRITNA	Long A	ITALIA	temperate_japonica	Plant/grain	
BORRACHO			temperate_japonica		
SAVIO	Long A	ITALIA	temperate_japonica	Plant/grain	
HAREM	Long A	PORTOGALLO	temperate_japonica		
ARTICO			indica		
SENATORE_NOVELLI	Long A	ITALIA	temperate_japonica		
RUBIDIO			temperate_japonica		
RUBI		PORTOGALLO	temperate_japonica	Plant/grain	
RIVA		ITALIA	temperate_japonica		
LORD	Long A	ITALIA	temperate_japonica	Plant/grain	Root
ZENA	Long B	ITALIA	tropical_japonica	Plant/grain	Root
SANDOCA	Long B	PORTOGALLO	temperate_japonica		
GALILEO	Long A	ITALIA	temperate_japonica	Plant/grain	
CT36	Long B	COLOMBIA	temperate_japonica		
ARTIGLIO	Long B	ITALIA	indica		
GHIBLI		ITALIA	temperate_japonica		
CRESO	Long A	ITALIA	temperate_japonica	Plant/grain	
UPLA_75	Long B	ARGENTINA	tropical_japonica	Plant/grain	Root
SCIROCCO	Medium	ITALIA	temperate_japonica		
UPLA_32	Long B	ARGENTINA	tropical_japonica	Plant/grain	Root
UPLA_80	Long B	ARGENTINA	tropical_japonica		
BASMATI_DETRADUNI	Long B	INDIA	indica		
BIANCA	Long A	ITALIA	temperate_japonica	Plant/grain	Root
VELA	Long A	ITALIA	temperate_japonica		
MELAS	Long B	GRECIA	temperate_japonica	Plant/grain	Root
UPLA_63	Long B	ARGENTINA	tropical_japonica	Plant/grain	
TARRISO	Long B	ITALIA	tropical_japonica		
UPLA_104	Long B	ARGENTINA	tropical_japonica		
PANDA		ITALIA	tropical_japonica	Plant/grain	Root
SAGRES	Long A	PORTOGALLO	temperate_japonica		
UPLA_60	Long B	ARGENTINA	tropical_japonica		
MIARA	Long B	ITALIA	temperate_japonica		
UPLA_68	Long B	ARGENTINA	tropical_japonica	Plant/grain	
CARIOCA	Long B	ITALIA	tropical_japonica		
SAMBA	Long A	ITALIA	tropical_japonica		
GRAAL	Long B	FRANCIA	tropical_japonica		
CHACARERO			temperate_japonica		
ALINANO		FRANCIA	temperate_japonica		
ELLEBI	Long B	ITALIA	tropical_japonica		
NANO	Round	ITALIA	temperate_japonica		
CT23		COLOMBIA	temperate_japonica		
MEJANES	Long B	FRANCIA	temperate_japonica		
GIANO	Long B	ITALIA	tropical_japonica	Plant/grain	Root
ITALPATNAxMILYANG	Long A	PORTOGALLO	temperate_japonica		
BAJANGxALLORIO			temperate_japonica		
SALVO	Long B	ITALIA	tropical_japonica	Plant/grain	Root
DARDO	Long A	ITALIA	tropical_japonica		
EUROSIS	Long A	ITALIA	temperate_japonica	Plant/grain	Root
SANTERNO	Long B	ITALIA	temperate_japonica	Plant/grain	Root
ALBATROS		ITALIA	temperate_japonica		
GREPPI	Round	ITALIA	tropical_japonica	Plant/grain	Root
ESTRELA	Long A	PORTOGALLO	temperate_japonica		
UPLA_64	Long B	ARGENTINA	tropical_japonica		
MAIORAL	Long A	PORTOGALLO	temperate_japonica		
PREVER	Long B	ITALIA	tropical_japonica		
ERMES	Long B	ITALIA	tropical_japonica	Plant/grain	
PECOS	Medium	USA	tropical_japonica	Plant/grain	Root
ANSEATICO	Long A	ITALIA	temperate_japonica		
ARTEMIDE	Long B	ITALIA	tropical_japonica	Plant/grain	Root
BENGAL	Long A	USA	temperate_japonica	Plant/grain	
MIDA			temperate_japonica		
GRALDO	Long B	ITALIA	tropical_japonica	Plant/grain	Root
RONALDO	Long A	ITALIA	tropical_japonica		
ROMBO	Long A	ITALIA	temperate_japonica		
OSCARxSUWEON	Long A	PORTOGALLO	temperate_japonica		
FIDJI	Long B	FILIPPINE	tropical_japonica		

S2 Table: Command lines. Command lines and parameters used to generate the imputed non-filtered SNP dataset with Tassel GBS pipeline 3.0. Command lines with the "Plugin" suffix are part of the TASSEL pipeline; filenames and directories are not listed. In order to run the "FastImputationBitFixedWindowPlugin", a tab-separated file with an inbreeding coefficient of 0,99 for each taxon was provided. When not reported, default parameters were used.

Command line	Parameters
FastqToTagCountPlugin	-Xmx16G -i fastq -e ApeKI -s 400000000
MergeMultipleTagCountPlugin	-c 5 -t
bwa index	-a is
bwa aln	-t 8
bwa samse	
SAMConverterPlugin	
FastqToTBTPlugin	-Xmx16G -i fastq -e ApeKI -y -s 400000000
MergeTagsByTaxaFilesPlugin	-Xmx16G -s 400000000
TagsToSNPByAlignmentPlugin	-mnF 0.99 -mnMAF 0.01 -mnMAC 10 -s 1 -e 12 -y
MergeDuplicateSNPsPlugin	-s 1 -e 12
FastImputationBitFixedWindowPlugin	

S3 Table: Probability of assignment. List of accessions used in the study with the taxonomical group they have been assigned to and the probability of assignment to each of the 5 major groups based on their SNP genotypes.

Accession	Group_Admixture	Score_Admixture	indica	trop_jap	temp_jap	aromatic	aus
9311	indica	0.99996	0.99996	0.00001	0.00001	0.00001	0.00001
A201	tropical_japonica	0.99996	0.00001	0.99996	0.00001	0.00001	0.00001
A301	tropical_japonica	0.99996	0.00001	0.99996	0.00001	0.00001	0.00001
ADAIR	tropical_japonica	0.99996	0.00001	0.99996	0.00001	0.00001	0.00001
AGOSTANO	temperate_japonica	0.99996	0.00001	0.00001	0.99996	0.00001	0.00001
AIACE	tropical_japonica	0.99996	0.00001	0.99996	0.00001	0.00001	0.00001
AKITAKOMACHI	temperate_japonica	0.99996	0.00001	0.00001	0.99996	0.00001	0.00001
ALAN	tropical_japonica	0.99996	0.00001	0.99996	0.00001	0.00001	0.00001
ALEXANDROS	tropical_japonica	0.99996	0.00001	0.99996	0.00001	0.00001	0.00001
ALLORIO	temperate_japonica	0.99996	0.00001	0.00001	0.99996	0.00001	0.00001
ALPE	temperate_japonica	0.99996	0.00001	0.00001	0.99996	0.00001	0.00001
ALPHA	temperate_japonica	0.99996	0.00001	0.00001	0.99996	0.00001	0.00001
AMBRA	temperate_japonica	0.99996	0.00001	0.00001	0.99996	0.00001	0.00001
AMERICANO_1600	temperate_japonica	0.99996	0.00001	0.00001	0.99996	0.00001	0.00001
ANTONI	temperate_japonica	0.99996	0.00001	0.00001	0.99996	0.00001	0.00001
APOLLO	tropical_japonica	0.99996	0.00001	0.99996	0.00001	0.00001	0.00001
ARBORIO	temperate_japonica	0.99996	0.00001	0.00001	0.99996	0.00001	0.00001
ARC_10352	aromatic	0.99996	0.00001	0.00001	0.00001	0.99996	0.00001
ARGO	temperate_japonica	0.99996	0.00001	0.00001	0.99996	0.00001	0.00001
ARIANA	indica	0.99996	0.99996	0.00001	0.00001	0.00001	0.00001
Arias	tropical_japonica	0.99996	0.00001	0.99996	0.00001	0.00001	0.00001
ARIETE	temperate_japonica	0.99996	0.00001	0.00001	0.99996	0.00001	0.00001
ARLESIENNE	temperate_japonica	0.99996	0.00001	0.00001	0.99996	0.00001	0.00001
ARSENAL	tropical_japonica	0.99996	0.00001	0.99996	0.00001	0.00001	0.00001
ASIA	aromatic	0.99996	0.00001	0.00001	0.00001	0.99996	0.00001
Asse_Y_Pung	tropical_japonica	0.99996	0.00001	0.99996	0.00001	0.00001	0.00001
AUGUSTO	temperate_japonica	0.99996	0.00001	0.00001	0.99996	0.00001	0.00001
AUZGUSTA	temperate_japonica	0.99996	0.00001	0.00001	0.99996	0.00001	0.00001
Azucena	tropical_japonica	0.99996	0.00001	0.99996	0.00001	0.00001	0.00001
BACCO	temperate_japonica	0.99996	0.00001	0.00001	0.99996	0.00001	0.00001
BAHIA	temperate_japonica	0.99996	0.00001	0.00001	0.99996	0.00001	0.00001
BAIXET	temperate_japonica	0.99996	0.00001	0.00001	0.99996	0.00001	0.00001
BALILLA	temperate_japonica	0.99996	0.00001	0.00001	0.99996	0.00001	0.00001
BALILLONE_R253	temperate_japonica	0.99996	0.00001	0.00001	0.99996	0.00001	0.00001
BALOCCO_G	temperate_japonica	0.99996	0.00001	0.00001	0.99996	0.00001	0.00001
BALZARETTI	temperate_japonica	0.99996	0.00001	0.00001	0.99996	0.00001	0.00001
BARAGGIA	temperate_japonica	0.99996	0.00001	0.00001	0.99996	0.00001	0.00001
Basmati	aromatic	0.99996	0.00001	0.00001	0.00001	0.99996	0.00001
Basmati_1	aus	0.99996	0.00001	0.00001	0.00001	0.00001	0.99996
BASMATI_C621	aromatic	0.99996	0.00001	0.00001	0.00001	0.99996	0.00001
BEIRAO	temperate_japonica	0.99996	0.00001	0.00001	0.99996	0.00001	0.00001
BELLE_PATNA	tropical_japonica	0.99996	0.00001	0.99996	0.00001	0.00001	0.00001
BERTONE	temperate_japonica	0.99996	0.00001	0.00001	0.99996	0.00001	0.00001
BETIS	temperate_japonica	0.99996	0.00001	0.00001	0.99996	0.00001	0.00001
Bico_Branco	aromatic	0.99996	0.00001	0.00001	0.00001	0.99996	0.00001
Binulawan-F	indica	0.99996	0.99996	0.00001	0.00001	0.00001	0.00001
BJ_1	aus	0.99996	0.00001	0.00001	0.00001	0.00001	0.99996
BLACK_EGYPT	temperate_japonica	0.99996	0.00001	0.00001	0.99996	0.00001	0.00001
Black_Gora	aus	0.99996	0.00001	0.00001	0.00001	0.00001	0.99996
BLUE_BONNET	tropical_japonica	0.99996	0.00001	0.99996	0.00001	0.00001	0.00001
BOMBA	temperate_japonica	0.99996	0.00001	0.00001	0.99996	0.00001	0.00001
BOMBILLA	temperate_japonica	0.99996	0.00001	0.00001	0.99996	0.00001	0.00001
BOMBON	temperate_japonica	0.99996	0.00001	0.00001	0.99996	0.00001	0.00001
BOND	tropical_japonica	0.99996	0.00001	0.99996	0.00001	0.00001	0.00001
BRAZOS	tropical_japonica	0.99996	0.00001	0.99996	0.00001	0.00001	0.00001
BRIO	temperate_japonica	0.99996	0.00001	0.00001	0.99996	0.00001	0.00001
BURMA	tropical_japonica	0.99996	0.00001	0.99996	0.00001	0.00001	0.00001
CADET	tropical_japonica	0.99996	0.00001	0.99996	0.00001	0.00001	0.00001
CALMOCHI_101	temperate_japonica	0.99996	0.00001	0.00001	0.99996	0.00001	0.00001
CAMPINO	temperate_japonica	0.99996	0.00001	0.00001	0.99996	0.00001	0.00001
CAPATAZ	temperate_japonica	0.99996	0.00001	0.00001	0.99996	0.00001	0.00001
CARINA	temperate_japonica	0.99996	0.00001	0.00001	0.99996	0.00001	0.00001
CARMEN	temperate_japonica	0.99996	0.00001	0.00001	0.99996	0.00001	0.00001
CARNAROLI	temperate_japonica	0.99996	0.00001	0.00001	0.99996	0.00001	0.00001
CARNISE	temperate_japonica	0.99996	0.00001	0.00001	0.99996	0.00001	0.00001
CARNISE_PRECOCE	temperate_japonica	0.99996	0.00001	0.00001	0.99996	0.00001	0.00001
CARRICO	temperate_japonica	0.99996	0.00001	0.00001	0.99996	0.00001	0.00001
CASTELMOCHI	temperate_japonica	0.99996	0.00001	0.00001	0.99996	0.00001	0.00001
CENTAURO	temperate_japonica	0.99996	0.00001	0.00001	0.99996	0.00001	0.00001
CENTURY_PATNA	tropical_japonica	0.99996	0.00001	0.99996	0.00001	0.00001	0.00001
Chau	indica	0.99996	0.99996	0.00001	0.00001	0.00001	0.00001
Chiem_Chanh	indica	0.99996	0.99996	0.00001	0.00001	0.00001	0.00001
Chinese	temperate_japonica	0.99996	0.00001	0.00001	0.99996	0.00001	0.00001
CHIPKA	temperate_japonica	0.99996	0.00001	0.00001	0.99996	0.00001	0.00001
CIGALON	temperate_japonica	0.99996	0.00001	0.00001	0.99996	0.00001	0.00001
CINIA_40	temperate_japonica	0.99996	0.00001	0.00001	0.99996	0.00001	0.00001
CLOT	temperate_japonica	0.99996	0.00001	0.00001	0.99996	0.00001	0.00001
CNA_4081	indica	0.99996	0.99996	0.00001	0.00001	0.00001	0.00001
COCODRIE	tropical_japonica	0.99996	0.00001	0.99996	0.00001	0.00001	0.00001
COLINA	temperate_japonica	0.99996	0.00001	0.00001	0.99996	0.00001	0.00001
CORBETTA	temperate_japonica	0.99996	0.00001	0.00001	0.99996	0.00001	0.00001
CRIPTO	temperate_japonica	0.99996	0.00001	0.00001	0.99996	0.00001	0.00001
CRLB1	tropical_japonica	0.99996	0.00001	0.99996	0.00001	0.00001	0.00001
CT58	temperate_japonica	0.99996	0.00001	0.00001	0.99996	0.00001	0.00001
Dee_Geo_Woo_Gen	indica	0.99996	0.99996	0.00001	0.00001	0.00001	0.00001
DELFINO	temperate_japonica	0.99996	0.00001	0.00001	0.99996	0.00001	0.00001
DELLROSE	tropical_japonica	0.99996	0.00001	0.99996	0.00001	0.00001	0.00001
DELTA	temperate_japonica	0.99996	0.00001	0.00001	0.99996	0.00001	0.00001
Dhala_Shaitta	aus	0.99996	0.00001	0.00001	0.00001	0.00001	0.99996
DIMITRA	temperate_japonica	0.99996	0.00001	0.00001	0.99996	0.00001	0.00001
DIXIEBELLE	tropical_japonica	0.99996	0.00001	0.99996	0.00001	0.00001	0.00001
Dom_Sofid	aromatic	0.99996	0.00001	0.00001	0.00001	0.99996	0.00001
DRAGO	temperate_japonica	0.99996	0.00001	0.00001	0.99996	0.00001	0.00001
DREW	tropical_japonica	0.99996	0.00001	0.99996	0.00001	0.00001	0.00001
DUCATO	temperate_japonica	0.99996	0.00001	0.00001	0.99996	0.00001	0.00001
DV85	aus	0.99996	0.00001	0.00001	0.00001	0.00001	0.99996
ELIO	temperate_japonica	0.99996	0.00001	0.00001	0.99996	0.00001	0.00001
EOLO	tropical_japonica	0.99996	0.00001	0.99996	0.00001	0.00001	0.00001
ERCOLE	temperate_japonica	0.99996	0.00001	0.00001	0.99996	0.00001	0.00001
ESCARLATE	temperate_japonica	0.99996	0.00001	0.00001	0.99996	0.00001	0.00001
ESTRELA_IRRADIADO	temperate_japonica	0.99996	0.00001	0.00001	0.99996	0.00001	0.00001
EUROPA	temperate_japonica	0.99996	0.00001	0.00001	0.99996	0.00001	0.00001
Firooz	aromatic	0.99996	0.00001	0.00001	0.00001	0.99996	0.00001
FLIPPER	temperate_japonica	0.99996	0.00001	0.00001	0.99996	0.00001	0.00001
FORTUNA	tropical_japonica	0.99996	0.00001	0.99996	0.00001	0.00001	0.00001
FRAGRANCE	aromatic	0.99996	0.00001	0.00001	0.00001	0.99996	0.00001
FRANCES	temperate_japonica	0.99996	0.00001	0.00001	0.99996	0.00001	0.00001
FULGENTE	temperate_japonica	0.99996	0.00001	0.00001	0.99996	0.00001	0.00001
GANGE	tropical_japonica	0.99996	0.00001	0.99996	0.00001	0.00001	0.00001
GARDE_SADRI	temperate_japonica	0.99996	0.00001	0.00001	0.99996	0.00001	0.00001
Geumbyeo	temperate_japonica	0.99996	0.00001	0.00001	0.99996	0.00001	0.00001
GIADA	tropical_japonica	0.99996	0.00001	0.99996	0.00001	0.00001	0.00001
GIGANTE_VERCELLI	temperate_japonica	0.99996	0.00001	0.00001	0.99996	0.00001	0.00001
GIOVANNI_MARCHETTI	temperate_japonica	0.99996	0.00001	0.00001	0.99996	0.00001	0.00001
GIZA_177	temperate_japonica	0.99996	0.00001	0.00001	0.99996	0.00001	0.00001
GIZA_178	indica	0.99996	0.99996	0.00001	0.00001	0.00001	0.00001
GLADIO	tropical_japonica	0.99996	0.00001	0.99996	0.00001	0.00001	0.00001
GUADIAMAR	temperate_japonica	0.99996	0.00001	0.00001	0.99996	0.00001	0.00001
Guan-Yin-Tsan	indica	0.99996	0.99996	0.00001	0.00001	0.00001	0.00001
GZ6296	indica	0.99996	0.99996	0.00001	0.00001	0.00001	0.00001
GZ8367	temperate_japonica	0.99996	0.00001	0.00001	0.99996	0.00001	0.00001
HANDAO_11	temperate_japonica	0.99996	0.00001	0.00001	0.99996	0.00001	0.00001
HANDAO_297	temperate_japonica	0.99996	0.00001	0.00001	0.99996	0.00001	0.00001
HARRA	temperate_japonica	0.99996	0.00001	0.00001	0.99996	0.00001	0.00001
HONDURAS	tropical_japonica	0.99996	0.00001	0.99996	0.00001	0.00001	0.00001
IAC32_52	tropical_japonica	0.99996	0.00001	0.99996	0.00001	0.00001	0.00001
IBO_380-33	temperate_japonica	0.99996	0.00001	0.00001	0.99996	0.00001	0.00001
IBO_400	temperate_japonica	0.99996	0.00001	0.00001	0.99996	0.00001	0.00001
ILANG_ILANG	tropical_japonica	0.99996	0.00001	0.99996	0.00001	0.00001	0.00001
IR64	indica	0.99996	0.99996	0.00001	0.00001	0.00001	0.00001
ITALMOCHI	temperate_japonica	0.99996	0.00001	0.00001	0.99996	0.00001	0.00001
ITALPATNA_48	temperate_japonica	0.99996	0.00001	0.00001	0.99996	0.00001	0.00001
JACINTO	tropical_japonica	0.99996	0.00001	0.99996	0.00001	0.00001	0.00001
Jaya	indica	0.99996	0.99996	0.00001	0.00001	0.00001	0.00001
JEFFERSON	tropical_japonica	0.99996	0.00001	0.99996	0.00001	0.00001	0.00001
Jhona_349	aus	0.99996	0.00001	0.00001	0.00001	0.00001	0.99996
JUBILIENI	temperate_japonica	0.99996	0.00001	0.00001	0.99996	0.00001	0.00001
KARNAK	temperate_japonica	0.99996	0.00001	0.00001	0.99996	0.00001	0.00001
Kasalath	aus	0.99996	0.00001	0.00001	0.00001	0.00001	0.99996
KATY	tropical_japonica	0.99996	0.00001	0.99996	0.00001	0.00001	0.00001
Khao_Gaew	aus	0.99996	0.00001	0.00001	0.00001	0.00001	0.99996
KING	tropical_japonica	0.99996	0.00001	0.99996	0.00001	0.00001	0.00001
Kitrana_508	aromatic	0.99996	0.00001	0.00001	0.00001	0.99996	0.00001
KORAL	temperate_japonica	0.99996	0.00001	0.00001	0.99996	0.00001	0.00001
Koshihikari	temperate_japonica	0.99996	0.00001	0.00001	0.99996	0.00001	0.00001
KULON	temperate_japonica	0.99996	0.00001	0.00001	0.99996	0.00001	0.00001
KYEEMA	tropical_japonica	0.99996	0.00001	0.99996	0.00001	0.00001	0.00001
L201	tropical_japonica	0.99996	0.00001	0.99996	0.00001	0.00001	0.00001
L202	tropical_japonica	0.99996	0.00001	0.99996	0.00001	0.00001	0.00001
L204	tropical_japonica	0.99996	0.00001	0.99996	0.00001	0.00001	0.00001
L205	tropical_japonica	0.99996	0.00001	0.99996	0.00001	0.00001	0.00001
LACASSINE	tropical_japonica	0.99996	0.00001	0.99996	0.00001	0.00001	0.00001
LADY_WRIGHT	tropical_japonica	0.99996	0.00001	0.99996	0.00001	0.00001	0.00001
LAGRUE	tropical_japonica	0.99996	0.00001	0.99996	0.00001	0.00001	0.00001
LAMONE	tropical_japonica	0.99996	0.00001	0.99996	0.00001	0.00001	0.00001
Lemont	tropical_japonica	0.99996	0.00001	0.99996	0.00001	0.00001	0.00001
LENCINO	temperate_japonica	0.99996	0.00001	0.00001	0.99996	0.00001	0.00001
LIBERO	tropical_japonica	0.99996	0.00001	0.99996	0.00001	0.00001	0.00001
LIDO	temperate_japonica	0.99996	0.00001	0.00001	0.99996	0.00001	0.00001
LOMELLINO	temperate_japonica	0.99996	0.00001	0.00001	0.99996	0.00001	0.00001
LOTO	temperate_japonica	0.99996	0.00001	0.00001	0.99996	0.00001	0.00001
LUCERO	temperate_japonica	0.99996	0.00001	0.00001	0.99996	0.00001	0.00001
LUNA	temperate_japonica	0.99996	0.00001	0.00001	0.99996	0.00001	0.00001
LUSITO_IRRADIADO	temperate_japonica	0.99996	0.00001	0.00001	0.99996	0.00001	0.00001
LUXOR	temperate_japonica	0.99996	0.00001	0.00001	0.99996	0.00001	0.00001
M202	temperate_japonica	0.99996	0.00001	0.00001	0.99996	0.00001	0.00001
M203	temperate_japonica	0.99996	0.00001	0.00001	0.99996	0.00001	0.00001
M204	temperate_japonica	0.99996	0.00001	0.00001	0.99996	0.00001	0.00001
Mansaku	temperate_japonica	0.99996	0.00001	0.00001	0.99996	0.00001	0.00001
MARATELLI	temperate_japonica	0.99996	0.00001	0.00001	0.99996	0.00001	0.00001
MARENY	temperate_japonica	0.99996	0.00001	0.00001	0.99996	0.00001	0.00001
MARISMA	temperate_japonica	0.99996	0.00001	0.00001	0.99996	0.00001	0.00001
MARTA	tropical_japonica	0.99996	0.00001	0.99996	0.00001	0.00001	0.00001
MARTE	temperate_japonica	0.99996	0.00001	0.00001	0.99996	0.00001	0.00001
MAYBELLE	tropical_japonica	0.99996	0.00001	0.99996	0.00001	0.00001	0.00001
MERCURIO	tropical_japonica	0.99996	0.00001	0.99996	0.00001	0.00001	0.00001
MERLE	indica	0.99996	0.99996	0.00001	0.00001	0.00001	0.00001
MILEV_21	temperate_japonica	0.99996	0.00001	0.00001	0.99996	0.00001	0.00001
Miriti	tropical_japonica	0.99996	0.00001	0.99996	0.00001	0.00001	0.00001
Miriti-S2	tropical_japonica	0.99996	0.00001	0.99996	0.00001	0.00001	0.00001
MONTICELLI	temperate_japonica	0.99996	0.00001	0.00001	0.99996	0.00001	0.00001
Moroberekan	tropical_japonica	0.99996	0.00001	0.99996	0.00001	0.00001	0.00001
Mudgo	indica	0.99996	0.99996	0.00001	0.00001	0.00001	0.00001
MUGA	temperate_japonica	0.99996	0.00001	0.00001	0.99996	0.00001	0.00001
NEMBO	temperate_japonica	0.99996	0.00001	0.00001	0.99996	0.00001	0.00001
NILO	temperate_japonica	0.99996	0.00001	0.00001	0.99996	0.00001	0.00001
Nipponbare	temperate_japonica	0.99996	0.00001	0.00001	0.99996	0.00001	0.00001
Norin_20	temperate_japonica	0.99996	0.00001	0.00001	0.99996	0.00001	0.00001
NOVARA	temperate_japonica	0.99996	0.00001	0.00001	0.99996	0.00001	0.00001
OLYMPIADA	indica	0.99996	0.99996	0.00001	0.00001	0.00001	0.00001
OPALE	temperate_japonica	0.99996	0.00001	0.00001	0.99996	0.00001	0.00001
ORIGINARIO	temperate_japonica	0.99996	0.00001	0.00001	0.99996	0.00001	0.00001
ORIONE	temperate_japonica	0.99996	0.00001	0.00001	0.99996	0.00001	0.00001
OSTIGLIA	temperate_japonica	0.99996	0.00001	0.00001	0.99996	0.00001	0.00001
OTA	temperate_japonica	0.99996	0.00001	0.00001	0.99996	0.00001	0.00001
P6	temperate_japonica	0.99996	0.00001	0.00001	0.99996	0.00001	0.00001
PADANO	temperate_japonica	0.99996	0.00001	0.00001	0.99996	0.00001	0.00001
PEGONIL	temperate_japonica	0.99996	0.00001	0.00001	0.99996	0.00001	0.00001
PELDE	temperate_japonica	0.99996	0.00001	0.00001	0.99996	0.00001	0.00001
PERLA	temperate_japonica	0.99996	0.00001	0.00001	0.99996	0.00001	0.00001
Phudugey	aus	0.99996	0.00001	0.00001	0.00001	0.00001	0.99996
PIEMONTE	temperate_japonica	0.99996	0.00001	0.00001	0.99996	0.00001	0.00001
PLOVDIV_22	temperate_japonica	0.99996	0.00001	0.00001	0.99996	0.00001	0.00001
PLOVDIV_24	temperate_japonica	0.99996	0.00001	0.00001	0.99996	0.00001	0.00001
PLUS	tropical_japonica	0.99996	0.00001	0.99996	0.00001	0.00001	0.00001
POLIZESTI_28	temperate_japonica	0.99996	0.00001	0.00001	0.99996	0.00001	0.00001
POSEIDONE	temperate_japonica	0.99996	0.00001	0.00001	0.99996	0.00001	0.00001
Pratao	tropical_japonica	0.99996	0.00001	0.99996	0.00001	0.00001	0.00001
PRECOCE_6	temperate_japonica	0.99996	0.00001	0.00001	0.99996	0.00001	0.00001
PRECOCE_ROSSI	temperate_japonica	0.99996	0.00001	0.00001	0.99996	0.00001	0.00001
PRECOZ_2FA	temperate_japonica	0.99996	0.00001	0.00001	0.99996	0.00001	0.00001
PROMETEO	temperate_japonica	0.99996	0.00001	0.00001	0.99996	0.00001	0.00001
PUNTAL	tropical_japonica	0.99996	0.00001	0.99996	0.00001	0.00001	0.00001
R_271	temperate_japonica	0.99996	0.00001	0.00001	0.99996	0.00001	0.00001
RADON	temperate_japonica	0.99996	0.00001	0.00001	0.99996	0.00001	0.00001
RANGHINO	temperate_japonica	0.99996	0.00001	0.00001	0.99996	0.00001	0.00001
Rathuwee	indica	0.99996	0.99996	0.00001	0.00001	0.00001	0.00001
RAZZA_77	temperate_japonica	0.99996	0.00001	0.00001	0.99996	0.00001	0.00001
RB_GAMMA	temperate_japonica	0.99996	0.00001	0.00001	0.99996	0.00001	0.00001
REXMONT	tropical_japonica	0.99996	0.00001	0.99996	0.00001	0.00001	0.00001
RIBE	temperate_japonica	0.99996	0.00001	0.00001	0.99996	0.00001	0.00001
RIBE_JAUNE	temperate_japonica	0.99996	0.00001	0.00001	0.99996	0.00001	0.00001
RINALDO_BERSANI	temperate_japonica	0.99996	0.00001	0.00001	0.99996	0.00001	0.00001
RINGO	temperate_japonica	0.99996	0.00001	0.00001	0.99996	0.00001	0.00001
ROBBIO_SEL1	temperate_japonica	0.99996	0.00001	0.00001	0.99996	0.00001	0.00001
RODEO	temperate_japonica	0.99996	0.00001	0.00001	0.99996	0.00001	0.00001

RODINA	temperate_japonica	0.99996	0.00001	0.00001	0.99996	0.00001	0.00001
ROMA	temperate_japonica	0.99996	0.00001	0.00001	0.99996	0.00001	0.00001
ROTUNDUS	temperate_japonica	0.99996	0.00001	0.00001	0.99996	0.00001	0.00001
ROXANI	temperate_japonica	0.99996	0.00001	0.00001	0.99996	0.00001	0.00001
RPC_12	temperate_japonica	0.99996	0.00001	0.00001	0.99996	0.00001	0.00001
RUSSO	temperate_japonica	0.99996	0.00001	0.00001	0.99996	0.00001	0.00001
S101	temperate_japonica	0.99996	0.00001	0.00001	0.99996	0.00001	0.00001
S102	temperate_japonica	0.99996	0.00001	0.00001	0.99996	0.00001	0.00001
S102_2	temperate_japonica	0.99996	0.00001	0.00001	0.99996	0.00001	0.00001
SAEDINENIE	temperate_japonica	0.99996	0.00001	0.00001	0.99996	0.00001	0.00001
SAFARI	temperate_japonica	0.99996	0.00001	0.00001	0.99996	0.00001	0.00001
SAKHA_102	temperate_japonica	0.99996	0.00001	0.00001	0.99996	0.00001	0.00001
SAKHA_103	temperate_japonica	0.99996	0.00001	0.00001	0.99996	0.00001	0.00001
SALOIO	temperate_japonica	0.99996	0.00001	0.00001	0.99996	0.00001	0.00001
SANDORA	temperate_japonica	0.99996	0.00001	0.00001	0.99996	0.00001	0.00001
SANT_ANDREA	temperate_japonica	0.99996	0.00001	0.00001	0.99996	0.00001	0.00001
SATURNO	tropical_japonica	0.99996	0.00001	0.99996	0.00001	0.00001	0.00001
SCUDO	tropical_japonica	0.99996	0.00001	0.99996	0.00001	0.00001	0.00001
SELENIO	temperate_japonica	0.99996	0.00001	0.00001	0.99996	0.00001	0.00001
SELN_244A620	temperate_japonica	0.99996	0.00001	0.00001	0.99996	0.00001	0.00001
SENIA	temperate_japonica	0.99996	0.00001	0.00001	0.99996	0.00001	0.00001
SEQUIAL	temperate_japonica	0.99996	0.00001	0.00001	0.99996	0.00001	0.00001
SESIAMOCHI	temperate_japonica	0.99996	0.00001	0.00001	0.99996	0.00001	0.00001
SETANTUNO	temperate_japonica	0.99996	0.00001	0.00001	0.99996	0.00001	0.00001
SHANGHAI	aus	0.99996	0.00001	0.00001	0.00001	0.00001	0.99996
Shinrike	temperate_japonica	0.99996	0.00001	0.00001	0.99996	0.00001	0.00001
Shoemed	temperate_japonica	0.99996	0.00001	0.00001	0.99996	0.00001	0.00001
SHSS_381	temperate_japonica	0.99996	0.00001	0.00001	0.99996	0.00001	0.00001
SHSS_53	temperate_japonica	0.99996	0.00001	0.00001	0.99996	0.00001	0.00001
SIRIO_CL	tropical_japonica	0.99996	0.00001	0.99996	0.00001	0.00001	0.00001
SIS_R215	tropical_japonica	0.99996	0.00001	0.99996	0.00001	0.00001	0.00001
SLAVA	temperate_japonica	0.99996	0.00001	0.00001	0.99996	0.00001	0.00001
SMERALDO	temperate_japonica	0.99996	0.00001	0.00001	0.99996	0.00001	0.00001
SP55	temperate_japonica	0.99996	0.00001	0.00001	0.99996	0.00001	0.00001
SPRINT	tropical_japonica	0.99996	0.00001	0.99996	0.00001	0.00001	0.00001
SR_113	temperate_japonica	0.99996	0.00001	0.00001	0.99996	0.00001	0.00001
Suweon_362	temperate_japonica	0.99996	0.00001	0.00001	0.99996	0.00001	0.00001
T_1	aus	0.99996	0.00001	0.00001	0.00001	0.00001	0.99996
T757	temperate_japonica	0.99996	0.00001	0.00001	0.99996	0.00001	0.00001
Taducan	indica	0.99996	0.99996	0.00001	0.00001	0.00001	0.00001
TAICHUNG_65	temperate_japonica	0.99996	0.00001	0.00001	0.99996	0.00001	0.00001
TEA	temperate_japonica	0.99996	0.00001	0.00001	0.99996	0.00001	0.00001
TEJO	temperate_japonica	0.99996	0.00001	0.00001	0.99996	0.00001	0.00001
TEQING	indica	0.99996	0.99996	0.00001	0.00001	0.00001	0.00001
TEXMONT	tropical_japonica	0.99996	0.00001	0.99996	0.00001	0.00001	0.00001
THAIBONNET	tropical_japonica	0.99996	0.00001	0.99996	0.00001	0.00001	0.00001
THAIPERLA	temperate_japonica	0.99996	0.00001	0.00001	0.99996	0.00001	0.00001
TIMICH_108	temperate_japonica	0.99996	0.00001	0.00001	0.99996	0.00001	0.00001
TITANIO	temperate_japonica	0.99996	0.00001	0.00001	0.99996	0.00001	0.00001
TITANO (ATLANTE)	tropical_japonica	0.99996	0.00001	0.99996	0.00001	0.00001	0.00001
TOPAZIO	temperate_japonica	0.99996	0.00001	0.00001	0.99996	0.00001	0.00001
TORIO	temperate_japonica	0.99996	0.00001	0.00001	0.99996	0.00001	0.00001
Trembese	tropical_japonica	0.99996	0.00001	0.99996	0.00001	0.00001	0.00001
ULISSE	temperate_japonica	0.99996	0.00001	0.00001	0.99996	0.00001	0.00001
ULLAL	temperate_japonica	0.99996	0.00001	0.00001	0.99996	0.00001	0.00001
UPLA_77	tropical_japonica	0.99996	0.00001	0.99996	0.00001	0.00001	0.00001
UPLA_79	tropical_japonica	0.99996	0.00001	0.99996	0.00001	0.00001	0.00001
UPLA_91	tropical_japonica	0.99996	0.00001	0.99996	0.00001	0.00001	0.00001
URANO	tropical_japonica	0.99996	0.00001	0.99996	0.00001	0.00001	0.00001
VALTEJO	temperate_japonica	0.99996	0.00001	0.00001	0.99996	0.00001	0.00001

VARIETA_16	temperate_japonica	0.99996	0.00001	0.00001	0.99996	0.00001	0.00001
VENERE	temperate_japonica	0.99996	0.00001	0.00001	0.99996	0.00001	0.00001
VIALE	temperate_japonica	0.99996	0.00001	0.00001	0.99996	0.00001	0.00001
VIALONE_190	temperate_japonica	0.99996	0.00001	0.00001	0.99996	0.00001	0.00001
VIALONE_NANO	temperate_japonica	0.99996	0.00001	0.00001	0.99996	0.00001	0.00001
VICTORIA	temperate_japonica	0.99996	0.00001	0.00001	0.99996	0.00001	0.00001
VIRGO	temperate_japonica	0.99996	0.00001	0.00001	0.99996	0.00001	0.00001
VOLANO	temperate_japonica	0.99996	0.00001	0.00001	0.99996	0.00001	0.00001
YRL_196	temperate_japonica	0.99996	0.00001	0.00001	0.99996	0.00001	0.00001
YRM_6_2	temperate_japonica	0.99996	0.00001	0.00001	0.99996	0.00001	0.00001
ZENITH	tropical_japonica	0.99996	0.00001	0.99996	0.00001	0.00001	0.00001
ZHENSHANG_97	indica	0.99996	0.99996	0.00001	0.00001	0.00001	0.00001
RUDINO	temperate_japonica	0.99996	0.00001	0.00001	0.99996	0.00001	0.00001
RONCOLO	temperate_japonica	0.99996	0.00001	0.00001	0.99996	0.00001	0.00001
OLCENENGO	temperate_japonica	0.99996	0.00001	0.00001	0.99996	0.00001	0.00001
SOURE	temperate_japonica	0.99996	0.00002	0.00001	0.99996	0.00001	0.00001
NIBBIO	temperate_japonica	0.99995	0.00002	0.00001	0.99995	0.00001	0.00001
REDI	temperate_japonica	0.99995	0.00001	0.00002	0.99995	0.00001	0.00001
VENERIA	temperate_japonica	0.99995	0.00002	0.00001	0.99995	0.00001	0.00001
BALDO	temperate_japonica	0.99538	0.00001	0.00459	0.99538	0.00001	0.00001
STRELLA	temperate_japonica	0.99356	0.00001	0.00641	0.99356	0.00001	0.00001
CERVO	temperate_japonica	0.98967	0.00001	0.01030	0.98967	0.00001	0.00001
ONDA	temperate_japonica	0.98727	0.00001	0.01270	0.98727	0.00001	0.00001
M6	temperate_japonica	0.97842	0.00001	0.02155	0.97842	0.00001	0.00001
ADELAIDE CHIAPPELLI	temperate_japonica	0.97784	0.00001	0.02213	0.97784	0.00001	0.00001
SILLA	temperate_japonica	0.97104	0.00001	0.02813	0.97184	0.00001	0.00001
MOLO	temperate_japonica	0.97161	0.00001	0.02836	0.97161	0.00001	0.00001
CALENDAL	temperate_japonica	0.96708	0.00001	0.03289	0.96708	0.00001	0.00001
TOSCA	temperate_japonica	0.96355	0.00002	0.03641	0.96355	0.00001	0.00001
PIERINA_MARCHETTI	temperate_japonica	0.95146	0.00001	0.04851	0.95146	0.00001	0.00001
RIZZOTTO_51_1	temperate_japonica	0.94355	0.00001	0.05642	0.94355	0.00001	0.00001
SESIA	temperate_japonica	0.91730	0.00001	0.08267	0.91730	0.00001	0.00001
ALICE	temperate_japonica	0.91206	0.00001	0.08791	0.91206	0.00001	0.00001
MANTOVA	temperate_japonica	0.90435	0.00001	0.09562	0.90435	0.00001	0.00001
FAMILIA_181	temperate_japonica	0.89653	0.00001	0.10344	0.89653	0.00001	0.00001
FAISCA	temperate_japonica	0.88146	0.00001	0.10962	0.88146	0.00001	0.00890
BONNI	temperate_japonica	0.87482	0.00001	0.12515	0.87482	0.00001	0.00001
SUPER	temperate_japonica	0.86903	0.00001	0.13094	0.86903	0.00001	0.00001
EUROSE	temperate_japonica	0.86032	0.00001	0.13965	0.86032	0.00001	0.00001
GRITNA	temperate_japonica	0.85951	0.00001	0.14046	0.85951	0.00001	0.00001
BORRACHO	temperate_japonica	0.84453	0.00001	0.14934	0.84453	0.00001	0.00611
SAVIO	temperate_japonica	0.84177	0.00001	0.15820	0.84177	0.00001	0.00001
HAREM	temperate_japonica	0.84094	0.00001	0.15292	0.84094	0.00001	0.00612
ARTICO	indica	0.83992	0.83992	0.16005	0.00001	0.00001	0.00001
SENATORE_NOVELLI	temperate_japonica	0.83754	0.00001	0.16243	0.83754	0.00001	0.00001
RUBIDIO	temperate_japonica	0.82836	0.00001	0.17161	0.82836	0.00001	0.00001
RUBI	temperate_japonica	0.82734	0.00808	0.15675	0.82734	0.00001	0.00782
RIVA	temperate_japonica	0.80991	0.00001	0.19006	0.80991	0.00001	0.00001
LORD	temperate_japonica	0.80443	0.00001	0.19554	0.80443	0.00001	0.00001
ZENA	tropical_japonica	0.80305	0.00001	0.80305	0.19692	0.00001	0.00001
SANDOCA	temperate_japonica	0.80035	0.00519	0.18896	0.80035	0.00001	0.00548
GALILEO	temperate_japonica	0.78458	0.00001	0.21539	0.78458	0.00001	0.00001
CT36	temperate_japonica	0.77946	0.00001	0.22051	0.77946	0.00001	0.00001
ARTIGLIO	indica	0.76156	0.76156	0.23629	0.00001	0.00214	0.00001
GHIBLI	temperate_japonica	0.76098	0.00001	0.23899	0.76098	0.00001	0.00001
CRESO	temperate_japonica	0.76047	0.00001	0.23950	0.76047	0.00001	0.00001
UPLA_75	tropical_japonica	0.75430	0.01624	0.75430	0.22943	0.00001	0.00001
SCIROCCO	temperate_japonica	0.74414	0.00001	0.25583	0.74414	0.00001	0.00001
UPLA_32	tropical_japonica	0.74356	0.00001	0.74356	0.21946	0.03696	0.00001
UPLA_80	tropical_japonica	0.73384	0.00001	0.73384	0.26613	0.00001	0.00001

BASMATI_DETRADUNI	indica	0.72967	0.72967	0.00001	0.00001	0.27030	0.00001
BIANCA	temperate_japonica	0.71994	0.00001	0.28003	0.71994	0.00001	0.00001
VELA	temperate_japonica	0.70343	0.00002	0.29653	0.70343	0.00001	0.00001
MELAS	temperate_japonica	0.70329	0.00001	0.29668	0.70329	0.00001	0.00001
UPLA_63	tropical_japonica	0.69475	0.03397	0.69475	0.27126	0.00001	0.00001
TARRISO	tropical_japonica	0.68340	0.00001	0.68340	0.31657	0.00001	0.00001
UPLA_104	tropical_japonica	0.68254	0.00001	0.68254	0.31744	0.00001	0.00001
PANDA	tropical_japonica	0.68064	0.00001	0.68064	0.31933	0.00001	0.00001
SAGRES	temperate_japonica	0.67600	0.00356	0.30580	0.67600	0.00001	0.01463
UPLA_66	tropical_japonica	0.67470	0.08384	0.67470	0.24144	0.00001	0.00001
MIARA	temperate_japonica	0.66760	0.00002	0.33236	0.66760	0.00001	0.00001
UPLA_68	tropical_japonica	0.65017	0.12995	0.65017	0.21986	0.00001	0.00001
CARIOCA	tropical_japonica	0.63819	0.00001	0.63819	0.36178	0.00001	0.00001
SAMBA	tropical_japonica	0.63752	0.00001	0.63752	0.36246	0.00001	0.00001
GRAAL	tropical_japonica	0.63547	0.00001	0.63547	0.36450	0.00001	0.00001
CHACARERO	temperate_japonica	0.63525	0.00001	0.36472	0.63525	0.00001	0.00001
ALINANO	temperate_japonica	0.62619	0.04813	0.18637	0.62619	0.04017	0.09914
ELLEBI	tropical_japonica	0.62329	0.00001	0.62329	0.37668	0.00001	0.00001
NANO	temperate_japonica	0.61901	0.00001	0.38095	0.61901	0.00001	0.00001
CT23	temperate_japonica	0.61659	0.00001	0.38338	0.61659	0.00001	0.00001
MEJANES	temperate_japonica	0.61435	0.00001	0.38561	0.61435	0.00001	0.00001
GIANO	tropical_japonica	0.61367	0.02008	0.61367	0.36623	0.00001	0.00001
ITALPATNAxMILYANG	temperate_japonica	0.61312	0.29224	0.09463	0.61312	0.00001	0.00001
BAJANGxALLORIO	temperate_japonica	0.60607	0.00001	0.39390	0.60607	0.00001	0.00001
SALVO	tropical_japonica	0.59880	0.00002	0.59880	0.36583	0.03534	0.00001
DARDO	tropical_japonica	0.58469	0.00001	0.58469	0.41528	0.00001	0.00001
EUROSIS	temperate_japonica	0.57250	0.00001	0.42747	0.57250	0.00001	0.00001
SANTERNO	temperate_japonica	0.56696	0.00001	0.43301	0.56696	0.00001	0.00001
ALBATROS	temperate_japonica	0.56127	0.00001	0.43871	0.56127	0.00001	0.00001
GREPPI	tropical_japonica	0.55341	0.00001	0.55341	0.44656	0.00001	0.00001
ESTRELA	temperate_japonica	0.54767	0.04722	0.34711	0.54767	0.00244	0.05556
UPLA_64	tropical_japonica	0.53701	0.11984	0.53701	0.34313	0.00001	0.00001
MAIORAL	temperate_japonica	0.53691	0.00001	0.44510	0.53691	0.00001	0.01797
PREVER	tropical_japonica	0.53110	0.00001	0.53110	0.46887	0.00001	0.00001
ERMES	tropical_japonica	0.53002	0.00001	0.53002	0.45102	0.01893	0.00001
PECOS	tropical_japonica	0.52785	0.00001	0.52785	0.47212	0.00001	0.00001
ANSEATICO	temperate_japonica	0.52538	0.00001	0.47459	0.52538	0.00001	0.00001
ARTEMIDE	tropical_japonica	0.51416	0.00001	0.51416	0.46085	0.00906	0.01592
BENGAL	temperate_japonica	0.50951	0.01473	0.47575	0.50951	0.00001	0.00001
MIDA	temperate_japonica	0.50621	0.00001	0.49377	0.50621	0.00001	0.00001
GRALDO	tropical_japonica	0.50375	0.00001	0.50375	0.49622	0.00001	0.00001
RONALDO	tropical_japonica	0.50135	0.00001	0.50135	0.49862	0.00001	0.00001
ROMBO	temperate_japonica	0.49691	0.38250	0.12057	0.49691	0.00001	0.00001
OSCARxSUWEON	temperate_japonica	0.49230	0.38791	0.11298	0.49230	0.00680	0.00001
FIDJI	tropical_japonica	0.40973	0.04323	0.40973	0.22448	0.32255	0.00001

S4 Table: SD1 without correcting for population structure. Associations between SNP genotypes and plant height in the region of the gene SD1 on chromosome 1, from a linear GWAS model without accounting for population structure.

Trait	Chr	Marker	Position	p.value	Candidate.Gene
PH	1	S1_330484	330.484	7.2100000000E-05	
PH	1	S1_23553310	23553.31	3.0400000000E-05	
PH	1	S1_30310666	30310.666	8.9500000000E-05	
PH	1	S1_35155442	35155.442	2.6600000000E-05	
PH	1	S1_35586856	35586.856	8.1900000000E-06	
PH	1	S1_35847644	35847.644	4.7000000000E-06	
PH	1	S1_36682877	36682.877	4.5300000000E-06	
PH	1	S1_37012118	37012.118	4.0500000000E-06	
PH	1	S1_37403082	37403.082	8.2200000000E-06	
PH	1	S1_38197794	38197.794	5.8400000000E-10	
PH	1	S1_38457643	38457.643	1.7700000000E-10	SD1 (Sasaki et al. 2002)
PH	1	S1_38718425	38718.425	2.4000000000E-10	
PH	1	S1_39286700	39286.7	7.1600000000E-09	
PH	1	S1_39711918	39711.918	1.2500000000E-07	
PH	1	S1_40254090	40254.09	2.0300000000E-07	
PH	1	S1_40964141	40964.141	6.8800000000E-08	
PH	1	S1_41298835	41298.835	1.4200000000E-07	
PH	1	S1_42690322	42690.322	3.5800000000E-06	
PH	1	S1_43143774	43143.774	4.2300000000E-06	

ACKNOWLEDGMENTS

The research leading to these results has received funding from the European research projects "GS-Ruse" (http://www.first-gsruse.net/) and "Eurigen" (Action 49 of the AGRI GEN RES PROGRAM) and from the Italian national research project "Risinnova" (http://risinnova. entecra.it/).

AUTHOR CONTRIBUTIONS

Conceived and designed the experiments: PP. Performed the experiments: RG L. Casella PR AV AT CB PA. Analyzed the data: FB PC RG JS. Contributed reagents/materials/analysis tools: JS PA GV. Wrote the paper: FB GO RP GT L. Cattivelli SM GV PP RG.

REFERENCES

1. Caicedo AL, Williamson SH, Hernandez RD, Boyko A, Fledel-Alon A, York TL, et al. Genome-wide patterns of nucleotide polymorphism in domesticated rice. PLoS genetics. 2007; 3(9):e163. doi: 10.1371/ journal.

pgen.0030163

2. FAOstat, Food;. Available from: http://faostat.fao.org/.
3. Temperate rice on the global stage; 2014. Available from: http://irri-news.blogspot.it/2014/07/ temperate-rice-on-global-stage.html.
4. Okumoto Y, Ichitani K, Inoue H, Tanisaka T. Photoperiod insensitivity gene essential to the varieties grown in the northern limit region of paddy rice (Oryza sativa L.) cultivation. Euphytica. 1996; 92(1- 2):63–66.
5. Ichitani K, Okumoto Y, Tanisaka T. Photoperiod sensitivity gene of Se-1 locus found in photoperiod insensitive rice cultivars of the northern limit region of rice cultivation. Breeding Science. 1997; 47:145– 152.
6. Tonini A, Cabrera E. Opportunities for Global Rice Research in a Changing World: Global Futures for Agriculture Project. International Rice Research Institute Los Baños; 2011.
7. Fitzgerald MA, McCouch SR, Hall RD. Not just a grain of rice: the quest for quality. Trends in plant science. 2009; 14(3):133–139. doi: 10.1016/j.tplants.2008.12.004 PMID: 19230745
8. Biselli C, Cavalluzzo D, Perrini R, Gianinetti A, Bagnaresi P, Urso S, et al. Improvement of markerbased predictability of Apparent Amylose Content in japonica rice through GBSSI allele mining. Rice. 2014; 7(1):1. doi: 10.1186/1939-8433-7-1 PMID: 24383761
9. El Nahry A, Ali R, El Baroudy A. An approach for precision farming under pivot irrigation system using remote sensing and GIS techniques. Agricultural Water Management. 2011; 98(4):517–531. doi: 10. 1016/j. agwat.2010.09.012
10. Serraj R, McNally KL, Slamet-Loedin I, Kohli A, Haefele SM, Atlin G, et al. Drought resistance improvement in rice: an integrated genetic and resource management strategy. Plant Production Science. 2011; 14(1):1–14. doi: 10.1626/pps.14.1
11. Jiao Y, Wang Y, Xue D, Wang J, Yan M, Liu G, et al. Regulation of OsSPL14 by OsmiR156 defines ideal plant architecture in rice. Nature genetics. 2010; 42(6):541–544.
12. Biselli C, Bagnaresi P, Cavalluzzo D, Urso S, Desiderio F, Orasen G, et al. Deep sequencing transcriptional fingerprinting of rice kernels for dissecting grain quality traits. BMC genomics. 2015; 16(1):1. doi: 10.1186/ s12864-015-2321-7
13. Comas LH, Becker SR, Von Mark VC, Byrne PF, Dierig DA. Root traits contributing to plant productivity under drought. Frontiers in Plant Science. 2013; 4:442. doi: 10.3389/fpls.2013.00442 PMID: 24204374
14. Khush GS. Green revolution: the way forward. Nature Reviews Genetics. 2001; 2(10):815–822. doi: 10. 1038/35093585 PMID: 11584298
15. Xing Y, Zhang Q. Genetic and molecular bases of rice yield. Annual review of plant biology. 2010; 61:421–442. doi: 10.1146/annurev-arplant-042809-112209 PMID: 20192739
16. Zuo J, Li J. Molecular genetic dissection of quantitative trait loci regulating rice grain size. Annual review of genetics. 2014; 48:99–118. doi: 10.1146/annurev-genet-120213-092138 PMID: 25149369
17. Korte A, Farlow A. The advantages and limitations of trait analysis with GWAS: a review. Plant methods. 2013; 9(1):29. doi: 10.1186/1746-4811-9-29 PMID: 23876160
18. Xu Y, Crouch JH. Marker-assisted selection in plant breeding: from publications to practice. Crop Science. 2008; 48(2):391–407. doi: 10.2135/cropsci2007.04.0191
19. Kaur S, Panesar PS, Bera MB, Kaur V. Simple Sequence Repeat Markers in Genetic Divergence and Marker Assisted Selection of Rice Cultivars: A Review. Critical Reviews in Food Science and Nutrition. 2013;(just-accepted).
20. Seck PA, Diagne A, Mohanty S, Wopereis MC. Crops that feed the world 7: Rice. Food Security. 2012; 4(1):7–24. doi: 10.1007/s12571-012-0168-1
21. Zhang Z, Ober U, Erbe M, Zhang H, Gao N, He J, et al. Improving the accuracy of whole genome prediction for complex traits using the results of genome wide association studies. PloS one. 2014; 9(3): e93017. doi: 10.1371/journal.pone.0093017 PMID: 24663104
22. Hirschhorn JN, Daly MJ. Genome-wide association studies for common diseases and complex traits. Nature Reviews Genetics. 2005; 6(2):95–108. doi: 10.1038/ nrg1521 PMID: 15716906
23. Brachi B, Morris GP, Borevitz JO. Genome-wide association studies in plants: the missing heritability is in the field. Genome biology. 2011; 12(10):1–8. doi: 10.1186/gb-2011-12-10-232
24. Elshire RJ, Glaubitz JC, Sun Q, Poland JA, Kawamoto K, Buckler ES, et al. A robust, simple genotyping-by-sequencing (GBS) approach for high diversity species. PloS one. 2011; 6(5):e19379. doi: 10. 1371/journal. pone.0019379 PMID: 21573248
25. Huang X, Wei X, Sang T, Zhao Q, Feng Q, Zhao Y, et al. Genome-wide association studies of 14 agronomic traits in rice landraces. Nature genetics. 2010; 42(11):961–967. doi: 10.1038/ng.695 PMID: 20972439
26. Zhao K, Tung CW, Eizenga GC, Wright MH, Ali ML, Price AH, et al. Genome-wide association mapping reveals a rich genetic architecture of complex traits in Oryza sativa. Nature communications. 2011; 2:467. doi: 10.1038/ncomms1467 PMID: 21915109
27. Huang R, Jiang L, Zheng J, Wang T, Wang H, Huang Y, et al. Genetic bases of rice grain shape: so many genes, so little known. Trends in plant science. 2013; 18(4):218–226. doi: 10.1016/j.tplants.2012. 11.001 PMID: 23218902
28. Gamuyao R, Chin JH, Pariasca-Tanaka J, Pesaresi P, Catausan S, Dalid C, et al. The protein kinase Pstol1 from traditional rice confers tolerance of phosphorus deficiency. Nature. 2012; 488(7412):535– 539. doi: 10.1038/nature11346 PMID: 22914168
29. Courtois B, Audebert A, Dardou A, Roques S, Ghneim-Herrera T, Droc G, et al. Genome-wide association mapping of root traits in a japonica rice panel. PloS one. 2013; 8(11):e78037. doi: 10.1371/journal. pone.0078037 PMID: 24223758

30. Faivre-Rampant O, Bruschi G, Abbruscato P, Cavigiolo S, Picco AM, Borgo L, et al. Assessment of genetic diversity in Italian rice germplasm related to agronomic traits and blast resistance (Magnaporthe oryzae). Molecular breeding. 2011; 27(2):233–246.

31. Garris AJ, Tai TH, Coburn J, Kresovich S, McCouch S. Genetic structure and diversity in Oryza sativa L. Genetics. 2005; 169(3):1631–1638. doi: 10.1534/genetics.104.035642 PMID: 15654106

32. Kawahara Y, de la Bastide M, Hamilton JP, Kanamori H, McCombie WR, Ouyang S, et al. Improvement of the Oryza sativa Nipponbare reference genome using next generation sequence and optical map data. Rice. 2013; 6(1):4. doi: 10.1186/1939-8433-6-4 PMID: 24280374

33. Glaubitz JC, Casstevens TM, Lu F, Harriman J, Elshire RJ, Sun Q, et al. TASSEL-GBS: a high capacity genotyping by sequencing analysis pipeline. PLoS One. 2014; 9(2):e90346. doi: 10.1371/journal.pone. 0090346 PMID: 24587335

34. Buckler Lab for Maize Genetics and Diversity;. Available from: http://www.maizegenetics.net.

35. Li H, Durbin R Fast and accurate short read alignment with Burrows-Wheeler transform. Bioinformatics. 2009; 25(14):1754–1760 doi: 10.1093/bioinformatics/btp324 PMID: 19451168

36. Alexander DH, Novembre J, Lange K. Fast model-based estimation of ancestry in unrelated individuals. Genome research. 2009; 19(9):1655–1664. doi: 10.1101/gr.094052.109 PMID: 19648217

37. Perrier X, Flori A, Bonnot F. Data analysis methods. In: Genetic diversity of cultivated tropical plants. Science Publishers; 2003.

38. VanLiere JM, Rosenberg NA. Mathematical properties of the r2 measure of linkage disequilibrium. Theoretical population biology. 2008; 74(1):130–137. doi: 10.1016/j.tpb.2008.05.006 PMID: 18572214

39. Astle W, Balding DJ. Population structure and cryptic relatedness in genetic association studies. Statistical Science. 2009;p. 451–471. doi: 10.1214/09-STS307

40. Benjamini Y, Hochberg Y. Controlling the false discovery rate: a practical and powerful approach to multiple testing. Journal of the Royal Statistical Society Series B (Methodological). 1995;p. 289–300.

41. Bradbury PJ, Zhang Z, Kroon DE, Casstevens TM, Ramdoss Y, Buckler ES. TASSEL: software for association mapping of complex traits in diverse samples. Bioinformatics. 2007; 23(19):2633–2635. doi: 10.1093/bioinformatics/btm308 PMID: 17586829

42. Purcell S, Neale B, Todd-Brown K, Thomas L, Ferreira MA, Bender D, et al. PLINK: a tool set for wholegenome association and population-based linkage analyses. The American Journal of Human Genetics. 2007; 81(3):559–575. doi: 10.1086/519795 PMID: 17701901

43. R Development Core Team. R: A Language and Environment for Statistical Computing. Vienna, Austria; 2008. ISBN 3-900051-07-0. Available from: http://www.R-project.org.

44. Sonah H, Bastien M, Iquira E, Tardivel A, Légaré G, Boyle B, et al. An improved genotyping by sequencing (GBS) approach offering increased versatility and efficiency of SNP discovery and genotyping. PloS one. 2013; 8(1):e54603. doi: 10.1371/journal.pone.0054603 PMID: 23372741

45. McCouch SR, Wright MH, Tung CW, Maron LG, McNally KL, Fitzgerald M, Singh N, DeClerck G, Agosto-Perez F, Korniliev P, Greenberg AJ Open access resources for genome-wide association mapping in rice. Nature communications. 2016; 7:10532.

46. Huang YF, Poland JA, Wight CP, Jackson EW, Tinker NA. Using genotyping-by-sequencing (GBS) for genomic discovery in cultivated oat. PloS one. 2014; 9(7):e102448. doi: 10.1371/journal.pone.0102448 PMID: 25047601

47. Courtois B, Frouin J, Greco R, Bruschi G, Droc G, Hamelin C, et al. Genetic diversity and population structure in a European collection of rice. Crop science. 2012; 52(4):1663–1675. doi: 10.2135/cropsci2011.11.0588

48. Mather KA, Caicedo AL, Polato NR, Olsen KM, McCouch S, Purugganan MD. The extent of linkage disequilibrium in rice (Oryza sativa L.). Genetics. 2007; 177(4):2223–2232. doi: 10.1534/genetics.107. 079616 PMID: 17947413

49. Khatkar MS, Nicholas FW, Collins AR, Zenger KR, Cavanagh JA, Barris W, et al. Extent of genomewide linkage disequilibrium in Australian Holstein-Friesian cattle based on a high-density SNP panel. BMC genomics. 2008; 9(1):1. doi: 10.1186/1471-2164-9-187

50. O'Brien AMP, Mészáros G, Utsunomiya YT, Sonstegard TS, Garcia JF, Van Tassell CP, et al. Linkage disequilibrium levels in Bos indicus and Bos taurus cattle using medium and high density SNP chip data and different minor allele frequency distributions. Livestock Science. 2014; 166:121–132. doi: 10. 1016/j.livsci.2014.05.007

51. Bacciu N, Bomba L, Nicolazzi E, Milanesi M, Ajmone-Marsan P. Linkage disequilibrium in BTA6 for the italian holstein breed using a 54k and 800k SNP-chip. In: Proceedings of the 16th, QTL-MAS Workshop; 2012. p. 24–25.

52. Bohmanova J, Sargolzaei M, Schenkel FS. Characteristics of linkage disequilibrium in North American Holsteins. BMC genomics. 2010; 11(1):1. doi: 10.1186/1471-2164-11-421

53. Burton PR, Clayton DG, Cardon LR, Craddock N, Deloukas P, Duncanson A, et al. Genome-wide association study of 14,000 cases of seven common diseases and 3,000 shared controls. Nature. 2007; 447 (7145):661–678. doi: 10.1038/nature05911

54. Lim JH, Yang HJ, Jung KH, Yoo SC, Paek NC. Quantitative trait locus mapping and candidate gene analysis for plant architecture traits using whole genome re-sequencing in rice. Molecules and cells. 2014; 37(2):149. doi: 10.14348/molcells.2014.2336 PMID: 24599000

55. Bing Y, Wei-Ya X, Li-Jun L, Yong-Zhong X. QTL analysis for flag leaf characteristics and their relationships with

yield and yield traits in rice. Acta Genetica Sinica. 2006; 33(9):824–832. doi: 10.1016/S0379- 4172(06)60116-9

56. Zhang GH, Li SY, Wang L, Ye WJ, Zeng DL, Rao YC, et al. LSCHL4 from japonica cultivar, which is allelic to NAL1, increases yield of indica super rice 93-11. Molecular plant. 2014;p. ssu055.

57. Dan Jiang, Fang Jingjing, Lou Lamei, Zhao Jinfeng, Yuan Shoujiang, Yin Liang, Sun Wei, Peng Lixiang, Guo Baotai, Li Xueyong. Characterization of a null allelic mutant of the rice nal1 gene reveals its role in regulating cell division. PloS One. 2015; 10(2):e0118169. doi: 10.1371/journal.pone.0118169

58. Shen G, Zhan W, Chen H, Xing Y. Dominance and epistasis are the main contributors to heterosis for plant height in rice. Plant Science. 2014; 215:11–18. doi: 10.1016/j.plantsci.2013.10.004 PMID: 24388510

59. Ys LIANG, Zq GAO, Xh SHEN,ZHAN Xd, ZHANG Yx, WU Wm, et al. Mapping and comparative analysis of QTL for rice plant height based on different sample sizes within a single line in RIL population. Rice science. 2011; 18(4):265–272. doi: 10.1016/S1672-6308(12)60004-3

60. Begum Hasina, Spindel JE, Lalusin A, Borromeo T, Gregorio G, Hernandez J, Virk P, Collard B, McCouch SR. Genome-wide association mapping for yield and other agronomic traits in an elite breeding population of tropical rice (Oryza sativa). PloS One. 2015; 10(3):e0119873. doi: 10.1371/journal. pone.0119873 PMID: 25785447

61. Yang C, Li D, Mao D, Liu X, Ji C, Li X, et al. Overexpression of microRNA319 impacts leaf morphogenesis and leads to enhanced cold tolerance in rice (Oryza sativa L.). Plant, cell & environment. 2013; 36 (12):2207–2218.

62. Itoh H, Tatsumi T, Sakamoto T, Otomo K, Toyomasu T, Kitano H, et al. A rice semi-dwarf gene, TanGinbozu (D35), encodes the gibberellin biosynthesis enzyme, ent-kaurene oxidase. Plant molecular biology. 2004; 54(4):533–547.

63. Matusmoto T, Yamada K, Yoshizawa Y, Oh K. Comparison of Effect of Brassinosteroid and Gibberellin Biosynthesis Inhibitors on Growth of Rice Seedlings. Rice Science. 2016; 23(1):51–5. doi: 10.1016/j.rsci.2016.01.006

64. Chung PJ, Kim YS, Jeong JS, Park SH, Nahm BH, Kim JK. The histone deacetylase OsHDAC1 epigenetically regulates the OsNAC6 gene that controls seedling root growth in rice. The Plant Journal. 2009; 59(5):764–76. doi: 10.1111/j.1365-313X.2009.03908.x PMID: 19453457

65. Hu Y, Qin F, Huang L, Sun Q, Li C, Zhao Y, et al. Rice histone deacetylase genes display specific expression patterns and developmental functions. Biochemical and biophysical research communications. 2009; 388(2):266–271. doi: 10.1016/j.bbrc.2009.07.162 PMID: 19664599

66. Hittalmani S, Huang N, Courtois B, Venuprasad R, Shashidhar H, Zhuang J, et al. Identification of QTL for growth-and grain yield-related traits in rice across nine locations of Asia. Theoretical and Applied Genetics. 2003; 107(4):679–690. doi: 10.1007/s00122-003-1269-1 PMID: 12920521

67. Septiningsih E, Prasetiyono J, Lubis E, Tai T, Tjubaryat T, Moeljopawiro S, et al. Identification of quantitative trait loci for yield and yield components in an advanced backcross population derived from the Oryza sativa variety IR64 and the wild relative O. rufipogon. Theoretical and applied genetics. 2003; 107(8):1419–1432. doi: 10.1007/s00122-003-1373-2 PMID: 14513215

68. Spielmeyer W, Ellis MH, Chandler PM. Semidwarf (sd-1) "green revolution" rice, contains a defective gibberellin 20-oxidase gene. Proceedings of the National Academy of Sciences. 2002; 99(13):9043– 9048. doi: 10.1073/pnas.132266399

69. Li N, Li Y. Ubiquitin-mediated control of seed size in plants. Advances in Seed Biology. 2015; 11:78.

70. Nagasawa N, Hibara KI, Heppard EP, Vander Velden KA, Luck S, Beatty M, et al. GIANT EMBRYO encodes CYP78A13, required for proper size balance between embryo and endosperm in rice. The Plant Journal. 2013; 75(4):592–605. doi: 10.1111/tpj.12223 PMID: 23621326

71. Weng J, Gu S, Wan X, Gao H, Guo T, Su N, et al. Isolation and initial characterization of GW5, a major QTL associated with rice grain width and weight. Cell research. 2008; 18(12):1199–1209. doi: 10.1038/cr.2008.307 PMID: 19015668

72. Wang L, Wang A, Huang X, Zhao Q, Dong G, Qian Q, et al. Mapping 49 quantitative trait loci at high resolution through sequencing-based genotyping of rice recombinant inbred lines. Theoretical and applied genetics. 2011; 122(2):327–340. doi: 10.1007/s00122-010-1449-8 PMID: 20878143

73. Amarawathi Y, Singh R, Singh AK, Singh VP, Mohapatra T, Sharma TR, et al. Mapping of quantitative trait loci for basmati quality traits in rice (Oryza sativa L.). Molecular Breeding. 2008; 21(1):49–65.

74. Chen Y, Liu L, Shen Y, Liu S, Huang J, Long Q, Wu W, Yang C, Chen H, Guo X, Cheng Z. Loss of function of the cytochrome P450 gene CYP78B5 causes giant embryos in rice. Plant Molecular Biology Reporter. 2015; 33(1):69–83. doi: 10.1007/s11105-014-0731-3

75. Werck-Reichhart D, Bak S, Paquette S. Cytochromes P450. The Arabidopsis Book. 2002;e0028. doi: 10.1199/tab.0028 PMID: 22303202

76. Miura K, Ikeda M, Matsubara A, Song XJ, Ito M, Asano K, et al. OsSPL14 promotes panicle branching and higher grain productivity in rice. Nature genetics. 2010; 42(6):545–549. doi: 10.1038/ng.592 PMID: 20495564

77. Shomura A, Izawa T, Ebana K, Ebitani T, Kanegae H, Konishi S, et al. Deletion in a gene associated with grain size increased yields during rice domestication. Nature genetics. 2008; 40(8):1023–1028. doi: 10.1038/ng.169 PMID: 18604208

78. Hemamalini G, Shashidhar H, Hittalmani S. Molecular marker assisted tagging of morphological and physiological traits under two contrasting moisture regimes at peak vegetative stage in rice (Oryza sativa

L.). Euphytica. 2000; 112(1):69–78.

79. Champoux M, Wang G, Sarkarung S, Mackill D, O'Toole J, Huang N, et al. Locating genes associated with root morphology and drought avoidance in rice via linkage to molecular markers. Theoretical and Applied Genetics. 1995; 90(7-8):969–981. doi: 10.1007/BF00222910 PMID: 24173051

80. Ahmadi N, Audebert A, Bennett MJ, Bishopp A, de Oliveira AC, Courtois B, et al. The roots of future rice harvests. Rice. 2014; 7(1):1–9. doi: 10.1186/s12284-014-0029-y

CHAPTER 4

VARIABILITY ASSESSMENT OF AROMATIC AND FINE RICE GERMPLASM

M. Z. Islam, M. Khalequzzaman, M. K. Bashar, N. A. Ivy, M. M. Haque, and M. A. K. Mian

INTRODUCTION

Bangladesh is mainly a country of rice based cropping system, where thousands of local rice varieties are being cultivated from the time immemorial [1]. To date, farmers across the country are used to cultivate different local varieties or landraces particularly in the unfavourable ecosystems. Local variety including aromatic rice genotypes occupied about 12.16% of the rice growing areas in Bangladesh [2]. Many of these local varieties have some special characteristics such as aroma, better taste, and higher cooking quality which also provide additional value in socioeconomic aspects. Moreover, aromatic rice constitutes a special group of rice genotypes well known in many countries across the world for their aroma and/or super fine grain quality [3]. Bangladesh has a stock of above 8,000 rice germplasms of which nearly 100 are aromatic [1, 4]. It is worthwhile to mention that aromatic rice is closely related to the social and cultural heritage in Bangladesh and it is consumed during weddings and other festivals [5]. Aromatic and fine rice germplasm native to Bangladesh generally have short bold and medium bold grain type with mild to strong aroma [6, 7]. In Bangladesh, among the different aromatic rice varieties, Chinigura is the predominant one that covers more than 70% of rice farms in the northern districts of Naogaon and Dinajpur. Other important aromatic rice varieties are Kalijira (predominantly grown in Mymensingh) and Kataribhog (mainly cultivated in Dinajpur) [8]. Most of the aromatic rice varieties in Bangladesh are of locally adapted, photoperiod-sensitive, and grown during Aman season under rainfed lowland ecosystem. The production cost of aromatic and fine rice is low compared to that of coarse rice. Therefore, the income potential is higher with aromatic fine rice cultivation, since its cultivation does not usually require additional expenditures on fertilizer, pesticides, and irrigation. However, the average yield of high yielding rainfed lowland rice is 3.4 t/ha, whereas that of aromatic rice is 2.0–2.3 t/ha [9].

Knowledge on genetic diversity among crop populations and its quantitative assessment usually helps a breeder to select suitable parents to be utilized in breeding programmes [10–15]. Among the different cereal crops, rice is (Oryza sativa) one of the best models to undertake the study of genome structure and genetic diversity. Its diploid genome is relatively smaller in size (430 Mb) with a significant level of genetic polymorphism and a large amount of well-conserved genetically diverse material [16–18].

In a breeding programme, genetic improvement primarily depends upon the amount of genetic variability present in the population. In many cases, characters are mostly governed by poly genes which are highly influenced by the environment. Therefore, it is difficult to predict whether the existing variability is heritable or not. Furthermore, heritability of a genetic trait is very important in determining the response to selection because it implies the extent of transmissibility of that trait into next generations [19, 20]. In addition, high genetic advance coupled with high heritability offers the most effective condition for selection for a specific character [21].

These days, plant breeders usually evaluate genetic diversity on the basis of morphological traits because they are more economic, faster, and easier to score compared to the molecular traits [22, 23]. Investigation to these traits also does not require any sophisticated procedure or advanced equipment. In addition, these traits can be transmitted without adapting any special biochemical or molecular techniques. The rice plant is morphologically diverse, especially in terms of

Citation: M. Z. Islam, M. Khalequzzaman, M. K. Bashar, N. A. Ivy, M. M. Haque, and M. A. K. Mian, "Variability Assessment of Aromatic and Fine Rice Germplasm," The Scientific World Journal, vol. 2016, Article ID 2796720, 14 pages, 2016. doi:10.1155/2016/2796720.

the vegetative traits such as plant height and leaf length. Our previous studies involving local aromatic and nonaromatic rice germplasm from Bangladesh using morphological, physicochemical, and molecular markers revealed high genetic diversity [6, 24–28]. However, such investigations on aromatic and fine rice genotypes are not yet to be conducted. Therefore, the present study was undertaken to assess the genetic diversity of aromatic and fine rice genotypes in Bangladesh.

MATERIALS AND METHODS

Experimental Site

The experiment was conducted at the research farm of Bangladesh Rice Research Institute (BRRI), Gazipur, during July to December (T. Aman season), 2011. Geographically, the place is located at about 24.00°N latitude and 90.25°E longitude with an elevation of 8.4 meters from the sea level and is characterized by subtropical climate. The soil of the experimental site was clay loam in texture.

Plant Materials

A total of 113 aromatic and fine rice genotypes (Table 1) were selected from BRRI genebank. Pregerminated seeds were sown in the seed bed.

Table 1: Information on place of collection, source, and local name of the aromatic and fine rice accessions.

Sl. No.	Genotypes	Acc. No./source	Place of collection	Kernel size and shape	1.7% KOH (aroma)
1	Sakor	197	Mymensingh	Short, bold	Lightly scented
2	Sagardana	229	Mymensingh	Short, medium	Lightly scented
3	Nunia	233	Mymensingh	Short, medium	Lightly scented
4	Chini Sagar (2)	245	Mymensingh	Short, medium	Scented
5	Meny	288	Gaibandha	Short, bold	Scented
6	Tilkapur	296	Gaibandha	Short, medium	Lightly scented
7	Binnaphul	315	Gaibandha	Short, medium	Lightly scented
8	Kalobhog	318	Gaibandha	Short, medium	Scented
9	Jabsiri	331	Gaibandha	Short, medium	Scented
10	Kalgochi	352	Gaibandha	Short, bold	Scented
11	Chinisakkor	387	Rajshahi	Short, medium	Scented
12	Chiniatob	399	Rajshahi	Short, medium	Scented
13	Noyonmoni	461	Rajshahi	Short, medium	Scented
14	Saubail	873	Sylhet	Short, medium	Scented
15	Chinniguri	1880	Kishoreganj	Short, bold	Scented
16	Kalomala	1886	Kishoreganj	Short, medium	Scented
17	Begunmala	1896	Kishoreganj	Short, medium	Scented
18	Gopalbhog	1938	Kishoreganj	Short, medium	Scented
19	Tulsimoni	1980	Jamalpur	Short, medium	Scented
20	Jirabuti	1984	Mymensingh	Short, bold	Scented
21	Khirshabuti	1996	Tangail	Short, medium	Scented
22	Rajbut	1999	Tangail	Short, medium	Scented
23	Soru kamina	2015	Satkhira	Short, bold	Lightly scented
24	Kamini soru	2027	Satkhira	Short, bold	Lightly scented
25	Doiarguru	2037	Khulna	Short, bold	Lightly scented
26	Premful	2041	Satkhira	Short, medium	Scented
27	Begun bichi	2073	Kishoreganj	Short, bold	Lightly scented
28	Elai	2423	Dhaka	Long, slender	Nonscented
29	Gua masuri	3666	Sherpur	Short, medium	Nonscented
30	Luina	3676	Netrokona	Short, medium	Scented
31	Lal Soru	4135	Dinajpur	Short, medium	Scented
32	Chini Kanai	4356	Khulna	Short, bold	Scented
33	Kalijira (short grain)	4357	Khulna	Short, bold	Scented
34	Rajbhog	4360	Khulna	Short, medium	Scented
35	Philippines Kataribhog	4365	Dinajpur	Short, medium	Scented
36	Baoibhog	4813	Kurigram	Short, medium	Scented
37	Baoijhaki	4826	Dinajpur	Short, medium	Lightly scented
38	Jirabhog (Bolder)	4828	Dinajpur	Short, bold	Lightly scented
39	Chinigura	4867	Mymensingh	Short, bold	Scented
40	Tulsimala	4870	Mymensingh	Short, bold	Scented
41	Bashmati 370	4904	Pakistan	Medium, slender	Scented
42	Uknimodhu	5083	Rangpur	Short, medium	Scented
43	Ranisalut	5286	Khulna	Short, bold	Lightly scented
44	Jira dhan	5313	Khulna	Short, bold	Scented
45	Gandhakusturi	5319	Bagerhat	Short, bold	Nonscented
46	Sakkorkhora	5347	Barguna	Short, bold	Scented
47	Badshabhog	5349	Bagerhat	Short, bold	Scented
48	Jirakatari	5975	Dinajpur	Short, bold	Scented
49	Desikatari	5978	Dinajpur	Short, medium	Scented
50	Thakurbhog	5983	Sylhet	Short, medium	Nonscented

51	Tulsimaloty	6638	Tangail	Short, bold	Scented
52	Raduni pagal	6711	Rajshahi	Short, medium	Scented
53	Sugandhi dhan	7063	Nawabganj	Short, medium	Nonscented
54	Kalijira (long grain)	4358	Khulna	Short, medium	Scented
55	Jesso balam TAPL-25	2454	GRSD, BRRI	Short, medium	Scented
56	Dakshahi	983	Khulna	Short, bold	Nonscented
57	Hatisail TAPL-101	2528	GRSD, BRRI	Short, bold	Scented
58	Khasa	682	Comilla	Short, medium	Scented
59	Buchi	369	Gaibandha	Short, bold	Scented
60	Awned TAPL-545	2939	GRSD, BRRI	Short, bold	Scented
61	Black TAPL-554	2947	GRSD, BRRI	Short, bold	Scented
62	Straw TAPL-500	2898	GRSD, BRRI	Long, slender	Scented
63	Dubsail	4840	Satkhira	Short, bold	Scented
64	Duksail	2028	Satkhira	Short, bold	Nonscented
65	Khaskani	4341	Jessore	Short, medium	Scented
66	Khazar	4921	Iran	Long, slender	Nonscented
67	Basmati Sufaid 106	4498	Pakistan	Medium, slender	Lightly scented
68	BR5	4343	GRSD, BRRI	Short, bold	Scented
69	BRRI dhan34	7093	GRSD, BRRI	Short, medium	Scented
70	BRRI dhan37	7094	GRSD, BRRI	Short, medium	Scented
71	BRRI dhan38	7095	GRSD, BRRI	Medium, slender	Scented
72	BRRI dhan50	6882	GRSD, BRRI	Long, slender	Lightly scented
73	Khasa Mukpura	7586	Khagrachhari	Short, medium	Scented
74	Uknimodhu	298	Gaibandha	Short, bold	Scented
75	Bawaibhog-2	301	Gaibandha	Short, medium	Scented
76	Chiniatob-2	398	Rajshahi	Short, bold	Scented
77	Tilokkachari	758	Chittagong	Short, bold	Scented
78	Begunbichi-2	508	Rangpur	Short, bold	Scented
79	Chinairri	764	Chittagong	Short, bold	Scented
80	Bhatir chikon	774	Chittagong	Short, medium	Scented
81	Gordoi	1908	Kishoreganj	Short, bold	Nonscented
82	Dolagocha	451	Rajshahi	Short, bold	Nonscented
83	Kalonunia	537	Rangpur	Short, medium	Lightly scented
84	Dhan chikon	538	Dinajpur	Short, medium	Lightly scented
85	Badshabhog-2	03	Dhaka	Short, bold	Scented
86	Thakurbhog-2	872	Sylhet	Short, bold	Nonscented
87	Khuti chikon	4107	Comilla	Short, bold	Lightly scented
88	Sunduri samba	4803	Rajshahi	Short, medium	Nonscented
89	Basmati	4754	Barguna	Short, bold	Scented
90	Basmati 37	4491	India	Long, slender	Lightly scented
91	Basnatu sufaid 187	4499	Pakistan	Long, slender	Lightly scented
92	Tulsimala-2	7342	Sherpur	Short, bold	Lightly scented
93	Chinisail	7343	Sherpur	Short, medium	Lightly scented
94	Malshira	7347	Sherpur	Short, bold	Lightly scented
95	Sadagura	—	Khagrachhari	Short, medium	Lightly scented
96	Modhumadab	7352	Habiganj	Short, medium	Lightly scented
97	Parbatjira	7351	Habiganj	Short, bold	Lightly scented
98	Chinikanai-2	7350	Dinajpur	Short, bold	Lightly scented
99	Meedhan	7537	Habiganj	Short, medium	Lightly scented
100	Gobindhabhog	—	Jessore	Short, medium	Lightly scented
101	Kataribhog	7082	Dinajpur	Short, medium	Scented
102	Fulkari	7531	Habiganj	Short, bold	Lightly scented
103	BU dhan2R	7413	GRSD, BRRI	Long, slender	Lightly scented
104	Padmabhog	4812	Kurigram	Short, medium	Lightly scented
105	Dudsail	4840	Satkhira	Short, medium	Lightly scented
106	Sakkorkhana	4761	Barguna	Short, medium	Scented
107	Maloti	169	Tangail	Short, medium	Lightly scented
108	Bashful	4215	Kishoreganj	Short, medium	Scented
109	Kalijira TAPL-64	2492	GRSD, BRRI	Short, medium	Scented
110	Oval TAPL-2990	2990	GRSD, BRRI	Short, medium	Lightly scented
111	Kalijira TAPL-68	2496	GRSD, BRRI	Short, medium	Scented
112	Kalijira TAPL-74	2501	GRSD, BRRI	Short, bold	Scented
113	Kalobakri	2108	Narsingdi	Short, bold	Scented

Experimental Design and Setting the Experiment

The experiment was conducted following a randomized complete block design with three replicates for each treatment. Thirty-day-old seedlings of each test genotypes were transplanted on the 15th August, 2011 using single seedling per hill in 2.4 m^2 plot with 25 cm and 20 cm space between rows and plants, respectively.

Intercultural Operations

Fertilizers were applied @ 80:60:40: 12 kg N:P:K:S per hectare. However, except N, the other fertilizers were applied at final land preparation. Nitrogen was applied in three equal splits, at 15 days after transplanting (DAT), at 35 DAT, and just before flowering. Intercultural operations and pest control measures were done as and when necessary.

Data Collection

Data were collected on culm diameter (mm), flag leaf area (cm^2), plant height (cm), days to flowering, days to maturity, effective tiller number (ET No.), panicle length (cm), primary branch length (cm), secondary branch length (cm), primary braches per panicle, secondary branches per panicle, number of filled grains per panicle, number of unfilled grains per panicle, grain length (mm), grain breadth (mm), grain length breadth ratio, 1000-grain weight (g), grain yield per hill (g), and harvest index (HI). Kernel quality was determined using dehusked grains. Kernels were classified on the basis of length (size) and for *L/B* ratio (shape) following classification described by Cruz and Khush [29] (Table 2).

Table 2: List of quantitative traits of 113 aromatic and fine rice genotypes

Traits	Method of evaluation
Culm diameter (CD, mm)	Outer diameter internodes of the 10 culms were measured and averaged
Flag leaf area (FLA, cm^2)	Flag leaf area (cm^2) = flag leaf length (cm) × maximum width (cm) × 0.75
Plant height (PH, cm)	The average of height from the base to the tip of last leaf (flag leaf)
Days to flowering (DF, days)	The number of days from seeding to flowering day
Days to maturity (DM, days)	The number of days from seeding to maturing day
Effective tiller number (ET No.)	Counting of effective tiller per hill
Panicle length (PL, cm)	Distance between apex of the panicle (excluding awn) and top most node (neck node) of the culm
Primary braches per panicle (PBP, no.)	Primary branches were counted from 5 randomly selected panicles and averaged
Primary branch length (PBL, cm)	Lengths of the primary branches present in a panicle were measured (cm) from five panicles and averaged
Secondary branches per panicle (SBP, no.)	Secondary branches were counted from 5 randomly selected panicles and averaged
Secondary branch length (SBL, cm)	Length (cm) of the 30 random secondary branches from five randomly selected panicles and averaged
Filled grains per panicle (FGP, no.)	Number of filled grains per panicle was counted from 10 randomly selected panicles and averaged
Unfilled grains per panicle (UFGP, no.)	Number of unfilled grains per panicle was recorded from 10 randomly selected panicles and averaged
Grain length (GL, mm)	Length (mm) of a grain was measured by a digital slide caliper from 10 randomly selected fertile grains excluding awn and averaged
Grain breadth (GB, mm)	Breadth of a grain (mm) was measured form 10 randomly selected fertile grains by a digital slide caliper and averaged
Grain length breadth ratio (GLBR)	Dividing grain length by grain breadth and averaged
1000-grain weight (TGW, g)	200 grains were weighted then 1000-weight grains were calculated from these weights
Yield per plant (GYP, g)	Ten randomly selected plants per replication and averaged
Harvest index (HI)	Ratio of grain yield to biological yield

Aroma Test

Aroma was detected by sniffing and was scored as nonscented, lightly scented, and scented following 1.7% KOH based method [30] (Table 2).

Statistical Analysis

Univariate analysis of the individual character (ANOVA) including the estimation of mean, range, and coefficient of variation (CV%) was conducted using a statistical software package MstatC. The test of significance was performed using Fisher's (*F*) test. Multivariate analysis was conducted using another statistical software package GENSTAT version 5.5. Genetic parameters were also estimated to understand genetic variations among the test genotypes and to determine genetic and environmental effects on different characters. These genetic parameters were calculated using the following formula [31–33]. These parameters include the following:

(a) Genotypic variance, $\sigma_g^2 = (\text{GMS} - \text{EMS})/r$, where GMS is the genotypic mean sum of squares, EMS is the error mean sum of squares, and *r* is number of replication.

(b) Phenotypic variance, $\sigma_p^2 = \sigma_g^2 + \sigma_e^2$, where σ_g^2 is the genotypic variance and σ_e^2 is the mean squares of error.

(c) Genotypic coefficient of variation (GCV)% = $\sqrt{\sigma_g^2}/\bar{x} \times 100,$ where σ_g^2 is the genotypic variance and $\bar{x}$ is the mean of character.

(d) Phenotypic coefficient of variation (PCV)% = $\sqrt{\sigma_p^2}/\bar{x} \times 100$, where σ_p^2 is the phenotypic variance and $\bar{x}$ is the mean of trait.

(e) Heritability (broad sense) $h_B^2\% = \sigma_g^2/\sigma_p^2 \times 100$, where σ_g^2 is the genotypic variance and σ_p^2 is the phenotypic variance.

(f) Expected genetic advance (GA): $\text{GA}(\%) = K \times \sigma_p^2 \times h_B^2 \times 100$, where GA is a percent of the mean assuming selection of the superior 5% of accession: $\text{GA}(\%) = K \times \sqrt{\sigma_p^2}/\bar{x} \times \quad h_B^2 \times 100$, where k is a constant, $\sqrt{\sigma_p^2}/\bar{x}$ is the phenotypic standard deviation, $h_B^2 \times 100$ is the heritability, and $\bar{x}$ is the mean of traits.

RESULT

Variation in Grain Diversity and Genetic Parameters among Accessions

The grain morphology varied considerably in genotypes collected from BRRI genebank (Figure 1) with respect to awning, colour and size of awns, lemma and palea with presence or absence of coloured furrows and spots, pubescence, and varied coloured apiculus and sterile lemma. Analysis of variance of 19 quantitative characters based on individual sample means showed highly significant ($P \leq 0.01$) variations among the genotypes for all the characters outlined in Table 3. The range, mean, standard error, coefficients of variation, and ($P \leq 0.01$) value of 19 characters are presented in Table 4. The coefficient of variation ranged from 1.69 to 35.81% which indicates considerable variation among the character studied. Out of 19 traits, unfilled grains per panicle, harvest index, yield per plant, filled grains per panicle, primary branches per panicle, and secondary branches per panicle found with relatively higher coefficient of variation (35.81, 20.45, 18.70, 16.79, 11.61, and 11.03%, resp.) than the other traits. These possibly occurred because of sampling error and/or characters were more influenced by the environmental factors. In this study, most of the growth traits showed higher PCV compared to yield and yield component traits. However, lower PCV belonged to days to maturity (5.79%) while unfilled grains per panicle (46.57%) were recorded with higher value. Secondary branches per panicle (34.95%), 1000 grain weight (34.20%), and filled grains per panicle (29.32%) were recorded with higher values of PCV. However, panicle length (6.31%), days to flowering (7.40%), and plant height (10.04%) were found with lower values. The higher GCV was associated with 1000-grain weight (33.18%) whereas the value was fairly low in case of panicle length (5.06%). Results also showed narrow differences between PCV and GCV for most of the traits. Heritability ranged from 29.03 to 97.44%. The highest and the lowest amount of heritability were recorded at grain length and yield per plant, respectively.

Table 3: Analysis of variance for different quantitative characters in 113 aromatic and fine rice genotypes

(a)

Source of variation	df	CD (mm)	FLA (cm^3)	DF	DM	PH (cm)	ET No.	PL (cm)	PB No.	PBL (cm)	SB No.
Replication	2	1.173	32.105	0.51	1.507	250.964	28.769	15.972	4.975	0.081	38.629
Genotype	112	1.422**	155.904**	170.559**	164.857**	615.038**	5.491**	7.455**	5.129**	3.825**	365.148**
Error	224	0.158	17.342	2.257	4.951	16.130	1.156	1.164	1.441	0.593	12.993
CV%		8.60	11.70	2.10	2.69	2.78	10.93	3.77	11.61	7.71	11.03

(b)

Source of variation	SBL (cm)	FGP	UFGP	GL (mm)	GB (mm)	GLBR	TGW (g)	YP (g)	HI
Replication	0.044	3274.094	544.179	0.234	0.039	0.107	35.785	18.735	0.016
Genotype	0.228**	4828.209**	741.844**	5.412	0.423**	1.48**	58.883**	11.961**	0.005**
Error	0.034	675.308	241.344	0.047	0.017	0.03	1.206	1.624	0.002
CV%	6.66	16.79	35.81	3.06	5.58	5.55	8.31	18.70	20.44

** indicates significance at 1% level of probability.
CD: culm diameter (mm), FLA: flag leaf area, DF: days to flowering, DM: days to maturity, PH: plant height, ET No.: effective tiller number, PL: panicle length, PB No.: primary branches per panicle, PBL: primary branch length, SB No.: secondary branches per panicle, SBL: secondary branch length, FGP: filled grains per panicle, UFGP: unfilled grains per panicle, GL: grain length, GB: grain breadth, GLBR: grain length breadth ratio, TGW: 1000-grain weight, Y/P: yield per plant, and HI: harvest index.

Table 4: Variability in different quantitative characters in 113 aromatic and fine rice genotypes

Characters	Range	Mean	SE	CV%	F-value
CD (mm)	3.14–7.50	4.63	0.065	8.60	8.98**
FLA (cm^3)	21.94–54.43	35.28	0.599	11.70	8.99**
DF	76.00–125	100.5	0.709	2.45	75.58**
DM	102–150	126.00	0.697	1.69	33.30**
PH (cm)	79.00–179.00	129.00	1.382	2.74	38.13**
ET No.	4.00–13.87	9.84	0.127	10.39	4.75**
PL (cm)	24.13–33.00	28.62	0.149	3.77	6.40**
PB No.	6.33–13.33	10.34	0.123	11.61	3.56**
PBL (cm)	8.84–14.62	10.75	0.106	7.71	6.45**
SB No.	12.20–65.67	32.67	1.038	11.03	28.10**
SBL (cm)	1.95–3.58	2.75	0.026	6.66	6.79**
FG/P	79.00–262.00	170.50	3.774	16.79	7.15**
UFG/P	4.00–74.00	39.00	1.479	35.81	3.07**
GL (mm)	5.62–12.24	7.11	0.126	3.06	114.34**
GB (mm)	1.68–3.60	2.33	0.035	5.58	24.95**
GLBR	1.97–5.60	3.10	0.066	5.55	49.59**
TGW (g)	7.70–28.33	13.22	0.417	8.31	48.81**
YP (g)	6.47–17.45	12.38	0.188	18.70	2.23**
HI	0.16–0.34	0.24	0.004	20.45	1.93**

** indicates significance at 1% level of probability.
CD: culm diameter (mm), FLA: flag leaf area, DF: days to flowering, DM: days to maturity, PH: plant height, ET No.: effective tiller number, PL: panicle length, PB No.: primary branches per panicle, PBL: primary branch length, SB No.: secondary branches per panicle, SBL: secondary branch length, FGP: filled grains per panicle, UFGP: unfilled grains per panicle, GL: grain length, GB: grain breadth, GLBR: grain length breadth ratio, TGW: 1000-grain weight, YP: yield per plant, and HI: harvest index.

Figure 1: Variation in grain morphology of some aromatic and fine rice genotypes.

Days to flowering, grain breadth, grain length breadth ratio, plant height, and days to maturity were highly heritable, all with an estimated $H^2 > 0.90$ whereas other characters showed relatively low heritability. GA ranged from 0.03% for harvest index to 48.19% for filled grains per panicle. The genetic advance as percent of mean (GA%) ranged from 6.41% in panicle length to 50.85% in 1000 grain weight. In this study, flag leaf area, secondary branches per panicle, filled grains per panicle, grain length, grain breadth, grain length breadth ratio, and 1000-grain weight showed high heritability and high genetic advance indicated the presence of additive genes controlling these characters (Table 5).

Table 5: Estimation of genetic parameters of different quantitative characters in 113 aromatic and fine rice genotypes

Character	σ^2_G	σ^2_P	GCV (%)	PCV (%)	h^2_B (%)	GA (%)	GA in % of mean
CD (mm)	0.42	0.58	14.03	16.45	72.73	0.87	18.91
FLA (cm^3)	46.19	63.53	19.10	22.40	72.70	9.16	25.73
DF	56.10	58.36	7.25	7.40	96.13	11.60	11.24
DM	53.30	58.25	5.53	5.79	91.50	11.03	8.36
PH (cm)	199.64	215.76	9.65	10.04	92.53	21.47	14.67
ET No.	1.45	2.60	12.22	16.39	55.56	1.42	14.39
PL (cm)	2.10	3.26	5.06	6.31	64.31	1.83	6.41
PB No.	1.23	2.67	10.72	15.80	46.04	1.19	11.50
PBL (cm)	1.08	1.67	9.66	12.03	64.50	1.32	12.26
SB No.	117.39	130.38	33.17	34.95	90.03	16.24	49.72
SBL (cm)	0.06	0.10	9.24	11.41	65.54	0.33	11.82
FG/P	1384.30	2059.61	24.04	29.32	67.21	48.19	31.14
UFG/P	166.83	408.18	29.77	46.57	40.87	13.05	30.08
GL (mm)	1.79	1.84	18.82	19.07	97.44	2.09	29.36
GB (mm)	0.14	0.15	15.78	16.74	88.84	0.55	23.50
GLBR	0.48	0.51	22.35	23.04	94.16	1.07	34.27
TGW (g)	19.23	20.43	33.18	34.20	94.10	6.72	50.85
YP (g)	2.19	7.55	11.96	22.20	29.03	1.26	10.18
HI	0.00	0.00	13.40	23.21	33.33	0.03	12.22

CD: culm diameter (mm), FLA: flag leaf area, DF: days to flowering, DM: days to maturity, PH: plant height, ET No.: effective tiller number, PL: panicle length, PB No.: primary branches per panicle, PBL: primary branch length, SB No.: secondary branches per panicle, SBL: secondary branch length, FGP: filled grains per panicle, UFGP: unfilled grains per panicle, GL: grain length, GB: grain breadth, GLBR: grain length breadth ratio, TGW = 1000 grain weight, YP: yield per plant, HI: harvest index, σ^2_G = genotypic variance, σ^2_P = phenotypic variance, (GCV) % = genotypic coefficient of variation,(PCV) % = phenotypic coefficient of variation, h^2_B (%) = heritability (broad sense), and GA (%) = genetic advance.

Association between Traits

Pearson's correlation coefficient was computed between 19 quantitative traits among 113 accessions of aromatic and fine rice genotypes (Table 6). Culm diameter was significantly and positively correlated with flag leaf area, days to flowering, days to maturity, plan height, and primary branches per panicle. Plant height showed highly significant positive correlation with culm diameter, days to flowering, days to maturity, and panicle length. Grain yield was highly significant ($P < 0.01$) and positively correlated with days to flowering ($r = 0.407$), days to maturity ($r = 0.431$), filled grains per panicle ($r = 0.267$), primary branch length ($r = 0.324$), secondary branch length ($r = 0.324$), and 1000-grain weight ($r = 0.258$) and positively correlated with panicle length ($r = 0.190$) and secondary branches per panicle ($r = 0.231$).

Table 6: Pearson's correlation coefficient among 19 quantitative traits of aromatic and fine rice genotypes

	CD	FLA	DF	DM	PL	ET No.	PH	FG	UFG	GL	GB	GLB	PB No.	PBL	SB No.	SBL	TGW	YP	HI
CD	1.00																		
FLA	0.471**	1.00																	
DF	0.270**	−0.061	1.00																
DM	0.223*	−0.129	0.978**	1.00															
PL	−0.036	0.004	0.123	0.187*	1.00														
ET No.	−0.441**	−0.242**	−0.057	−0.029	0.069	1.00													
PH	0.266**	−0.073	0.268**	0.306**	0.265**	−0.007	1.00												
FG	−0.007	−0.267**	0.230*	0.272**	0.178	0.012	0.123	1.00											
UFG	−0.103	−0.242**	−0.143	−0.118	−0.012	0.017	0.062	0.288**	1.00										
GL	0.153	0.342**	−0.270**	−0.334**	−0.068	−0.012	−0.244**	−0.595**	−0.258**	1.00									
GB	0.152	0.409**	0.014	0.004	0.204*	−0.101	0.064	−0.335**	−0.018	0.148	1.00								
GLB	0.047	0.048	−0.237*	−0.286**	−0.190*	0.049	−0.264**	−0.312**	−0.239*	0.763**	−0.511**	1.00							
PB No.	0.413**	0.115	0.168	0.160	−0.140	−0.231*	0.135	0.081	0.053	−0.060	−0.020	−0.055	1.00						
PBL	−0.285**	0.085	−0.204*	−0.171	0.359**	0.243**	−0.229*	0.218*	−0.071	0.039	−0.043	0.083	−0.383**	1.00					
SB No.	−0.116	−0.218*	0.038	0.075	0.107	0.044	−0.013	0.784**	0.351**	−0.534**	−0.301**	−0.267**	0.011	0.345**	1.00				
SBL	−0.109	0.125	−0.299**	−0.316**	0.093	0.268**	−0.198*	0.115	0.032	0.286**	−0.087	0.295**	−0.319**	0.598**	0.283**	1.00			
TGW	0.108	0.479**	−0.041	−0.058	0.200*	−0.066	−0.154	−0.491**	−0.421**	0.651**	0.501**	0.249**	−0.245**	0.266**	−0.445**	0.250**	1.00		
YP	0.143	0.042	0.407**	0.431**	0.190*	0.118	−0.023	0.267**	−0.141	−0.001	0.016	−0.003	−0.073	0.324**	0.231*	0.324**	0.258**	1.00	
HI	−0.140	−0.152	0.025	0.043	0.056	0.048	−0.216*	0.343**	0.008	−0.194*	−0.282**	0.034	−0.120	0.390**	0.436**	0.324**	−0.112	0.545**	1.00

** and * indicate significance at 1% and 5% level of probability, respectively.
CD: culm diameter (mm), FLA: flag leaf area, DF: days to flowering, DM: days to maturity, PL: panicle length, ET No.: effective tiller number, PH: plant height, FGP: filled grains per panicle, UFGP: unfilled grains per panicle, GL: grain length, GB: grain breadth, GLBR: grain length breadth ratio, PB No.: primary branches per panicle, PBL: primary branch length, SB No.: secondary branches per panicle, SBL: secondary branch length, TGW: 1000-grain weight, YP: yield per plant, and HI: harvest index.

Principal Component Analysis (PCA)

Eigen values (latent roots) of 19 principal component axes and percentage of total variation accounted for them obtained from component analysis are presented in Table 7. The result revealed that the first axis largely accounted for the variations observed among the genotypes (48.8%) followed by the second axis (10.37%). The first nine axes accounted for about 90% of the total variations among the 19 characters describing 113 aromatic and fine rice genotypes where only 59.17% variation was accounted for the first two axes.

Table 7: Latent roots (eigen values) and their variation in 19 quantitative characters in 113 aromatic and fine rice genotypes

Principal component axes	Latent roots	Variation (%)	Cumulative% of variation
I	2.731	13	48.8
II	2.178	10.37	59.17
III	1.613	7.68	66.85
IV	1.219	5.81	72.66
V	1.01	4.81	77.47
VI	0.791	3.76	81.23
VII	0.669	3.19	84.42
VIII	0.654	3.12	87.54
IX	0.551	2.62	90.16
X	0.476	2.26	92.42
XI	0.372	1.77	94.19
XII	0.293	1.4	95.59
XIII	0.258	1.23	96.82
XIV	0.22	1.05	97.87
XV	0.176	0.84	98.71
XVI	0.151	0.72	99.43
XVII	0.096	0.46	99.89
XVIII	0.014	0.08	99.97
XIX	0.007	0.03	100

Cluster Analysis

The pattern of distribution of 113 aromatic and fine rice genotypes were grouped into 10 clusters shown in Table 8. The number of genotypes ranged from 3 to 19 in different cluster. The distribution pattern indicated that the maximum number of test genotypes (19) was grouped into the cluster I followed by 18 in clusters VIII, 17 in clusters III, 13 in clusters IV, 11 in clusters II, 10 in clusters X, 8 in clusters V and IX, and 6 in cluster VII. Cluster VI contained the lowest (3) number of genotypes.

Table 8: Distribution of 113 aromatic and fine rice genotypes into ten clusters

Cluster	Number of genotypes	% total	Name of genotypes
I	19	16.81	Nunia, Chini Sagar (2), Tilkapur, Kalobhog, Jabsiri, Chinisakkor, Noyonmoni, Tulsimoni, Khirshabuti, Gua masuri, Rajbhog, Baoijhaki, Tulsimala, Desikatari, Thakurbhog, Tulsimaloty, Radunipagal, Khasa, and Kataribhog
II	11	9.73	Begun bichi, Elai, Bashmati 370, Sugandhi dhan, Khazar, Basmati Sufaid 106, BRRI dhan50, Basmati 37, Basnatu sufaid 187, BU dhan2R, and Bashful
III	17	15.04	Jirabuti, Soru kamina, Kamini soru, Doiarguru, Luina, Kalijira (short grain), Philliphine kataribhog, Jirabhog (Bolder), Uknimodhu, Jira dhan, Badshabhog, Kalijira (long grain), Jesso balam, Dakshahi, Straw, Dubsail, and Sunduri samba
IV	13	11.50	Sagardana, Kalgochi, Chiniatob, Gopalbhog, Hatisail, Buchi, BRRI dhan38, Gordoi, Basmati, Padmabhog, Oval TAPL-2990, Kalijira TAPL-74, and Kalobakri
V	8	7.08	Saubail, Begunmala, Rajbut, Ranisalut, Gandhakusturi, Jirakatari, Awned TAPL-545, and Black TAPL-554
VI	3	2.65	Sakkorkhana, Kalijira TAPL-64, and Kalijira TAPL-68
VII	6	5.31	Chini Kanai, Chinigura, Khaskani, BRRI dhan34, BRRI dhan37, and Chinisail
VIII	18	15.92	Chinniguri, Premful, Baoibhog, Sakkorkhora, BR5, Uknimodhu, Chiniatob-2, Begunbichi-2, Bhatir cikon, Dolagocha, Dhan chikon, Badshabhog-2, Malshira, Sadagura, Chinikanai, Meedhan, Gobindhabhog, and Fulkari
IX	8	7.08	Sakor, Binnaphul, Lal Soru, Duksail, Tilokkachari, Chinairr, Kalonunia, and Thakurbhog-2
X	10	8.85	Meny, Kalomala, Khasa Mukpura, Bawaibhog-2, Khuti chikon, Tulsimala-2, Modhumadab, Parbatjira, Dudsail, and Maloti

Results of 10 higher and 10 lower intergenotypic distances estimated from distant matrix of Principal Coordinate Analysis are shown in Table 9. Highest intergenotypic distance was 2.274 observed between Gopalbhog and Kalobakri followed by the distance of 2.126 observed between Haitsail TAPL101 and Kalobakri. The 10th highest distance of 1.522 was observed between Jirabuti and Straw TAPL554 followed by 1.528 observed between Ranisaluit and Jirakatari. The lowest distance was calculated (0.299) between Doiagura and Jiradhan followed by the distance of 0.301 observed between Kamini soru and Jirabhog (bolder).

Table 9: Ten higher and ten lower intergenotypic distances among the 113 aromatic and fine rice genotypes

Sl. No.	Genotypic combination	Distances
	(a) Ten higher intergenotypic distance	
01	Gopalbhog-Kalobakri	2.274
02	Haitsail TAPL101-Kalobakri	2.126
03	Buchi-BU dhan 2R	1.836
04	Kalgochi-Kutichikon	1.832
05	Begun Mala-Elai	1.779
06	BRRI dhan50-Bashful	1.632
07	Elai-Khazar	1.589
08	Khasa Mukpura-Dudsail	1.580
09	Ranisalut-Jirakatari	1.528
10	Jirabuti-Straw TAPL-554	1.522
	(b) Ten lower intergenotypic distance	
01	Doiagura-Jiradhan	0.299
02	Kamini soru-Jirabhog (bolder)	0.301
03	Kutichikon-Parbatjira	0.318
04	Chinisagor (2)-Khasa	0.321
05	Tilkapur-Guamasuri	0.333
06	Kalobhog-Noyonmoni	0.341
07	Noyonmoni-Rajbhog	0.356
08	Chinisagor (2)-Jabsiri	0.361
09	Jabsiri-Chinisakkor	0.362
10	Tilkapur-Kalobhog	0.374

Intra- and intercluster distances value are presented in Table 10. There were marked variations in intracluster distances which ranged from 0.61 in cluster VI to 1.27 in cluster II indicating homogeneous nature of the genotypes within the cluster. The highest intracluster distance was computed for cluster II (1.27) which was comprised of eleven genotypes followed by cluster IV (1.01) with thirteen genotypes. The genotypes under cluster II (with the highest intracluster mean) were most heterogeneous and genotypes under cluster VI (with the lowest intracluster mean) were comparatively homogenous.

Table 10: Average intra- (bold) and intercluster distances (D^2) for 113 aromatic and fine rice genotypes

Cluster	I	II	III	IV	V	VI	VII	VIII	IX	X
I	**0.68**	7.499	3.710	4.274	4.807	7.332	3.975	7.592	4.895	12.119
II		**1.27**	9.438	7.161	8.645	10.77	9.681	12.206	9.087	15.791
III			**0.65**	6.288	7.713	6.876	4.332	4.799	3.775	9.087
IV				**1.01**	4.46	6.946	5.365	8.983	4.908	13.609
V					**0.96**	9.198	6.713	11.435	7.951	16.116
VI						**0.61**	7.524	7.896	6.498	10.93
VII							**0.75**	6.461	5.036	11.064
VIII								**0.71**	4.78	5.294
IX									**0.79**	9.256
X										**0.90**

The intercluster distances ranged from 3.710 to 16.116. Regarding the intercluster distance, the highest value was found between clusters V and X (16.116) followed by clusters II and X (15.791) and so on. On the other hand, the lowest intercluster distance was observed between clusters I and III (3.710) followed by clusters III and IX (3.775) indicating that genotypes of these clusters were genetically closed.

The mean values for all of 19 characters along with the marking of the highest (*H*) and lowest (*L*) for each of the cluster are presented in Table 11. Differences in cluster means existed for almost all the characters. Genotypes of cluster VI produced the highest mean for days to flowering (DF), days to maturity (DM), plant height (PH), and yield per plant (Y/P). Genotypes in cluster II had higher mean values for flag leaf area (FLA), secondary branch length (SBL), grain length (GL), grain length breadth ratio (GLBR), and 1000-grain weight (1000). Higher mean values for panicle length (PL), secondary branches per panicle (SB No.), filled grains per panicle (FG/P), and harvest index (HI) were recorded in cluster X whereas those for effective tiller (ET No.) per plant and grain breadth (GB) were recorded in cluster IV.

Table 11: Cluster means for 19 quantitative characters in 113 aromatic and fine rice genotypes

	I	II	III	IV	V	VI	VII	VIII	IX	X
CD (mm)	4.57	4.75	4.97	4.53	4.61	**5.86 (H)**	4.42	**4.16 (L)**	4.2	5.08
FLA (cm^2)	33.37	**43.25 (H)**	35.12	35.77	36.82	40.16	33.45	33.89	**30.95 (L)**	34.2
DF	101	**90.33 (L)**	102.5	107.56	106.75	**120.11 (H)**	107.8	104.7	103.7	104.03
DM	130.1	**118.2 (L)**	131.1	136.13	135.12	**146.67 (H)**	135.8	133.7	132.9	133.07
PH (cm)	152	**125.2 (L)**	158.4	138.5	157.56	**168 (H)**	131.2	143.7	142.1	146.39
ET No.	9.99	9.15	**9.08 (L)**	**10.44 (H)**	10.33	9.19	9.92	10.21	9.96	9.8
PL (cm)	28.67	27.57	28.56	28.99	28.81	27.95	**27.4 (L)**	28.98	28.85	**29.28 (H)**
PB No.	10.67	**9.80 (L)**	**10.85 (H)**	10.15	9.82	11.22	10.58	9.83	10.61	10.4
PBL (cm)	10.25	11.28	**10.11 (L)**	10.85	10.26	10.15	10.45	**11.42 (H)**	11.02	11.40
SB No.	28.38	22.96	35.44	23.83	**18.97 (L)**	25.78	37.56	42.53	34.06	**49.5 (H)**
SBL (cm)	2.72	**3.03 (H)**	2.65	2.61	**2.60 (L)**	2.58	2.74	2.83	2.65	2.99
FG/P	135.5	**98.84 (L)**	162.9	125.96	102.69	158.11	156.3	198.3	162.2	**232.2 (H)**
UFG/P	53.41	28.3	51.78	27.46	43.45	**13.89 (L)**	**65.44 (H)**	47.91	28.91	46.31
GL (mm)	7.36	**9.67 (H)**	6.81	7.26	7.71	6.34	6.57	**6.13 (L)**	6.61	6.32
GB (mm)	2.31	2.33	2.36	**2.42 (H)**	2.84	**1.89 (L)**	2.36	2.21	2.34	2.13
GLBR	3.21	**4.28 (H)**	2.91	3.12	2.81	3.36	2.8	**2.79 (L)**	2.91	3.0
TGW (g)	11.83	**19.03 (H)**	11.59	15.49	17.46	12.28	**11.02 (L)**	11.16	13.17	11.25
YP (g)	**11.02 (L)**	11.59	11.96	13.29	11.78	**14.06 (H)**	13.94	12.77	12.42	13.60
HI	0.22	0.22	0.23	0.24	**0.21 (L)**	0.23	0.26	0.25	0.24	**0.27 (H)**

CD: culm diameter (mm), FLA: flag leaf area, DF: days to flowering, DM: days to maturity, PH: plant height (cm), ET No.: effective tiller number, PL: panicle length (cm), PB No.: primary branches per panicle, PBL: primary branch length (cm), SB No.: secondary branches per panicle, SBL: secondary branch length (cm), FGP: filled grains per panicle, UFGP: unfilled grains per panicle, GL: grain length (mm), GB: grain breadth (mm), GLBR: grain length breadth ratio, TGW: 1000-grain weight (g), YP: yield per plant (g), and HI: harvest index.

DISCUSSION

During the current study, all traits showed highly significant ($P < 0.01$) variations among 113 accessions, which originated in Bangladesh except Khazar, Basmati 37, Basmati 370, Basmati Sufaid 106, and Basmati Sufaid 187 genotypes. Our results are in close agreement with those of Pandey et al. [34] who recorded highly significant variability among the different rice genotypes. Similarly the finding of Wang et al. [35] also gives support to the current findings. The findings of Chandra et al. [36] and Abarshahr et al. [37] further strengthen the current findings, who also found valuable and highly significant and positive variability among their studied genotypes.

The dependence of grain yield on other traits has been reported for many crops [38]. As mentioned, in this study, yield of plant had positive correlation with 8 quantitative traits. Lasalita-Zapico et al. [39] studied correlation coefficient of 10 quantitative traits for 32 upland rice varieties. In this distinguished significant positive correlation the majority of the morphological traits was recorded except flag leaf angle that had negative correlation with most of characters such as panicle length, leaf length, leaf width, ligule length, leaf area, and culm length. In our studies, grain yield positively correlated with panicle length. The findings indicate that plants with high panicles have high number of filled grains thereby increasing rice yield. Similar correlations were reported by Zafar et al. [40].

The calculation of heritability and genetic advance are used to help the breeder to select traits that are highly heritable as compared to a trait which is less heritable [33]. Both high heritability and genetic advance value obtained in this study, flag leaf area, secondary branches per panicle, filled grains per panicle, grain length, grain breadth, length breadth ratio, and 1000-grain weight indicated reasonable variation for this traits. This suggests that selection can be easily practiced by using these traits to improve grain yield in aromatic rice genotypes. The results support the findings of Sedeek et al. [41], Laxuman et al. [42], and Pandey et al. [34] who reported such type of heritability in rice.

In the present study, 113 aromatic and fine rice genotypes were clustered into ten groups based on 19 quantitative traits. This result supports the findings of Singh et al. [43] and Rao et al. [44] who reported ten clusters in rice genotypes. Ahmadikhah et al. [38] clustered 58 rice varieties into four groups based on 18 morphological traits and genetic distance was around 0.75. Group A was comprised of only one genotype and groups B, C, and D contained 14, 20, and 23 genotypes, respectively. Veasey et al. [45] computed clustering for 23 populations of rice by 20 morphological characteristics. So the varieties were clustered into 10 groups; the last group was the biggest group with seven members and groups 1, 2, 7, and 8 were the smallest groups including only one variety. So, genotypes having distant clusters could be hybridized to get the higher heterotic responses. The similar findings were also reported in a number of previous studies [18, 46–48].

Principal component analysis indicated diversity among 113 aromatic and fine rice genotypes. "Proper values" measure the importance and contribution of each component to total variance, whereas each coefficient of proper vectors indicates the degree of contribution of every original variable with which each principal component is associated. The higher the coefficients are, regardless of the direction (positive or negative), the more effective they will be in discriminating between accessions. In the present study, the first three axes accounted for about 66% of the total variations. Lasalita-Zapico et al. [39] computed approximately 82.7% of total variation among 32 upland rice varieties, 66.9% variation for PC1, and 15.87% for PC2. Rajiv et al. [49] reported the first two principal components accounting for 82.1% of total variation in control and 68.6% in the stress induced genotypes. To obtain greater heterosis, genotypes having distant clusters could be used as parents for hybridization program. In Bangladesh, most of the aromatic rice genotypes are traditional, photoperiod-sensitive, tall stature, and lower yields with mild to strong aroma and also they showed high variability (6, 28). In the present study, it was observed that the genotypes in clusters V and X (16.116) were more diverse than the genotypes of clusters I and III

(3.710). Considering cluster distance and cluster mean, the highest mean value for panicle length (cm), secondary branch length (cm), filled grains per panicle, and harvest index was observed in cluster X, which means that those traits might be selected for their high heterosis. Therefore, selection of parents for hybridization program from clusters V and X may result in the desirable heterosis for heterotic rice hybrids. Genotypes under cluster II may also give higher heterosis, if crossing is done within the genotypes of this cluster due to high value of intracluster distance.

CONCLUSION

In the present study, flag leaf area, secondary branches per panicle, filled grains per panicle, grain length, grain breadth, grain length breadth ratio, and 1000-grain weight showed high heritability and high genetic advance in percent of mean had high heritability and high genetic advance. Yield of plant had positively correlated with days to flowering, days to maturity, panicle length, filled grains per panicle, and 1000-grain weight. The cluster analysis placed 113 aromatic and fine rice genotypes into ten groups. The highest intercluster distance was observed between clusters V and X followed by clusters II and X. The maximum value of intercluster distance indicated that the genotypes belonging to cluster V were far diverged from those of cluster X. So, it is expected in our results that parent's selection for hybridization from the clusters V and X may give the desirable heterosis for heterotic rice hybrids. Finally, molecular characterizations of the studied germplasm are required for high resolution QTL mapping and validating the presence of candidate genes responsible for valuable characters.

ACKNOWLEDGMENTS

The authors are highly grateful to the collaborative research project entitled "Genetic Enhancement of Local Rice Germplasm towards Aromatic Hybrid Rice Variety Development in Bangladesh" funded by the NATP: Phase I of PIU, Bangladesh Agricultural Research Council (BARC), for providing all necessary supports.

REFERENCES

1. A. Hamid, N. Uddin, M. Haque, and E. Haque, Deshi Dhaner Jat, Publication no. 59, Bangladesh Rice Research Institute, Gazipur, Bangladesh, 1982 (Bangla).
2. BBS (Bangladesh Bureau of Statistics), Statistical Pocket Book of Bangladesh, Government of Peoples Republic of Bangladesh, Dhaka, Bangladesh, 2011.
3. R. K. Singh, U. S. Singh, and G. S. Khush, Aromatic Rices, Oxford and IBH Publishing, New Delhi, India, 2000.
4. M. Khalequzzaman, M. A. Siddique, and M. K. Bashar, "Rice genetic resources conservation and utilization in Bangladesh," in Proceedings of the National Workshop on Plant Genetic Resources for Nutritional Food Security, pp. 50–60, BARC, Dhaka, Bangladesh, May 2012.
5. M. A. H. Sarker, "Indigenous fine aromatic rice production: Bangladesh perspective," in Development of Basic Standard for Organic Rice Cultivation, pp. 1–9, RDA (Rural Development Administration) and Dankook University, Yongin, South Korea, 2002.
6. S. M. Shahidullah, M. M. Hanafi, M. Ashrafuzzaman, M. R. Ismail, and A. Khair, "Genetic diversity in grain quality and nutrition of aromatic rices," African Journal of Biotechnology, vol. 8, no. 7, pp. 1238–1246, 2009. ·
7. M. Z. Islam, S. Banik, M. A. Haque, M. A. Siddiquee, M. Khalequzzaman, and M. A. K. Mian, "Physico-chemical and cooking properties of local aromatic rice germplasm in Bangladesh," Eco-Friendly Agriculture Journal, vol. 6, no. 10, pp. 243–248, 2013.
8. M. A. Baqui, M. E. Ham, D. Jones, and R. Straingfellow, The Export Potential of Traditional Varieties of Rice from Bangladesh, Bangladesh Rice Research Institute, Gazipur, Bangladesh, 1997.
9. T. Das, Rices in Bangladesh, Dhanmondi, Dhaka-1205, Bangladesh, India, 2005.
10. B. K. Babu, V. Meena, V. Agarwal, and P. K. Agrawal, "Population structure and genetic diversity analysis of Indian and exotic rice (Oryza sativa L.) accessions using SSR markers," Molecular Biology Reports, vol. 41, no. 7, pp. 4329–4339, 2014. · ·
11. M. Allhgholipour, E. Farshdfar, and B. Rabiei, "Molecular characterization and genetic diversity analysis of different rice cultivars by microsatellite markers," Genetika, vol. 46, no. 1, pp. 187–198, 2014. · ·
12. V. Pachauri, N. Taneja, P. Vikram, N. K. Singh, and S. Singh, "Molecular and morphological characterization of Indian farmers rice varieties (Oryza sativa L.)," Australian Journal of Crop Science, vol. 7, no. 7, pp. 923–932, 2013. ·
13. C.-H. Wang, X.-M. Zheng, Q. Xu et al., "Genetic diversity and classification of Oryza sativa with emphasis on Chinese rice germplasm," Heredity, vol. 112, no. 5, pp. 489–496, 2014. · ·
14. R. K. Varshney, T. Thiel, T. Sretenovic-Rajicic et al., "Identification and validation of a core set of informative genic SSR and SNP markers for assaying functional diversity in barley," Molecular Breeding, vol. 22, no. 1, pp. 1–13, 2008. · ·
15. International Rice Genome Sequencing Project, "The map-based sequence of the rice genome," Nature, vol. 436, no. 7052, pp. 793–800, 2005. ·
16. M. J. Thomson, E. M. Septiningsih, F. Suwardjo, T. J. Santoso, T. S. Silitonga, and S. R. McCouch, "Genetic diversity analysis of traditional and improved Indonesian rice (Oryza sativa L.) germplasm using microsatellite markers," Theoretical and Applied Genetics, vol. 114, no. 3, pp. 559–568, 2007. · ·
17. Z. Y. Wang, G. Second, and S. D. Tanksley, "Polymorphism and phylogenetic relationships among species in the genus Oryza as determined by analysis of nuclear RFLPs," Theoretical and Applied Genetics, vol. 83, no. 5, pp. 565–581, 1995. · ·
18. M. A. Latif, M. M. Rahman, M. S. Kabir, M. A. Ali, M. T. Islam, and M. Y. Rafii, "Genetic diversity analyzed by quantitative traits among rice (Oryza sativa L.) genotypes resistant to blast disease," African Journal of Microbiology Research, vol. 5, no. 25, pp. 4383–4391,

2011. · ·

19. T. Zhang, X.-L. Ni, K.-F. Jiang et al., "Relationship between heterosis and parental genetic distance based on molecular markers for functional genes related to yield traits in rice," Rice Science, vol. 17, no. 4, pp. 288–295, 2010. · ·
20. H. Surek and N. Beser, "Selection for grain yield and yield components in early generations for temperate rice," Philippine Journal of Crop Science, vol. 28, no. 3, pp. 3–15, 2003.
21. A. S. Larik and L. S. Rajput, "Estimation of selection indices in Brassica juncea L. and Brassica napus L.,"Pakistan Journal of Botany, vol. 32, no. 2, pp. 323–330, 2000. ·
22. L. S. Kumar, "DNA markers in plant improvement: an overview," Biotechnology Advances, vol. 17, no. 2-3, pp. 143–182, 1999. · ·
23. M. Khalid, Farahatullah, N. U. Khan, A. Din, and M. Y. Khan, "Linkage of morphological markers in Brassica," Pakistan Journal of Botany, vol. 42, no. 5, pp. 2995–3000, 2010.
24. M. Khalequzzaman, M. Z. Islam, K. Akter, and M. K. Bashar, "genetic diversity in local rainfed lowland rice (Oryza sativa L.) in Bangladesh," Bangladesh Journal of Plant Breeding and Genetics, vol. 21, no. 1, pp. 49–54, 2013. ·
25. M. Z. Islam, K. Akter, E. S. M. H. Rashid, M. Khalequzzaman, and M. K. Bashar, "Genetic diversity of local irrigated rice (Oryza sativa L.)," Bangladesh Journal of Plant Breeding and Genetics, vol. 22, no. 2, pp. 51–56, 2009. ·
26. M. A. Siddique, M. Z. Islam, M. Khalequzzaman, and M. S. Ahmed, "Genetic diversity in rice (Oryza sativa L.) landraces of hilly areas in Bangladesh," Bangladesh Journal of Plant Breeding and Genetics, vol. 24, no. 2, pp. 25–30, 2013. ·
27. M. Z. Islam, M. A. Siddique, E. S. M. H. Rashid, M. S. Ahmed, and M. Khalequzzaman, "Genetic diversity in sadajra rice (Oryzasativa L.) germplasm," The Agriculturists, vol. 12, no. 1, pp. 26–32, 2014. ·
28. M. A. Rahman, M. J. Thomson, M. S. Alam, M. de Ocampo, J. Egdane, and A. M. Ismail, "Exploring novel genetic sources of salinity tolerance in rice through molecular and physiological characterization,"Annals of Botany, In press.
29. D. N. Cruz and G. S. Khush, "Rice grain quality evaluation procedures," in Aromatic Rices, R. K. Singh, U. S. Singh, and G. S. Khush, Eds., pp. 15–28, Kalyani, New Delhi, India, 2004.
30. B. C. Sood and E. A. Siddiq, "A rapid technique for scent determination in rice," Indian Journal of Genetics and Plant Breeding, vol. 38, pp. 268–271, 1978.
31. G. W. Burton, "Quantitative inheritance in grasses," in Proceedings of the International Grassland Congress, vol. 1, pp. 277–283, State College, Pa, USA, August 1952.
32. G. W. Burton and E. H. DeVane, "Estimating heritability in tall fescue (Festuca arundinacea) from replicated clonal material," Agronomy Journal, vol. 45, no. 10, pp. 478–481, 1953. ·
33. H. W. Johnson, H. F. Robinson, and R. E. Comstock, "Estimates of genetic and environmental variability in soybeans," Agronomy Journal, vol. 47, no. 7, pp. 314–318, 1955. ·
34. P. Pandey, P. John Anurag, D. K. Tiwari, S. K. Yadav, and B. Kumar, "Genetic variability, diversity and association of quantitative traits with grain yield in rice (Oryza sativa L.)," Journal of Bio-Science, vol. 17, no. 1, pp. 77–82, 2009. ·
35. J.-L. Wang, Y.-B. Gao, N.-X. Zhao et al., "Morphological and RAPD analysis of the dominant species Stipa krylovii Roshev. in Inner Mongolia steppe," Botanical Studies, vol. 47, no. 1, pp. 23–35, 2006. ·
36. R. Chandra, S. Pradhan, S. Singh, L. Bose, and O. Singh, "Multivariate analysis in upland rice genotypes,"World Journal of Agricultural Sciences, vol. 3, no. 3, pp. 295–300, 2007.
37. M. Abarshahr, B. Rabiei, and H. S. Lahigi, "Assessing genetic diversity of rice varieties under drought stress conditions," Notulae Scientia Biologicae, vol. 3, no. 1, pp. 114–123, 2011.
38. A. Ahmadikhah, S. Nasrollanejad, and O. Alishah, "Quantitative studies for investigating variation and its effect on heterosis of rice," International Journal of Plant Production, vol. 2, no. 4, pp. 297–308, 2008.
39. F. C. Lasalita-Zapico, J. A. Namocatcat, and J. L. Cariño-Turner, "Genetic diversity analysis of traditional upland rice cultivars in Kihan, Malapatan, Sarangani Province, Philippines using morphometric markers," Philippine Journal of Science, vol. 139, no. 2, pp. 177–180, 2010. ·
40. N. Zafar, S. Aziz, and S. Masood, "Phenotypic divergence for agro-morphological traits among landrace genotypes of rice (Oryza sativa L.) from Pakistan," International Journal of Biological Sciences, vol. 6, no. 2, pp. 335–339, 2006.
41. S. Sedeek, S. Hammoud, M. Ammar, and T. Metwally, "Genetic variability, heritability, genetic advance and cluster analysis for for some physiological traits and grain yield nad its components in rice (Oryza sativa L.)," Journal of Agricultural Sciences, vol. 35, no. 3, pp. 858–878, 2009.
42. L. Laxuman, P. Salimath, H. Shashidhar, et al., "Analysis of genetic variability in inter specific backcross inbred lines in rice (Oryza sativa L.)," Karnataka Journal of Agricultural Sciences, vol. 23, no. 4, pp. 563–565, 2010.
43. S. K. Singh, R. S. Singh, D. M. Maurya, and O. P. Verma, "Genetic divergence among low land rice cultivars," The Indian Journal of Genetics and Plant Breeding, vol. 47, no. 1, pp. 11–14, 1987.
44. A. V. Rao, A. S. R. Prashad, T. Sai Krishna, D. V. Seshu, and T. E. Srinivasan, "Genetic divergence among some brown plant hopper resistant rice varieties," Indian Journal of Genetics and Plant Breeding, vol. 41, no. 2, pp. 179–185, 1981.
45. E. A. Veasey, E. F. Da Silva, E. A. Schammass, G. C. X. Oliveira, and A. Ando, "Morphoagronomic genetic

diversity in American wild rice species," Brazilian Archives of Biology and Technology, vol. 51, no. 1, pp. 95–104, 2008. ·

46. M. A. Latif, M. Rafii Yusop, M. Motiur Rahman, and M. R. Bashar Talukdar, "Microsatellite and minisatellite markers based DNA fingerprinting and genetic diversity of blast and ufra resistant genotypes," Comptes Rendus-Biologies, vol. 334, no. 4, pp. 282–289, 2011. · ·

47. N. Abdullah, M. Rafii Yusop, M. Ithnin, G. Saleh, and M. A. Latif, "Genetic variability of oil palm parental genotypes and performance of its› progenies as revealed by molecular markers and quantitative traits," Comptes Rendus Biologies, vol. 334, no. 4, pp. 290–299, 2011. · ·

48. M. Sohrabi, M. Y. Rafii, M. M. Hanafi, A. S. N. Akmar, and M. A. Latif, "Genetic diversity of upland rice germplasm in malaysia based on quantitative traits," The Scientific World Journal, vol. 2012, Article ID 416291, 9 pages, 2012. · ·

49. S. Rajiv, P. Thivendran, and S. Deivanai, "Genetic divergence of rice on some morphological and physicochemical responses to drought stress," Pertinaka Journal, vol. 32, no. 2, pp. 315–328, 2010.

CHAPTER 5

VARIATION IN COOKING AND EATING QUALITY TRAITS IN JAPANESE RICE GERMPLASM ACCESSIONS

Kiyosumi Hori, Keitaro Suzuki, Ken Iijima, and Kaworu Ebana

Keywords: Eating quality, natural variation, *Oryza sativa* L., *temperate japonica* rice accessions

INTRODUCTION

Rice (*Oryza sativa* L.) is the most important food crop in the world, as it is a staple food for more than half of the world's population. Rice eating quality largely determines its market price and consumer acceptance, because consumers pay particular attention to high eating quality (Anacleto *et al.* 2015, Hori and Yano 2013, Juliano *et al.*1964).

Grain components affect eating quality that are determined by many physicochemical properties and cooking characteristics of rice. Amylose content is widely recognized as an important determinant of eating quality, because it reflects the functionality of the rice grain's starch (Juliano *et al.* 1993). The granule-bound starch synthase I enzyme (GBSSI) is required for amylose synthesis in rice, and the gene encoding GBSSI was named *Waxy* (*Wx*). Glutinous rice cultivars and *waxy* mutant lines carry a deletion in the *Wx* gene that prevents the activity of GBSSI (the *wx* allele), and consequently they contain no amylose. In non-glutinous rice cultivars, a single-nucleotide polymorphism (SNP) at the splice site of intron 1 of *Wx* defines two alleles, Wx^a and Wx^b, affects differences of amylose content (Cai *et al.*1998, Hirano *et al.* 1998, Issiki *et al.* 1998, Wang *et al.* 1995, Yamanaka *et al.* 2004). The varietal differentiation of *Wx*gene, Wx^a and Wx^b, alleles was reported by Sano (1984). Wx^a is distributed predominantly in *indica* accessions, andWx^b in *japonica* accessions (Sano *et al.* 1986). Protein content also plays an important role in determining eating quality, as does the grain texture and the surface hardness of cooked rice (Okadome 2005). By focusing on physicochemical properties, such as gel consistency, gelatinization temperature, and pasting properties, several evaluation methods have been developed to estimate the eating quality of rice accessions. The Rapid Visco Analyzer (RVA) is often used to measure the gelatinization and viscosity changes that occur during heating, holding, and cooling of rice flour (Bergman *et al.* 2004). Because of the small sample size required for RVA evaluation, this method is suitable for screening the pasting properties of grains from the accessions produced in breeding programs. The Tensipresser is another evaluation instrument for scoring the texture of cooked rice grains (Suzuki *et al.* 2006). These methods have let researchers clearly differentiate among rice accessions on the basis of the physicochemical properties of their grains (Nakamura *et al.* 2012a, Okadome *et al.* 1999).

In current Japanese breeding programs, the eating quality of cooked rice is usually measured through sensory analysis by well-trained panelists of at least 20 people (Matsue 1992, Yamamoto *et al.* 1996). Sensory tests evaluate the glossiness, stickiness, hardness, and taste of the cooked rice grains and provide a score for overall eating quality. Sensory tests are the most effective method to determine eating quality. However, they make it difficult to evaluate many samples simultaneously and to select superior accessions in early generations such as F_3 to F_5 populations in breeding programs, because of the requirement for large amounts of grain (*e.g.*, at least several hundred grams of polished rice; Takeuchi *et al.* 2008). To solve this problem, evaluation instruments such as the Cooked Rice Taste Analyzer and the Mido Meter have been developed to predict the eating quality that would be determined by a sensory test (Goto *et al.* 2014, Kwon *et al.* 2011). These instrumental methods showed significant correlations with the eating quality scores by sensory tests (Okadome 2005). More than 580 QTLs for eating-quality traits have been detected

Citation: Kiyosumi Hori, Keitaro Suzuki, Ken Iijima, and Kaworu Ebana, Variation in cooking and eating quality traits in Japanese rice germplasm accessions, doi:10.1270/jsbbs.66.309.

on all of rice chromosomes by using segregating populations derived from crosses among rice accessions in the Q-TARO QTL database (http://qtaro.abr.affrc.go.jp/;Yonemaru *et al.* 2010) and the Gramene QTL database (http://archive.gramene.org/qtl/; Monaco *et al.* 2014). Most previous QTL studies used segregating populations derived from crosses between accessions of the *indica* and*japonica* rice variety groups, and the studies found QTLs with a large effect near the *wx* locus on the short arm of chromosome 6 (Hsu *et al.* 2014, Takeuchi *et al.* 2007, Tan *et al.* 1999). Recently, QTL studies have been performed in segregating populations derived from crosses within non-glutinous *temperate japonica* rice that are homozygous for the Wx^b allele. These studies commonly detected eating-quality QTLs that were evaluated by sensory tests on the short arm of chromosome 3 (Hori and Yano 2013, Kobayashi and Tomita 2008, Takeuchi *et al.* 2008, Wada *et al.*2008, 2013). However, we do not yet fully understand the genetic basis of the variation in eating quality among non-glutinous *temperate japonica* rice, which has been a staple food for many years in northern parts of rice cultivation areas such as Japan, Korea, and China. It is therefore necessary to comprehensively elucidate the variation of eating quality in non-glutinous *temperate japonica* rice accessions.

In this study, we evaluated 33 eating-quality traits and heading date in 183 Japanese rice germplasm accessions, comprising 146 non-glutinous *temperate japonica* rice accessions and 40 accessions that did not belong to this category (see the Methods section for details). The present study therefore represents the first study of a wide range of eating-quality traits based on a large number of Japanese rice germplasm accessions. Principal-components analysis (PCA) revealed that alleles of the *Wx* gene were significantly associated with a wide range of phenotypic variation in eating quality among *indica*, *tropical japonica*, and *temperate japonica* rice accessions. In the non-glutinous *temperate japonica* rice accessions, which all had the Wx^b allele, a QTL on chromosome 3 was significantly associated with phenotypic variations in the eating-quality traits. These results suggest the substantial importance of the *Wx* gene and the other loci in future breeding programs of non-glutinous *temperate japonica* rice.

MATERIALS AND METHODS

Plant materials

Among the genetic resources conserved in the Genebank of the National Institute of Agrobiological Sciences (NIAS), we chose 183 rice germplasm accessions. Landraces and improved rice accessions were selected as representative Japanese rice cultivars on the basis of their geographical origins and the results of cluster analysis by means of genome-wide DNA marker analysis (Ebana *et al.* 2008, Yamasaki and Ideta 2013). The rice accessions consisted of 164 *temperate japonica*, 14 *tropical japonica*, and 5 *indica* accessions; these comprised 160 paddy and 23 upland accessions; 164 non-glutinous and 19 glutinous accessions; and 125 improved and 58 landrace accessions. We raised 48 plants of each accession in the paddy fields at NIAS in Tsukuba, Japan (36.03°N, 140.11°E), in 2010 and 2011. Seeds were sown in April, and seedlings were transplanted into the fields in May in plots with a double row per line, with 18 cm between plants and 30 cm between rows. Cultivation management followed the standard procedures used at NIAS.

Evaluation of eating quality

We evaluated 33 eating-quality traits plus the heading date of the 183 accessions (Table 1). We compared the evaluation scores of 22 accessions in 2010 and 2011 to investigate year-to-year correlations and the stability of the quality characteristics, because eating quality of cooked rice was influenced by environmental factors such as the air temperature during the ripening period. The 22 accessions showed a wide range of eating quality phenotypes, and included ancestral cultivars of Koshihikari, and cultivars derived from the same cross with Koshihikari. Rice seeds were harvested at the full maturity stage. Brown rice was polished to 90% of the original grain weight by a Pearlest machine (Kett Electric Laboratory, Tokyo, Japan). Polished rice grains were milled in an SRG10A mill (Satake Co. Ltd., Tokyo, Japan) to allow an evaluation of the amylose content, protein content, and pasting properties.

Table 1: Ranges of values for the 33 eating-quality traits and for heading date in the 183 Japanese rice germplasm accessions

Trait name	Units of measurement or score	Range of values
Protein content	(%)	4.9–8.5
Amylose content	(%)	0.4–30.9
Cooked Rice Taste Analyzer		
Eating quality	(–)	30.0–91.0
Appearance	(–)	0.2–9.8
Hardness	(–)	3.8–9.8
Stickiness	(–)	1.0–9.5
Balance degree	(–)	0.0–9.7
Rapid Visco Analyzer		
Maximum viscosity	(RVU)	90.6–576.8
Minimum viscosity	(RVU)	31.8–264.7
Breakdown	(RVU)	58.8–391.3

Final viscosity	(RVU)	49.3–452.3
Setback	(RVU)	17.5–224.1
Peak time	(min)	3.5–7.0
Gelatinization temperature	(°C)	67.3–82.0
Tensipresser		
on the grain surface		
Hardness H1	($N/m^2 \times 10^2$)	33.9–136.6
Stickiness S1	($N/m^2 \times 10^2$)	0.3–21.2
Adhered mass L3	(mm)	0.0–2.0
Adhesiveness A3	($N/m^2m^1 \times 10^2$)	0.0–118.0
Balance degree S1/H1	(–)	0.0–0.4
Balance degree L3/L1	(–)	0.0–3.2
Balance degree A3/A1	(–)	0.0–1.1
for the whole grain		
Hardness H2	($N/m^2 \times 10^4$)	13.6–37.1
Stickiness S2	($N/m^2 \times 10^4$)	0.1–0.6
Adhered mass L6	(mm)	0.5–4.3
Adhesiveness A6	($N/m^2m^1 \times 10^4$)	3.0–27.9
Balance degree S2/H2	(–)	0.04–0.34
Balance degree L6/L4	(–)	0.20–2.14
Balance degree A6/A4	(–)	0.01–0.34
Sample thickness	(mm)	2.0–3.0
Tenderness	($N/m^2m^1 \times 10^3$)	5.0–126.1
Pliability	(–)	0.5–1.3
Toughness	($N/m^2m^1 \times 10^4$)	3.1–198.4
Brittleness	(–)	1.4–7.2
Heading date	(month/day)	6/12–9/1

The apparent amylose content was estimated according to the method reported by Ando *et al.* (2010). Rice flour was dissolved with 0.5 N aqueous NaOH and left overnight at room temperature. The solution was then diluted to 0.05 N with H_2O, and the amylose content was determined by using an Auto Analyzer II (Bran + Luebbe Co. Ltd., Norderstedt, Germany).

The crude protein content was determined by the combustion method (American Association of Cereal Chemists International [AACCI] International Approved Method 46-30.01 (2000)). Rice flour was placed into a quartz combustion tube in an induction furnace at 900°C. Protein content was calculated from the nitrogen content using the conversion coefficient for nitrogen in protein, as N × 5.95 (Nakamura *et al.* 2012a).

The viscosity of the rice flour was analyzed by using a Rapid Visco Analyzer (RVA; Model No. RVA-4; Newport Scientific, Warriewood, NSW, Australia), according to the method reported by Toyoshima *et al.* 1997). Sample temperatures were 50°C for 1 min, followed by 93°C for 4 min, 93°C for 7 min, decreasing from 93°C to 50°C for 4 min, and 50°C for 3 min. RVA profiles were characterized by the parameters shown in Table 1.

To evaluate the characteristics of the cooked rice grains, polished rice is cooked following the method of Okadome*et al.* (1999). Each sample (20 g) of polished rice was placed in a pudding cup with 25 mL of water, soaked for 1 h at 20°C, and then cooked in an electric rice cooker (SR-ULH18, Panasonic, Kadoma Japan). Physical properties of the cooked rice grains were measured by using a Tensipresser MyBoy texture analyzer (Takemoto Electric, Co., Tokyo, Japan) with the high-compression/low-compression method under the conditions used by Okadome *et al.* (1999), and the continuous progressive compression method used by Okadome *et al.* (1995) and Takeyama *et al.* (1998). The Tensipresser physical properties in Table 1 were obtained under both the high-compression/low-compression method and the continuous progressive compression method.

The eating quality characteristics of the rice (Table 1) were as measured by using a Cooked Rice Taste Analyzer STA1A (Satake Co. Ltd., Tokyo, Japan), according to the method reported by Mikami (2009). Eating quality was calculated by the estimation formulae based on the amount of light reflected and transmitted by the cooked rice grains.

PCA for the eating-quality traits

We used Pearson's correlation coefficient (*r*) to establish relationships among all eating-quality traits. We then performed PCA using version 11 of JMP software (SAS Institute, Cary, NC, USA). PCA was based on the correlation matrix and the calculated Eigenvalues and Eigenvectors. Principal components

(PCs) were extracted until they accounted for more than 80% of the cumulative contribution of the Eigenvalues.

DNA marker analysis for the Wx gene and simple sequence repeats

Total DNA of each rice accession was extracted from leaves by the CTAB method reported previously (Hori *et al.*2012). The alleles of the *Wx* gene were identified by means of derived cleaved amplified polymorphic sequences analysis following the method of Yamanaka *et al.* (2004). We amplified a DNA fragment containing the first exon–intron junction of the *Wx* gene. PCR amplification was performed using the primer set WP-B (5′-TTA ATT TCC AGC CCA ACA CC-3′) and WP-CAPS (5′-TGT TGT TCA TCA GGA AGA ACA TCT CCA AG-3′). Each PCR product was digested with *Eco*T14I at 37°C overnight. Digestion products were then electrophoresed in 3% agarose gels.

Simple sequence repeat (SSR) marker RM4108 on chromosome 3 was amplified by using the primer set 5′-GTC CCT CGC TTT ATA TCT AG-3′ and 5′-CAA CTC TGC TAA ACG AAT TA-3′. PCR products were fluorescently labeled and their fragment sizes were determined by using an ABI 3130xl system and version 3.6 of GeneMapper software (Applied Biosystems, Foster City, CA, USA), according to the method of Hori *et al.* (2012).

RESULTS

Variation of eating-quality traits in the 183 Japanese rice accessions

All 33 eating-quality traits and heading date varied widely among the accessions (Table 1). For example, amylose content ranged from 0.4% to 30.9%, protein content ranged from 4.9% to 8.5%, the eating quality estimated by the Cooked Rice Taste Analyzer ranged from 30.0 to 91.0, maximum viscosity by the RVA ranged from 90.6 to 576.8, and stickiness on grain surface by the Tensipresser ranged from 0.3 to 21.2. The accessions with the lowest amylose content were the glutinous rice Meguromochi (0.4%), Okabo (0.5%), and Gaisenmochi (0.8%). Of the non-glutinous rice accessions, the lowest amylose content was in Sakihikari (14.2%), Shimahikari (14.4%), Nanatsuboshi (14.4%), Sorachi (14.4%), and Fukoku (146%). The accessions with low protein content were Kabashiko (4.9%), Nikomaru (5.0%), Yumetsukushi (5.0%), Aichinokaori (5.1%), and Koshihikari (5.1%). The accessions with the highest eating quality were found in the glutinous rice Hiyokumochi (91.0), Koganemochi (90.0), and Himenomochi (89.7). In the non-glutinous rice accessions, the highest eating qualities were those of Koshihikari (88.5), Itadaki (88.3), Hitomebore (88.0), Yumetsukushi (88.0), and Nikomaru (87.0). The accessions with the highest maximum viscosity were Joushuu (576.8), Kameji (548.3), Hananomai (528.7), Norin 1 (524.9), and Yamasenishiki (513.8). The accessions with the highest stickiness on the grain surface were Hinohikari (21.2), Akiho (19.9), Norin 22 (19.3), Notohikari (19.2), and Fusaotome (19.1). Year-to-year correlations between the values in 2010 and 2011 were significant for all 33 eating-quality traits except hardness of the grain surface, and the tenderness and toughness in the Tensipresser measurements (Table 2), indicating the consistency of these traits among years.

Table 2: Between-year correlation coefficients and correlations with heading date for the 33 eating quality traits

Trait name	Year-to year correlation	Correlation with heading date
Protein content	0.87 ***	−0.39 ***
Amylose content	0.71 ***	0.19 **
Cooked Rice Taste Analyzer		
Eating quality	0.78 ***	0.17 *
Appearance	0.80 ***	0.15 *
Hardness	0.84 ***	−0.06
Stickiness	0.82 ***	0.23 **
Balance degree	0.76 ***	0.17 *
Rapid Visco Analyzer		
Maximum viscosity	0.65 ***	−0.18 *
Minimum viscosity	0.44 **	−0.33 ***
Breakdown	0.62 ***	−0.01
Final viscosity	0.69 **	−0.19 *
Setback	0.49 **	0.04
Peak time	0.31 *	0.00
Gelatinization temperature	−0.30 *	−0.50 ***
Tensipresser		

on the grain surface		
Hardness H1	0.07	0.00
Stickiness S1	0.55 ***	0.16 *
Adhered mass L3	0.32 **	0.04
Adhesiveness A3	0.38 **	0.11
Balance degree S1/H1	0.36 **	0.12
Balance degree L3/L1	0.35 **	0.05
Balance degree A3/A1	0.47 ***	0.09
for the whole grain		
Hardness H2	0.23 *	0.19 **
Stickiness S2	0.38 **	−0.26 ***
Adhered mass L6	0.58 ***	0.00
Adhesiveness A6	0.27 *	0.07
Balance degree S2/H2	0.28 *	−0.31 ***
Balance degree L6/L4	0.57 ***	0.02
Balance degree A6/A4	0.29 *	−0.10
Sample thickness	0.61 ***	−0.39 ***
Tenderness	0.07	0.07
Pliability	0.22 *	−0.06
Toughness	0.09	0.03
Brittleness	0.23 *	0.10
Heading date	0.98 ***	–

Significance levels: *$P < 0.05$, **$P < 0.01$, and **$P < 0.001$.

The heading date also varied widely, from 12 June to 1 September (Table 1). Significant correlations with heading date were observed for 15 of the 33 eating-quality traits (Table 2). Heading date was positively correlated with amylose content; with eating quality comprosing appearance, stickiness and balance degree evaluated using the Cooked Rice Taste Analyzer; and stickiness on the grain surface and hardness of the whole grain evaluated using the Tensipresser. On the other hand, it was negatively correlated with the protein content; with the maximum viscosity, minimum viscosity, final viscosity, and gelatinization temperature evaluated using the RVA and stickiness, balance degrees, and sample thickness of the whole grain evaluated using the Tensipresser. Thus, although the correlations were generally weak to moderate ($|r| \leq 0.5$), these traits were influenced to some extent by the heading date.

There were significant correlations between many pairs of the 33 eating-quality traits. The accessions with high eating quality showed an overall trend toward low protein content, low amylose content, high maximum viscosity, low minimum viscosity, low final viscosity, low setback, low peak time, low gelatinization temperature, low hardness, high stickiness, high adhered mass, high adhesiveness, and a high balance degree. Strong and statistically significant correlations ($P < 0.001$) were often observed between traits evaluated using the same instrument. The absolute values of the correlations among the five traits evaluated using the Cooked Rice Taste Analyzer were at least 0.9, so we selected eating quality as representative of all traits evaluated by the Cooked Rice Taste Analyzer, and did not subject the other four traits (appearance, hardness, stickiness and balance degree) to additional statistical analysis to avoid an excessive influence of autocorrelation on the results.

PCA results for the 183 Japanese rice accessions

To elucidate the main components of variation in the eating-quality traits, we subjected the 29 traits (excluding appearance, hardness, stickiness, and balance degree) to PCA. The results revealed that the contributions of the first and second principal components (PC1 and PC2) were 51.2% and 18.6%, respectively. The sum of the contributions of the top four PCs accounted for 84.1% of the total variance. For PC1, the corresponding loadings were negative for amylose content and setback and positive for eating quality, adhered mass, adhesiveness, the three balance degrees on the grain surface, the adhered mass of the whole grain, and two of the three balance degrees of the whole grain. For PC2, the loadings were negative for protein content and positive for eating quality, maximum viscosity, minimum viscosity, breakdown, final viscosity, peak time, and stickiness on the grain surface. For PC3, the loadings were an equal mixture of positive and negative. For PC4, the loadings were almost all positive.

PCA for the 183 accessions revealed that PC1 clearly distinguished among alleles at *wx* locus (Fig. 1A). PC1 identified three groups: the Wx^a, Wx^b, and *wx* alleles were separate from each other, but members of each group were clustered together in the ordination, without differences among accessions from the *indica*, *tropical japonica*, and *temperate japonica* variety groups. The average PC scores for the eating-quality traits differed significantly among the three alleles at *wx* locus (Table 3). Compared with the rice accessions that were homozygous for the Wx^aallele, rice accessions that were homozygous for the Wx^b allele had a significantly lower amylose content, higher eating quality, higher maximum viscosity, and higher stickiness both on the grain surface and of the whole grain. Compared with the rice accessions that were homozygous for the *Wx*^{b} allele, rice accessions that were homozygous for the *wx* allele had significantly lower amylose content, higher eating quality, and lower stickiness on the grain surface. These results indicate that differences in the amylose and amylopectin super long chains content, which are caused by differences in the *Wx* alleles, strongly affect the eating quality of the rice in the overall collection.

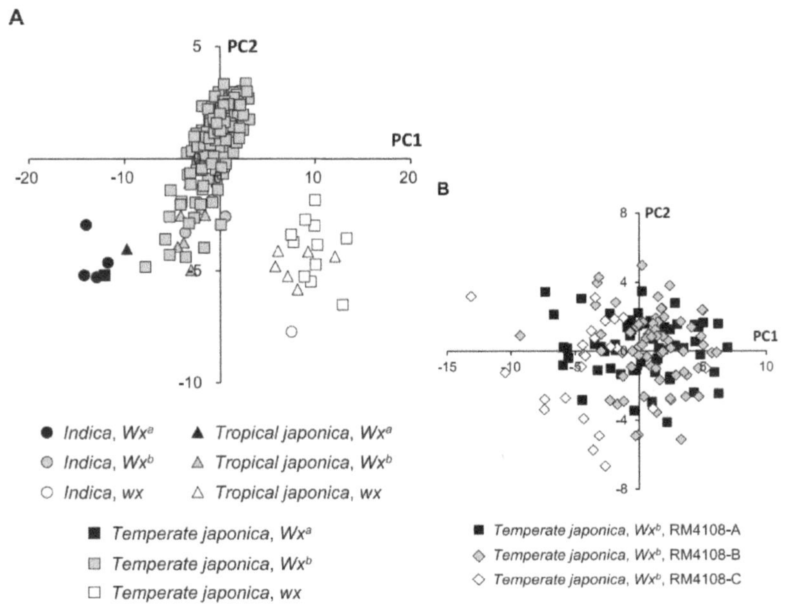

Figure 1: (A) Ordination for variations of the eating-quality traits of 183 Japanese rice accessions consisting of 5 *indica*, 14 *tropical japonica*, and 164 *temperate japonica* rice accessions using the first and second principal components (PC1 and PC2). (B) Ordination for variations of the eating-quality traits of the 146 non-glutinous *temperate japonica* rice accessions for PC1 and PC2

Table 3: Statistical significance of differences among the three *Wx* alleles (*Wx*^{a}, *Wx*^{b}, and *wx*) based on the four strongest principal components (PC1 to PC4) and six of the key eating-quality traits. The *Wx*^{a}, *Wx*^{b}, and *wx* alleles include 17, 160, and 6 rice accessions, respectively

Trait name		*Wx*^{a} allele	*Wx*^{b} allele	*wx* allele
PC1	(–)	9.1 ± 0.5 a	−12.5 ± 0.8 b	−0.5 ± 0.2 c
PC2	(–)	−4.5 ± 0.4 a	−4.5 ± 0.7 a	0.6 ± 0.1 b
PC3	(–)	1.3 ± 0.4 a	4.7 ± 0.6 b	−0.3 ± 0.1 c
PC4	(–)	0.2 ± 0.3 a	−1.3 ± 0.5 b	0.0 ± 0.1 a
Protein content	(%)	6.9 ± 0.3 a	6.0 ± 0.1 a	6.9 ± 0.2 a
Amylose content	(%)	28.7 ± 0.6 a	16.7 ± 0.1 b	2.2 ± 0.3 c
Eating quality	(–)	33.1 ± 3.8 a	73.9 ± 0.7 b	84.1 ± 2.2 c
Maximum viscosity	(RVU)	356.0 ± 15.3 a	448.7 ± 2.9 b	213.6 ± 8.8 c
Stickiness on the grain surface S1	($N/m^2 \times 10^2$)	2.0 ± 1.3 a	13.8 ± 0.3 b	11.6 ± 0.8 c
Stickiness for the whole grain S2	($N/m^2 \times 10^4$)	0.21 ± 0.02 a	0.48 ± 0.00 b	0.49 ± 0.01 b

Values for a trait labeled with different letters differ significantly ($P < 0.05$, Tukey-Kramer HSD test).

PCA for the 146 accessions of non-glutinous temperate japonica rice that was homozygous for the same Wx allele (Wx^b)

PC1 and PC2 accounted for 38.7% and 13.9%, respectively, of the cumulative variance. The contributions of the top six PCs accounted for 80.2% of the total variance. For PC1, the loadings were negative for protein content but positive for eating quality, stickiness, adhered mass, adhesiveness, and the three balance degrees on the grain surface. For PC2, the loadings were negative for brittleness but positive for amylose content, sample thickness, tenderness, pliability, and toughness. For PC3 to PC6, the loadings were largely associated with individual components of the RVA profile and the whole-grain Tensipresser parameters.

To test for the significance of genetic factors other than *Wx* for the eating-quality traits of the non-glutinous*temperate japonica* rice accessions, we focused on SSR marker RM4108, which was located near eating-quality QTLs revealed on the short arm of chromosome 3 by sensory tests. We divided the 146 accessions into three groups (A, B, and C) based on their RM4108 alleles. The group with the RM4108-C allele was somewhat distinct from the groups with the RM4108-A and RM4108-B alleles, although the distributions of the three alleles overlapped somewhat in the ordination (Fig. 1B). The RM4108-A and RM4108-B allele groups consisted mainly of improved accessions of non-glutinous *temperate japonica* rice, whereas the RM4108-C allele group included many landrace accessions (Fig. 2A). For example, the RM4108-A allele group included Koshihikari, Hitomebore, Aichinokaori, Nanatsuboshi, and Itadaki, whereas the RM4108-B allele group included Hinohikari, Nikomaru, Sakihikari, Fusaotome, and Notohikari. The mean PC1 and eating-quality values differed significantly among the three RM4108 allele groups (Fig. 2B, Table 4). Compared with the accessions with the RM4108-C allele, accessions with the RM4108-A and RM4108-B alleles showed significantly lower protein content but significantly higher eating quality and stickiness on the grain surface. There were no significant differences among the allele groups in amylose content, maximum viscosity, or stickiness of the whole grain.

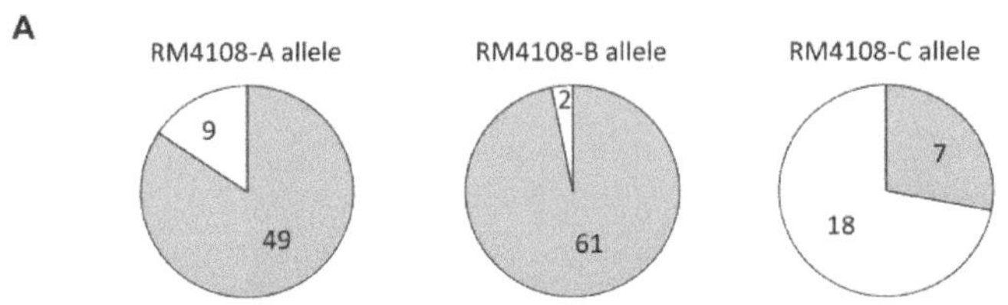

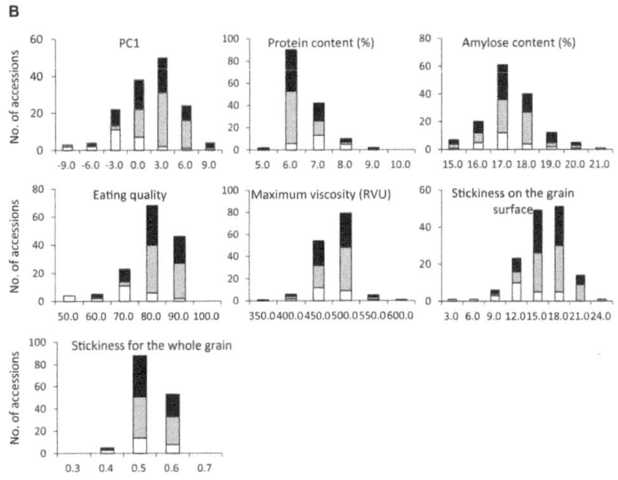

Figure 2: (A) The proportions of landrace and improved rice accessions in the three RM4108 allele groups. Landraces are shown in *white* and improved rice accessions in *gray*. Numbers in each circle indicate the number of accessions in each group. (B) Frequency distributions for principal component 1 (PC1), protein content, amylose content, eating quality, maximum viscosity, and stickiness on the grain surface and of the whole grain for the 146 non-glutinous *temperate japonica* rice accessions. Rice accessions of the RM4108-A, B, and C allele groups are shown in *black*, *gray* and *white*, respectively.

Table 4: Statistical significance of differences among the three RM4108 genotypes (A, B, and C) for the top four principal components (PC1 to PC4) and six key eating-quality traits. Genotypes A, B, and C include 56, 68, and 22 non-glutinous accessions of *temperate japonica* rice, respectively

Trait name		RM4108-A allele	RM4108-B allele	RM4108-C allele
PC1	(–)	0.2 ± 0.4 a	1.0 ± 0.4 b	−3.9 ± 0.5 c
PC2	(–)	0.2 ± 0.3 a	0.1 ± 0.3 a	−1.0 ± 0.4 b
PC3	( )	0.1 ± 0.2 a	0.1 ± 0.2 a	−0.5 ± 0.3 b
PC4	(–)	−0.2 ± 0.2 a	0.1 ± 0.2 a	0.1 ± 0.3 a
Protein content	(%)	5.9 ± 0.1 a	5.8 ± 0.1 a	6.5 ± 0.1 b
Amylose content	(%)	16.8 ± 0.1 a	16.9 ± 0.1 a	16.6 ± 0.2 a
Eating quality	(–)	75.9 ± 1.0 a	77.4 ± 0.9 a	65.3 ± 1.6 b
Maximum viscosity	(RVU)	450.9 ± 4.5 a	452.9 ± 4.2 a	443.0 ± 7.4 a
Stickiness on the grain surface S1	($N/m^2 \times 10^2$)	14.4 ± 0.4 a	14.9 ± 0.4 a	12.1 ± 0.6 b
Stickiness for the whole grain S2	($N/m^2 \times 10^4$)	0.48 ± 0.01 a	0.48 ± 0.01 a	0.48 ± 0.01 a

Values for a trait labeled with different letters differ significantly ($P < 0.05$, Tukey-Kramer HSD test).

DISCUSSION

Researchers have recognized the wide range of eating-quality characteristics in rice germplasm accessions since the early 20th century (Ikeno 1914, Warth and Darabsett 1914). Nakamura *et al.* (2004, 2012b) reported varietal differences of eating quality traits in world-wide rice germplasm accessions including *indica* and *japonica* rice cultivars. Germplasm accessions provide basic knowledge that supports the discovery of genes that control eating quality and their introduction into novel rice cultivars. The present study represents a wide range of eating-quality traits using a large number of Japanese rice germplasm accessions. We evaluated 33 eating-quality traits and heading date in 183 Japanese rice germplasm accessions, and revealed that the difference of amylose content between Wx^a and Wx^b alleles was a primary factors involved in determining the phenotypic differences in eating-quality traits. In the 146 accessions of non-glutinous *temperate japonica* rice that were homozygous for Wx^b allele, the protein content and surface texture of the cooked rice grains were significantly associated with phenotypic differences in eating-quality. The phenotypic variation in eating quality was controlled by one or more QTLs near SSR marker locus, RM4108, on the short arm of chromosome 3. These results suggest that genetic factors other than the*Wx* gene are involved in creating the natural variation of eating-quality traits in the non-glutinous *temperate japonica*rice accessions, and that the genetic factors have been consciously selected for eating quality during rice breeding programs in Japan. Our study confirmed previous QTL studies that eating-quality QTLs have been commonly detected near SSR marker locus, RM4108, in non-glutinous *temperate japonica* rice accessions (Hori and Yano 2013, Kobayashi and Tomita 2008, Takeuchi *et al.* 2008, Wada *et al.* 2008). These previous studies used Koshihikari or one of its progeny as a crossing parent to develop mapping populations. Koshihikari has been a top cultivar in Japan since 1979, and possesses superior eating quality among the Japanese non-glutinous *temperate japonica* cultivars. Koshihikari and other cultivars such as Hitomebore, Itadaki and Nikomaru had high eating quality, and had the A or B alleles at SSR marker RM4108. Eating quality differed significantly among the accession groups identified on the basis of differences at marker RM4108. Therefore, the QTL or QTLs on the short arm of chromosome 3 might be associated mainly with variations of the eating-quality traits in the Japanese non-glutinous *temperate japonica* rice accessions. However, we found a few inconsistencies based on pedigree information of Japanese rice cultivars. For example, Koshihikari (RM4108-A allele) was derived from the cross between Norin 1 (RM4108-A allele) and Norin 22 (RM4108-A allele), but Norin 22 was derived from the cross between Norin 6 (RM41085-B allele) and Norin 8 (RM4108-C allele). These results suggested the need to add genotypes of other DNA markers located on the short arm of chromosome 3.

The surface texture of cooked rice grains is one of important traits that determines the eating quality of non-glutinous *temperate japonica* rice accessions (Okabe 1979, Okadome *et al.* 1999, Suzuki *et al.* 2006). The accumulation of storage proteins in the outer layer of the rice endosperm is one factor that influences the surface texture of cooked rice grains (Okadome 2005). Our results also indicate a substantial effect of surface texture on the protein content and eating quality of rice grains in the non-glutinous *temperate japonica* rice accessions. In this study, accessions with high eating quality had a low protein content and strong stickiness on the surface of the cooked rice grain. These results reflect the preferences of Japanese consumers for rice grain characteristics.

Amylose content has been considered an important factor that affects eating quality in rice (Juliano *et al.* 1993). The 146 non-glutinous *temperate japonica* accessions that were homozygous for the same Wx^b allele showed amylose contents ranging from 14.2% to 20.4%. Varietal difference of the amylose content might be due to alteration of heading date and ambient temperature in the grain filling period among rice cultivars. However, Biselli *et al.* (2014)reported 21 alleles of the *Wx* gene in 125 *japonica* rice accessions with a wide range of amylose content (from <5% to >25%). Therefore, some of the *Wx* gene alleles that we did not investigate in the present study are likely to be associated with differences in amylose contents among rice accessions that were homozygous for the same Wx^ballele. Amylose content in the non-glutinous *temperate japonica* rice accessions would also be controlled by genetic factors other than the *Wx* gene. Recently, a QTL for amylose content has been detected and fine-mapped on chromosomes 2 in populations derived from crosses between Koshihikari and Akihikari (Kobayashi *et al.* 2008). Another QTL for amylose content has been detected and fine-mapped on chromosomes 9 in populations derived from crosses between Itadaki and Kuiku 162 (Ando *et al.* 2010, Takemoto-Kuno *et al.* 2015). Koshihikari and Kuiku 162 alleles at these QTLs showed a relatively small decrease (2% to 3%) in the amylose content. These QTLs may explain some of the variation of amylose content among the non-glutinous *temperate japonica* rice accessions in the present study.

Several previous studies evaluated the phenotypic variations of grain characteristics associated with eating quality, and elucidated genetic variation of these traits by means of PCA in several dozen non-glutinous *japonica* accessions. Kamara *et al.* (2010) evaluated the free amino acid contents in 49 accessions. Matsuzaki *et al.* (1992) carried out PCA for the contents of amylose, amino acids, and essential elements (N, P, K, Mg, and Ca) in 18 accessions. Kim *et al.* (2010) investigated the flavonoid and carotenoid contents in five accessions. Mabashi *et al.* (2010), Nakamura *et al.* (2012b) and Tsujii and Takano (2015) measured activities of endogenous enzymes of rice grains in five, 42 and nine accessions, respectively. In the present study, PCA let us summarize the relationships among the many eating-quality traits. Our results for 33 eating-quality traits in 146 accessions will complement these previous studies and enhance knowledge of the natural variation in eating-quality traits in non-glutinous *japonica* accessions.

We found that accessions with high eating quality tended to have high maximum viscosity but low levels of the other RVA traits. In contrast, eating quality was negatively correlated with amylose content. Previous studies reported that the amylase content had a significant negative correlation with maximum viscosity (and, thus, a significant positive correlation with eating quality) and significant positive correlations with minimum viscosity, final viscosity, and setback (Nakamura *et al.* 2012a, Tong *et al.* 2014). In the Tensipresser evaluation scores, accessions with high eating quality tended toward low hardness, high stickiness, high adhered mass and adhesiveness, and high balance degrees of both the grain surface and the whole cooked grain. These observations agree with those in previous studies (Kobayashi *et al.* 2008, Okadome 2005, Wada *et al.* 2013). Therefore, the RVA and Tensipresser evaluations can provide a good estimate of the eating quality of rice germplasm accessions.

QTL detection and gene cloning are necessary to dissect the genetic variation of eating-quality traits. Recent advances in molecular genomics technologies that have permitted genome-wide SNP detection and large-scale SNP genotyping allowed the detection of QTLs involved in various agronomic traits, even in genetically close groups of accessions such as non-glutinous *temperate japonica* rice (Hori *et al.* 2012). However, these QTL studies examined too few eating-quality traits. Genome-wide association studies (GWAS) make it possible to directly detect QTLs in germplasms without the need to develop genetically distinct populations, and has already been successfully applied in analyses of heading date and grain yield, two agronomically important traits in rice (Huang *et al.* 2012, Zhao *et al.*2011). GWAS would also be a suitable way to detect QTLs for eating-quality traits in the non-glutinous *temperate japonica* rice accessions because of the relatively simple population structure within this group (Yamasaki and Ideta 2013, Yonemaru *et al.* 2014). Additional research will be needed to enhance the identification of genes that control eating-quality traits in the non-glutinous *temperate japonica* accessions. These future studies will support the development of novel rice cultivars with high eating quality.

ACKNOWLEDGMENTS

We are grateful to the technical staff of the Field Management Division at the NIAS for their technical assistance and management of the rice fields. This work was supported by grants from the Ministry of Agriculture, Forestry and Fisheries of Japan (Genomics for Agricultural Innovation QTL4010; Research Project for Genomics-based Technology for Agricultural Improvement, RBS2011).

REFERENCES

1. American Association of Cereal Chemists Approved Methods Committee (2000) Approved methods of the American Association of Cereal Chemists, 10th edn. AACC, St. Paul, pp. 1–2.
2. Anacleto, R., R.P. Cuevas, R. Jimenez, C. Llorente, E. Nissila, R. Henry and N. Sreenivasulu (2015) Prospects of breeding high-quality rice using post-genomic tools. Theor. Appl. Genet. 128: 1449–1466.
3. Ando, I., H. Sato, N. Aoki, Y. Suzuki, H. Hirabayashi, M. Kuroki, H. Shimizu, T. Ando and Y. Takeuchi (2010) Genetic analysis of the low-amylose characteristics of rice cultivars Oborozuki and Hokkai-PL9. Breed. Sci. 60: 187–194.
4. Bergman, C.J., K.R. Bhattacharya and K. Ohtsubo (2004) Rice end-use quality analysis. *In*: Champagne, E.T. (ed.) Rice: Chemistry and Technology, 3rd edn., AACCI, MN, USA, pp. 415–472.
5. Biselli, C., D. Cavalluzzo, R. Perrini, A. Gianinetti, P. Bagnaresi, S. Urso, G. Orasen, F. Desiderio, E. Lupotto, L. Cattivelli et al. (2014) Improvement of marker-based predictability of Apparent Amylose Content in *japonica* rice through *GBSSI* allele mining. Rice 7: 1.
6. Cai, X.L., Z.Y. Wang, Y.Y. Xing, J.L. Zhang and M.M. Hong (1998) Aberrant splicing of intron 1 leads to the heterogeneous 5′ UTR and decreased expression of *waxy* gene in rice cultivars of intermediate amylose content. Plant J. 14: 459–465.
7. Ebana, K., Y. Kojima, S. Fukuoka, T. Nagamine and M. Kawase (2008) Development of mini core collection of Japanese rice landrace. Breed. Sci. 58: 281–291.
8. Goto, H., N. Asanome, K. Suzuki, T. Sano, H. Saito, Y. Abe, M. Chuba and T. Nishio (2014) Objective evaluation of whiteness of cooked rice and rice cakes using a portable spectrophotometer. Breed. Sci. 63: 489–494.
9. Hirano, H.Y., M. Eiguchi and Y. Sano (1998) A single base change altered the regulation of the *Waxy* gene at the posttranscriptional level during the domestication of rice. Mol. Biol. Evol. 15: 978–987.
10. Hori, K., T. Kataoka, K. Miura, M. Yamaguchi, N. Saka, T. Nakahara, Y. Sunohara, K. Ebana and M. Yano (2012) Variation in heading date conceals quantitative trait loci for other traits of importance in breeding selection of rice. Breed. Sci. 62: 223–234.
11. Hori, K. and M. Yano (2013) Genetic improvement of grain quality in *Japonica* rice. *In*: Varshney, R.K. and R. Yuberosa (eds.) Translational Genomics for Crop Breeding Vol. II Abiotic Stress, Yield and Quality, Wiley Blackwell, Iowa, USA, pp. 143–160.
12. Hsu, Y.C., M.C. Tseng, Y.P. Wu, M.Y. Lin, F.J. Wei, K.K. Hwu, Y.I. Hsing and Y.R. Lin (2014) Genetic factors responsible for eating and cooking qualities of rice grains in a recombinant inbred population of an inter-subspecific cross. Mol. Breed. 34: 655–673.
13. Huang, X., Y. Zhao, X. Wei, C. Li, A. Wang, Q. Zhao, W. Li, Y. Guo, L. Deng, C. Zhu et al. (2012) Genome-wide association study of flowering time and grain yield traits in a worldwide collection of rice germplasm. Nat. Genet.44: 32–39.
14. Ikeno, S. (1914) Ueber die Bestaubung und die Bastardierung von Reis. Zeit. f. Pflanzenzucht. 2: 495–503.
15. Issiki, M., K. Morino, M. Nakajima, R.J. Okagaki, S.R. Wessler, T. Izawa and K. Shimamoto (1998) A naturally occurring functional allele of the rice *waxy* locus has a GT to TT mutation at the 5′ splice site of the first intron. Plant J. 15: 133–138.
16. Juliano, B.O., G.M. Bautista, J.C. Lugay and A.C. Reyes (1964) Rice quality, studies on physicochemical properties of rice. J. Agr. Food Chem. 12: 131–138.
17. Juliano, B.O., C.M. Perez and F. Cuevas-Perez (1993) Screening for stable high head rice yields in rough rice. Cereal Chem. J. 70: 650–655.
18. Kamara, J.S., S. Konishi, T. Sasanuma and T. Abe (2010) Variation in free amino acid profile among some rice (*Oryza sativa* L.) cultivars. Breed. Sci. 60: 46–54.
19. Kim, J.K., S.Y. Lee, S.M. Chu, S.H. Lim, S.C. Suh, Y.T. Lee, H.S. Cho and S.H. Ha (2010) Variation and correlation analysis of flavonoids and carotenoids in Korean pigmented rice (*Oryza sativa* L.) cultivars. J. Agr. Food Chem. 58: 12804–12809.
20. Kobayashi, A. and K. Tomita (2008) QTL detection for stickiness of cooked rice using recombinant inbred lines derived from crosses between *japonica* rice cultivars. Breed. Sci. 58: 419–426.
21. Kobayashi, A., K. Tomita, F. Yu, Y. Takeuchi and M. Yano (2008) Verification of quantitative trait locus for stickiness of cooked rice and amylose content by developing near-isogenic lines. Breed. Sci. 58: 235–242.
22. Kwon, S.W., Y.C. Cho, J.H. Lee, J.P. Suh, J.J. Kim, M.K. Kim, I.S. Choi, H.G. Hwang, H.J. Koh and Y.G. Kim (2011) Identification of quantitative trait loci associated with rice eating-quality traits using a population of recombinant inbred lines derived from a cross between two temperate *japonica* cultivars. Mol. Cells 31: 437–445.
23. Mabashi, Y., Y. Miwa, T. Ookura and M. Kasai (2010) Role of endogenous enzymes in milled rice of various cultivars in the accumulation of chemical components in rice grain during cooking. J. Cookery Sci. Jpn. 4: 228–236.
24. Matsue, Y. (1992) On a sensory test of cooked rice in which 10 cultivars were evaluated by 13 panel members. J. Home Econ. Jpn. 43: 1027–1032.
25. Matsuzaki, A., T. Takano, S. Sakamoto and T. Kuboyama (1992) Relation between eating quality and chemical

components in milled rice and amino acid contents in cooked rice. Jpn. J. Crop Sci. 61: 561–567.

26. Mikami, T. (2009) Development of evaluation systems for rice taste quality. Jpn. J. Food Eng. 4: 191–197.

27. Monaco, M.K., J. Stein, S. Naithani, S. Wei, P. Dharmawardhana, S. Kumari, V. Amarasinghe, K. Youens-Clark, J. Thomason, J. Preece et al. (2014) Gramene 2013: comparative plant genomics resources. Nucleic Acids Res. 42: D1193–D1199.

28. Nakamura, S., H. Okadome, K. Yoza, K. Haraguchi, T. Okunishi, K. Suzuki, H. Sato and K. Ohtsubo (2004) Differentiation and search for palatability-factors of world-wide rice grains by PCR method. J. Jpn. Soc. Food Sci.78: 764–779.

29. Nakamura, S., D. Suzuki, R. Kitadume and K. Ohtsubo (2012a) Quality evaluation of rice crackers based on physicochemical measurements. Biosci. Biotechnol. Biochem. 76: 794–804.

30. Nakamura, S., K. Machida and K. Ohtsubo (2012b) Search for cell-wall-degrading enzymes of world-wide rice grains by PCR and their effects on the palatability of rice. Biosci. Biotechnol. Biochem. 76: 1645–1654.

31. Okabe, M. (1979) Texture measurement of cooked rice and its relationship to the eating quality. J. Texture Stud. 10: 131–152.

32. Okadome, H., H. Toyoshima and K. Ohtsubo (1995) Many-sided evaluation of physical properties of cooked rice grains with a single apparatus. J. Jpn. Soc. Food Sci. 43: 1004–1011.

33. Okadome, H., H. Toyoshima and K. Ohtsubo (1999) Multiple measurements of physical properties of individual cooked rice grains with a single apparatus. Cereal Chem. 76: 855–860.

34. Okadome, H. (2005) Application of instrument-based multiple texture measurement of cooked milled-rice grains to rice quality evaluation. JARQ 39: 261–268.

35. Sano, Y. (1984) Differential regulation of waxy gene expression in rice endosperm. Theor. Appl. Genet. 68: 467–473.

36. Sano, Y., M. Katsumata and K. Okuno (1986) Genetic studies of speciation in cultivated rice. 5. Inter- and intraspecific differentiation in the *waxy* gene expression of rice. Euphytica 35: 1–9.

37. Suzuki, K., H. Okadome, S. Nakamura and K. Ohtsubo (2006) Quality evaluation of various "new characteristic rice" varieties based on physicochemical measurements. J. Jpn. Soc. Food Sci. 53: 287–295.

38. Takemoto-Kuno, Y., H. Mitsueda, K. Suzuki, H. Hirabayashi, O. Ideta, N. Aoki, T. Umemoto, T. Ishii, I. Ando, H. Kato et al. (2015) *qAC2*, a novel QTL that interacts with *Wx* and controls the low amylose content in rice (*Oryza sativa* L.). Theor. Appl. Genet. 128: 563–573.

39. Takeuchi, Y., Y. Nonoue, T. Ebitani, K. Suzuki, N. Aoki, H. Sato, O. Ideta, H. Hirabayashi, M. Hirayama, H. Ohta et al. (2007) QTL detection for eating quality including glossiness, stickiness, taste and hardness of cooked rice. Breed. Sci. 57: 231–242.

40. Takeuchi, Y., K. Hori, K. Suzuki, Y. Nonoue, Y. Takemoto-Kuno, H. Maeda, H. Sato, H. Hirabayashi, H. Ohta, T. Ishii et al. (2008) Major QTLs for eating quality of an elite Japanese rice cultivar, Koshihikari, on the short arm of chromosome 3. Breed. Sci. 58: 437–445.

41. Takeyama, S., R. Toyama, Y. Fujiwara and Y. Arakawa (1998) Physical properties of steamed cakes made of miscellaneous cereals. Res. Bull. Iwate Ind. Res. Inst. 5: 28.

42. Tan, Y.F., J.X. Li, S.B. Yu, Y.Z. Xing, C.G. Xu and Q. Zhang (1999) The three important traits for cooking and eating quality of rice grains are controlled by a single locus in an elite rice hybrid, Shanyou 63. Theor. Appl. Genet. 99: 642–648.

43. Tong, C., Y. Chen, F. Tang, F. Xu, Y. Huang, H. Chen and J. Bao (2014) Genetic diversity of amylose content and RVA pasting parameters in 20 rice accessions grown in Hainan, China. Food Chem. 161: 239–245.

44. Toyoshima, H., H. Okadome, K. Ohtsubo, M. Suto, N. Horisue, O. Inatsu, A. Narizuka, M. Aizaki, T. Okawa, N. Inouchi et al. (1997) Cooperative test on the small-scale rapid method for the gelatinization properties test of rice flours with a rapid-visco-analyser (RVA). J. Jpn. Soc. Food Sci. 44: 579–584.

45. Tsujii, Y. and K. Takano (2015) Effects of cultivar and cultivation region on endosperm enzyme activity involved in the palatability of cooked rice. J. Jpn. Soc. Food Sci. 62: 34–40.

46. Wada, T., T. Ogata, M. Tsubone, Y. Uchimura and Y. Matsue (2008) Mapping of QTLs for eating quality and physicochemical properties of the *japonica* rice 'Koshihikari'. Breed. Sci. 58: 427–435.

47. Wada, T., H. Yasui, T. Inoue, M. Tsubone, T. Ogata, K. Doi, A. Yoshimura and Y. Matsue (2013) Validation of QTLs for eating quality of *japonica* rice 'Koshihikari' using backcross inbred lines. Plant Prod. Sci. 16: 131–140.

48. Wang, Z.Y., F.Q. Zheng, G.Z. Shen, J.P. Gao, D.P. Snustad, M.G. Li, J.L. Zhang and M.M. Hong (1995) The amylose content in rice endosperm is related to the post-transcriptional regulation of the *waxy* gene. Plant J. 7: 613–622.

49. Warth, F.J. and D.B. Darabsett (1914) Disintegration of rice grains by means of alkali. Bull. Agr. Res. Inst. 38: 1–9.

50. Yamamoto, T., N. Horisue and R. Ikeda (1996) Rice breeding manual. Yokendo Ltd., Tokyo, pp. 74–124.

51. Yamanaka, S., I. Nakamura, K.N. Watanabe and Y.I. Sato (2004) Identification of SNPs in the *waxy* gene among glutinous rice cultivars and their evolutionary significance during the domestication process of rice. Theor. Appl. Genet. 108: 1200–1204.

52. Yamasaki, M. and O. Ideta (2013) Population structure in Japanese rice population. Breed. Sci. 63: 49–57.

53. Yonemaru, J., T. Yamamoto, S. Fukuoka, Y. Uga, K. Hori and M. Yano (2010) Q-TARO: QTL Annotation Rice Online Database. Rice 3: 194–203.

54. Yonemaru, J., R. Mizobuchi, H. Kato, T. Yamamoto, E.

Yamamoto, K. Matsubara, H. Hirabayashi, Y. Takeuchi, H. Tsunematsu, T. Ishii et al. (2014) Genomic regions involved in yield potential detected by genome-wide association analysis in Japanese high-yielding rice cultivars. BMC Genomics 15: 346.

55. Zhao, K., C.W. Tung, G.C. Eizenga, M.H. Wright, M.L. Ali, A.H. Price, G.J. Norton, M.R. Islam, A. Reynolds, J. Mezey et al. (2011) Genome-wide association mapping reveals a rich genetic architecture of complex traits in *Oryza sativa*. Nat. Commun. 2: 467.

CHAPTER 6

ASSOCIATION OF SNP HAPLOTYPES OF HKT FAMILY GENES WITH SALT TOLERANCE IN INDIAN WILD RICE GERMPLASM

Mishra S, Singh B, Panda K, Singh BP, Singh N, Misra P, Rai V,and Singh NK

Keywords: Allele mining, HKT, Association, Salt stress, NaCl, Wild rice

BACKGROUND

Rice (*Oryza sativa* L.) is cultivated around the world and consumed by more than 50 % of the global human population (Mohanty 2013). The current status of rice production is 495.63 million tonnes (MT) which increased by only 9 MT in the preceding 4 year block (2011–2014) as compared to 80 MT increase in two such 4 year blocks during 2004–2011 (http://faostat3.fao.org/browse/Q/*/E). In contrast, the global population is projected to increase by 25 % to 9.2 billion by 2050 (Schroeder et al. 2013). Rice production must increase correspondingly by 70 % to fulfil the growing demand (FAOSTAT 2009). Fifteen percent of the land currently used for agricultural practices is at saturation point to maintain environmentally sustainable production, therefore innovations are needed to enhance crop productivity to meet the projected demand (Rockström et al. 2009). However, multiple abiotic stresses specifically salinity changes the soil texture and creates unfavourable conditions for rice production by severe inhibition of plant growth and development (Miller and Donahue 1995). Globally 45 Mha of irrigated and 32 Mha of rain fed agriculture are affected by salinity (http://www.fao.org/soils-portal/en/). Furthermore, irrigation with brackish water, tidal waves and tsunami continue to increase the soil salinity (Schroeder et al. 2013). According to one estimate up to 50 % of the present arable land may be affected by salinity by 2050 (Wang et al. 2003). Therefore, we need to explore natural genetic resources to find novel genes and alleles that can help withstand high salt concentration and maintain crop productivity.

Salinity is characterised by presence of exchangeable sodium ions (Na^+) which imposes both ion toxicity and osmotic stress on the plant and alters the physiological status and ionic homeostasis of the cell. Crop plants are vulnerable to salinity and further their level of sensitivity and mechanism of tolerance depends on the growth stage of the plant (Hossain et al. 2015). Rice is most sensitive at the seedling stage and beyond 3 dSm^{-1} of electrical conductivity, seedlings are considered under stress and yield loss due to salinity is measured at twelve percent reduction per dSm^{-1} above threshold level (Lutts et al. 1995; Maas and Hoffman 1977). However, in most of the modern high yielding rice varieties, 50 % yield reduction occurs at 6 dSm^{-1} and they become totally unproductive beyond 12 dSm^{-1} (Linh et al. 2012). A limited number of large scale screening of germplasm for salinity stress has been conducted (Platten et al. 2013). Most of the genetic studies have focused on *Saltol* QTL derived from Indian rice landrace Pokkali which provides seedling stage salt tolerance (Thomson et al. 2010). There is very little work on the genetics of reproductive stage salt tolerance in rice (Mohammadi et al. 2013). The present understanding of salt tolerance mechanisms has facilitated exploitation of spatially located membrane transporters (Zhu 2001; Munns and Tester 2008). The known mechanisms include sequestration of ions in vacuole and exclusion of ions from root and leaves (Munns and Tester 2008). Thus, allelic variations in the sequence of ion transporter genes are likely to play a significant role in providing effective tolerance to salt stress.

In the present study we selected eight different transporters of the *HKT* gene family. These genes are further classified into two subfamilies based on their amino acid sequence similarity and differences in their Na^+ and K^+selectivity (Horie et al. 2001), though product of each of these is a transmembrane

Citation: Mishra S, Singh B, Panda K, et al. Association of SNP Haplotypes of HKT Family Genes with Salt Tolerance in Indian Wild Rice Germplasm. Rice. 2016;9:15. doi:10.1186/s12284-016-0083-8.

protein (Horie et al. 2009; Mäser et al. 2001; Platten et al. 2006). Functionally, products of subfamily 1 [HKT1;1, HKT1;2, HKT1;3, HKT1;4 and HKT1;5] are Na+specific transporters and have S-G-G-G signature while products of subfamily 2 [HKT2;1, HKT2;2, HKT2;3 and HKT2;4] are Na^+-K^+ co-transporters or Na^+-K^+ uniporters with G-G-G-G signature (Jabnoune et al.2009; Mäser et al. 2002). Oomen et al. 2012 has reported a hybrid of HKT2;1 and HKT2;2 gene as HKT2;2/1 from Nona Bokra which has strong permeability to Na^+ and K^+ even at high external Na^+ concentrations. Spatial localization and differential expression of these genes further enhances their importance. Horie et al. (2001) reported two isoforms of HKT transporters, a Na+transporter OsHKT1 and a Na^+- K^+-coupled transporter OsHKT2, which may act harmoniously in the salt tolerant indica rice. Further studies reported that OsHKT1;4 is expressed around xylem in the leaf sheath while OsHKT1;5 is expressed around the root xylem (Cotsaftis et al. 2011). Similarly, OsHKT2;1 and OsHKT2;4 are expressed in the outer part of the root and in the root hairs and may provide an entry point for Na^+ into plant roots from the soil (Lan et al. 2010; Schachtman and Schroeder 1994). HKT2;1 expression is significantly upregulated in the root cortex under K^+ starvation and high Na^+ concentration (Almeida et al. 2013; Horie et al. 2001). Further, it was proven that the Na^+ influx into K^+-starved rice roots was primarily OsHKT2;1-dependent while at concentration >30 mM NaCl, it was permeable to Na^+ only (Jabnoune et al. 2009). Another class 2 transporter gene *OsHKT2;4* shows 93 % similarity with *OsHKT2;3* and mediates K^+ transport independent of Na^+ concentration (Horie et al. 2011).

The level of salt tolerance provided by known transporter genes in the modern rice cultivars is insufficient and hence new genes or novel allelic variants of these transporters are needed for enhanced salinity tolerance. Crop domestication has selected only a few agronomically desirable alleles and left behind vast pool of genetic diversity in the wild progenitor species due to domestication bottleneck (Tanksley and McCouch 1997). Potential implication of germplasm and crop wild relatives under extreme environment conditions has been reviewed (Mickelbart et al.2015). Allelic variant of members of HKT transporters such TmHKT1;5-A (Munns et al. 2012), TaKHT1;5-D (Yang et al. 2014) has been introgressed from wild relatives that led to increase in yield of the plant. Allele mining identifies superior alleles from related genotypes that may have been the effect of mutations in the process of evolution. The superior alleles can be used to develop allele specific markers and use them in marker assisted selection and also in tracing the evolution of alleles. Sequencing based allele mining and association analysis is an effective strategy to unravel the hidden potential of wild rice germplasm. Allele mining has been used across the crop species and novel alleles have been identified for abiotic stress tolerance genes in rice (Latha et al. 2004; Negrão et al. 2013; Platten et al. 2013; Singh et al. 2015a), maize (Yu et al. 2010) and barley (Cseri et al. 2011). However, a collection of untapped germplasm is required to mine novel desirable alleles and identify nucleotide sequence variations associated with these alleles (Kumar et al. 2010). India has a wealth of untapped wild rice germplasm that requires hasty expeditions to collect and exploit this fast depleting genetic resource (Singh et al. 2013). The genes already exploited from Indian wild rice include grassy stunt virus resistance from *Oryza nivara* (Khush and Ling 1974), *Bph 19(t)* from *Oryza rufipogon* (Li et al. 2010), *Xa38* from *Oryza nivara* (Bhasin et al. 2012), salinity tolerance genes, inositol methyl transferase (Sengupta et al. 2008) and L-myo-inositol 1-phosphate synthase from wild rice *Porteresia coarctata* (Das-Chatterjee et al. 2006). The male sterility (MS) gene from *O. rufipogon* was introduced into the cultivated rice, leading to development of high yielding hybrid rice (Yuan et al. 1993). The beneficial alleles derived from wild relatives of rice have been transferred into elite genetic backgrounds leading to enhanced trait performance in rice (McCouch et al. 2007; Xiao et al. 1996; Xiao et al. 1998).

In the present study, 299 accessions of rice comprising *O. rufipogon* and *O. nivara* were screened for their level of salt tolerance. A candidate gene based allele mining was used to find natural allelic variants of agronomically important HKT family genes. Genes were re-sequenced with their promoters from 103 accessions of wild and cultivated rice varieties and their nucleotide and haplotype diversity was analysed. SNP based association analysis was done to link traits for salt tolerance e.g., SPAD for chlorophyll, Na^+ and K^+ concentration in the shoot and root of rice. Further, merit of *HKT1;5* and *HKT2;3* were discussed for their potentiality in salt tolerance of rice.

RESULTS

Phenotypic Screening of the Wild Rice Germplasm for Salt Tolerance

Based on standard evaluation system (SES) of IRRI (Gregorio et al. 1997), 299 wild rice accessions were screened and classified according to their level of salt tolerance (Fig. 1). Two accessions NKSWR173 and NKSWR202 were found highly tolerant with SES score 1 as compared to the tolerant check FL478 with SES score 3 after 15 days of salt stress, 30 accessions were found tolerant at 10 days of salt stress, of which 14 accessions maintained the tolerance comparable to FL478 till fifteenth day. Total 84 accessions were found moderately tolerant, 115 sensitive and 67 highly sensitive at 10 days of salt treatment. However, at fifteenth day only 28 accessions were moderately tolerant, while 69 accessions became sensitive and 186 highly sensitive. One accession, *O. nivara* 336676 displayed interesting result showing high tolerance up to tenth day and thereafter it became highly sensitive and died within the next 2 days in all three years of the experiment. After this primary screening, 45 selected accessions, including all tolerant lines and representative lines from the other sensitive classes, were further evaluated for salt tolerance parameters such as biomass, chlorophyll content (using SPAD meter) and Na+and K+ concentrations in replicated tests. The Na^+/K^+ ratio in shoot under stress ranged from 0.8 to 2.1 with the maximum value in *O. rufipogon* 336679 and the minimum in CSR27. The lowest Na^+/K^+ ratio (1.3) in the root was observed in a tolerant line NKSWR232 and the highest ratio (3.3) was in NKSWR144. FL478 had highest chlorophyll content while VSR156 and NKS074 had no chlorophyll as the plants died.

Figure 1: Response of wild rice accessions after 15 days of 150 mM of salt stress in hydroponics. **a** Border; **b** NKSWR149; **c**NKSWR112; **d** NKSWR104; **e** NKSWR092; **f** NKSWR119; **g** NKSWR085; **h** NKSWR115; **i** NKSWR132; **j** NKSWR097; **k**Oryza nivara330646; **l** NKSWR143; **m** NKSWR101; **n** NKSWR079; **o** FL478; **p** VSR156; **q** Border (NKSWR143 is tolerant other accessions are sensitive, FL478- tolerant check, VSR156- sensitive check)

Nucleotide Diversity, Haplotype Diversity and Haplotype Networks of the *HKT* Genes

Eight HKT transporter family genes along with their promoters were re-sequenced from 103 accessions of wild rice and cultivated rice varieties and their nucleotide and haplotype diversity was analysed (Table 1). Nucleotide diversity (π) was the highest (0.00169) for *HKT2;3* and the lowest (0.00012) for *HKT1;1* gene. Maximum haplotype diversity (HD) of 0.891 was found in *HKT2;3* gene followed by *HKT1;4* (0.692) and the lowest in *HKT1;1* (0.084) gene (Table 1). Haplotype analysis showed the maximum 23 haplotypes each for the *HKT1;5* and *HKT2;3* genes with 45 and 28 SNP sites, respectively. The *HKT2;4* gene showed the least number of 5 haplotypes. The Tajima's D test, Fu and Li's D* test and Fu and Li's F* statistics were performed to distinguish between random versus non-random mutations. All the test values were analysed for entire gene, as well as separately for the coding and non-coding regions. Except for the coding region of all accessions, which had positive values of Tajima's D test, all other regions of the genes among each class individually and in group deviated significantly from neutrality and Tajima's D test values were negative.

Table 1: Nucleotide and haplotype diversity and tests of neutrality for eight high affinity potassium transporter (HKT) family genes from 95 Indian wild rice accessions and 8 cultivated varieties

Candidate gene (Locus ID)	Region	S	π	θω	H	HD	D	D*	F*
HKT1;1	Coding	8	0.00013	0.00084	10	0.217	−2.10133	−3.97252	−3.94393
(LOC_Os04g51820)	Non-coding	3	0.0001	0.00059	4	0.076	−1.48732	−2.01292	−2.16727
	All	11	0.00012	0.00076	13	0.084	−2.22874	−4.21517	−4.17222
HKT1;2	Coding	0	0	0	1	0	0	0	0
	Non-coding	20	0.00027	0.00135	12	0.387	−2.30631	−4.55086	−4.42942
Chr4:30548314..30545885	All	20	0.00026	0.00131	12	0.387	−2.30631	−4.55086	−4.42942
HKT1;3	Coding	8	0.00019	0.00108	9	0.25	−1.99519	−2.80081	−2.99198
(LOC_Os02g07830)	Non-coding	6	0.00067	0.00137	4	0.164	−1.14077	0.15513	−0.32717
	All	14	0.00037	0.00118	10	0.29	−1.88141	−1.8801	−2.24114
HKT1;4	Coding	2	0.00031	0.00091	3	0.128	−1.01249	−1.0804	−1.23948
(LOC_Os04g51830)	Non-coding	42	0.00059	0.00191	19	0.69	−2.1536	3.07065	−3.2462
	All	44	0.00057	0.00182	20	0.692	−2.15775	−3.12193	−3.27713
HKT1;5	Coding	8	0.00025	0.00183	8	0.166	−2.08122	−3.61227	−3.65909
(LOC_Os01g20160)	Non-coding	37	0.0012	0.00322	18	0.571	−1.94371	−2.70539	−2.87895
	All	45	0.00094	0.00284	23	0.631	−2.10258	−3.40294	−3.44926
HKT2;1	Coding	5	0.00015	0.00073	5	0.18	−1.66589	−3.23816	−3.20982
(LOC_Os06g48810)	Non-coding	46	0.00042	0.00241	17	0.333	−2.60346	−4.60994	−4.55342
	All	51	0.00035	0.00196	18	0.421	−2.61003	−4.91693	−4.76765
HKT2;3	Coding	24	0.00238	0.00353	23	0.891	−0.96284	−1.72038	−1.71118
(LOC_Os01g34850)	Non-coding	4	0.00013	0.00132	3	0.038	−1.77201	−3.90552	−3.78567
	All	28	0.00169	0.00284	23	0.891	−1.22676	−2.67757	−2.5308
HKT2;4	Coding	0	0	0	1	0	0	0	0
(LOC_Os06g48800)	Non-coding	4	0.0012	0.00448	5	0.198	−1.44156	−1.48409	−1.72999
	All	4	0.0012	0.00448	5	0.198	−1.44156	−1.48409	−1.72999

(*S* SNP/Indel sites, *π* nucleotide diversity, θω nucleotide diversity with Watterson's estimator, *H* number of haplotypes, *HD* haplotype diversity, *D* Tajima's D statistic, *D** Fu and Li's D* statistic, *F**: Fu and Li's F* statistic.)

Haplotype analysis of *HKT1;1* gene showed 13 haplotypes with haplotype number H12 having a single highly tolerant accession and a distinguishing guanine residue instead of adenine in the first intron. *HKT1;2* gene has diversified to 13 haplotypes, *HKT1;3* into 10 haplotypes and *HKT1;4* to 20 haplotypes. Network analysis of *HKT1;5*gene showed that out of the total 23 haplotypes of this gene two haplotypes, H5 and H22 included all the tolerant accessions. Haplotype H5 had 13 accessions, including tolerant cultivated varieties FL478 and CSR27 along with sensitive variety VSR156 and five sensitive accessions from Gujrat. A unique haplotype H22 was represented by salt tolerant cultivated rice variety CSR11, which also matched with the published sequence of tolerant Indian rice variety Nona Bokra (Ren et al. 2005). The members of HKT sub-family 2 genes, *HKT2;1* and *HKT2;4* showed 18 and 5 haplotypes, respectively. Further, haplotype analysis of *HKT2;3* gene showed that its haplotype H1 included four tolerant wild rice accessions along with the salt tolerant check varieties FL478 and CSR27. However, another salt tolerant variety CSR11 possessing haplotype H10 grouped separately along with tolerant *O. nivara* accession 330641. A phenotypically unique accession, showing tolerance only up to 10 days of stress followed by extreme sensitivity, had a unique haplotype H5 of the *HKT2;3* gene (Fig. 2).

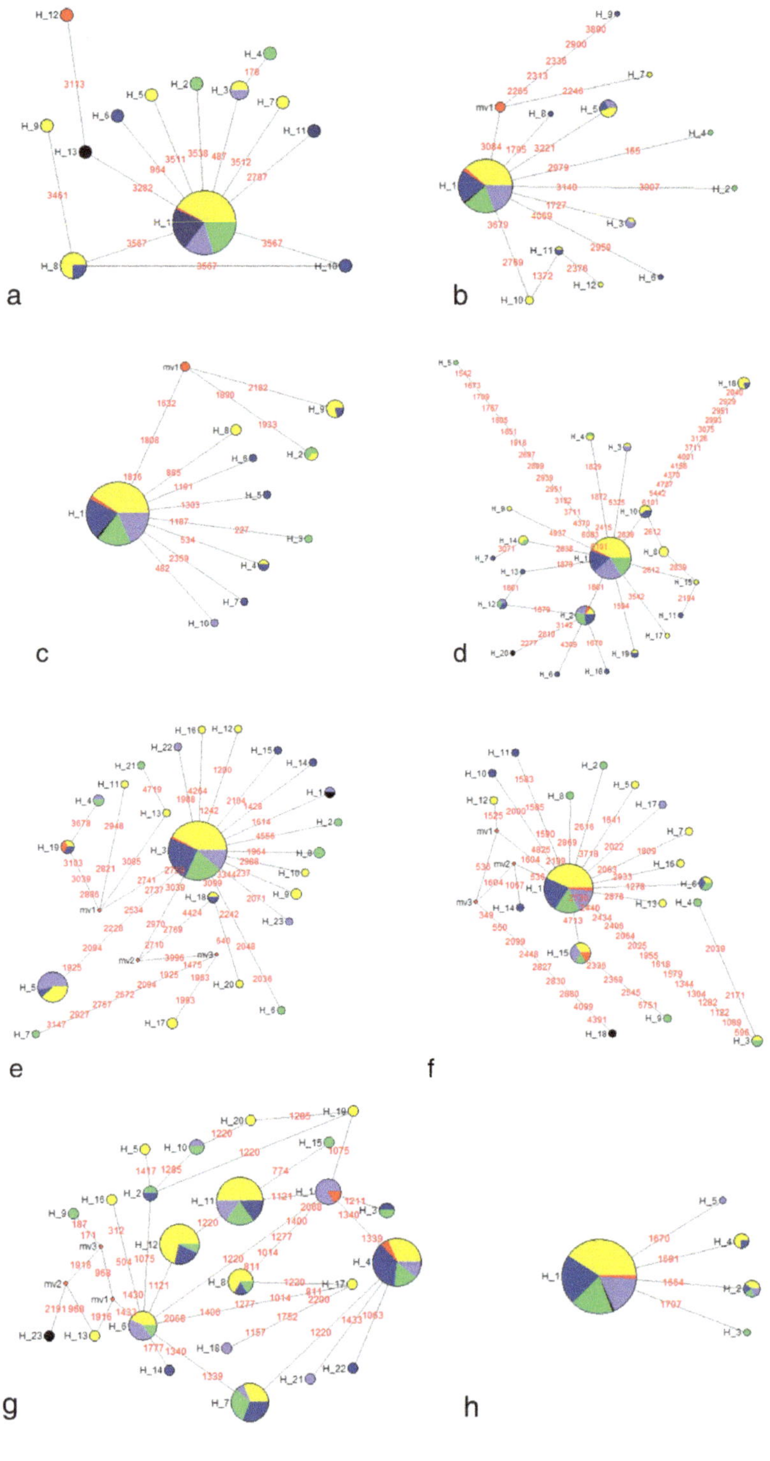

Figure 2: Haplotype networks of eight HKT transporter genes including (**a**) *HKT1;1* (**b**) *HKT1;2* (**c**) *HKT1;3* (**d**) *HKT1;4* (**e**) *HKT1;5*(**f**) *HKT2;1* (**g**) *HKT2;3* (**h**) *HKT2;4*. Each circle represents a haplotype and size of each circle is proportional to haplotype (allele) frequency. Colour coding represents phenotype class based on standard evaluation system (SES) score. (Red-highly tolerant, Purple-tolerant, Green moderately tolerant, Blue-Sensitive, Yellow-highly sensitive)

Evolution of the HKT Family Genes

A phylogenetic tree was constructed using nucleotide sequences of the eight HKT family genes using NJ method. Sorghum *HKT* gene taken from sorghum gene database was included as outgroup. As expected the tree showed total eight clades representing the individual HKT genes that clustered into two major groups. The phylogenetic tree revealed a somewhat different relationship among the HKT transporter genes as compared with their functional classification. The *HKT1;3* gene is at the root of divergence as it was most closely related to the sorghum *HKT*gene. A close relationship was observed between the *HKT1;3, HKT1;2* and *HKT2;3.* The remaining five genes were clustered in group I and showed that *HKT1;4* was most primitive followed by *HKT1;5* and *HKT1;2.* The recently evolved genes *HKT2;4* and *HKT2;1* are located in tandem on chromosome 6 and very closely related to each other diverging from a common branching point (Fig. 3).

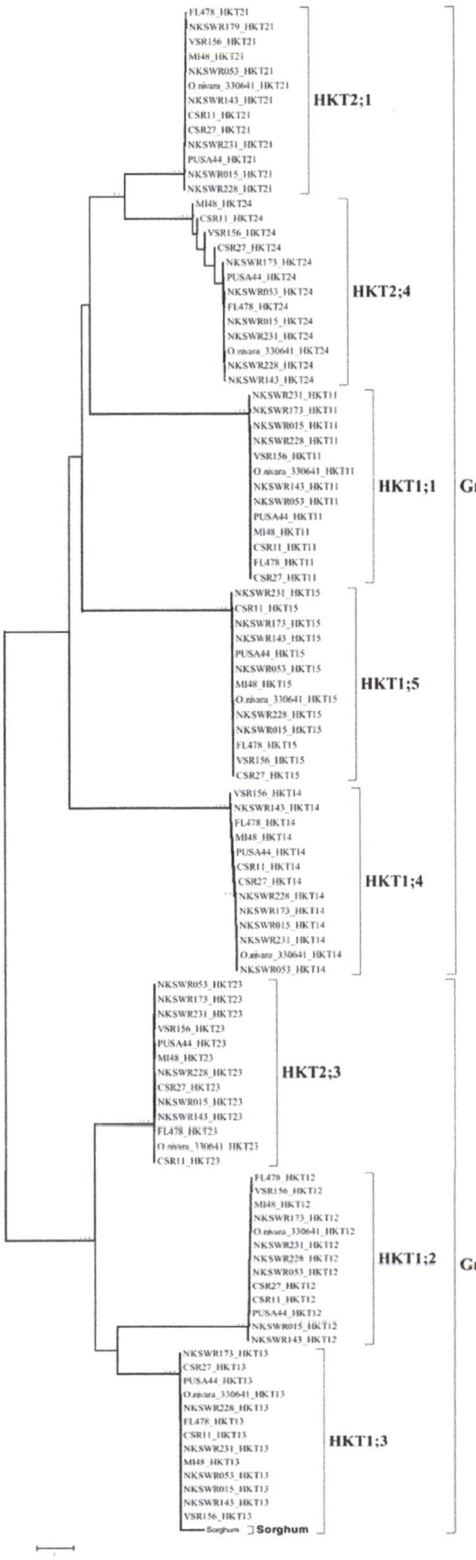

Figure 3: Minimum evolution phylogenetic tree of eight *HKT* genes re-sequenced from representative wild rice accessions and cultivar checks constructed using MEGA 5.1 software

Association of HKT Gene SNPs with Salt Tolerance

The association study was conducted on 45 accessions including 3 check rice varieties, of which, 18 belonged to tolerant class, 9 moderately tolerant, 7 sensitive and 11 highly sensitive. For association study a Bayesian based analysis of population structure was conducted which used an ad hoc statistics (ΔK), the rate of change in log probability of data between successive K values. It showed that the highest log likelihood was at $K=3$ suggesting three major sub-population in the analysed Indian wild rice accessions. The individual assignments and Q-matrix in the three different sub-population revealed that sub-population I included 51.4 %, sub-population II 33.8 % and sub-population III 14.8 % of the analysed accessions. The mean Fst values were 0.8981, 0.6154 and 0.7291 for the sub-populations I, II and III, respectively (Fig. 4).

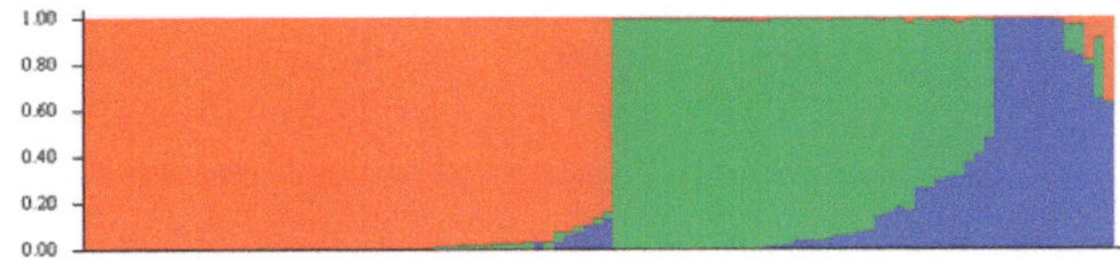

Figure 4: Distribution of 103 wild rice accessions and cultivated rice into three different sub-populations based on 48 genome wide SNP markers. X-axis shows rice accessions and y-axis represents Fst values (*Blue*: Population I; *Green*: Population II; *Red*: Population III)

SNP based association analysis showed significant associations with the salt tolerance traits using MLM_Q+K statistical method. Thus, the results were filtered with association values greater than 5 % and p value less than 0.01. A total of 50 SNPs from all the eight *HKT* genes showed significant associations with the salt tolerance traits (Table 2).

Table 2: Association of SNPs with different salt related traits at 10 days of treatment with HKT genes (with R2>5 % and $p<0.01$)

Gene	Trait	SNP	Site	*p*	R2 (%)	SNP type	MAF	Gene	Trait	SNP	Site	*p*	R2 (%)	SNP type	MAF
HKT 1;1	N_R	S0231	30724332	0.003	34.43	Int	37	HKT 1;5	N_S	S0077	11459042	0.002	31.57	Int	36
	K_S	S1955	30726056	0.003	30.75	Syn	5		N_S	S3893	11462858	0.005	33.00	P/A_{140}	20
	SPAD	S2094	30726195	0.003	31.90	S/N_{258}	32		K_S	S0077	11459042	0.003	30.92	Int	36
	SPAD	S2331	30726432	0.003	31.90	S/F_{179}	30		N_R	S0887	11459852	0.003	34.12	Int	13
	SPAD	S2546	30726647	0.003	31.90	Syn	30		N_R	S0919	11459884	0.009	34.42	Int	35
HKT 1;2	K_S	S0221	30545988	0.003	31.83	5' NC	36		N_R	S0958	11459923	0.009	34.65	Int	35
	K_S	S1430	30547197	0.009	25.16	Int	21		N_R	S0959	11459924	0.003	34.12	Int	13
	K_S	S2282	30548049	0.003	31.54	Int	31		N_R	S0974	11459939	0.008	35.09	Int	6
HKT 1;3	N_R	S0337	4103487	0.005	33.93	C/R_{16}	30		N_R	S1036	11460001	0.002	37.37	Int	5
	SPAD	S0091	4103241	0.009	27.83	5' NC	12		K_R	S1036	11460001	0.008	26.80	Int	5
	N_R	S0744	4103894	0.008	38.19	Syn	22		N_R	S1379	11460344	0.003	34.44	K/Q_{429}	49
	SPAD	S0744	4103894	0.008	35.21	Syn	22		N_R	S1913	11460878	0.003	34.62	Int	43
	SPAD	S1682	4104832	0.009	27.83	Int	10		N_R	S2043	11461008	0.004	33.62	Int	29
	SPAD	S1762	4104912	0.005	31.75	Int	45		K_S	S2146	11461111	0.002	34.93	Int	45
	SPAD	S2347	4105497	0.003	34.78	A/V_{488}	23		K_S	S2347	11461312	0.006	27.43	Int	28
	SPAD	S2374	4105524	0.004	32.49	R/Q_{497}	46		K_S	S2865	11461830	0.001	36.63	Int	14
	SPAD	S2392	4105542	0.009	27.68	T/M_{503}	13		K_S	S3893	11462858	0.009	31.17	Int	20
HKT 1;4	SPAD	S0886	30734275	0.003	34.72	3'NC	32	HKT 2;1	K_S	S3062	29541671	0.008	32.35	5' NC	7
	SPAD	S1267	30734656	0.003	34.72	3'NC	32		N_R	S3062	29541671	0.004	39.70	5' NC	7
	SPAD	S2284	30735673	0.002	37.42	Int	32		SPAD	S4153	29542762	0.009	25.80	5' NC	38
	SPAD	S2804	30736193	0.003	34.72	Int	32		SPAD	S4377	29542986	0.002	33.96	5' NC	36
	SPAD	S2974	30736363	0.002	37.03	Int	33		K_S	S5815	29544424	0.002	40.52	5' NC	21
	SPAD	S3185	30736574	0.002	37.03	Int	32		K_S	S5817	29544426	0.007	32.66	5' NC	7
	SPAD	S3228	30736617	0.007	36.36	Int	33	HKT 2;3	K_S	S0356	19241884	0.002	34.84	Int	10
	SPAD	S3307	30736696	0.003	34.87	Int	28		SPAD	S0761	19242289	0.003	31.88	I/T_{77}	34
	SPAD	S3544	30736933	0.003	34.72	Int	32		K_S	S1001	19242529	0.009	25.08	I/T_{157}	5
	SPAD	S4678	30738067	0.003	35.09	Int	35		SPAD	S2053	19243581	0.003	31.79	Syn	34
								HKT 2;4	K_S	S0334	29534433	0.004	37.20	Int	37

(*N_S* sodium concentration in shoot, *K_S* potassium concentration in shoot, *N_R* sodium concentration in root, *K_S* potassium concentration in root, *Int* intron, *Syn* synonymous, *NC* non-coding, *MAF* minor allele frequency in %)

Among the HKT subfamily I members, analysis of associated SNPs of *OsHKT1;1* gene revealed two non-synonymous substitutions, (i) S2094 (Ser_{258} to Asp_{258}) and (ii) S2331 (Ser_{179} to Phe_{179}) that were significantly associated with SPAD chlorophyll content. These two SNPs are spatially located, one (S2094) on the extracellular side and the other on the cytoplasmic side (S2331). In addition, one SNP (S0231) in the intronic region showed highly significant association with sodium concentration in the root. Further, two synonymous SNPs were associated with potassium concentration in the shoot (S1955) and SPAD chlorophyll content (S2546). In case of *HKT1;2* gene three SNPs were associated with potassium concentration in the shoot, two of these (S1430 and S2282) were present in the intronic region while third one (S221) was in the upstream promoter region. In case of *OsHKT1;3* a non-synonymous SNP S337 (Cys_{16} to Arg_{16}) present in the extracellular domain of the protein explained 34 % of phenotypic variance (R^2 value) and was associated with sodium concentration in the root while, three non-synonymous SNPs in the cytoplasmic domain, S2374 (Ala_{488} to Val_{488}), S2374 (Arg_{479} to Glu_{479}) and S2392 (Thr_{503} to Met_{503}) showed association with SPAD chlorophyll content. A synonymous SNP (S744) was associated with potassium concentration in the shoot and SPAD chlorophyll content. Interestingly, the SNP in the promoter region (S91) and those in the intronic region (S1682 and S1762) were also associated with the SPAD chlorophyll value. All SNPs identified in the *OsHKT1;4* gene were associated with SPAD chlorophyll content, among them, two were in the downstream promoter region while eight were in the intronic region. Of the two SNPs in the downstream promoter region (S886 and S1267), one was in the consensus GT-1 binding site and the other one in the ACGT sequence. In the *OsHKT1;5* gene total 14 SNPs were associated with the salt tolerance traits. Of which 12 were in the intronic region while remaining 2, S1379 (Lys_{429} Glu_{429}) and S3893 (Pro_{140} Ala_{140}), were non-synonymous. Both the non-synonymous SNPs were associated with sodium concentration: S1379 was located on extracellular side and showed significant association with the concentration of sodium in the root while, the other S3893 was on the cytoplasmic side and showed association with concentration of sodium in shoot, explaining 32 % (R^2 value) of the phenotypic variance (Table 2).

From the HKT subfamily II, *OsHKT2;1* gene had five SNPs in the promoter region showing association with the salt tolerance traits. Three of these (S3062, S5815, and S5817) were associated with shoot potassium concentration, of which S3062 was also associated with root sodium concentration, while the other two, S4153, and S4377, were associated with SPAD chlorophyll content. In *OsHKT2;3* a total of 4 SNPs were associated with the salt tolerance traits. Of the two non-synonymous SNPs, S1001 was associated with potassium concentration in shoot (extracellular side) while S761 was associated with SPAD chlorophyll content (cytoplasmic side), both the SNPs resulted in Isoleucine to Threonine substitution at amino acid positions 157 and 77, respectively. In addition, one synonymous SNP S2053 was associated with SPAD chlorophyll content and another SNP S356 in the intronic region was associated with potassium concentration in shoot. Only one SNP S334 present in the intronic region of*OsHKT2;4* gene was associated with potassium concentration in the shoot (Table 2).

Discussion

Wild rice is expected to play important role in rice improvement in the coming years. In this study we explored untapped diversity of Indian wild rice to identify natural alleles of the HKT transporter family genes. Analysis of nucleotide sequence variations for eight HKT family genes in wild rice showed higher nucleotide and haplotypic diversity as compared to the cultivated rice varieties (Table 1, Fig. 2). This supports the notion that wild relatives are genetically much more diverse than their cultivated counterparts (Hoisington et al. 1999). However, nucleotide variations of a gene is also associated with respect to position in the gene (McNally et al. 2006). Platten et al. (2013) observed comparatively lower nucleotide diversity and haplotypic diversity in cultivated rice and identified relationship between different haplotypes and salt tolerance for *HKT1;5* gene. Our results indicate that nucleotide diversity was quite different from haplotype diversity as only effective SNPs participated in haplotypic group determination (Goodall-Copestake et al. 2012). Differences between nucleotide and haplotype diversity has been measured across different genes such as Sucrose Synthase 3 in rice (Lestari et al. 2011) and *OsDREB1F* in wild and cultivated rice (Singh et al. 2015a).

To understand the natural selection process in the evolutions of HKT genes, Tajima's D test, Fu and Li's D* test and Fu and Li's F* statistics were calculated. Negative values obtained for wild rice population shows an excess of rare polymorphisms which still undergoes some population expansion and positive selection pressure and were further validated by Ka/Ks ratio and Jukes and Counter correction values (>1). However, positive value for coding region in cultivated rice revealed that alleles had evolved either by balancing selection or population bottlenecks (Table 1, Akey et al. 2004).

Among the eight genes only two *HKT1;5* and *HKT2;3* showed tolerant haplotypes. Functionally, HKT1;5 is a Na^+specific transporter that maintains Na^+/K^+ homeostasis in the leaves under salt stress (Ren et al. 2005; Mickelbart et al. 2015), whereas HKT2;3 regulates K^+ transport independent of Na^+ concentration (Horie et al. 2011). The genotypes having tolerant allele of *HKT1;5* showed lower Na^+/K^+ concentration in the shoot as compared to root (Thomson et al. 2010). Six wild rice accessions that showed lower shoot Na^+ grouped into four separate haplotypic group. In addition, four other accessions that showed low Na^+ concentration in shoot comparable to the tolerant checks fell into separate haplotypic groups indicating allelic variation for the *HKT1;5* gene among the accessions (Fig. 2). An exceptionally higher Na^+ and K^+ ratio was observed in the shoots of a tolerant accession *O. rufipogon* 336679 indicating presence of an alternate mechanism to sustain such a high level of salt concentration in shoot. Mian et al. (2011) identified role of *HKT 2;1* gene in barley and reported increased shoot Na+concentrations and improved biomass production under salt stress. Haplotype analysis of *HKT2;3* gene showed almost even distribution of accessions into different haplotypic groups, but it also formed a separate haplotype for the reference sequence. Of the total 23 observed haplotypes, one was associated with tolerant and two were associated with moderately tolerant genotypes. A unique accession of *O. nivara 336676* showing tolerance only up to tenth day of applying the stress separated in a different haplotype indicating that it is possibly following some different tolerance mechanism which breaks down beyond 10 days of stress. Salt susceptibility of this accessions after 10 days might be due to higher accumulation of Na^+ in root. Another plausible explanation for the tolerance in this accession is presence of root specific tonoplast transporters with some accessory factors associated with the genes for providing energy to transport ions into vacuole but with limited accumulation capacity of ions in vacuole. Functionally, HKT2;3 mediates K^+ transport independent of Na^+concentration hence, even at high salt concentration the physiological functions that are vital to crop plants relating to K^+ e.g., stomatal movement, CO_2 uptake and cofactor to make enzymes etc. are maintained inside the cell. It also maintains ionic homeostasis (Horie et al. 2007; Roy et al. 2014). However, haplotype analysis revealed uneven distribution of tolerant genotypes among different haplotypic group for different genes. It shows that an unknown type of mechanism might be present among the Indian wild rice that could help to enhance our understanding towards salt tolerance mechanism and be used for improvement of cultivated rice.

Phylogenetic analysis describes the evolutionary relationship among genes and different clustering patterns indicate the level of their functional divergence (Mäser et al. 2001). Here, two genes of phylogenetic group I (*HKT2;1* and*HKT 2;4*) showed high nucleotide similarity with each other. They most likely have evolved from recent duplication of a single gene as also suggested by their tandem location in the rice chromosome 6. Interestingly, in the phylogenetic tree functionally diverse, *HKT1;3* and *HKT1;2* genes grouped together along with *HKT2;3*. This suggests that members of functionally different groups might have evolved from a common ancestor through duplication and subsequent functional divergence.

Population structure analysis revealed three distinct subpopulations in the Indian wild rice accessions with extensive genetic variation between and also within the populations as was also shown recently using a high density 50 K SNP chip (Singh et al. 2015b). Higher average values of Fst indicated that a high level of unshared allelic variation was present in the population (Holsinger and Weir 2009). Here sub-population I comprising more than 50 % of the wild rice accessions was the largest of the three sub-populations.. Linkage disequilibrium (LD) based association analysis is supposed to be the best method for association analysis (Yu and Buckler 2006). Here, LD based association analysis showed strong associations of multiple SNPs in different candidate *HKT* genes with the salt tolerance traits.

Haplotype based association analysis for identification of allelic variants associated with specific traits is considered better over single-allele studies (Johnson et al. 2001). Overall, only three tolerant haplotypes were identified, one for the *HKT1;5* gene and two for the *HKT 2;3* gene, but, multiple SNPs were found to be associated with different salt tolerance traits. It may be because of less number of accessions grouping in many haplotypes resulting in low minor allele frequencies (<2 %), and synonymous or noncoding nature of the SNPs. Negrão et al. (2013) identified allelic variants for *HKT1;5* gene by setting filter for associated SNPs and individually assessing the results. Here total 14 filtered SNPs were found to be associated with traits for *HKT1;5*, and among them 12 SNPs were in the intronic regions, one was synonymous and two were non- synonymous, S1379 (Lys_{429} to Glu_{429}) and S3893 (Pro_{140} to Ala_{140}) leading to amino acid substitutions. Replacement of proline causes changes in the protein backbone that is responsible to introduce tight turns or kinks into alpha helices (Betts and Russell 2003). Conversion of proline to alanine was also found in Pokkali and Nona Bokra alleles of the *HKT1;5* gene which is supposed to increase the substrate specificity (Cotsaftis et al. 2012; Negrão et al. 2013). Cotsaftis et al. (2012) reported change in conformation of protein with change in overall charge.

Two SNPs in the promoter region of *OsHKT1;4* gene, S886 (GT-1 binding site region) and S1267 (downstream to the promoter ACGT sequence), were significantly associated with SPAD chlorophyll content value. It has been reported that in many light-regulated genes like *PHYA* in oat and rice are regulated by GT-1 binding site (Villain et al.1996). Further, Park et al. (2004) has reported this site to be salt induced. Downstream to the promoter region is the ACGT sequence that is required for etiolation induced expression of *erd1* (early responsive to dehydration) gene in*Arabidopsis* (Simpson et al. 2003), hence it is speculated that *OsHKT1;4* gene may regulate dehydration stress imposed by salinity.

Two haplotypes (H1 and H10) of the *OsHKT2;3* were associated with salt tolerance SES score. In addition, total 4 SNPs were associated with the analysed salt tolerance traits, one synonymous, one intronic and two non-synonymous SNPs, S761 and S1001, both leading to Iso to Thr substitution at amino acid positions 77 and 157, respectively (Table 2). The isoleucine to threonine substitution is known to impact post translation modifications such as phosphorylation. An increase in the phosphorylation level has been observed with increase in Thr in a protein sequence (Vlad et al. 2008). The non-synonymous mutations outside functional domain of genes may alter structure of the protein and consequently its function (Negrão et al. 2013). Both synonymous and non-synonymous SNPs showed significant association with salt tolerance traits, perhaps affecting the RNA splicing, mRNA stability, and post-translational modification of protein function (Negrão et al. 2013). Some of this could be due to close linkage of a non-functional SNP with the functional SNP. A large number of rare alleles and haplotypes were observed for different HKT genes, whose association with trait could not be studied due to low minor allele frequency. Rare alleles contributing to the gain or enhancement of the trait value may be useful for future adaptability of the rice crop, these may involve novel mechanisms of salt tolerance. Introgression of rare alleles through marker-assisted backcross breeding (MABB) techniques may help develop new genetic resources for breeding of rice for tolerance to extreme salt stress. To find out effective rare alleles, bi-parental mapping populations involving these lines will be needed to validate the function of rare alleles and also to understand their genetic control mechanisms (Semagn et al. 2010).

CONCLUSIONS

The wild rice accessions screened in this study have been collected from different ecological habitats including salt-affected areas. An accessions (NKSWR173) from upland and another accession (NKSWR202) from upper gangetic plain region showed high level of tolerance phenotypic patterns and may have different salt tolerance mechanisms. Haplotype analysis indicated a substantial level of natural diversity for the HKT family of genes, especially for the HKT1;5 and HKT2;3 genes among the Indian wild rice accessions. The novel haplotypes showing association with salt tolerance may have great impact on rice salinity breeding. A programme for introgression of the identified haplotypes into high yielding but salt sensitive rice varieties has been initiated. It will help breed rice genotypes with higher level of salt tolerance beyond the existing salt tolerant varieties.

METHODS

Plant Material

A total of 299 wild rice accessions including 244 new accessions collected from different geographical regions of India along with their passport data and 58 accessions from NBPGR gene bank, New Delhi, India were analysed. In addition, 6 cultivated *O. sativa* (salt tolerant FL478, CSR27, CSR11 with different tolerance mechanisms, MI48 as a moderate check while VSR156 and Pusa 44 as sensitive checks) were used. The collection sites and other detailed information on each wild rice accession is available in a database at http://nksingh.nationalprof.in/.

Phenotyping for Salt Tolerance

Phenotyping for all the accessions was done at National Phytotron Facility, IARI, New Delhi, India for three times. Experiment was planned according to IRRI protocol with minor modifications (Gregorio et al. 1997). Each set of experiment had three replicates with 10 plants of each accession per replicate. The seeds were germinated in petri plates and then transferred into thermocol trays after two days of germination. Each tray had a positive check and a negative check. The seedlings were allowed to grow for 14 days in Hoagland's solution and on 15th day, they were transferred to Hoagland's solution with 150 mM NaCl concentrations. Rice seedlings were allowed to grow and scoring was done after 10 days and 15 days of stress according to the 1–9 scale of Standard Evaluation System (SES) developed by Gregorio et al. (1997). Most of the SES were same for all the replicates but in case of variation mode was taken. Based on the above data and geographical location 45 representative accessions from each class were selected to evaluate physiological parameters to reduce the cost and labour. Sodium and Potassium ion concentrations from roots and shoots were estimated using a flame photometer LABTRONICS model LT66 after digestion of 0.1 gm of dried plant sample with 1:3 ratio of perchloric acid and nitric acid. Chlorophyll content of leaves was determined using SPAD meter.

PCR Amplification and re-Sequencing

After phenotypic screening of 299 accessions, 103 were selected based on their response to salt stress. All the salt tolerant lines and representative accessions from moderately tolerant and sensitive classes were taken for re-sequencing of *HKT* genes. Genomic DNA was extracted from leaf tissue using the CTAB method described by Murray and Thompson (1980). Eight *HKT* genes were amplified using primer walking method. Nucleotide sequence of the genes were retrieved from NCBI database (http://www.ncbi.nlm.nih.gov/) and primers were designed using Primer3 software (http://bioinfo.ut.ee/primer3-0.4.0/). Specific amplification and validation of the primers was done by NCBI Primer BLAST (http://www.ncbi.nlm.nih.gov/tools/primer-blast/) against*Oryza* taxon and high stringency conditions during PCR amplification. The PCR amplifications were carried out in 25 μl reaction consisting of 1 Unit SpeedStar™HS DNA Polymerase from TAKARA BIO INC, 1x Fast Buffer, 2 μl of dNTP mixture, 0.5 Pico moles of each primers and 80 ng of template DNA. The PCR reaction was carried out in BIORAD Thermal cycler under following conditions: Initial denaturation at 98 °C for 3 min followed by 38 cycles of denaturation for 10 s at 98 °C, annealing for1 min at 64 °C and extension for 1 min at 68 °C, and for final extension for 10 min at 68 °C. The amplified products were further checked by electrophoresis in 1 % agarose gel in 1X TBE buffer.

PCR products were directly sequenced by Ion Torrent PGM sequencer (Life Technologies) after fragmentation, library preparation, purification and cycle sequencing according to manufacturer's instructions. Briefly, the amplicons of different genes of the same genotype were pooled in equimolar concentration, sheared to size of ~200 bp and then barcoded to identify individual accessions. Since each gene has unique sequence no barcoding was needed for individual gene. The sheared and barcoded products were now size selected, pooled together in equi-molar ratio and then PCR sequenced. At each step purification was done using Agencourt AMPure XP reagent using Kingfisher Flex. Ion OneTouch2™ was used for emulsion PCR to clone the library on the beads and thereafter for enrichment. The enriched library was loaded on Ion PGM 316 chip and sequencing was performed on Ion PGM™ sequencer.

Sequence Data Analysis

The depth of sequencing obtained was approximately 88x. Coverage analysis and variant caller plugins were run and the sequences were viewed in IGV (Robinson et al. 2011; Thorvaldsdóttir et al. 2012). SamTools mpileup command was used to generate consensus sequences against the reference for each gene and alignment of*OsHKT* genes was done by ClustalW (Thompson et al. 1994) in BioEdit (Hall 2011). Sequences were deposited to NCBI GenBank database (KT795544-KT796361). Nucleotide polymorphisms were analysed using the DnaSP software version 5.10 (Rozas et al. 2003). Level of silent-site nucleotide diversities per site (π) (Nei 1987) and population mutation parameter (θ) (Watterson 1975) was estimated. Sliding window analysis was performed to examine nucleotide polymorphism across the genes in all accessions using DnaSP software. Statistical tests of neutrality such as Tajima's D (Tajima 1989), Fu and Li's D* and F were calculated to examine the selection pressure at this locus. A haplotype network was constructed for comparison of genealogical relationships among the haplotypes using Network software (Bandelt et al. 1999) (http://www.fluxus-engineering.com). The nucleotide sequences were translated into amino acid sequences, and the protein variants were identified as compared with the reference protein. The bootstrap consensus tree inferred from 500 replicates was constructed to represent the evolutionary history of the HKT genes by using the Minimum Evolution Method in MEGA5 (Tamura et al. 2011).

Population Structure and SNP-Trait Association Analysis

Association analysis was performed with the MLM model, considering both kinship (K) and population structure (Q), implemented in TASSEL software. The kinship (K) and population structure (Q) were generated from a genome wide 48-plex Illumina GoldenGate SNP genotyping assay

(B. Singh, unpublished) and analysing data with Structure software (Pritchard et al. 2000). To overcome the problems of interpreting the real value of K, a range of ad hoc K values were tested and analysed using Evanno plot (Evanno et al. 2005). The results of estimated likelihood values for a given K (from 2 to 10) in five independent runs were harvested with structure harvester an online program (Earl2012). For association mapping filtered sites within *OsHKT* genes were used to determine linkage disequilibrium (LD) by correlation between alleles at two loci in TASSEL 5.0 (Bradbury et al. 2007) software and significance of LD among SNPs was determined by Fisher's exact test. The mixed model showed least deviation of observed *P*-values from expected *P*-values in Q-Q plot when compared with that of Q (population structure) or K (kinship) model only. A probability value of 0.01 was used as the threshold for significance of SNP–trait associations. Functions of associated sites in the promoter region was identified by PLACE (Higo et al. 1999).

ACKNOWLEDGEMENTS

The financial assistance from Indian Council of Agricultural Research through National Professor B.P. Pal Chair project is gratefully acknowledged. We are thankful to National Bureau of Plant Genetic Resources, New Delhi, India for providing seeds of 58 wild rice accessions.

AUTHORS' CONTRIBUTIONS

SM: Design and conduct of experiments, Sequencing, Data Analysis and Manuscript preparation; BS: Collection and characterization of wild rice, Conduct of experiments and manuscript drafting; KP: Database maintenance; BPS: Collection of Wild Rice; NS: Design of 48-plex SNP genotyping assay; PM: Manuscript editing; VR: Planning of experiment, Manuscript editing; NKS: Wild rice collection, planning and supervision of the experiments, manuscript editing and finalization. All authors read and approved the final manuscript.

REFERENCES

1. Akey JM, Eberle MA, Rieder MJ, Carlson CS, Shriver MD, Nickerson DA, Kruglyak L (2004) Population history and natural selection shape patterns of genetic variation in 132 genes. PLoS Biol 2:1591–1599
2. Almeida P, Katschnig D, de Boer AH (2013) HKT transporters—state of the art. Int J Mol Sci 14:20359–20385
3. Bandelt H-J, Forster P, Röhl A (1999) Median-joining networks for inferring intraspecific phylogenies. Mol Biol Evol 16:37–48
4. Betts MJ, Russell RB (2003) Amino acid properties and consequences of substitutions. Bioinformatics Genet 317:289
5. Bhasin H, Bhatia D, Raghuvanshi S, Lore JS, Sahi GK, Kaur B, Vikal Y, Singh K (2012) New PCR-based sequence-tagged site marker for bacterial blight resistance gene Xa38 of rice. Mol Breed 30:607–611
6. Bradbury PJ, Zhang Z, Kroon DE, Casstevens TM, Ramdoss Y, Buckler ES (2007) TASSEL: software for association mapping of complex traits in diverse samples. Bioinformatics 23:2633–2635
7. Cotsaftis O, Plett D, Johnson AAT, Walia H, Wilson C, Ismail AM, Close TJ, Tester M, Baumann U (2011) Root-specific transcript profiling of contrasting rice genotypes in response to salinity stress. Mol Plant 4:25–41
8. Cotsaftis O, Plett D, Shirley N, Tester M, Hrmova M (2012) A Two-staged model of Na (+) exclusion in rice explained by 3D modeling of HKT transporters and alternative splicing. PLoS One 7, e39865
9. Cseri A, Cserháti M, Von KM, Nagy B, Horváth GV, Palágyi A, Pauk J, Dudits D, Törjék O (2011) Allele mining and haplotype discovery in barley candidate genes for drought tolerance. Euphytica 181:341–356
10. Das-Chatterjee A, Goswami L, Maitra S, Dastidar KG, Ray S, Majumder AL (2006) Introgression of a novel salt-tolerant L-myo-inositol 1-phosphate synthase from *Porteresia coarctata* (Roxb.) Tateoka (PcINO1) confers salt tolerance to evolutionary diverse organisms. FEBS Lett 580:3980–3988
11. Earl DA (2012) Structure harvester: a website and program for visualizing structure output and implementing the Evanno method. Conserv Genet Resour 4:359–361
12. Evanno G, Regnaut S, Goudet J (2005) Detecting the number of clusters of individuals using the software STRUCTURE: a simulation study. Mol Ecol 14:2611–2620
13. FAOSTAT, D. (2009). Food and Agriculture Organization of the United Nations Year book.
14. Goodall-Copestake W, Tarling G, Murphy E (2012) On the comparison of population-level estimates of haplotype and nucleotide diversity: a case study using the gene cox1 in animals. Heredity 109:50–56
15. Gregorio GB, Senadhira D, Mendoza RD (1997) Screening rice for salinity tolerance. International Rice Research Institute discussion paper series, Metro Manila
16. Hall T (2011) BioEdit: an important software for molecular biology. GERF Bull Biosci 2:60–61
17. Higo K, Ugawa Y, Iwamoto M, Korenaga T (1999) Plant cis-acting regulatory DNA elements (PLACE) database. Nucleic Acids Res 27:297–300
18. Hoisington D, Khairallah M, Reeves T, Ribaut J-M, Skovmand B, Taba S, Warburton M (1999) Plant genetic resources: what can they contribute toward increased crop productivity? Proc Natl Acad Sci 96:5937–5943
19. Holsinger KE, Weir BS (2009) Genetics in geographically structured populations: defining, estimating and interpreting FST. Nat Rev Genet 10:639–650
20. Horie T, Yoshida K, Nakayama H, Yamada K, Oiki S, Shinmyo A (2001) Two types of HKT transporters with different properties of Na+and K+ transport in *Oryza sativa*. Plant J 27:129–138
21. Horie T, Costa A, Kim TH, Han MJ, Horie R, Leung HY, Miyao A, Hirochika H, An G, Schroeder JI (2007) Rice OsHKT2; 1 transporter mediates large Na+influx component into K+−starved roots for growth. EMBO J 26:3003–3014
22. Horie T, Hauser F, Schroeder JI (2009) HKT transporter-mediated salinity resistance mechanisms in Arabidopsis

and monocot crop plants. Trends Plant Sci 14:660–668

23. Horie T, Brodsky DE, Costa A, Kaneko T, Schiavo FL, Katsuhara M, Schroeder JI (2011) K+ transport by the OsHKT2; 4 transporter from rice with atypical Na+transport properties and competition in permeation of K+ over Mg2+ and Ca2+ ions. Plant Physiol 156:1493–1507
24. Hossain H, Rahman M, Alam M, Singh R (2015) Mapping of quantitative trait loci associated with reproductive-stage salt tolerance in rice. J Agron Crop Sci 201:17–31
25. Jabnoune M, Espeout S, Mieulet D, Fizames C, Verdeil J-L, Conéjéro G, Rodríguez-Navarro A, Sentenac H, Guiderdoni E, Abdelly C (2009) Diversity in expression patterns and functional properties in the rice HKT transporter family. Plant Physiol 150:1955–1971
26. Johnson GC, Esposito L, Barratt BJ, Smith AN, Heward J, Di Genova G, Ueda H, Cordell HJ, Eaves IA, Dudbridge F (2001) Haplotype tagging for the identification of common disease genes. Nat Genet 29:233–237
27. Khush GS, Ling K (1974) Inheritance of resistance to grassy stunt virus and its vector in rice. J Hered 65:135–136
28. Kumar GR, Sakthivel K, Sundaram RM, Neeraja CN, Balachandran SM, Rani NS, Viraktamath BC, Madhav MS (2010) Allele mining in crops: prospects and potentials. Biotechnol Adv 28:451–461
29. Lan W-Z, Wang W, Wang S-M, Li L-G, Buchanan BB, Lin H-X, Gao J-P, Luan S (2010) A rice high-affinity potassium transporter (HKT) conceals a calcium-permeable cation channel. Proc Natl Acad Sci 107:7089–7094
30. Latha R, Rubia L, Bennett J, Swaminathan M (2004) Allele mining for stress tolerance genes in *Oryza* species and related germplasm. Mol Biol 27:101–108
31. Lestari P, Lee G, Ham TH, Woo MO, Piao R, Jiang W, Chu SH, Lee J, Koh HJ (2011) Single nucleotide polymorphisms and haplotype diversity in rice sucrose synthase 3. J Hered 102:735–746
32. Li R, Li L, Wei S, Wei Y, Chen Y, Bai D, Yang L, Huang F, Lu W, Zhang X (2010) The evaluation and utilization of new genes for brown plant hopper resistance in common wild rice (*Oryza rufipogon* Griff.). Mol Entomol. doi:10.5376/me.2010.01.0001
33. Linh LH, Linh TH, Xuan TD, Ham LH, Ismail AM, Khanh TD (2012) Molecular breeding to improve salt tolerance of rice (*Oryza sativa* L.) in the Red River Delta of Vietnam. Intern J Plant genom 2012, 949038. doi:10.1155/2012/949038
34. Lutts S, Kinet J, Bouharmont J (1995) Changes in plant response to NaCl during development of rice (*Oryza sativa* L.) varieties differing in salinity resistance. J Exp Bot 46:1843–1852
35. Maas E, Hoffman G (1977) Crop salt tolerance: evaluation of existing data. J Irrigat Drainage 15:187
36. Mäser P, Thomine S, Schroeder JI, Ward JM, Hirschi K, Sze H, Talke IN, Amtmann A, Maathuis FJM, Sanders D (2001) Phylogenetic relationships within cation transporter families of Arabidopsis. Plant Physiol 126:1646–1667
37. Mäser P, Eckelman B, Vaidyanathan R, Horie T, Fairbairn DJ, Kubo M, Yamagami M, Yamaguchi K, Nishimura M, Uozumi N (2002) Altered shoot/root Na+distribution and bifurcating salt sensitivity in Arabidopsis by genetic disruption of the Na+transporter AtHKT1. FEBS Lett 531:157–161
38. McCouch SR, Sweeney M, Li J, Jiang H, Thomson M, Septiningsih E, Edwards J, Moncada P, Xiao J, Garris A (2007) Through the genetic bottleneck: *O. rufipogon* as a source of trait-enhancing alleles for O. *sativa*. Euphytica 154:317–339
39. McNally KL, Bruskiewich R, Mackill D, Buell CR, Leach JE, Leung H (2006) Sequencing multiple and diverse rice varieties. Connecting whole-genome variation with phenotypes. Plant Physiol 141:26–31
40. Mian A, Oomen RJ, Isayenkov S, Sentenac H, Maathuis FJ, Véry AA (2011) Over-expression of an Na+−and K+−permeable HKT transporter in barley improves salt tolerance. Plant J 68:468–479
41. Mickelbart MV, Hasegawa PM, Bailey-Serres J (2015) Genetic mechanisms of abiotic stress tolerance that translate to crop yield stability. Nat Rev Genet 16(4):237–251. doi:10.1038/nrg3901
42. Miller RW, Donahue RL (1995) Soils in our environment. Prentice hall, Upper Saddle River
43. Mohammadi R, Mendioro MS, Diaz GQ, Gregorio GB, Singh RK (2013) Mapping quantitative trait loci associated with yield and yield components under reproductive stage salinity stress in rice (*Oryza sativa* L.). J Genet 92:433–443
44. Mohanty S (2013) Trends in global rice consumption. Rice Today 12(1):44–45
45. Munns R, Tester M (2008) Mechanisms of salinity tolerance. Annu Rev Plant Biol 59:651–681
46. Munns R, James RA, Xu B, Athman A, Conn SJ, Jordans C, Byrt CS, Hare RA, Tyerman SD, Tester M, Plett D (2012) Wheat grain yield on saline soils is improved by an ancestral Na+transporter gene. Nat Biotechnol 30:360–364
47. Murray M, Thompson WF (1980) Rapid isolation of high molecular weight plant DNA. Nucleic Acids Res 8:4321–4326
48. Negrão S, Cecília AM, Pires IS, Abreu IA, Maroco J, Courtois B, Gregorio GB, McNally KL, Margarida OM (2013) New allelic variants found in key rice salt-tolerance genes: an association study. Plant Biotechnol J 11:87–100
49. Nei M (1987) Molecular evolutionary genetics. Columbia University Press, New York
50. Oomen RJ, Benito B, Sentenac H, Rodríguez-Navarro A, Talón M, Véry AA, Domingo C (2012) HKT2; 2/1, a K+−permeable transporter identified in a salt-tolerant rice cultivar through surveys of natural genetic polymorphism. Plant J 71:750–762
51. Park HC, Kim ML, Kang YH, Jeon JM, Yoo JH, Kim MC, Park CY, Jeong JC, Moon BC, Lee JH (2004)

Pathogen-and NaCl-induced expression of the SCaM-4 promoter is mediated in part by a GT-1 box that interacts with a GT-1-like transcription factor. Plant Physiol 135:2150–2161

52. Platten JD, Cotsaftis O, Berthomieu P, Bohnert H, Davenport RJ, Fairbairn DJ, Horie T, Leigh RA, Lin H-X, Luan S (2006) Nomenclature for HKT transporters, key determinants of plant salinity tolerance. Trends Plant Sci 11:372–374

53. Platten JD, Egdane JA, Ismail AM (2013) Salinity tolerance, Na+exclusion and allele mining of HKT1; 5 in *Oryza sativa* and O. glaberrima: many sources, many genes, one mechanism? BMC Plant Biol 13:32

54. Pritchard JK, Stephens M, Donnelly P (2000) Inference of population structure using multilocus genotype data. Genetics 155:945–959

55. Ren Z-H, Gao J-P, Li L-G, Cai X-L, Huang W, Chao D-Y, Zhu M-Z, Wang Z-Y, Luan S, Lin H-X (2005) A rice quantitative trait locus for salt tolerance encodes a sodium transporter. Nature Genet 37:1141–1146

56. Robinson JT, Thorvaldsdóttir H, Winckler W, Guttman M, Lander ES, Getz G, Mesirov JP (2011) Integrative genomics viewer. Nat Biotechnol 29:24–26

57. Rockström J, Falkenmark M, Karlberg L, Hoff H, Rost S, Gerten D (2009) Future water availability for global food production: the potential of green water for increasing resilience to global change. Water Resour Res 45, W00A12. doi:10.1029/2007WR006767

58. Roy SJ, Negrão S, Tester M (2014) Salt resistant crop plants. Curr Opin Biotechnol 26:115–124

59. Rozas J, Sánchez-DelBarrio JC, Messeguer X, Rozas R (2003) DnaSP, DNA polymorphism analyses by the coalescent and other methods. Bioinformatics 19:2496–2497

60. Schachtman DP, Schroeder JI (1994) Structure and transport mechanism of a high-affinity potassium uptake transporter from higher plants. Nature 370(6491):655–658

61. Schroeder JI, Delhaize E, Frommer WB, Guerinot ML, Harrison MJ, Herrera-Estrella L, Horie T, Kochian LV, Munns R, Nishizawa NK (2013) Using membrane transporters to improve crops for sustainable food production. Nature 497:60–66

62. Semagn K, Bjørnstad Å, Xu Y (2010) The genetic dissection of quantitative traits in crops. Electron J Biotechnol 13:16–17

63. Sengupta S, Patra B, Ray S, Majumder AL (2008) Inositol methyl tranferase from a halophytic wild rice, Porteresia coarctata Roxb. (Tateoka): regulation of pinitol synthesis under abiotic stress. Plant Cell Environ 31:1442–1459

64. Simpson SD, Nakashima K, Narusaka Y, Seki M, Shinozaki K, Yamaguchi-Shinozaki K (2003) Two different novel cis-acting elements of erd1, a clpA homologous Arabidopsis gene function in induction by dehydration stress and dark-induced senescence. Plant J 33:259–270

65. Singh A, Singh B, Panda K, Rai VP, Singh AK, Singh SP, Chouhan SK, Rai V, Singh PK, Singh NK (2013) Wild rices of Eastern Indo-Gangetic plains of India constitute two sub-populations harbouring rich genetic diversity. POJ 6(2):121–127

66. Singh BP, Jayaswal PK, Singh B, Singh PK, Kumar V, Mishra S, Singh N, Panda K, Singh NK (2015a) Natural allelic diversity in OsDREB1F gene in the Indian wild rice germplasm led to ascertain its association with drought tolerance. Plant Cell Rep 34:993–1004

67. Singh N, Jayaswal PK, Panda K, Mandal P, Kumar V, Singh B, Mishra S, Singh Y, Singh R, Rai V, Gupta A, Sharma TR, Singh NK (2015b) Single-copy gene based 50 K SNP chip for genetic studies and molecular breeding in rice. Sci Rep 5, 11600. doi:10.1038/srep11600

68. Tajima F (1989) Statistical method for testing the neutral mutation hypothesis by DNA polymorphism. Genetics 123:585–595

69. Tamura K, Peterson D, Peterson N, Stecher G, Nei M, Kumar S (2011) MEGA5: molecular evolutionary genetics analysis using maximum likelihood, evolutionary distance, and maximum parsimony methods. Mol Biol Evol 28:2731–2739

70. Tanksley SD, McCouch SR (1997) Seed banks and molecular maps: unlocking genetic potential from the wild. Science 277:1063–1066

71. Thompson JD, Higgins DG, Gibson TJ (1994) CLUSTAL W: improving the sensitivity of progressive multiple sequence alignment through sequence weighting, position-specific gap penalties and weight matrix choice. Nucleic Acids Res 22:4673–4680

72. Thomson MJ, de Ocampo M, Egdane J, Rahman MA, Sajise AG, Adorada DL, Tumimbang-Raiz E, Blumwald E, Seraj ZI, Singh RK (2010) Characterizing the Saltol quantitative trait locus for salinity tolerance in rice. Rice 3:148–160

73. Thorvaldsdóttir H, Robinson JT, Mesirov JP (2012) Integrative Genomics Viewer (IGV): high-performance genomics data visualization and exploration. Brief Bioinform 14(2):178–192. doi:10.1093/bib/bbs017

74. Villain P, Mache R, Zhou D-X (1996) The mechanism of GT element mediated cell type-specific transcriptional control. J Biol Chem 271:32593–32598

75. Vlad F, Turk BE, Peynot P, Leung J, Merlot S (2008) A versatile strategy to define the phosphorylation preferences of plant protein kinases and screen for putative substrates. Plant J 55:104–117

76. Wang W, Vinocur B, Altman A (2003) Plant responses to drought, salinity and extreme temperatures: towards genetic engineering for stress tolerance. Planta 218:1–14

77. Watterson G (1975) On the number of segregating sites in genetical models without recombination. Theor Popul Biol 7:256–276

78. Xiao J, Li J, Yuan L, Tanksley S (1996) Identification of QTLs affecting traits of agronomic importance in a recombinant inbred population derived from a subspecific rice cross. Theor Appl Genet 92:230–244

79. Xiao J, Li J, Grandillo S, Ahn SN, Yuan L, Tanksley SD, McCouch SR (1998) Identification of trait-improving quantitative trait loci alleles from a wild rice relative,

Oryza rufipogon. Genetics 150:899–909

80. Yang C, Zhao L, Zhang H, Yang Z, Wang H, Wen S, Zhang C, Rustgi S, von Wettstein D, Liu B (2014) Evolution of physiological responses to salt stress in hexaploid wheat. Proc Natl Acad Sci 111:11882–11887

81. Yu J, Buckler ES (2006) Genetic association mapping and genome organization of maize. Curr Opin Biotechnol 17:155–160

82. Yu Y-T, Wang R-H, Shi Y-S, Song Y-C, Wang T-Y, Yu L (2010) Haplotypic structure and allelic variation of rab17, an ABA-responsive gene, in a mini core set of Chinese diversified maize inbred lines. Agric Sci China 9:1726–1738

83. Yuan S-C, Zhang Z-G, He H-H, Zen H-L, Lu K-Y, Lian J-H, Wang B-X (1993) Two photoperiodic-reactions in photoperiod-sensitive genic male-sterile rice. Crop Sci 33:651–660

84. Zhu J-K (2001) Plant salt tolerance. Trends Plant Sci 6:66–71

CHAPTER 7

INDICA RICE GENOME ASSEMBLY, ANNOTATION AND MINING OF BLAST DISEASE RESISTANCE GENES

H. B. Mahesh, Meghana Deepak Shirke, Siddarth Singh, Anantharamanan Rajamani, Shailaja Hittalmani, Guo-Liang Wang and Malali Gowda

Keywords: *O. sativa*, Genome sequencing, *Indica*, Illumina, Pacific Biosciences, SSRs, SNPs, Blast resistance

BACKGROUND

Rice (*Oryza sativa L.*) is a staple food for more than half of the world's population. India is believed to be a center of origin and diversity of rice. India is the second most rice producer in the world and 65 % of its population depends on rice as a staple food. Rice is a model cereal crop with small genome size, short generation time, diploid (2n=24) and amenable for genetic manipulation. Due to its global importance, several genomes of rice cultigen including *japonica* (Nipponbare [1]), *indica* (93-11 [2] and IR64 [3]) and *aus*(Kasalath [4]) have been sequenced. Currently gold standard assembly and annotation are available for *japonica* rice. Although over 80 % of rice cultivation in the world under *indica* rice cultivation however, genomic resources are lacking for this subspecies. Due to non-availability of proper genome assembly, *indica* genome studies are still using Nipponbare genome as a reference. This will introduce the potential bias in analysis and may not capture conclusive results at the nucleotide and chromosomal level for *indica* subspecies [3]. Thus, we and other researchers [3] believe that creating an *indica* reference rice genome is essential for genome-wide studies, which will enable genome assisted *indica* rice breeding program.

Here we report the *de novo* genome assembly and annotation of *indica* cultivar, HR-12. This cultivar was bred and released in India (http://www.drricar.org). It has good agronomic traits, but highly susceptible to blast disease [5], caused by *Magnaporthe oryzae*. Rice blast is a serious constraint in rice production and utilization of resistant (R) genes in variety development has become the most effective method of blast disease management. Sequencing of highly resistant and susceptible varieties will enable the identification of novel R genes and their deployment in rice breeding programme. In addition to HR-12, we resequenced whole genomes of three *indica* cultivars, which are highly susceptible (Co-39) and resistant (Tetep and Tadukan) to rice blast. Our *indica* rice sequencing efforts have complemented the global rice genomic resources, which eventually will help to characterize *indica* rice germplasm to identify genes for agronomically important traits including disease, pest and yield attributing traits.

Citation: Mahesh HB, Shirke MD, Singh S, et al. Indica rice genome assembly, annotation and mining of blast disease resistance genes. BMC Genomics. 2016;17:242. doi:10.1186/s12864-016-2523-7.

METHODS

Indica rice cultivars and genome size estimation

The HR-12 (Himmatsagar Rice-12) was derived from Raja Hansa (http://inger.irri.org). Subsequently HR-12 was used to develop Hamsa (HR-12 x TN-1) and Tellahamsa (HR-12 x TN-1) rice varieties. The Co-39 variety was developed by crossing Culture340 and Kannagi. These two varieties have been widely used as susceptible checks in rice blast screening nursery. Tetep and Tadukan are used as resistance checks, and donor parents for blast resistance breeding, however the genealogy information is not available. The seeds of HR-12, Co-39, Tetep and Tadukan were sown in a PVC pot containing red earth and fertilizers. Twenty one days old leaves from Co-39, HR-12, Tetep and Tadukan were collected and chopped into pieces in nuclear isolation buffer (Hypotonic Propidium Iodide, 50 µg/mL in 3 g/L TriSodium citrate Dihydride containing 0.05 % (v/v) of Nonidet P-40 containing 2 mg/mL RNase A) and samples were processed as per the protocol suggested by Krishna [6]. Debris was filtered and stained nuclei were analysed using BD FACS at Central Imaging and Flow Cytometry Facility (CIFF), C-CAMP, NCBS, Bengaluru, India. Values of rice nuclear DNA was estimated by comparing rice nuclear peak on the linear scale with the peak for *Pisum sativum* included as an internal standard.

Nucleic acid isolation

Genomic DNA was isolated from four (HR-12, Co-39, Tetep and Tadukan) varieties as per the manufacturer's instruction (DNAeasy Plant Mini Kit, Cat # 69104, USA). DNA quality was assessed by Nanodrop and DNA was quantified using Qubit (Applied Biosystems).

Paired-end (PE) and matepair (MP) library preparation, and Illumina sequencing

One micro gram of genomic DNA was fragmented in the range of 300 to 400 bases using ultra Sonicator (S220, Covaris, USA). Then PE library of HR-12, Co-39, Tetep and Tadukan samples were prepared using Tru Seq DNA sample preparation kit v2 (Catalog No: FC-121-2001, Illumina) as per the manufacture's instruction. PE libraries were sequenced using Illumina HiSeq1000 and the length of sequence was 101 nts from both ends of the fragment. The MP library (insert size upto 12Kb) was prepared only for HR-12 sample by Nextera MP sample preparation kit (catalog No.: FC-132-1001, Illumina) and sequenced 2×51 nts by Illumina HiSeq1000.

Strand-specific RNA seq library preparation and sequencing

Total RNA was isolated from HR-12 leaves using Direct-zol RNA MiniPrep kit (Catalog No. R2050, Zymo Research) and RNA integrity and quantity was assessed by Bioanalyzer using Agilent RNA 6000 nano chip. TruSeq stranded total RNA library preparation kit v2 (Catalog No.: RS-122-2201) from Illumina was used to prepare strand-specific RNA sequencing (ssRNA-seq) library by following manufacture's instruction. The ssRNA-seq library was sequenced 2x101 nts by Illumina HiSeq1000.

PacBio library preparation and sequencing

Around 20 micrograms of high quality genomic DNA was sheared using Hydroshear. The Bluepippin was used to select 20Kb double-stranded DNA fragments. Then, DNA fragments were end repaired and ligated with universal hairpin adapters. Subsequent steps were followed as per the manufacture's instruction to prepare SMRTbell library. The library was sequenced in PacBio RS SMRT instrument.

Illumina data output and data preprocessing

The low quality bases (quality less than Q30 or the accuracy less than 99.99 % of the base called) and adapter sequence contaminations in raw reads of Illumina sequencing (PE, MP and ssRNA-seq) was processed using FASTX-Toolkit (http://hannonlab.cshl.edu/fastx_toolkit/index.html).

Short read (Illumina) *de novo* genome assembly of HR-12

The quality processed Illumina reads (PE and MP reads) were used for *de novo* assembly using three *de novo*genome assemblers viz., Velvet [7], SOAPDenovo2 [8] and MaSuRCA [9]. The quality of genome assemblies were assessed using QUAST [10].

Gap filling of short read assembly using PacBio long reads, contigs scaffolding and anchoring

Raw PacBio reads were used for gap filling and to upgrade short read assembly using PBJelly pipeline [11]. Gap filled assembly was further scaffolded by L_RNA_scaffolder [12] using transcripts assembled by Trinity [13]. The pseudomolecules of HR-12 were constructed by anchoring HR-12 contigs on to pseudomolecules of Nipponbare genome (version 7.0) with ABACAS [14]. The synteny map of rice genomes was generated with default parameters by SyMAP [15]. The genome completeness of short read and gap filled assemblies were checked by CEGMA [16].

Gene prediction and functional annotation of HR-12 genome

The HR-12 pseudomolecules were subjected for gene prediction with MAKER-P [17] version 2.31.6 by providing expressed sequences (trinity assembled HR-12 transcripts, ESTs, cDNA and mRNA) of *Oryza* (NCBI). Protein domain structures and gene ontology (GO) terms were assigned using InterProScan5 software [18]. Functional annotation of genes was done by searching homology against rice protein sequences of SwissProt (http://www.uniprot.org) using BLASTp alignments with an e-value threshold of 1^{e-10}. The synonymous (Ks) and non-synonymous (Ka) substitution rate was calculated using NG [19] method in Ka/Ks calculator [20].

Annotation of HR-12 specific genes

Nipponbare (version 7.0) and 93–11 gene models were downloaded from MSU rice genome FTP site (ftp://ftp.plantbiology.msu.edu/pub/data/Eukaryotic_Projects/o_sativa/annotation_dbs/pseudomolecules/version_7.0/all.dir/) and BGI (http://rice.genomics.org.cn/rice/index2.jsp) site, respectively. The homology search was carried out with BLASTp using Nipponbare and 93–11 proteins as subject and HR-12 proteins as a query; the e-value cutoff was set to 1^{e-10}.

Repeat identification and prediction of Simple Sequence Repeats (SSRs)

De novo repeat prediction was performed using Repeat Masker 4.0.5 (http://www.repeatmasker.org) using*Oryza sativa* repeat library in Repbase as a reference. The SSRs were predicted using Microsatellite Identification tool [21]. Rice SSR markers in Gramene website (http://archive.gramene.org/markers/microsat/all-ssr.tab) were used for performing an electronic PCR [22] (e-PCR) to check polymorphic and novel SSRs in HR-12 genome.

Identification of SNPs, InDels and functional annotation of variants

We used Illumina data of HR-12, Co-39, Tetep and Tadukan for variant analysis. The high quality (Q30) Illumina reads were mapped to Nipponbare genome using Burrows-Wheeler Aligner (BWA) V0.7.9a [23]. Alignments with mapping quality ('q' option in samtools) <60 (Phred-scaled) were filtered using SAMtools [24]. Duplicate reads were removed using Picard tool v1.115 (http://broadinstitute.github.io/picard/). We performed local realignment around InDels to correct mapping related artifacts using InDelRealigner tool in GATK V3.3-0. Then base quality recalibration was performed using BaseRecalibrator tool in GATK V3.3-0. Variant calling was performed with a minimum Phred-scaled confidence threshold of 30, and a minimum Phred-scaled confidence threshold for emitting variants at 10 using HaplotypeCaller in GATK V3.3-0 across all four samples [25, 26]. Functional annotation and genetic consequences (effect on gene) of common variants (SNPs and InDels) of susceptible and resistant varieties was annotated using SnpEff tool V4.1b [27]. The Nipponbare genome version 7.0 was used as a reference to annotate SNPs and InDels.

Mining of blast disease resistance genes in *indica* cultivars

The sequences of twenty-two cloned blast *R*-genes (NCBI) were subjected to BLASTp alignment (e-value cutoff 1^{e-10}, minimum 70 % identity and query coverage) with proteins sequences of HR-12, Co-39, IR64 (http://schatzlab.cshl.edu/data/rice/), Tetep, and Tadukan.

RESULTS AND DISCUSSION

De novo short read HR-12 genome assembly

The genome of *indica* rice cultivar, HR-12 was assembled using combination of short reads (PE and MP libraries) from Illumina and long reads from Pacific Biosciences. Initially, the Illumina short reads were assembled using three *de novo* assemblers Velvet, SOAPdenovo2 and MaSuRCA. Among three assemblers, MaSuRCA covered highest genome size of 340.12 Mb and it had least number of N's as compared to other two assemblers. All three assemblers resulted comparatively lower contig N50, which might result due to inherent disadvantages of short read assemblies, such as poor repeat resolution [3, 28], missing exons and genes and genes split between scaffolds [29]. Less number of contigs in Velvet [7] and SOAPDenovo2 [8] assemblies were because of higher number of 'N's in the assembly, which were used to merge adjacent contigs into scaffolds. Based on quality assessment by QUAST, MaSuRCA assembly was chosen for further analysis since it showed more improvement with respect to genome coverage (74.80 %), lower number of gaps (47479) and higher number of genes covered (48428).

Improvement of short read *indica* rice genome assembly using PacBio long reads

To improve assembly quality and overcome the limitations of short read assembly, HR-12 genome was sequenced using PacBio RS SMRT platform with 20x coverage. The PacBio reads were used to fill the gaps in the short read assembly (Fig. 1) which improved HR-12 genome with respect to genome size, N50 and number of contigs/scaffolds. The genome assembly size was increased by 49.52 Mb which accounts for over 86 % of estimated HR-12 genome (454 Mb). The average genome size of four *indica* cultivars was found to be higher (457 Mb) as compared to previous report [30]. The N50 of gap filled assembly was increased from 6.82 to 26.46 Kb. Similarly, the number of scaffolds was reduced from 98939 to 61001. The Core Eukaryotic Genes (CEG) mapping approach has resulted 82.26 % and 94.35 % genes in short read and gap filled assemblies, respectively. According to our data, the gap filled *indica* rice, HR-12 genome assembly is much better with respect to genome size, number of genes, scaffold N50 and genome completeness as compared to other rice genomes including *indica* (93–11 and IR64), *aus* (Kasalath) and*japonica* (Nipponbare) (Table 1). The overall assembly quality comparison with other published genomes reiterated the importance of PacBio long reads in generating gold standard genomes for complex eukaryotic organisms. In a recent study, reference mapping of sequence reads of 50 rice accessions to *japonica*(Nipponbare), *indica* (93–11) and *aus* (Kasalath) genomes showed higher rate of mapping to Kasalath followed by 93–11 and Nipponbare genomes [4]. This indicates that 93–11 genome is not well assembled and annotated [3, 4].

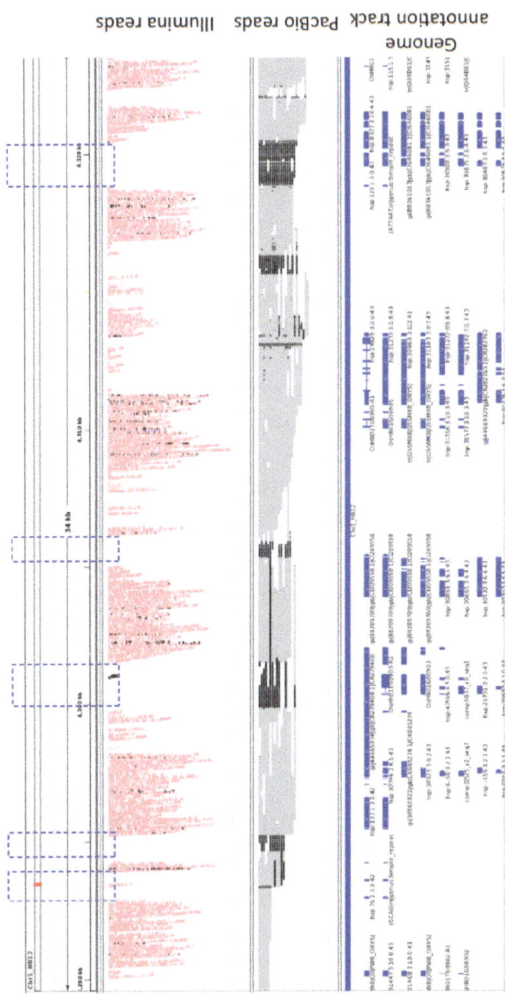

Figure 1: Diagram depicting the assembly gaps closed by PacBio long reads. Gaps in short read assembly are shown in broken rectangles and gene models are represented in genome annotation track

Table 1: Comparison of whole genome sequencing of *Oryza sativa* subspecies

Details	HR-12 (current study)	93–11 [2]	IR-64 [3]	Kasalath [4]	Nipponbare [1]
Subspecies	*indica*	*indica*	*indica*	*aus*	*japonica*
Sequence depth	73x (Illumina) and 20x (PacBio)	4.2x (Sanger) and 36x (Illumina)	110x (Illumina)	6x (Roche 454) and 148X (Illumina)	6x (Sanger), 62x (Illumina) and 2.57x (Roche 454)
Sequencing technologies	Illumina and PacBio	Sanger and Illumina	Illumina	Illumina and Roche 454	Sanger, Illumina and Roche 454
Genome assembler(s)	MaSuRCA and PBJelly	RePS and SOAPDenovo	ALLPATHS-LG	Celera Assembler and String Graph Assembler	TIGR Assembler
Genome size (Mb)	389.77 (389.60)	374.55 (359.41)	345.9 (321.2)	401.14 (328.25)	374.47 (374.31)
No. of N's (Mb)	0.17	15.14	24.7	72.89	0.17
Total no. of contigs/scaffolds	59,692	50,231	26,160	51,550	NA
a. Anchored contigs/scaffolds	43,781	35,415	NA	36,932	NA
b. Unanchored contigs/scaffolds	15,911	14,816	NA	14,618	NA
N50 (Kb)	28.5	6.69	22.2	13.73	NA
No. of genes	56,284	40,464	37,758	53,662	55,986
CEG mapping (%)	94.35	95.97	94.35	95.97	95.97

Value in parenthesis indicate genome size without 'N' content
NA Not available, *No.* Number, *Mb* Mega bases, *Kb* Kilo bases
N50 50 % of the contigs represent this contig length
CEG Core Eukaryotic Genes

HR-12 contigs ordering on Nipponbare chromosomes

Ordering and orienting of contigs/scaffolds onto pseudomolecules has facilitated identification of gaps, closure of gaps and also comparative whole genome analyses. We scaffolded HR-12 contigs by utilizing stranded RNA sequencing data, which reduced scaffold numbers from 61001 to 59692. Over 73 % of HR-12 contigs (43781 out of 59692 contigs) anchored onto pseudomolecules of Nipponbare with 80 % sequence identity and remaining 15911 were unanchored. The unanchored contigs could be part of structural variants like insertions, inversions and translocations. The minimum and maximum contig length of unanchored contigs was 226 and 392500 bp, respectively. The N50 was 36469 bp with an average contigs length of 6642 bp. Nearly 55.40 % (8815) and 60.34 % (9600) of HR-12 contigs aligned to Nipponbare and 93–11 genomes with 100 % query coverage and 80 % sequence similarity, which confirmed that unaligned contigs were part of rice genome. Around 1.55 % (247) of contigs did not show any alignment to Nipponbare genome indicating their absence in the reference genome.

The synteny map of HR-12, Nipponbare and 93–11 showed that most of the genomic blocks conserved across all three genomes with few translocations. Around 79 % and 74 % of HR-12 genome is in syntenic with Nipponbare and 93–11 genomes, respectively which restate that 93–11 genome is not well assembled.

Gene prediction and functional annotation

The gene prediction using MAKER-P in gap filled assembly of HR-12 genome yielded 56,284 protein-coding transcripts. We performed two-way comparison of genes from short read and gap filled assemblies, resulting into 31,933 genes with 100 % identity. In addition, 2615 genes were annotated which were unique to gap filled assembly but absent in short read assembly. About 38.6 % (21,736) of genes were fragmented in short read assembly as compared to gap filled assembly. This indicates that use of PacBio reads has significantly improved the *indica* rice genome annotation.

Genes of gap filled assembly were also compared with genes of Nipponbare and 93–11. There were 54,849 genes (e-value of 1^{e-10}) commonly present in both HR-12 and Nipponbare. Similarly 54,130 genes were commonly present in HR-12 and 93–11. This indicates that over ~97 % of annotated rice genes from *indica*(93–11) and *japonica* (Nipponbare) were annotated in HR-12 gap filled assembly. Overall, 1950 genes were unique to HR-12 genome in comparison with Nipponbare and 93–11 genomes. These unique genes in HR-12 were annotated and classified as proteins of unknown function (96), uncharacterized proteins (1349) and proteins with known function (475). There were 30 resistance genes with NBS-LRR, LRR and kinase domains.

Ka/Ks analysis for annotated proteins

To determine the evolutionary selection pressure on proteome between *indica* (HR-12 and 93–11) and*japonica* (Nipponbare) genomes, synonymous amino acids substitution rates (Ks) and non-synonymous amino acids substitution rates (Ka) were calculated. The Ka/Ks ratio can reflect the selection pressure between gene pairs (homologs) caused by evolutionary forces like natural mutations. This homolog group analyses resulted 3424 and 5527 gene pairs between *indica-indica* (HR-12 v/s 93–11), and *indica-japonica* (HR-12 v/s Nipponbare), respectively. Homolog groups were classified into three categories based on Ka/Ks ratio with probability value of <0.05 (Fisher exact test). The *indica-indica* homologs comparison yielded 32 and 3392 genes being positively and negatively selected with probability value of <0.05 (Fisher exact test), respectively. Similarly, *indica-japonica* homologs comparison showed 29 and 5498 genes were under positive (Ka/Ks ratio >1) and negative (Ka/Ks ratio <1) selection, respectively.

Genome-wide comparison of SSRs and SNPs in sequenced genomes

The SSRs are repetitive DNA sequences used as co-dominant molecular markers to determine genetic diversity and mapping of genes/QTLs. Identification of SSRs in genome sequences will increase the availability of more genomic resources. In total, 114508, 145371, 135501 and 141177 SSRs were identified in Illumina HR-12 assembly, long read HR-12 assembly, Nipponbare and 93–11 genomes, respectively. Mono and di-nucleotide repeats were more in HR-12 genome as compared to Nipponbare and 93–11. The 'AT' (di-repeats) and AAG, AGG (tri-repeats) repeats were more predominant in HR-12 genome. The 'AT' rich di-nucleotide repeats are reported to be most abundant in rice genome as compared to other SSRs [31]. Among tetra- repeats; AAAC, AAAG, AAAT, AACC, AACG, AAGG, ACAT, ACGC, ACGG, ACGT and AGAT types were more abundant in HR-12 genome as compared to Nipponbare and 93–11 genomes.

a

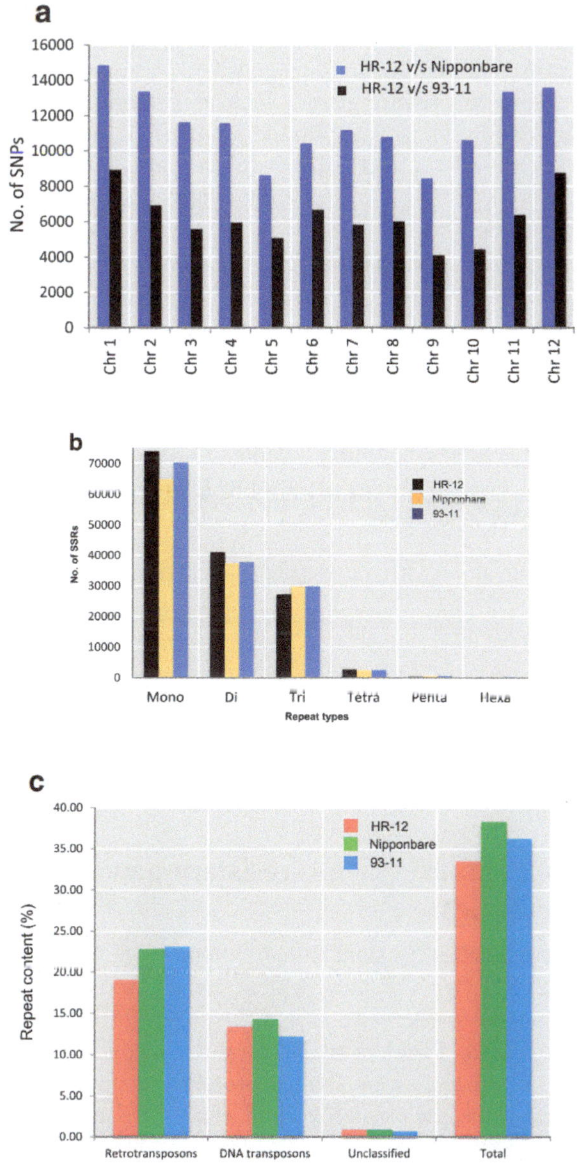

d

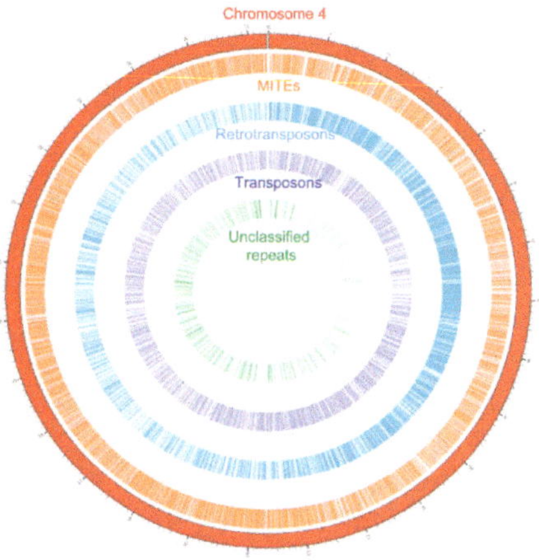

Figure 2: Genome-wide comparison of HR-12, 93–11 and Nipponbare genomes for SNPs, SSRs and repeats. **a** Number of SNPs, (**b**) number of SSRs, (**c**) distribution of repeat elements and (**d**) density of repeats on chromosome 4 of HR-12 genome

More than 19000 SSRs are available in the Gramene database (http://archive.gramene.org/markers/microsat/all-ssr.tab). These sequences were used for e-PCR analysis in HR-12, Nipponbare and 93–11 genomes. There were 11296 (58.55), 16144 (83.68) and 13429 (69.61) SSRs mapped to HR-12, Nipponbare and 93–11 genomes, respectively. This indicates that publicly available SSR markers were skewed towards *japonica* genome. Among 11296 SSRs from HR-12 assembly, 1095 SSRs were monomorphic and 10201 SSRs were polymorphic across three genomes. The pair-wise comparison of polymorphic SSRs has yielded 8914 polymorphic SSRs between HR-12 and Nipponbare, and 8445 SSRs polymorphic between HR-12 and 93–11. As expected, SSRs were less polymorphic within *indica* rice (HR-12 and 93–11) (*indica* type) and highly polymorphic between *japonica* rice (Nipponbare). Among 11296 SSRs in HR-12, 1617 SSRs were unique to HR-12 genome. These unique SSRs (1617) were compared with sequenced genomes of other *indica* varieties, Tetep and Tadukan. Out of these, 1260 SSRs were found in both Tetep and Tadukan, which resulted 209 SSR markers unique to HR-12 genome. Out of remaining 148 SSRs, 128 SSRs were common between HR-12 and Tetep but absent in Tadukan genome and 20 were common between HR 12 and Tadukan.

The comparison of Gramene SSRs in HR-12 (11296) and predicted SSRs (56560) revealed that all 11296 public SSRs were present in predicted SSRs. Elimination of publicly available SSRs in predicted SSRs of HR-12 yielded 47378 SSRs to be novel from *in silico* prediction for *indica* genome. Among these, 20547 di, 18760 tri, 1040 tetra, 332 penta and 6699 were complex SSRs. The conserved flanking regions (100 bp upstream and downstream from SSR motif) of 43834 and 44123 SSRs of HR-12 (47378) were mapped to Nipponbare and 93–11 genomes, respectively. Whereas 3533, 3255 HR-12 SSRs were unique to HR-12 and did not map to Nipponbare and 93–11 genomes.

Genome-wide comparison of SNPs between HR-12 and 93–11, HR-12 and Nipponbare discerned higher level of polymorphism between *indica-japonica* comparison (Fig. 2a). These species-specific variations can be used for marker assisted breeding, positional cloning and evolutionary studies.

Repeats in HR-12 genome

Repetitive DNA sequence can account for substantial portion of the many eukaryotic genomes and lead into genome expansion, gene disruption and gene duplication. Total 34.9 % repeats were found in HR-12 genome (gap filled assembly) as compared to Nipponbare (39.64 %) and 93–11 (37.62 %) (Fig. 2c). Total interspersed repeats were the major elements (33.57 %), consisting of retro-elements (19.14 %), DNA transposons (13.44 %) and unclassified repeats (1 %). In case of retro-elements, long terminal repeat (LTR) had a highest fraction (17.72 %), followed by LINEs (0.98 %) and SINEs (0.43 %). In case of DNA transposons, tourist/harbinger elements content was more (2.84 %), followed by Tc1-IS630-Pogo (2.61 %), hobo-Activator (0.52 %), En-Spm (0.43 %) and MuDR-IS905 (0.14 %). Similar trend of repeat distribution was observed in case of Nipponbare and 93–11 genomes. Chromosome-wise repeat distribution displayed highest repeat content of 39.60 % on chromosome 4 and the lowest of 31.74 % on chromosome 3. The overall distribution of retro-transposons, DNA transposons and Miniature Inverted-repeat Transposable Elements (MITEs) in HR-12 genome for chromosome 4 is represented in Fig. 2d.

Whole genome sequencing of blast susceptible and resistant *indica* rice varieties

We sequenced *indica* cultivars, which are highly resistant (Tetep and Tadukan) and susceptible (HR-12 and Co-39) to rice blast disease caused by *Ascomycetes* pathogen *Magnaporthe oryzae* (Fig. 3a). Mapping of sequence reads to Nipponbare genome enabled us to identify large number of variants in Co-39 (211066 SNPs and 305048 InDels), HR-12 (138412 SNPs and 256296 InDels), Tadukan (158161 SNPs and 262294 InDels) and Tetep (133583 SNPs and 383048 InDels) (Fig. 3c). Large fraction of SNPs and InDels were found in upstream and downstream regions of annotated genes. The transition to transversion (Ts/Tv) ratio was 1.69, 1.58, 1.61 and 1.58 in Co-39, HR-12, Tetep and Tadukan, respectively. The SNPs were located at the intervals of 1764, 2692, 2791 and 2357 nts in the genomes of Co-39, HR-12, Tetep and Tadukan, respectively. Similarly, the InDel rate was per 1207, 1449, 1493 and 1419 bases in Co-39, HR-12, Tetep and Tadukan, respectively. One base insertions (+1) and deletions (−1) were more [32] as compared to other types of InDels (Fig. 3d).

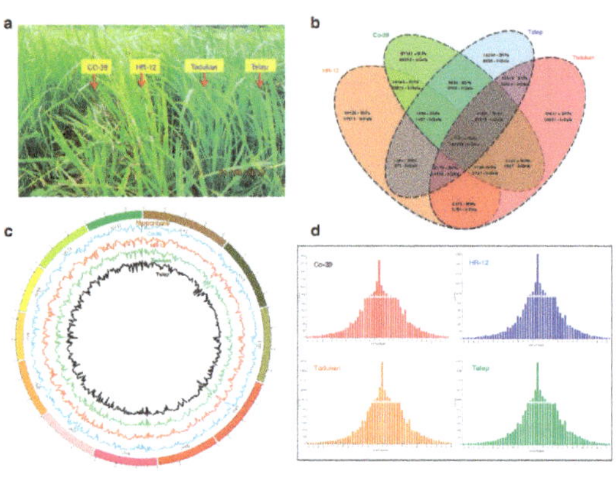

Figure 3: Genome-wide variants identified in HR-12, Co-39, Tadukan and Tetep *indica* cultivars. **a** Blast symptoms, (**b**) Venn diagram showing unique and shared SNPs and InDels between and among susceptible and resistant varieties, **c**) Genome-wide distribution of SNPs in Co-39, HR-12, Tadukan and Tetep and **d**) InDels distribution in Co-39, HR-12, Tadukan and Tetep

Among blast disease susceptible varieties (HR-12 and Co-39), 29566 SNPs and 53372 InDels and among resistant varieties (Tetep and Tadukan), 32545 SNPs and 50824 InDels were common (Fig. 3b). In blast susceptible varieties, 44192 SNPs and 44914 InDels were unique to HR-12 and 97181 SNPs and 69692 InDels were unique to Co-39. Similarly in blast resistant varieties, 36227 SNPs and 16824 InDels were unique to Tadukan where as 18290 SNPs and 8608 InDels were unique to Tetep.

Functional annotation of variants in blast resistant and susceptible varieties

Commonly occurring variants (29566 SNPs and 53372 InDels in susceptible; 32545 SNPs and 50824 InDels in resistant varieties) were annotated to know their functionality in the genome. In total, 24431 and 27562 genes were mutated (SNPs and InDels) in susceptible and resistant varieties, respectively. We further classified these genes based on their protein domains and focused on genes involved in host defense mechanism (resistance genes) such as protein kinase (PK), nucleotide-binding adapter shared by APAF-1, R-proteins and CED-4 (NB-ARC), and leucine-rich repeats (LRR) domains. We observed defense related genes have accumulated more mutations (non-synonymous) in resistant varieties as compared to susceptible varieties (Table 2). There were 40 NB-ARC, 20 LRR, 93 PK genes were mutated (SNPs and InDels) in resistant varieties. In susceptible varieties, 16 NB-ARC, 17 LRR and 67 PK genes were mutated. This functional annotation of variants is highly useful to develop PCR-based functional markers [33] to screen large set of rice germplasm and identify novel alleles of R genes, enabling breeders to rapid introgression of resistance genes and gene pyramiding [34] to elite cultivars for durable blast resistance.

Table 2: Mutation in defense related genes in *indica* rice varieties

Domain Name	Domain ID	No. of genes mutated			
		Blast resistant varieties (Tetep and Tadukan)		Blast susceptible varieties (HR-12 and Co-39)	
		SNPs	InDels	SNPs	InDels
NB-ARC	PF00931	15	25	6	10
LRR	PF00560, PF08263, PF12799	3	17	2	15
Protein Kinase	PF00069, PF07714, PF13947, PF03727, PF00406, PF00781, PF00485	31	62	17	50
TOTAL		49	104	25	75

NB-ARC Nucleotide-binding adaptor shared by APAF-1, R proteins, and CED-4
LRR Leucine Rich Repeats

Mining of blast disease resistant genes in *indica* rice varieties

Till-date, 22 blast resistant *R*-genes cloned [35, 36, 37, 38, 39] from several rice varieties, which confers resistance against *Magnaporthe* isolates. To assess the spectrum of R-genes in resistant (Tetep, Tadukan and IR64) and susceptible (Co-39 and HR-12) cultivars, we performed protein-protein alignment of R-genes. The R-genes such as *Pi37, Pid2, Pid3, Pi25, Pish and Pi64* were conserved at structure level in all *indica* varieties, however, several SNPs and InDels were interrupted these genes. The *Pi54* (*Pikh*) gene was present in Tadukan and Tetep but absent in Co-39, HR-12 and IR64. *Pi-ta*, a broad spectrum resistant gene was found in all four*indica* varieties. However, single amino acid substitution (Ala to Ser) was observed at 918 in *Pi-ta* protein, which reported to be responsible for determining resistance specificity [40]. We observed similar Ser substitution in *Pi-ta* protein in blast susceptible varieties (Co-39 and HR-12). In addition, we identified novel amino acid (Phe to Ser) substitution at 641 residues in all *indica (*Co-39, HR-12, IR64, Tadukan, Tetep) varieties (Fig. 4). Other genes such as *Pib, Piz-t, Pik-m, Pi5, pi21, Pb1, Pik, Pik-p and Pi1* were absent in all *indica* varieties (Table 3). Majority of R-genes were either mutated or fragmented in case of susceptible varieties. The R genes in host and AVR genes in pathogen follow gene -for -gene hypothesis [41]. Survey of avirulent genes (AVR) in*Magnaporthe* population isolated from HR-12 from Southern India showed predominance of *AVR-Pizt*, *AVR-Pita*, *AVR-Pik*, *AVR-Pii*, and *AVR-Pia* (Unpublished). The cognate R-genes like *Piz-t*, *Pi-ta*, *Pik* and *Pia* were absent in HR-12. Similarly, *Magnaporthe* isolates from Co-39 showed presence of *AVR-Pizt*, *AVR-Pii* and *AVR-Pik* and the cognate R-genes, *Piz-t* and *Pik* genes were absent in Co-39 genome. Thus, study of R and AVR genes in the crop ecosystems play an important role to understand the evolution of new pathotypes and to design better plant breeding strategies.

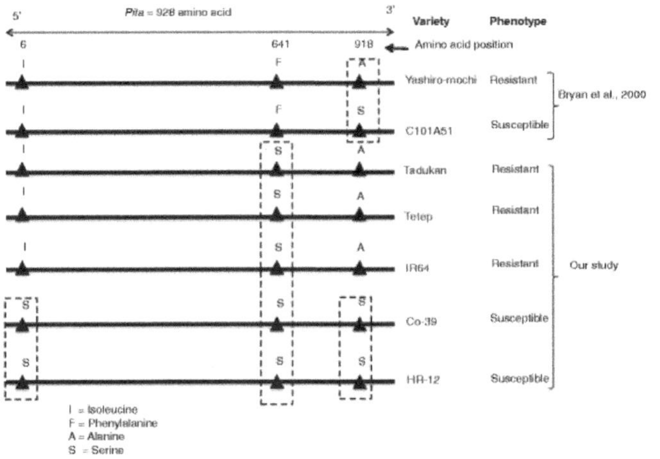

Figure 4: Structure of *Pi-ta* gene in *indica* cultivars as compared to reference gene

Table 3: Distribution of cloned blast resistance genes in sequenced rice varieties

'R' gene	Donor Variety	Group	HR-12	Co-39	Tetep	Tadukan	IR-64
Pib	Tohoku IL-9	*Japonica*	-	-	-	-	M
Pi-ta	Yashiro-mochi	*Japonica*	-	-	+	+	+
Pi54 (Pik-h)	Tetep	*Indica*	-	-	+	+	-
Pid2	Digu	*Indica*	M	M	M	M	M
Pi9	75-1-127	*Indica*	M	-	-	-	M
Piz-t	Toride 1	*Japonica*	-	-	-	-	-
Pi37	St.No. 1	*Japonica*	M	M	M	M	M
Pi36	Kasalath	*Indica*	M	-	-	M	M
Pik-m-TS1	Tsuyuake	*Japonica*	-	-	-	-	-
Pik-m-TS2	Tsuyuake	*Japonica*	-	-	-	-	M
pi21	Owarihatamochi	*Japonica*	-	-	-	-	-
Pit	Nipponbare	*Japonica*	M	-	M	-	M
Pi5-1	RIL260	*Japonica*	-	-	-	-	M
Pi5-2	RIL260	*Japonica*	-	+	-	-	-
Pid3	Digu	*Indica*	M	M	M	M	M
Pb1	St.No. 1	*Japonica*	-	-	-	-	-
Pish	Shin 2, Norin 22	*Japonica*	M	M	M	M	M
Pi25	Gumei 2	*Japonica*	M	M	M	M	M
Pia (RGA4)	Sasanishiki, Aichi-asahi	*Japonica*	-	M	M	M	M
Pik-p-1	K60	*Japonica*	-	-	-	-	M
Pik-p-2	K60	*Japonica*	-	-	-	-	M
Pik-1	Kusabue	*Japonica*	-	-	-	-	M
Pik-2	Kusabue	*Japonica*	-	-	-	-	M
Pi54rh	*O. rhizomatis*	Wild species	-	-	M	M	-
Pi1-5	LAC23, C101LAC	*Indica*	-	-	-	-	M
Pi1-6	LAC23, C101LAC	*Indica*	-	-	-	-	M
Pi64	Yangmaogu (YMG)	*Japonica*	M	M	M	M	M

+ = Present
- = Absent
M Mutated but protein structure retained

CONCLUSIONS

High quality genomic resources are prerequisite for rice breeding. The available *indica* reference genomes 93–11 and IR64 were sequenced and assembled using short read sequencing technology. Inherent drawbacks of short read genome assemblies may cause alignment problem during reference mapping and study of structural variations. Due to non-availability of high quality *indica* reference genome, Illumina reads from 3000 rice accessions (*indica*, *aus*, *tropical japonica*, *temperate japonica* and *aromatic*) were mapped to Nipponbare genome to identify variants [42]. Although another *indica* genome IR64 [3] is sequenced, its chromosomes consist of complex recombination of fragments (mosaic) from the genealogy of more than 38 parents [43] including *indica*, *japonica* and wild species. With advent of third generation sequencing technology, it is possible to sequence plant genomes with higher accuracy and coverage. Thus, we report improved *de novo*assembly of *indica* cultivar HR-12 using combinatorial approach of short and long reads which covers over 86 % of estimated genome size. The gap filling strategy with the help of long reads has improved the short read assembly with respect to genome size, repeat content, and number of protein coding genes. Whole genome comparison of HR-12, 93–11 and Nipponbare genomes revealed 1950 genes and 1617 SSRs unique to HR-12 genome with similar level of repeat content.

Rice blast is a major biotic stress in rice, which reduces yield significantly. To understand disease resistance genes, we sequenced highly resistant (Tetep and Tadukan) and susceptible (Co-39, HR-12) *indica* cultivars. Functional annotation of SNPs and InDels in *indica* cultivars showed higher non-synonymous substitutions in defense related genes [44, 45] containing NB-ARC, LRR and PK domains among resistant varieties indicating strong diversifying selection to confer resistance to fast evolving blast pathogen. Allele mining for resistance genes in all sequenced genomes showed presence/absence polymorphism and large number of structural variations. Most of the R genes were conserved in resistant cultivars with point mutations and InDels whereas loss of R-gene structure was noticed in susceptible cultivars. The broad-spectrum resistance in Tadukan, Tetep and IR64 could be attributed to presence of intact *Pi-ta* and *Pi54* (absent in IR64*)*. The rice cultivars, Tadukan and Tetep were found to be resistant against most of blast races across the globe [46, 47] and these have been used in IRRI for developing IR64. Previously, high rate of mutation in R genes have been reported in Rice (*Xa21*) [48] and Maize *(Rp1)* [49]. These allelic variations created by mutations will result in evolution of novel R genes/alleles and selection of genes that can recognize pathogen avirulence gene products. Identification of R genes/alleles is a prerequisite for effective utilization of genetic and genomic resources in modern plant breeding, which is driven by new genomics tools.

Data availability

The raw sequence reads deposited under NCBI Sequence Read Archive (SRA) accession numbers SRP067809 (HR-12), SRP067775 (Co-39), SRP067810 (Tetep), and SRP067808 (Tadukan). The whole genome shotgun projects have been deposited at DDBJ/EMBL/Genbank under the accessions AZTA02000000 (HR-12), LQHE01000000 (Co-39), LQHG01000000 (Tetep), and LQHF01000000 (Tadukan).

ACKNOWLEDGEMENTS

This work was supported by Department of Biotechnology (DBT), Government of India. We would like to acknowledge Texas A&M, USA for awarding Monsanto's Beachell-Borlaug International Fellowship to H. B, Mahesh for his Ph.D programme. We are grateful to Genomics facility members; Chandana S, Shilpa S, Sahana S, Ashwini M, Sheetal A and Goutham for their help in sample preparation, sequencing and bioinformatics analyses. We thank Ram Laxman, Pacific Biosciences for helping in PacBio sequencing. We acknowledge Indian Institute of Rice Research (IIRR, formerly known as DRR) for providing Co-39, HR-12, Tetep and Tadukan seed material.

AUTHORS' CONTRIBUTIONS

HBM performed isolation of DNA and RNA, genome estimation, genome sequencing, assembly, annotation, variant calling and R-gene analyses of all *indica* cultivars, prepared all tables, figures and wrote the manuscript. MDS prepared the DNA for PacBio sequencing, KaKs and Repeat analysis and helped in writing manuscript. SS coordinated PacBio sequencing and performed gap filling of Illumina short read assembly and helped in writing manuscript. AR prepared CIRCOS figures. SH helped in writing manuscript. G-LW helped in writing manuscript. MG conceived the project, designed the experiment and wrote the manuscript. All authors read and approved the final manuscript

REFERENCES

1. Goff SA, Ricke D, Lan T-H, Presting G, Wang R, Dunn M, Glazebrook J, Sessions A, Oeller P, Varma H, et al. A Draft Sequence of the Rice Genome (*Oryza sativa* L. ssp. japonica. Science. 2002;296(5565):92–100.
2. Yu J, Hu S, Wang J, Wong GK, Li S, Liu B, Deng Y, Dai L, Zhou Y, Zhang X, et al. A draft sequence of the rice genome (*Oryza sativa* L. ssp. indica). Science. 2002;296(5565):79–92.
3. Schatz MC, Maron LG, Stein JC, Wences AH, Gurtowski J, Biggers E, Lee H, Kramer M, Antoniou E, Ghiban E, et al. Whole genome de novo assemblies of three divergent strains of rice, Oryza sativa, document novel gene space of aus and indica. Genome Biol. 2014;15(11):506
4. Sakai H, Kanamori H, Arai-Kichise Y, Shibata-Hatta M, Ebana K, Oono Y, Kurita K, Fujisawa H, Katagiri S, Mukai Y. Construction of pseudomolecule sequences of the aus rice cultivar kasalath for comparative genomics of Asian cultivated rice. DNA Res. 2014;21(4):397–405.
5. Gowda M, Shirke MD, Mahesh HB, Chandarana P, Rajamani A, Chattoo BB. Genome analysis of rice-blast fungus Magnaporthe oryzae field isolates from southern India. Genomics Data. 2015;5:284–91.
6. Krishan A. Rapid flow cytofluorometric analysis of mammalian cell cycle by propidium iodide staining. J Cell Biol. 1975;66(1):188–93.
7. Zerbino DR, Birney E. Velvet: algorithms for de novo short read assembly using de Bruijn graphs. Genome Res. 2008;18(5):821–9.
8. Luo R, Liu B, Xie Y, Li Z, Huang W, Yuan J, He G, Chen Y, Pan Q, Liu Y. SOAPdenovo2: an empirically improved memory-efficient short-read de novo assembler. Gigascience. 2012;1(1):18.
9. Zimin AV, Marçais G, Puiu D, Roberts M, Salzberg SL, Yorke JA. The MaSuRCA genome assembler. Bioinformatics. 2013;29(21):2669–77.
10. Gurevich A, Saveliev V, Vyahhi N, Tesler G. QUAST: quality assessment tool for genome assemblies. Bioinformatics. 2013;29(8):1072–5.
11. English AC, Richards S, Han Y, Wang M, Vee V, Qu J, Qin X, Muzny DM, Reid JG, Worley KC. Mind the gap: upgrading genomes with Pacific Biosciences RS long-read sequencing technology. PLoS One. 2012;7(11), e47768.
12. Xue W, Li J-T, Zhu Y-P, Hou G-Y, Kong X-F, Kuang Y-Y, Sun X-W. L_RNA_scaffolder: scaffolding genomes with transcripts. BMC Genomics. 2013;14(1):604.
13. Grabherr MG, Haas BJ, Yassour M, Levin JZ, Thompson DA, Amit I, Adiconis X, Fan L, Raychowdhury R, Zeng Q, et al. Trinity: reconstructing a full-length transcriptome without a genome from RNA-Seq data. Nat Biotechnol. 2011;29(7):644–52.
14. Assefa S, Keane TM, Otto TD, Newbold C, Berriman M. ABACAS: algorithm-based automatic contiguation of assembled sequences. Bioinformatics. 2009;25(15):1968–9.
15. Soderlund C, Nelson W, Shoemaker A, Paterson A. SyMAP: A system for discovering and viewing syntenic regions of FPC maps. Genome Res. 2006;16(9):1159–68.
16. Parra G, Bradnam K, Korf I. CEGMA: a pipeline to accurately annotate core genes in eukaryotic genomes. Bioinformatics. 2007;23(9):1061–7.
17. Campbell MS, Law M, Holt C, Stein JC, Moghe GD, Hufnagel DE, Lei J, Achawanantakun R, Jiao D, Lawrence CJ. MAKER-P: A tool kit for the rapid creation, management, and quality control of plant genome annotations. Plant Physiol. 2014;164(2):513–24.
18. Jones P, Binns D, Chang H-Y, Fraser M, Li W, McAnulla C, McWilliam H, Maslen J, Mitchell A, Nuka G. InterProScan 5: genome-scale protein function classification. Bioinformatics. 2014;30(9):1236–40.
19. Nei M, Gojobori T. Simple methods for estimating the numbers of synonymous and nonsynonymous nucleotide substitutions. Mol Biol Evol. 1986;3(5):418–26.
20. Zhang Z, Li J, Zhao X-Q, Wang J, Wong GK-S, Yu J. KaKs_Calculator: Calculating Ka and Ks Through Model Selection and Model Averaging. Genomics Proteomics Bioinformatics. 2006;4(4):259–63.
21. Thiel T, Michalek W, Varshney R, Graner A. Exploiting EST databases for the development and characterization of gene-derived SSR-markers in barley (Hordeum vulgare L.). Theor Appl Genet. 2003;106(3):411–22.

22. Schuler GD. Sequence mapping by electronic PCR. Genome Res. 1997;7(5):541–50.

23. Li H, Durbin R. Fast and accurate short read alignment with Burrows–Wheeler transform. Bioinformatics. 2009;25(14):1754–60.

24. Li H, Handsaker B, Wysoker A, Fennell T, Ruan J, Homer N, Marth G, Abecasis G, Durbin R, Genome Project Data Processing S. The Sequence Alignment/Map format and SAMtools. Bioinformatics. 2009;25(16):2078–9.

25. Auwera GA, Carneiro MO, Hartl C, Poplin R, del Angel G, Levy-Moonshine A, Jordan T, Shakir K, Roazen D, Thibault J. From FastQ Data to High-Confidence Variant Calls: The Genome Analysis Toolkit Best Practices Pipeline. Curr. Protoc. Bioinformatics. 2013:11.10. 11–11.10. 33.

26. DePristo MA, Banks E, Poplin R, Garimella KV, Maguire JR, Hartl C, Philippakis AA, del Angel G, Rivas MA, Hanna M. A framework for variation discovery and genotyping using next-generation DNA sequencing data. Nat Genet. 2011;43(5):491–8.

27. Cingolani P, Platts A, le Wang L, Coon M, Nguyen T, Wang L, Land SJ, Lu X, Ruden DM. A program for annotating and predicting the effects of single nucleotide polymorphisms, SnpEff: SNPs in the genome of *Drosophila melanogaster*strain w1118; iso-2; iso-3. Fly. 2012;6(2):80–92.

28. Miller JR, Koren S, Sutton G. Assembly algorithms for next-generation sequencing data. Genomics. 2010;95(6):315–27.

29. Koren S, Phillippy AM. One chromosome, one contig: complete microbial genomes from long-read sequencing and assembly. Curr Opin Microbiol. 2015;23:110–20.

30. Arumuganathan K, Earle ED. Nuclear DNA content of some important plant species. Plant Mol Biol Rep. 1991;9(3):208–18.

31. Temnykh S, DeClerck G, Lukashova A, Lipovich L, Cartinhour S, McCouch S. Computational and experimental analysis of microsatellites in rice (Oryza sativa L.): frequency, length variation, transposon associations, and genetic marker potential. Genome Res. 2001;11(8):1441–52.

32. Ma J, Bennetzen JL. Rapid recent growth and divergence of rice nuclear genomes. Proc Natl Acad Sci U S A. 2004;101(34):12404–10.

33. Hayashi K, Yoshida H, Ashikawa I. Development of PCR-based allele-specific and InDel marker sets for nine rice blast resistance genes. Theor Appl Genet. 2006;113(2):251 60.

34. Hittalmani S, Parco A, Mew T, Zeigler R, Huang N. Fine mapping and DNA marker-assisted pyramiding of the three major genes for blast resistance in rice. Theor Appl Genet. 2000;100(7):1121–8.

35. Sharma T, Rai A, Gupta S, Vijayan J, Devanna B, Ray S. Rice blast management through host-plant resistance: retrospect and prospects. Agricultural Research. 2012;1(1):37–52.

36. Chen J, Shi Y, Liu W, Chai R, Fu Y, Zhuang J, Wu J. A Pid3 allele from rice cultivar Gumei2 confers resistance to Magnaporthe oryzae. J Genet Genomics. 2011;38(5):209–16.

37. Das A, Soubam D, Singh P, Thakur S, Singh N, Sharma T. A novel blast resistance gene, Pi54rh cloned from wild species of rice, Oryza rhizomatis confers broad spectrum resistance to Magnaporthe oryzae. Funct Integr Genomics. 2012;12(2):215–28.

38. Hua L, Wu J, Chen C, Wu W, He X, Lin F, Wang L, Ashikawa I, Matsumoto T, Wang L. The isolation of Pi1, an allele at the Pik locus which confers broad spectrum resistance to rice blast. Theor Appl Genet. 2012;125(5):1047–55.

39. Ma J, Lei C, Xu X, Hao K, Wang J, Cheng Z, Ma X, Ma J, Zhou K, Zhang X. Pi64, encoding a novel CC-NBS-LRR protein, confers resistance to leaf and neck blast in rice. Mol Plant-Microbe Interact. 2015;28:558–568.

40. Bryan GT, Wu K-S, Farrall L, Jia Y, Hershey HP, McAdams SA, Faulk KN, Donaldson GK, Tarchini R, Valent B. A single amino acid difference distinguishes resistant and susceptible alleles of the rice blast resistance gene Pi-ta. The Plant Cell Online. 2000;12(11):2033–45.

41. Flor HH. Current status of the gene-for-gene concept. Annu Rev Phytopathol. 1971;9(1):275–96.

42. Li J-Y, Wang J, Zeigler RS. The 3,000 rice genomes project: new opportunities and challenges for future rice research. GigaScience. 2014;3(1):1–3.

43. Ballini E, Berruyer R, Morel JB, Lebrun MH, Nottéghem JL, Tharreau D. Modern elite rice varieties of the 'Green Revolution'have retained a large introgression from wild rice around the Pi33 rice blast resistance locus. New Phytol. 2007;175(2):340–50.

44. Hulbert SH, Webb CA, Smith SM, Sun Q. Resistance gene complexes: evolution and utilization. Annu Rev Phytopathol. 2001;39(1):285–312.

45. Yang S, Li J, Zhang X, Zhang Q, Huang J, Chen J-Q, Hartl DL, Tian D. Rapidly evolving R genes in diverse grass species confer resistance to rice blast disease. Proc Natl Acad Sci. 2013;110(46):18572–7.

46. Padmanabhan S, Mathur S, Misra R. Breeding for blast resistance in India: genetics of blast resistance. Indian J Genet Plant Breed. 1974;34:424–9.

47. Ou SH. Rice diseases. Philippines: IRRI; 1985.

48. Wang G-L, Ruan D-L, Song W-Y, Sideris S, Chen L, Pi L-Y, Zhang S, Zhang Z, Fauquet C, Gaut BS. Xa21D encodes a receptor-like molecule with a leucine-rich repeat domain that determines race-specific recognition and is subject to adaptive evolution. Plant Cell. 1998;10(5):765–79.

49. Sun Q, Collins NC, Ayliffe M, Smith SM, Drake J, Pryor T, Hulbert SH. Recombination between paralogues at the rp1 rust resistance locus in maize. Genetics. 2001;158(1):423–38.

CHAPTER 8

GENOME-WIDE ASSOCIATION STUDY OF GRAIN APPEARANCE AND MILLING QUALITY IN A WORLDWIDE COLLECTION OF INDICA RICE GERMPLASM

Xianjin Qiu, Yunlong Pang, Zhihua Yuan, Danying Xing, Jianlong Xu, Michael Dingkuhn, Zhikang Li, and Guoyou Ye

INTRODUCTION

Rice (*Oryza sativa* L.), a staple cereal crop, feeds more than half of the world's population. Improvement of rice yield and grain quality is the major objective of rice breeding worldwide. Grain quality primarily includes grain appearance, milling, eating and cooking and nutrition qualities. Grain appearance quality mainly includes grain shape and chalkiness. Grain shape, described by grain length (GL), grain width (GW), grain thickness (GT) and grain length to width ratio (GLWR), contributes to grain weight and yield and has a great impact on the market values of rice grain products. A short and round rice grain is generally preferred by consumers in Northern China, Korea and Japan, whereas consumers in Southern China, the USA, and South and Southeast Asian countries prefer a long and slender rice grain [1]. Chalkiness, according to its location within the endosperm, can be divided into white belly, white core and white back in rice grain [2]. Chalky grains, filled with loosely packed, round and large compound of starch granules, are more prone to breakage during milling and reduce head milled rice rate (HMRR). Furthermore, when chalky grains are cooked, cracks occur easily and reduce the palatability of the cooked rice [3]. Degree of endosperm chalkiness (DEC) and percentage of grains with chalkiness (PGWC) are commonly used to measure grain chalkiness. Milling quality determines the final yield and the broken kernel rate of the milled rice, which is of concern for consumers and farmers. Milling quality is measured by brown rice rate (BRR), milled rice rate (MRR) and HMRR. Brown rice is the de-hulled rice with the palea and lemma removed that can be used for cooking and eating. Milled rice is the result of brown rice after removing all of the bran, which consists of aleurone and pericarp, and germ or embryo. Head milled rice is kernel longer than or equal to 3/4 full length of a kernel. Among the above-mentioned three milling quality parameters, HMRR is the most important and greatly affects market value. HMRR depends on varietal characteristics, production factors, and harvesting, drying and milling processes.

Most of the rice quality determining traits are quantitatively inherited, controlled by multiple genes/QTL [2] and affected by growing environment [4]. As for other complex traits, QTL mapping using bi-parental populations derived from parental lines of contrast performance has been widely used to identify QTL for various rice quality traits. Many genes/QTL for rice grain appearance quality and milling quality traits have been reported in the last decades. The *GW1-1* and *GW1-2* [5], *qGRL1.1* [6], *GS2* [7], *GW3* and *GW6* [8], *qGL-4b* [9], *qPGWC-7* [10] *qGL-7*[11], *qGRL7.1* [6], *gw8.1* [12], *gw9.1* [13], *tgw11* [14] have been fine mapped. The *GW2* [15],*GS3* [16], *qGL-3* [17], *qSW5* [18], *GS5* [19], *Chalk5* [20], *TGW6* [21], *GW6a* [22], *SRS1* [23],*GL7*/*GW7* [24, 25], *GW8* [26] and *CycT1;3* [27] have been cloned. The usefulness of some of the well characterized genes/QTL was proven in an *indica* population of diverse breeding lines [4].

However, QTL mapping using a single bi-parental population has a few important limitations including the need to create a population segregating for the target trait, the ability to assess only two alleles per locus, and the limited number of meiosis. Only a limited number of QTL can be identified, since only QTL for which the two parents differ can be detected. Given the limited scope of each study, mapping the same trait in different bi-parental populations may yield different QTL. More than two alleles are likely to segregate per locus, and the

Citation: Qiu X, Pang Y, Yuan Z, Xing D, Xu J, Dingkuhn M, et al. (2015) Genome-Wide Association Study of Grain Appearance and Milling Quality in a Worldwide Collection of Indica Rice Germplasm. PLoS ONE 10(12): e0145577. doi:10.1371/journal.pone.0145577.

directions of QTL effects may vary depending on the genetic background because of epistasis, pleiotropy and QTL- by-environment interaction (QEI) [28].

An alternative QTL mapping approach is to conduct association analysis in a large germplasm collection, known as association mapping. Association mapping is based on linkage disequilibrium (LD) or the non-independence of alleles in a population. Association mapping studies using sparse SSR markers has been proven to be successful in identifying marker-trait associations (MTA) in rice [29–32]. With the development of high-throughput sequencing and SNP chip techniques, genome-wide association study (GWAS) using high density markers becomes more and more popular in rice genetics. The first GWAS in rice was conducted by Huang et al. [33] using 517 rice landraces genotyped by sallow sequencing. Among the QTL identified two and five were for GW and GL, respectively [33]. Applying the same method to a large panel of 950 worldwide rice accessions, Huang et al. [34] identified two, four and two QTL for GW, GL and thousand grain weight (TGW), respectively. Based on genotyping 44,100 SNP across 413 diverse accessions from 82 countries, Zhao et al. [35] identified eight,10 and 15 QTL for GW, GL and GLWR, respectively. Begum et al. [36] identified seven, 10 and 11 QTL for GW, GL and GLWR, respectively, using advanced lines from IRRI's irrigated breeding program genotyped with 71,710 SNP generated by Genotyping-by-sequencing (GBS). Similarly, using 533 *O. sativa* landrace and elite accessions genotyped with 4,358,600 SNP derived from GBS, Yang et al. [37] identified 10, 11, 19 and 21 QTL for GLWR, TGW, GW and GL, respectively. These studies clearly demonstrated that GWAS offers higher resolution for QTL mapping and can be used as a powerful complementary strategy to the classical linkage mapping using bi-parental segregation populations.

Many of the reported association studies in rice included grain traits, such as GL, GW, GLWR, GT and TGW. However, rice chalkiness and milling quality traits were less studied. The objective of this study is to identify markers associated with 10 grain appearance quality and milling quality traits using an *indica* collection of 272 accessions. The panel was phenotyped in four locations representing major *indica* rice production environments in China. GBS and Diversity Arrays Technology (DArT) based on next-generation sequencing technique called DArTseq™ was used to generate genome-wide markers. The association panel was characterized for population structure using three different methods and for LD pattern by estimating basal LD and LD decay in the whole population and subpopulations. Sixteen association mapping models were tested to choose the best model for each of the trait-environment combinations. A number of known QTL and novel QTL were identified.

MATERIALS AND METHODS

Association mapping panel

A total of 272 *indica* rice accessions from 31 countries or regions (S1 Table) were used in this study. More accessions came from China (39), Philippines (36), Madagascar (29), India (28), Senegal (24), Sri Lanka (24) and Bangladesh (15). For other countries or regions the number of accessions was fewer than eight.

Phenotypic evaluation

Field trials were conducted using a randomized complete block design with two replicates in four environments including Hangzhou (HZ), Jingzhou (JZ), Sanya (SY) and Shenzhen (SZ) in China. The sowing and transplanting dates varied for the testing environments to fit into the local planting season. In all environments, the plot size was three rows of 10 plants planted at a spacing of 20cm x 25cm. Local farmers' management practices were followed. At maturity, eight plants in the middle row were harvested. The institutions including Yangtze University, Zhejiang Academy of Agricultural Sciences, Shenzhen Institute of Breeding & Innovation permitted the field trials were conducted in their experimental fields.

The nature dried seeds stored in room temperature for three months were used to measure quality traits. GL (mm), GW (mm), GLWR, GT (mm), TGW (g), BRR (%), MRR (%) and HMRR (%) were measured according to the National Rice Grain Quality Assessment Standard of China (GB/T17891-1999). The grain chalkiness characteristics were measured using a JMWT12 Rice Appearance Quality Detector (Dong Fu Jiu Heng, Beijing). PGWC (%) was the percentage of head milled grains with chalkiness. DEC (%) was calculated as the product of PGWC and chalkiness size (the area of chalkiness divided by the area of whole grain). All measurements were conducted for two independent samples.

Phenotypic analysis

Due to various reasons not all accessions had phenotypic data in all the four testing environments. The final population size for each of the trait-environment combinations varied greatly (Table 1). Phenotypic analysis was conducted using linear mixed models to properly handle the unbalance data. For single-site analysis, accession (genotype) was regarded as fixed effect and replicate as random effect. The best linear unbiased estimates (BLUE) of accessions were obtained. For multi-site analysis, all effects including accession (genotype), environment and replicate within environment were regarded as random to estimate variance components. The best linear unbiased predictions (BLUP) for each of the genotype-environment combinations were predicted. All analyses were conducted using the PBTools package of R [38] developed by IRRI (bbi.irri.org). Phenotypic correlations were computed from the BLUE using the "rcorr" function implemented in the R package Hmisc [39]. Narrow-sense heritability (h^2) based on genotypic means was computed using the estimated variance components as $V_G / (V_G + V_{GEI} /s + V_e /sr)$. Where, V_G, V_{GEI}, V_e are the variance of genotype, genotype by environment interaction (GEI) and residual error, respectively, s is the number of environments and r is the number of replicates.

Table 1: Popoation size for 10 traits in four environments

Trait	HZ	JZ	SY	SZ
GL	220	131	216	228
GW	218	132	213	229
GLWR	218	131	215	229
GT	198	132	217	221
TGW	167	130	214	206
DEC	147	128	206	197
PGWC	147	131	213	200
BRR	136	130	201	206
MRR	136	130	213	205
HMRR	136	130	213	205

doi:10.1371/journal.pone.0145577.t001

Genotyping

Genomic DNA was extracted for each accession from the leaf tissues of a single plant using the MATAB method described in Risterucci et al. [40] and then diluted to 100 ng/µl. Genotyping was conducted at Diversity Arrays Technology Pty Ltd. (DArT P/L), Australia, using a method of DArTseq™. The method involves genome complexity reduction using *Pst*I/*Taq*I restriction enzymes followed by Illumina short-read sequencing. *Pst*I specific adapters tagged with 96 different barcodes to encode a plate of DNA samples were ligated to the restriction fragments. The resulting products were amplified and checked for quality. The 96 samples were pooled and run in a single lane on an Illumina Hiseq2000 instrument. The *Pst*I adapters included a sequencing primer so that the tags generated were always read from the *Pst*I sites. As detailed by Courtois et al. [41], the resulting sequences were filtered and split into their respective target datasets, and the barcode sequences were trimmed. The sequences were trimmed at 69 bp (5 bp of the restriction site plus 64 bases with a minimum Q score of 10). A proprietary analytical pipeline developed by DArT P/L was used to produce DArT score tables and SNP tables. The remaining 69 bp sequences were aligned to the Os-Nipponbare-Reference-IRGSP-1.0 [42] pseudomolecule assembly using Bowtie v0.12 [43] with a maximum of three mismatches to recover the position of the restriction site for the DArT markers and the position of the polymorphism(s) within the 69 bp sequences for the SNPs. For the DArT markers, the position given is that of the second base of the 6 base *Pst*I restriction site (5'-C|TGCAG-3') because the mutated base is unknown and can be any of the six. The same sequences were then aligned to the pseudomolecules using BLAST (e-value <1.0 e-20) to assess whether additional sequences could be positioned. The sequences that had only one hit on the pseudomolecules or had more than one hit but with a difference of at least 1.0 e-5 between the first and the second hits were retained for further analyses. When the marker position fell within a Michigan State University-annotated gene (http://rice.plantbiology.msu.edu/), the feature was determined (intron, exon, 3' or 5' UTR), and the name and function of the gene were retrieved. Call rates were measured for all markers, and markers with call rates below 80% were discarded. The allele frequency of the remaining markers was then calculated, and markers for which the minor allele had a frequency below 2.5% were also discarded.

Imputation of missing genotypic data

Markers for which the minor allele frequency (MAF) is below 2.5% were discarded before imputing missing data. Missing data were estimated using Beagle v4.0 [44]. Beagle uses a localized haplotype cluster model. It is a special class of directed acyclic graph which empirically models haplotype frequencies on a local scale and therefore adapts to local structure in the data. The model determines a hidden Markov model that can be used to find the most likely haplotype pair for each individual, given the genotype data for that individual and the graphical haplotype frequency model. The method works iteratively using an expectation maximization type approach. The imputed missing data, probabilities of missing genotypes and inferred haplotypes are calculated from the model that is fitted at the final iteration.

Analysis of population structure

Different statistical methods were employed to infer the number of subgroups in the panel. First, a model based Bayesian clustering analysis method implemented in STRUCTURE software version 2.3.4 [45] was used. The program was run with the following parameters: k, the number of groups in the panel varying from 1 to 9; 10 runs per k value; for each run, 10,000 burnin iterations followed by 10,000 MCMC (Markov Chain Monte Carlo) iterations. The optimal number of groups (k) was determined by Δk following the method described in Evanno et al. [46]. STRUCTURE analysis was conducted for different numbers of markers evenly distributed on the whole genome. Very similar results were obtained when the number of markers was more than 1000. Second, a multi-dimensional scaling and cluster analysis method implemented in Awclust package [47] was used. Third, Tdiscriminant analysis of principal components (DAPC) method implemented in adegenet package [48] was used. The "find.cluster" function was used to identify clusters (k) and the optimal k value was determined according to the Bayesian Information Criterion (BIC), then the "dapc" function was used to check the classification quality.

Kinship coefficient

Three methods were used to calculate kinship coefficient matrix (K). (1) Pairwise_IBS: the distance between two accessions is computed as the proportion of SNP which are different. The IBS is computed by subtracting all values of the distance matrix from 2 and then scaling the resultant matrix so that the minimum value is 0 and the maximum value is 2. It is implemented in TASSEL 5.2.6 [49]. (2) The scaled_IBS: the IBS was scaled to have the mean diagonal element equal to 1+F, where F is the inbreeding coefficient of the current population. It is the default method of the TASSEL 5.2.6 [49]. (3) The VanRaden method: the realized relatedness between individuals is computed according to $ZZ'/(2\sum p_i(1-p_i))$ where Z = W-P, W is the marker matrix, P contains the allele frequencies multiplied by 2, p_i is the allele frequency of marker i, and the sum is over all loci [50]. It is the default method of GAPIT [51].

Linkage disequilibrium

LD was measured by squared allele frequency correlations (r^2) values between the pairs of markers using TASSEL 5.2.6 [49]. Only markers with MAF greater than 10% were used. Marker pairs were discretized into bins of 10 kb and the median r^2 value was used as the estimate of r^2 of a bin. The estimated r^2 values were formulated as a function of physical distances between markers. A power law curve ($y = ax^k$) was fitted to determine the physical position (x) corresponding to a given r^2 value (y) [52]. The following scheme was used to estimate the background LD beyond which LD was assumed to be caused by genetic linkage: One marker per chromosome was randomly sampled and the maximum of all the pairwise r^2values was taken. The median of the maximum r^2 values of 10,000 samples was taken as the basal r^2 value. The physical distance at which r^2 reached to the basal value was determined as the LD decay distance of the population. All analysis was done for the whole population and the three subpopulations inferred by STRUCTURE.

Association analysis

All association models under the unified model for association mapping [53] can be described by considering how the two factors, the population structure (Q) and genetic relatedness between genotypes (K), are considered. We used four options for handling Q and four options for handling K to create 16 models. The four options for Q were: no Q, Q3 derived from STRUCTURE, Q7 derived from STRUCTURE and C7 derived from adegenet. The four options for K were: no K, K computed as pairwise_IBS (Kp scaled_IBS (Ks) and VanRaden method (Kv). All analyses were conducted using TASSEL 5.2.6 [49]. For models without K, known as general linear model (GLM) 1,000 permutations were used. For models with K, known as mixed liner model (MLM), the compressed mixed linear model approach [53, 54] and P3D [54] algorithms were adopted to reduce computing time. The best model for each trait-environment combination was chosen using the mean square difference (MSD) between observed and expected p-values of all marker loci, as suggested by Stich et al. [55]. MSD was a measure for the deviation of the observed p-values from the uniform distribution. Model with a smaller MSD is more appropriate. The critical p-value for declaring significant MTA was set to 0.0001.

RESULTS

Basic statistics of markers

A total of 22,266 polymorphic markers were detected in the panel including 9,340 SNP and 12,926 DArT markers. By removing markers with MAF less than 5%, 18,824 high quality markers (7,885 SNP and 10,939 DArT) were used in association analysis. The number of markers per chromosome ranged from 891 on chromosome 10 to 2,361 on chromosome 1. The size of chromosome varied from 22.8Mb for chromosome 9 to 43.2 Mb for chromosome 1. The whole genome size was 371.7 Mb and the average distance between neighboring markers (marker spacing) was 20.2 kb. The average marker spacing ranged from 16.3 kb for chromosome 11 to 25.9 kb for chromosome 10 (S2 Table). About 70.4% of the neighboring markers distances were below the average value and 97.4% were less than 100 Kb. There were eight gaps devoid of markers larger than 500 kb on chromosomes 1 (D01_26116212—S01_26770440), 2 (D02_13852683—D02_15083642), 4 (D04_8765494—D04_9302397 and D04_16774867—D04_17319136), 6 (D06_18215124—D06_18803206), 7 (D07_11700243—S07_12785983 and S07_13836518—D07_14505070) and 11 (D11_11988263—S11_12639397). More than half of the markers (55.7%) had MAF lower than 0.20 (Fig 1).

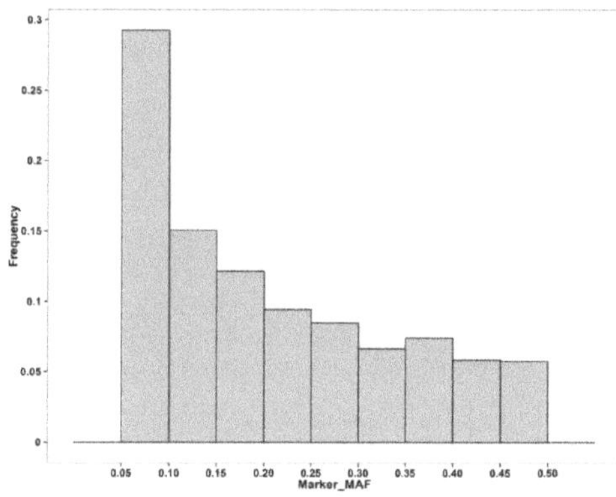

Figure 1: Frequency of markers in different MAF classes. SNP with MAF < 0.05 were excluded.

Phenotypic variation and trait correlation

In general, the panel revealed a wide range variation for all the evaluated traits (Fig 2). Most of the traits appeared to be normally distributed, but some trait-environment combinations showed skewed distributions. For instance, the distributions of DEC in all environments, HMRR and PGWC in HZ and BRR in JZ were seriously skewed (Fig 2). Variance components derived from across-site analysis were given in Table 2. For appearance quality traits, the V_G was much larger than V_{GEI}, indicating that GEI was negligible. For BRR, MRR and HMRR, the V_{GEI} was much larger than V_G, suggesting that GEI was very important. However, the V_e for HMRR was very large, indicating phenotyping precision was low. The h^2 for the appearance quality traits were high ranging from 0.82 for PGWC to 0.98 for GL and GLWR, while were low for the milling quality traits ranging from 0.19 for MRR to 0.27 for HMRR.

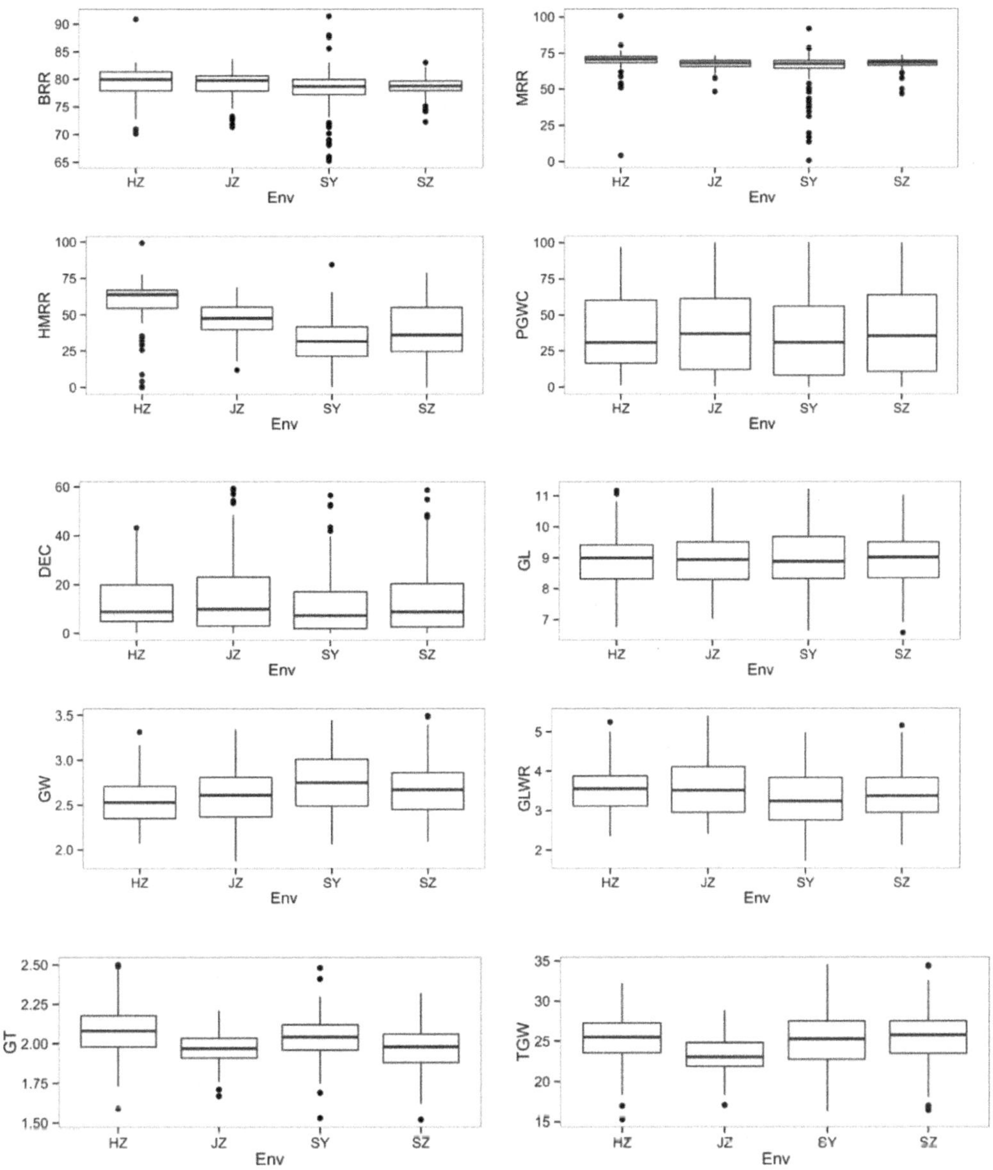

Figure 2: Box plot of 10 rice grain appearance and milling quality traits in four environments. HZ: Hangzhou. JZ: Jingzhou. SY: Sanya. SZ: Shenzhen. GL: Grain Length. GW: Grain Width. GLWR: Grain Length and Width Ratio. GT: Grain Thickness. TGW: Thousand Grain Weight. BRR: Brown Rice Rate. MRR: Milled Rice Rate. HMRR: Head Milled Rice Rate. PGWC. Percentage of Grains With Chalkiness. DEC: Degree of Endosperm chalkiness.

Table 2: Variance components and heritabilty estimated for 10 quality traits by across-site analysis

Trait	V_E	V_G	V_{GEI}	V_e	Rep(Env)	h^2
GL	0.00	0.62	0.00	0.08	0.01	0.98
GW	0.00	0.07	0.00	0.02	0.01	0.97
GLWR	0.00	0.32	0.00	0.05	0.01	0.98
GT	0.00	0.02	0.00	0.01	0.01	0.94
TGW	0.97	8.72	1.92	0.29	0.01	0.94
PGWC	0.62	86.04	59.81	27.7	1.72	0.82
DEC	0.00	465.84	277.44	138.17	6.72	0.84
BRR	0.08	0.57	5.57	0.97	0.31	0.27
MRR	3.01	2.75	44.98	4.71	0.48	0.19
HMRR	108.45	13.64	67.41	230.5	37.29	0.23

V_E: environment variance; V_G: genotype variance; V_{GEI}: genotype- by- environment interaction variance; V_e: residual variance; Rep(Env): replication variance within environment; h^2: narrow-sense heritability.

doi:10.1371/journal.pone.0145577.t002

The phenotypic correlations between traits were similar in the four testing environments. Low correlations were found between grain appearance quality traits and milling quality traits in all environments. GL showed a moderate and negative correlation with GW and a low and negative correlation with GT. Correlation between GW and GT was moderate and positive. TGW positively correlated with GL, GW and GT. High correlation was found between PGWC and DEC in all environments with coefficient ranging from 0.90 in SZ to 0.94 in HZ. The correlation between TGW and GT was moderate in all environments ranging from 0.46 in SY to 0.57 in HZ. Negative and moderate correlation was found between PGWC and HMRR. The correlation between PGWC and DED was strong and positive. The correlations between the three milling quality traits were positive ranging from low to moderate (Fig 3).

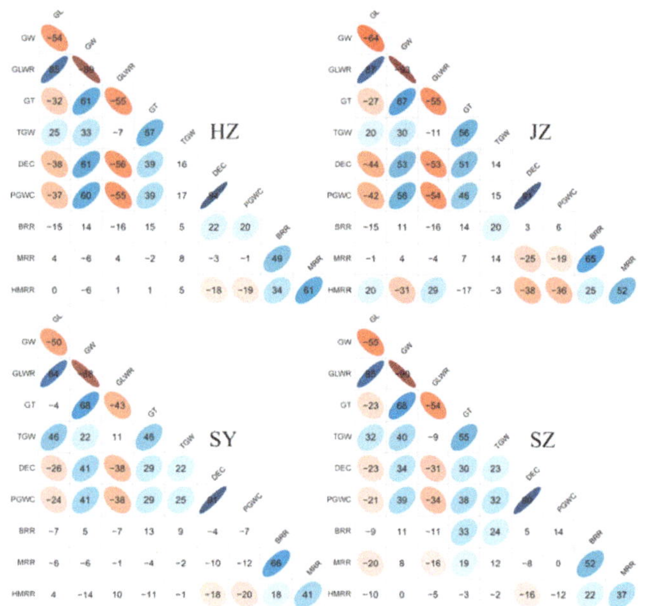

Figure 3: Correlations between 10 gain quality traits measured in HZ, JZ, SY and SZ. The values given were correlation coefficients multiplied by 100. The values without glyphs indicated insignificant at the level of 0.05.

Population structure

According to the value of Δk, there were three groups in the current panel (S1 Fig). Group-I consisted of 88 accessions from 21 countries with 36 from China (36). Group-II had 96 accessions from 19 countries with 30 and 23 from Philippines and Senegal. Group-III had 88 accessions from 13 countries with more accessions from Madagascar (28), India (20) and Sri Lanka (19) (S2 Fig).

Hierarchical cluster analysis using Awclust suggested that the best number of clusters was also three based on Gap stastics (S1 Fig). The classification of accessions into groups was similar to that derived from STRUCTURE. Fifteen accessions of the group Q3-2 of STRUCTURE were assigned to the C3-3 of Awclust (S3 Fig).

The "find.clusters" function of adegenet showed a clear decrease of BIC until $k = 7$ clusters, after which BIC increased (S1 Fig). In this case, the elbow in the curve also matched the smallest BIC, and clearly indicated that seven clusters should be retained. Roughly speaking, the three groups detected by STRUCTURE were divided into two, two and three clusters by adegenet (S3 Fig). If we also chose seven groups for STRUCTURE analysis the Q3-1, Q3-2 and Q3-3 were further divided into two, two and four groups (S3 Fig). There were some differences in the assignments of accessions between $k = 7$ in STRUCTURE and adegent. The biggest difference was that the C7-4 of adegenet was distributed over five of the seven groups of STRUCTURE (S3 Fig).

Whole genome patterns of LD decay

The decay of LD along physical distances was computed for both the whole population (272 accessions) and three subpopulations inferred by STRUCTURE $k = 3$. The determination coefficients (R^2) of regression models were 0.87, 0.93, 0.88 and 0.95 for the whole population, Q3-1, Q3-2 and Q3-3, respectively (S3 Table). The maximum r^2 was observed in the 0–10 kb marker interval for the whole population and subpopulations. The maximum r^2 was lower in the whole population (0.65) than in the subpopulations. Q3-2 had the largest maximum r^2 than Q3-1 (0.76) and Q3-3 (0.71). The basal r^2 was 0.11, 0.16, 0.16 and 0.19 for the whole population, Q3-1, Q3-2 and Q3-3, respectively (S3 Table). The LD decay distance was 150 kb for the whole population, which was longer (slower) than those for Q3-1 (110 kb) and Q3-3 (80 kb) but shorter (faster) than that for Q3-2 (240 kb) (Fig 4 and S3 Table).

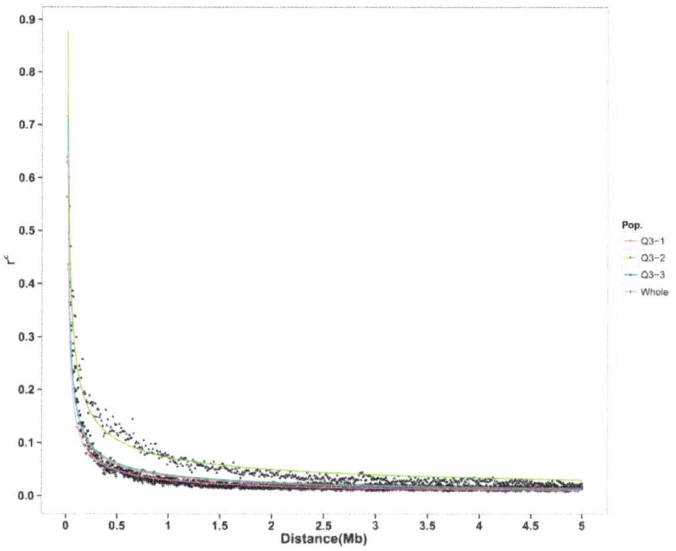

Figure 4: LD decay plot in the whole population and three subpopulations inferred by STRUCTURE. Y axis is the average r^2 value of each 10 kb region and X axis is physical distance between markers in unit of Mb. A power law curve ($y = ax^k$) was fitted to determine the physical position (x) corresponding to a given r^2 value (y)

Association analysis

Comparison of models.

A total of 16 models for detecting associations were compared to choose the best model for each of the trait-environment combinations using MSD. The complete MSD results were given in S4 Table. Fig 5 gave the quantile-quantile plots for the 16 models for GL measured in HZ to illustrate the relative differences between models. The best model varied with traits and environments (S4 Table). For all the trait-environment combinations except BRR in SY, the naïve model was always inferior to other models (S4 Table), indicating complex population structure and genetic relatedness between accessions present in our panel affected the detection of

MTA. According to the average MSD value across all trait-environment combinations, the Q models (Q3, Q7 and C7) that corrected for population structure only, were inferior to the K models (Kp, Ks and Kv) that corrected for the genetic relationships between accessions and the QK models (Q3Kp, Q3Ks, Q3Kv, Q7Kp, Q7Ks, Q7Kv, C7Kp, C7Ks and C7Kv) that corrected for both of population structure and genetic relationship (S4 Table). The QK model was the best model for most of the trait-environment combinations. The Q7 model was slightly better than the other two Q models. The Kp model was better than the Ks model and Kv model. The C7Kp model was better than the other QK models. For the models with the same Q but different K, the Q-Kp and Q-Ks models were similar and were slightly better than the Q-Kv model. The models with the same K but different Q had very similar MSD values. In the following sections only the MTA identified using the best model for each of the trait-environment combinations were presented and discussed.

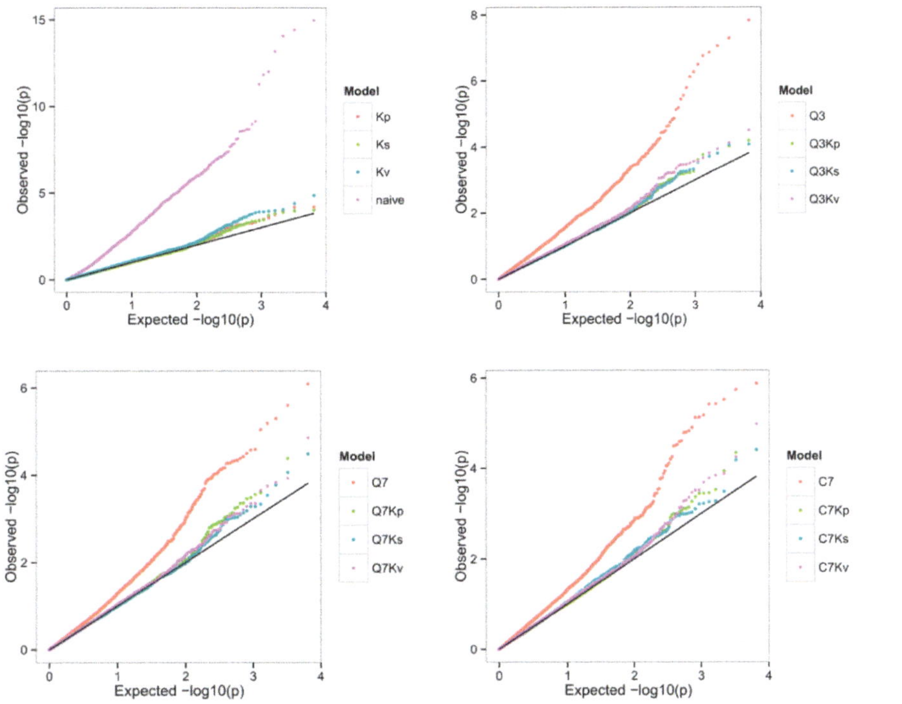

Figure 5: Quantile-quantile plots of 16 models for GL measured in HZ. The horizontal and vertical axes are -log10 transformed expected p-values and observed p-values, respectively. Model with more uniformly distributed p-values is better.

Marker-trait associations

A total of 74 markers on all 12 chromosomes were found to be significantly associated with the 10 measured traits. By delineating significant markers with LD higher than 0.2 into a single QTL, the 74 markers were grouped into 38 QTL (Table 3).

Table 3: QTL identified for 10 traits in four environments.

QTL	Env	Marker	Alleles[a]	MAF	p	Effect[b]	R^2 (%)[c]	Known gene/QTL
qGL2.1	SY	D02_25608957	T/A	0.16	5.93E-05	0.43	7.5	
qGL2.2	SY	S02_30573143	T/A	0.36	7.03E-05	0.5	7.4	
	SZ	S02_30573143	T/A	0.35	5.04E-05	0.43	7.0	
	HZ	S02_30664746	G/T	0.30	2.63E-05	0.51	8.1	
	SZ	S02_30664746	G/T	0.29	7.44E-05	0.44	6.7	
	HZ	S02_30664811	C/T	0.30	2.63E-05	0.51	8.1	
	SZ	S02_30664811	C/T	0.29	7.44E-05	0.44	6.7	
qGL3.1	HZ	S03_16663793	T/C	0.30	2.97E-05	0.79	8.0	*GS3*
	SY	S03_16663793	T/C	0.28	3.26E-06	0.98	10.2	
	SZ	S03_16663793	T/C	0.29	2.19E-05	0.69	7.7	
	SZ	S03_16731182	C/T	0.41	2.42E-05	-0.46	7.6	
	SZ	S03_16748937	C/T	0.47	3.92E-06	0.87	9.2	
	SZ	S03_16762099	A/G	0.47	1.13E-05	0.53	8.3	
	HZ	S03_16858510	C/G	0.45	7.31E-05	0.61	7.2	
	SY	S03_16858510	C/G	0.42	2.00E-05	0.69	8.5	
	SZ	S03_16858510	C/G	0.43	8.03E-07	0.65	10.5	
	SZ	S03_16996600	C/A	0.38	4.74E-05	0.48	7.0	
	SZ	S03_17000111	C/T	0.39	1.80E-05	0.5	7.9	
qGL5.1	SY	S05_5368086	C/T	0.38	8.44E-05	-0.48	7.2	*qSW5*
	SY	S05_5368151	A/G	0.38	8.10E-05	-0.48	7.2	
	SY	S05_5369527	A/T	0.37	5.71E-05	-0.49	7.5	
qGL7.1	HZ	S07_22019132	G/A	0.05	3.51E-05	-1.04	7.8	
	SY	S07_22019132	G/A	0.05	4.85E-05	-1.11	7.7	
	SZ	S07_22019132	G/A	0.05	9.95E-05	-0.81	6.4	
	SY	S07_22087092	T/A	0.13	6.54E-05	-0.66	7.4	
	HZ	S07_22099651	A/G	0.05	3.51E-05	-1.04	7.8	
	SY	S07_22099651	A/G	0.05	4.85E-05	-1.11	7.7	
	SZ	S07_22099651	A/G	0.05	9.95E-05	-0.81	6.4	

qGL7.2	SY	S07_22844850	C/T	0.18	6.30E-05	-0.63	7.5	*qGRL7.1*
qGL9.1	JZ	S09_10774239	A/T	0.50	7.39E-05	0.53	11.7	
qGL11.1	JZ	S11_2576141	T/G	0.35	5.76E-05	-0.57	12.0	*CycT1;3*
	SY	S11_2576141	T/G	0.35	6.41E-06	-0.63	9.6	
qGW4.1	SZ	S04_7167748	C/T	0.42	5.91E-05	0.16	6.8	
qGW5.1	SY	S05_5368086	C/T	0.38	1.94E-05	0.19	8.7	*qSW5*
	SZ	S05_5368086	C/T	0.35	3.26E-07	0.21	11.2	
	JZ	S05_5368151	A/G	0.39	1.06E-05	0.23	16.0	
	SY	S05_5368151	A/G	0.37	8.43E-06	0.19	9.5	
	SZ	S05_5368151	A/G	0.35	1.57E-07	0.22	11.9	
	JZ	S05_5369527	A/T	0.38	9.45E-06	0.24	16.2	
	SY	S05_5369527	A/T	0.37	5.09E-06	0.2	10.0	
	SZ	S05_5369527	A/T	0.34	5.54E-08	0.22	12.8	
qGW7.1	JZ	S07_19590107	G/C	0.08	4.41E-05	0.29	13.6	
	JZ	S07_19592078	T/A	0.08	4.41E-05	0.29	13.6	
qGW10.1	SZ	S10_14195109	A/T	0.05	7.54E-05	0.36	6.6	
	SZ	S10_14262980	A/G	0.05	7.54E-05	0.36	6.6	
qGW11.1	HZ	D11_7124485	T/A	0.17	8.76E-05	0.12	7.0	
qGLWR1.1	HZ	S01 5811755	A/T	0.08	4.83E-05	0.52	7.4	
	SZ	S01_5811755	A/T	0.08	9.32E-05	0.48	5.8	
qGLWR3.1	SZ	D03_14077712	T/A	0.15	9.18E-05	-0.4	5.8	
qGLWR3.2	SY	S03_16858510	C/G	0.42	9.44E-05	0.38	6.3	*GS3*
	SZ	S03_16858510	C/G	0.43	4.06E-05	0.36	6.4	
qGLWR5.1	JZ	S05_5368086	C/T	0.40	4.03E-05	-0.46	12.9	*qSW5*
	SY	S05_5368086	C/T	0.39	1.47E-06	-0.35	9.7	
	SZ	S05_5368086	C/T	0.35	3.09E-05	-0.31	6.6	
	JZ	S05_5368151	A/G	0.39	2.30E-05	-0.48	13.8	
	SY	S05_5368151	A/G	0.38	4.80E-07	-0.37	10.7	
	SZ	S05_5368151	A/G	0.35	1.42E-05	-0.33	7.2	
	HZ	S05_5369527	A/T	0.36	4.95E-05	-0.32	7.4	
	JZ	S05_5369527	A/T	0.38	1.96E-05	-0.48	14.1	
	SY	S05_5369527	A/T	0.38	1.97E-07	-0.38	11.4	
	SZ	S05_5369527	A/T	0.34	9.05E-06	-0.34	7.6	
qGLWR7.1	SY	S07_25383179	T/C	0.07	5.93E-05	0.5	6.6	*Srs1*
qGLWR11.1	JZ	S11_2576141	T/G	0.35	1.03E-05	-0.42	15.1	*CycT1;3*
qGT1.1	HZ	D01_2405901	A/T	0.10	2.77E-05	0.14	9.4	
qGT4.1	SZ	S04_29442031	T/C	0.16	6.57E-05	-0.13	7.0	
	SZ	S04_29495556	T/C	0.16	6.57E-05	-0.13	7.0	
qGT5.1	HZ	S05_5368086	C/T	0.33	2.37E-06	0.13	12.0	*qSW5*
	SZ	S05_5368086	C/T	0.35	1.13E-08	0.13	14.9	
	HZ	S05_5368151	A/G	0.33	9.02E-07	0.14	13.1	
	JZ	S05_5368151	A/G	0.39	8.98E-05	0.09	8.6	
	SZ	S05_5368151	A/G	0.35	4.47E-09	0.14	15.8	
	HZ	S05_5369527	A/T	0.32	8.77E-07	0.13	13.1	
	SY	S05_5369527	A/T	0.37	1.70E-05	0.09	8.5	
	SZ	S05_5369527	A/T	0.34	4.91E-09	0.14	15.7	
qGT5.2	SZ	S05_18307877	G/C	0.21	8.96E-05	-0.11	6.7	
	SZ	S05_18314309	C/T	0.21	8.96E-05	-0.11	6.7	
	SZ	S05_18321742	A/G	0.21	8.96E-05	-0.11	6.7	
qTGW2.1	SZ	S02_30688426	C/A	0.30	8.86E-05	1.78	7.7	
qTGW5.1	HZ	S05_4808850	T/A	0.48	6.59E-05	-2.06	10.1	
	HZ	S05_4856967	A/T	0.47	6.88E-05	-2.07	10.0	
	HZ	S05_4880400	A/G	0.47	6.88E-05	-2.07	10.0	
	HZ	S05_4889259	C/T	0.47	6.88E-05	-2.07	10.0	
qTGW5.2	SZ	S05_5368086	C/T	0.33	5.38E-05	2.07	8.2	*qSW5*
	SZ	S05_5369527	A/T	0.32	2.66E-05	2.14	8.9	
qTGW7.1	HZ	S07_21628162	C/T	0.08	5.77E-05	-3.72	10.3	
qTGW7.2	SZ	S07_22684516	C/A	0.05	9.17E-05	-4.26	7.7	*qGRL7.1*
qPGWC5.1	SZ	S05_5368086	C/T	0.32	5.16E-08	28.21	15.9	*qSW5*
	SZ	S05_5368151	A/G	0.32	2.97E-07	26.78	14.0	
	JZ	S05_5369527	A/T	0.38	6.52E-05	24.33	10.9	
	SZ	S05_5369527	A/T	0.31	1.72E-08	29.41	17.1	
qPGWC6.1	SZ	S06_6372444	A/C	0.09	5.76E-06	35.63	10.8	
qDEC1.1	SZ	S01_5811836	A/C	0.17	8.66E-05	11.18	7.3	
	SZ	S01_5972775	T/G	0.23	7.96E-05	10.37	7.4	
qDEC5.1	HZ	S05_4180511	T/C	0.10	7.25E-05	11.03	10.3	
qDEC5.2	SY	S05_5368086	C/T	0.37	7.27E-05	7.89	7.5	*qSW5*
	SY	S05_5368151	A/G	0.36	4.18E-05	8.2	8.0	
	SY	S05_5369527	A/T	0.36	2.05E-05	8.59	8.7	
	SZ	S05_5369527	A/T	0.29	3.23E-05	9.75	8.3	
qDEC5.3	SZ	S05_23647280	A/G	0.35	4.93E-05	-8.5	7.9	
qDEC9.1	HZ	S09_16262697	T/C	0.07	4.02E-05	-16.93	11.1	
qBRR2.1	SZ	D02_8060607	A/T	0.26	2.96E-05	-1.02	8.2	
	SZ	D02_8075892	T/A	0.27	2.24E-05	-1.03	8.5	
qBRR2.2	JZ	D02_25652984	A/T	0.06	1.96E-05	-4.75	15.6	
qBRR3.1	HZ	S03_22356459	T/C	0.07	8.52E-05	3.84	10.6	
qBRR4.1	HZ	S04_32002470	C/T	0.09	3.59E-06	3.65	14.5	
qBRR7.1	HZ	S07_29497149	T/A	0.21	8.03E-05	2.31	10.7	
qMRR1.1	HZ	S01_41957924	G/A	0.07	4.44E-05	10.06	6.6	
	HZ	S01_41957989	G/T	0.07	4.44E-05	10.06	6.6	
	HZ	S01_41970208	T/A	0.07	4.44E-05	10.06	6.6	
qMRR2.1	JZ	D02_26022390	A/T	0.07	1.61E-06	-8.71	19.9	
	JZ	D02_26022455	A/T	0.07	1.61E-06	-8.71	19.9	
	JZ	D02_26163208	T/A	0.05	3.29E-07	-7.59	22.8	
	JZ	D02_26265084	A/T	0.05	5.22E-06	-7.04	17.8	
	JZ	D02_26312559	A/T	0.06	4.14E-05	-7.86	14.1	
qMRR3.1	SZ	S03_4636350	A/G	0.21	9.69E-05	-1.91	7.3	
qMRR7.1	JZ	S07_5075055	A/G	0.06	9.29E-05	-5.45	12.8	
	JZ	S07_5081323	T/A	0.06	9.29E-05	-5.45	12.8	
qMRR11.1	SY	D11_6895700	T/A	0.06	4.05E-05	-12.33	7.8	
qHMRR3.1	SZ	D03_31372694	T/A	0.40	2.85E-05	12.34	9.0	
	SZ	D03_31380514	A/T	0.48	1.67E-05	12.53	9.5	
qHMRR4.1	HZ	S04_28018950	T/C	0.05	3.92E-06	-26.22	14.9	
qHMRR5.1	SZ	S05_12426396	T/C	0.18	8.01E-05	15.42	7.9	
qHMRR6.1	HZ	D06_2712642	A/T	0.12	9.37E-05	-14.26	11.0	
qHMRR8.1	SZ	S08_2638229	G/A	0.12	3.16E-05	17.8	8.9	
qHMRR12.1	SZ	S12_26692818	C/T	0.36	1.79E-05	12.81	9.5	
	SZ	S12_26695253	G/A	0.32	3.18E-05	12.65	8.9	

[a]Major/Minor allele.
[b]Effect: Allelic effect with respect to the minor allele.
[c]R^2: Phenotypic variation explained.

doi:10.1371/journal.pone.0145577.t003

For GL, eight QTL were identified on chromosomes 2, 3, 5, 7, 9 and 11. *qGL2.1*(D02_25608957) was detected in SY and explained 7.5% of the phenotypic variation (R^2 = 7.5%). *qGL2.2* (S02_305731433, S02_30664746 and S02_30664811) was detected in HZ, SY and SZ with R^2 being 8.1%, 7.4% and 7.0%, respectively. *qGL3.1* (S03_16663793, S03_16731182, S03_16748937, S03_16762099, S03_16858510, S03_16996600 and S03_17000111) was identified in HZ, SY and SZ with R^2 being 8.0%, 10.2% and 10.5%, respectively. *qGL5.1* (S05_5368086, S05_5368151 and S05_5369527) was detected in SY with a R^2 of 7.5%. The *qGL7.1* (S07_22019132, S07_22087092 and S07_22099651) was detected in HZ, SY and SZ and explained 7.8%, 7.7% and 6.4% of the phenotypic variance.*qGL7.2* (S07_22844850) was detected in SY with a R^2 of 7.5%. *qGL9.1* (S09_10774239) was detected in JZ and explained 11.7% of the phenotypic variance. *qGL11.1* (S011_2576141) was detected in JZ and SY with R^2 being 12.0% and 9.6%, respectively (Table 3).

For GW, five QTL on chromosomes 4, 5, 7, 10 and 11 were identified. *qGW4.1* (S04_7167748) was detected in SZ with a R^2 of 6.8%. *qGW5.1* (S05_5368086, S05_5368151 and S05_5369527) was detected in JZ, SY and SZ with R^2 being 16.2%, 10.0% and 12.8%, respectively. *qGW7.1* (S07_19590107 and S07_19592078) was detected in JZ with a R^2 of 13.6%. *qGW10.1* (S010_14195109 and S010_14262980) was detected in SZ with a R^2 of 6.6%. *qGW11.1* (D011_7124485) was detected in SZ with a R^2 of 7.0% (Table 3).

For GLWR, six QTL were identified on chromosomes 1, 3, 5, 7 and 11. *qGLWR1.1*(S01_5811755) was detected in HZ and SZ with R^2 being 7.4% and 5.8%, respectively.*qGLWR3.1* (D03_14077712) was detected in SZ with a R^2 of 5.8%. *qGLWR3.2*(S03_16858510) was detected in SY and SZ with R^2 being 6.3% and 6.4%, respectively.*qGLWR5.1* (S05_5368086, S05_5368151 and S05_5369527) was detected in HZ, JZ, SY and SZ with R^2 being 7.4%, 14.1%, 11.4% and 7.6%, respectively. *qGLWR7.1* (S07_25383179) was detected in SY with a R^2 of 6.6%. *qGLWR11.1* (S011_2576141) was detected in JZ and explained 6.6% of the phenotypic variance (Table 3).

For GT, four QTL were found on chromosome 1, 4 and 5. *qGT1.1* (D01_2405901) was detected in HZ with a R^2 of 9.4%. *qGT4.1* (S04_29442031 and S04_29495556) was detected in SZ with a R^2 of 7.0%. *qGT5.1* (S05_5368086, S05_5368151 and S05_5369527) was detected in HZ, JZ, SY and SZ with R^2 being 13.1%, 8.6%, 8.5% and 15.8%, respectively.*qGT5.2* (S05_18307877, S05_18314309 and S05_18321742) was detected in SZ with a R^2 of 6.7% (Table 3).

For TGW, five QTL were found on chromosome 2, 5 and 7. *qTGW2.1* (S02_30688426) was detected in SZ with a R^2 of 7.7%. *qTGW5.1* (S05_4808850, S05_4856967, S05_4880400 and S05_4889259) was detected in HZ with a R^2 of 10.1%. *qTGW5.2* (S05_5368086 and S05_5369527) was detected in SZ with a R^2 of 8.9%. *qTGW7.1* (S07_21628162) was detected in HZ with a R^2 of 10.3%. *qTGW7.2* (S07_22684516) was detected in SZ with a R^2 of 7.7% (Table 3).

For PGWC, two QTL on chromosomes 5 and 6 were identified. *qPGWC5.1* (S05_5368086, S05_5368151 and S05_5369527) was detected in JZ and SZ with R^2 being 10.9% and 17.1%, respectively. *qPGWC6.1* (S06_6372444) was detected in SZ with a R^2 of 10.8% (Table 3).

For DEC, five QTL were identified on chromosomes 1, 5 and 9. *qDEC1.1* (S01_5811836 and S01_5972775) was detected in SZ with a R^2 of 7.4%. *qDEC5.1* (S05_4180511) was detected in HZ with a R^2 of 10.3%. *qDEC5.2* (S05_5368086, S05_5368151 and S05_5369527) was detected in SY and SZ with R^2 being 8.7% and 8.3%, respectively. *qDEC5.3* (S05_23647280) was detected in SZ with a R^2 of 7.9%. *qDEC9.1* (S09_16262697) was detected in HZ with a $R^2$of 11.1% (Table 3).

For BRR, five QTL located on chromosomes 2, 3, 4 and 7 were detected. *qBRR2.1*(D02_8060607 and D02_8075892) was identified in SZ with a R^2 of 8.5%. *qBRR2.2*(D02_25652984) was identified in SZ with a R^2 of 15.6%. *qBRR3.1* (S03_22356459) was identified in HZ with a R^2 of 10.6%. *qBRR4.1* (S04_32002470) was identified in HZ with a R^2 of 14.5% (Table 3).

For MRR, five QTL located on chromosomes 1, 2, 3, 7 and 11 were detected. *qMRR1.1*(S01_41957924, S01_41957989 and S01_41970208) was identified in HZ with a R^2 of 6.6%. *qMRR2.1* (D02_26022390, D02_26022455, D02_26163208, D02_26265084 and D02_26312559) was identified in JZ with a R^2 of 22.8%. *qMRR3.1* (S03_4636350) was identified in SZ with a R^2 of 7.3%. *qMRR7.1* (S07_5075055 and S07_5081323) was identified in JZ with a R^2 of12.8%. *qMRR11.1* (D11_6895700) was identified in SY with a R^2 of 7.8% (Table 3).

Six QTL were detected for HMRR. They were on chromosomes 3, 4, 5, 6, 8 and 12. *qHMRR3.1*(D03_31372694 and D03_31380514) was identified in SZ with a R^2 of 9.5%. *qHMRR4.1*(S04_28018950) was identified in HZ with a R^2 of 14.9%. *qHMRR5.1* (S05_12426396) was identified in SZ with a R^2 of 7.9%. *qHMRR6.1* (D06_2712642) was identified in HZ with a R^2 of 11.0%. *qHMRR8.1* (S08_2638229) was identified in SZ with a R^2 of 8.9%. *qHMRR12.1*(S012_26692818 and S012_26695253) was identified in SZ with a R^2 of 9.5% (Table 3).

Some of the QTL for different traits were in the same chromosomal regions and were regarded as the same QTL when the number of QTL was counted across traits. They were *qDEC1.1* and*qGLWR1.1* on chromosome 1; *qBRR2.2* and *qGL2.1* on chromosome 2; *qGL2.2* and *qTGW2.1*on chromosome 2; *qGLWR3.2* and *qGL3.1* on chromosome 3; *qGL5.1*, *qGW5.1*, *qGLWR5.1*,*qGT5.1*, *qTGW5.2*, *qPGWC5.1* and *qDEC5.2* on chromosome 5; *qGL7.2* and *qTGW7.2* on chromosome 7; *qGW11.1* and *qMRR11.1* on chromosome 11; and *qGLWR11.1* and *qGL11.1*on chromosome 11.

DISCUSSION

Phenotypic variation and trait correlation

Great variation was observed for all traits in all testing environments in the present study, suggesting that the association panel can be effectively used for GWAS of various quality traits. Grain appearance quality traits, particularly grain size and weight traits had high heritability and showed only small GEI (Table 2), suggesting that breeding for a wide range of growing environments could be feasible and effective. Grain chalkiness and milling quality traits had low heritability and large contribution of GEI and as a result will be more difficult to improve through conventional selection breeding. The relationship between grain quality traits in rice is complex. In the present study, positive correlations were observed between TGW and GL, GW and GT, which was consistent with previous studies [2, 26, 56]. The coefficients were moderate, which might be the result of negative correlations between GL and GW and GT. The correlation between GL and GW were -0.54, -0.64, -0.50 and -0.55 in HZ, JZ, SY and SZ, respectively. However, a weak and positive (0.11 in SY) or negative (-0.11, -0.07 and -0.09 in JZ, HZ and SZ, respectively) correlation was observed between TGW and GLWR, which was different with the result of Xu et al. [57]. They found TGW and GLWR had a moderate and positive correlation (0.44). Consistent with previous reports, BRR was negatively correlated with GL but positively correlated with GW, even though the correlation coefficients were small. In the process of milling, the long and slender rice grain is easier to break than short and round one. Chalkiness is also an important factor influencing rice milling traits especially HMRR. Chalky grains are also more prone to breakage at chalky area. We found HMRR was negatively correlated with PGWC and DEC, which was consistent with the study of Zheng et al. [58].

LD pattern

The basal r^2 was 0.11 and estimated LD decay distance was 150 kb for the whole population (S3 Table). The LD decay distance ranged from 80 kb to 240 kb in three subpopulations. This was in agreement with previous finding that cultivated rice has a long range LD from close to 100 kb to over 200 kb [59, 60]. Compared with the whole population, LD dropped much faster in Q3-1 and Q3-3 but more slowly in Q3-2. The effect of population structure on LD pattern has been widely reported [52, 60–62]. Varying patterns of LD decay in different subpopulations were likely reflecting their breeding histories and the origins of accessions in subpopulations [63] and may influence the QTL mapping resolution of the panel. Given that our average marker density (one marker per 20.2 kb) was higher than the r^2 decay, we expected to have reasonable resolution and power to identify common variants of large effect associated with traits of interest.

Model comparison

It is well-known that the results of association analysis using different models would be very different. However, there is no generally acceptable best model. It is likely that the best model differs among populations, traits and growing environments. The most appropriate model should be chosen by comparing different models. In the present study, MSD proposed by Stich et al. [55] was used to determine the best model for each of the 40 trait-environment combinations. The results showed that the naïve model and the Q models were almost always inferior to the K and QK models (S4 Table). The average MSD value across all trait-environment combinations of the naïve model was much higher compared to those of the Q models, indicating introduction of population structure could tremendously reduce false positives. Q derived from different numbers of subpopulations (*k*) also leaded to different results. According to MSD, increasing *k* from three to seven resulted in better control of false positives (S4 Table). When population structure was detected by different methods, the resultant Q and as a result the association results were different. For instance, the Q7 model, which used the Bayesian clustering analysis method implemented in STRUCTURE, was better than the C7 model, which used the Q from the DAPC method (S4 Table). It was clear from our results that the three Q models were rarely the best model and the K and QK models were better (S4 Table). This was consistent with most of the previous studies that showed that GLM method (Q model) was inferior to MLM method (K and QK models). A recent example in rice was the study ofCourtois et al. [41]. They performed association study for 16 root depth and associated traits in a panel of japonica accessions and chose the best model by comparing the likelihoods of each model using the BIC [64]. Their results showed that the QK model was the best model for 13 measured traits, the K model was the best for the other 3 traits, and the Q model was always inferior to the other two models.

For the models with the same Q but different K, the Q-Kp and Q-Ks models were similar and were slightly better than the Q-Kv model (S4 Table). The models with the same K but different Q had very similar MSD values, indicating that once the genetic relatedness between genotypes was well accounted, the differences between methods for inferring Q became less important. Therefore, when both Q and K were considered simultaneously, K played a more important role than Q and different K should be tried to determine the best model.

The best model differed for the same trait measured in different environments and different traits measured in the same environment (S4 Table). For traits with low heritability, the best model changed greatly across environments. For instance, BRR (h^2 = 0.27), MRR (h^2 = 0.19) and HMRR (h^2 = 0.23), there was no common best model in the four environments. However, for traits with high heritability the same best model tended to be found across environments. For instance, Q7Kv was the best model for GL (h^2 = 0.98) in HZ, JZ and SY and C7Kv was the best model for GW (h^2 = 0.97) in HZ and SZ and Kv was the best model for GW in JZ and SY.

In the present study, we selected the model with the smallest MSD value as the best model. It was observed that models with very similar MSD values had very similar association results, indicating that MSD is a simple but effective method for choosing the appropriate model(s) in GWAS.

The effects of environment and QEI

In the present study, the total number of identified QTL varied greatly across environments, indicating that testing environment had important influence on GWAS. Out of the 38 identified QTL only one (*qGLWR5.1* (*qGT5.1*)), four (*qGL2.2*, *qGL3.1*, *qGL7.1* and qGW5.1) and four QTL (*qGLWR1.1*, *qGL11.1*, *qPGWC5.1* (*qDEC5.2*) and qGLWR3.2) were detected in four, three and two environments, respectively (Table 3). Furthermore, even for *qGLWR5.1*, which was detected in all the four environments, the R^2 varied from 7.4% to 14.1% across environments. These results may be partially attributed to QEI. For BRR, MRR and HMRR that showed significant GEI, all QTL were detected in only one of the four environments (Table 3). However, the big differences in population size across environments may also be partially responsible for the observed differences between environments. For most of the traits the population size was much larger in SY and SZ than in HZ and JZ (Table 1).

GWAS is effective

In the present study, five out of the 38 identified QTL were adjacent to previously known genes or fine mapped QTL, suggesting that association mapping was an effective way to map QTL for rice grain quality traits. On chromosome 5, *qGL5.1*, *qGW5.1*, *qGLWR5.1*, *qGT5.1*, *qTGW5.2*,*qPGWC5.1* and *qDEC5.2* were located in genomic region (S05_5368086, S05_5368151 and S05_5369527) close to the cloned gene *qSW5* affecting GW and TGW [18]. *qSW5* was in 5362625bp - 5370506bp in the Os-Nipponbare-Reference- IRGSP-1.0 according to the complete cds sequence of *qSW5* in *indica*. The *qSW5* functioned as negative regulator of grain width and weight and a deletion in *qSW5* resulted in a significant increase in sink size owing to an increase in cell number in the outer glume of the rice flower and finally increased the grain width and weight [18]. The *qSW5* was also reported to regulate GL, GLWR and GT [65]. However, the effects of *qSW5* on PGWC and DEC have not been reported before. Our results suggested that *qSW5* could be used to increase GL and reduce DEC and PGWC simultaneously. The *qGL3.1* (S03_16663793, S03_16731182, S03_16748937, S03_16762099, S03_16858510, S03_16996600 and S03_17000111) and *qGLWR3.2* (S03_16858510) associated with GL and GLWR, were close to *GS3* (16729228–16735641) affecting grain shape. *GS3* functions as a negative regulator for grain size which is a major QTL for GL and TGW and a minor QTL for GW and GT [16]. The *qGL7.2* (S07_22844850) and *qTGW7.2*(S07_22684516) were located in the region harboring the previously fine mapped QTL affecting GL, GW and GLWR, *qGRL7.1* (22127.4Kb—24526.7Kb) [6]. The *qGLWR7.1*(S07_25383179) was located in the region of *Srs1* (25381698–25389547) affecting grain size [23]. The *srs1* results in small and round seed due to the reduction in both cell length and cell number in the longitudinal direction, and the elongation of the cells in the lateral direction of the lemma. The *qGL11.1* (S11_2576141) and *qGLWR11.1* (S11_2576141) were close to *CycT1;3*(2730143–2736649) whose down regulation results in a shorter grain.

Thirty-three of the identified QTL were located on chromosomal regions where no QTL for the measured traits have been reported and can be regarded as novel QTL. Five of the QTL including *qDEC1.1* (*qGLWR1.1*), *qBRR2.2* (*qGL2.1*), *qTGW2.1* (*qGL2.2*), *qGW11.1*(*qMRR11.1*) and *qGL7.1* affected multiple traits with relatively large effects and/or were detected in multiple environments.

The number of QTL identified is small

Compared to the previous GWAS studies in rice the number of QTLs detected was small in the present study. Many of the known genes affecting traits measured were not detected in the present study. The most likely reason is the small population size. Most of the association panels had more than 400 accessions while the present panel only had 230 or fewer accessions. The effect of population size on GWAS is well known. It is particularly important for detecting loci with low MAF to use a large panel. Because the panel size was small markers with MAF <0.05 were filleted out before association analysis. The trait-affecting genes were missed out if markers in strong LD with them had low MAF in our panel. For example, the gene*PGL2* (LOC_Os02g51320) was involved in controlling grain length and weight of rice through interaction with a typical bHLH protein APG. Marker S02_31422564 very close to this gene (1.4 kb) was removed before association analysis since its MAF was only 0.02. The big differences between environments in population size could also explain why most of the QTL (53%) were detected in SZ. Nevertheless, loci with large effects were also detected for some of the traits with small population size. For instance, the population size of GW in JZ is only 132, but*qGW5.1* was identified with a high R^2 of 16.2%. Therefore, provided that the MAF is high enough, major loci would still be detected even if the population size was small. Another possible reason is that the present panel consists of only *indica* accessions while most of the previous studies used panels with a few subspecicies/ecotypes. It is expected that genes that determine the major trait differences between subspecies/ecotypes could not be detected in our study but might be detectable in other studies. We also expected that more false positives were present when a panel with very diverse accessions was used. The use of a single subspecies allows controlling population structure at finer scale and as results reduces false positives. The critical significant threshold value used to declare significant association also influenced the analysis results. For instance, the *qGW5.1* was found to be significant in JZ, SY and SZ but not in HZ. By inspecting the p values in HZ, it could be seen that the p values of S05_5368086, S05_5368151 and S05_5369527, were 0.00065, 0.00035 and 0.00014, respectively, which were only slightly higher than the applied critical significant threshold 10^{-4}. It might be worth pointing out that the number of markers used in the present study was smaller than or similar to the reported GWAS in rice. However, the marker density was sufficient for most of the genomic regions based on LD decay pattern. Furthermore, the eight gaps devoid of markers below critical LD did not harbor any known genes affecting the studied traits. Therefore, the relatively low marker density was unlikely a major contributor to the small number of detected QTL. Nevertheless, the marker density might be too low in some chromosomal regions where LD decays more rapidly or breaks down due to recombination events [41]. For instance, the *GW2*(LOC_Os02g14720) controlling grain width and weight, was in the marker interval of S02_8069581—S01_8181973. However, we could not identify any markers associated with GW or TGW in this region, since LD was very low (about 0.02). Additional markers in such regions are clearly needed.

CONCLUSIONS

Wide ranges of genetic variations were present in the *indica* association panel for all 10 measured grain quality traits. The accessions could serve as sources of novel genes and alleles for improving rice appearance quality and milling quality traits. GWAS conducted using proper statistical models is a powerful method for elucidating the genetic basis of rice quality traits. Although it is generally true that MLM is better than the naïve model and GLM (Q) model for GWAS, the most appropriate model varies across traits and environments and must be chosen by comparing different models. MSD was found to be a simple but effective method for choosing the appropriate model in GWAS. Based on the selected best model for each of the trait-environment combinations, 38 QTL were identified for 10 rice grain appearance quality and milling quality traits including GL, GW, GLWR, GT, TGW, PGWC, DEC, BRR,

MRR and HMRR. Eight genomic regions harbored QTL for more than one trait. Five QTL were found to concur with previously cloned genes or fine mapped QTL. Thirty-three were novel. Five of the novel QTL including *qDEC1.1* (*qGLWR1.1*), *qBRR2.2* (*qGL2.1*), *qTGW2.1* (*qGL2.2*), *qGW11.1*(*qMRR11.1*) and *qGL7.1* affected multiple traits with relatively large effects and/or were detected in multiple environments. Our results provided an insight into the genetic architecture of rice grain appearance quality and milling quality traits. The novel QTL are important candidates for future functional characterization studies.

SUPPORTING INFORMATION

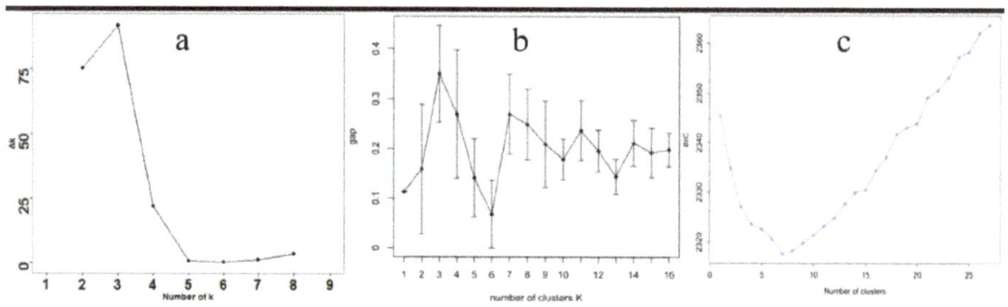

S1 Figure: Population structure analysis of 272 indica accessions. (a) Log probability change (Δk) as function of number of subpopulations (STRUCTURE). (b) Gap statistics as function of number of clusters (Awclust). (c) Value of BIC versus number of clusters (adegenet).

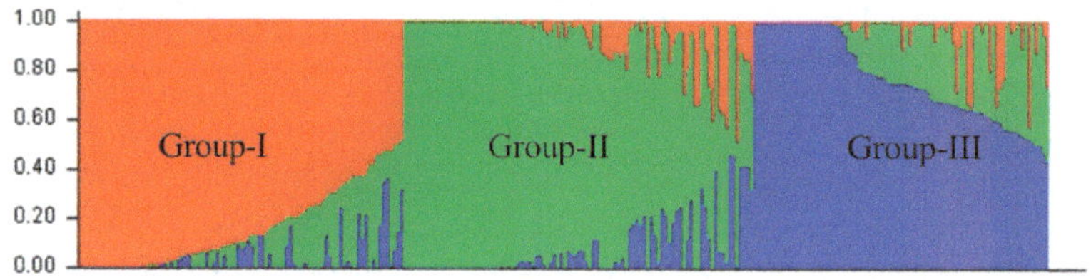

S2 Figure: Population sub-structuring of 272 *indica* accessions based on 1083 SNP.

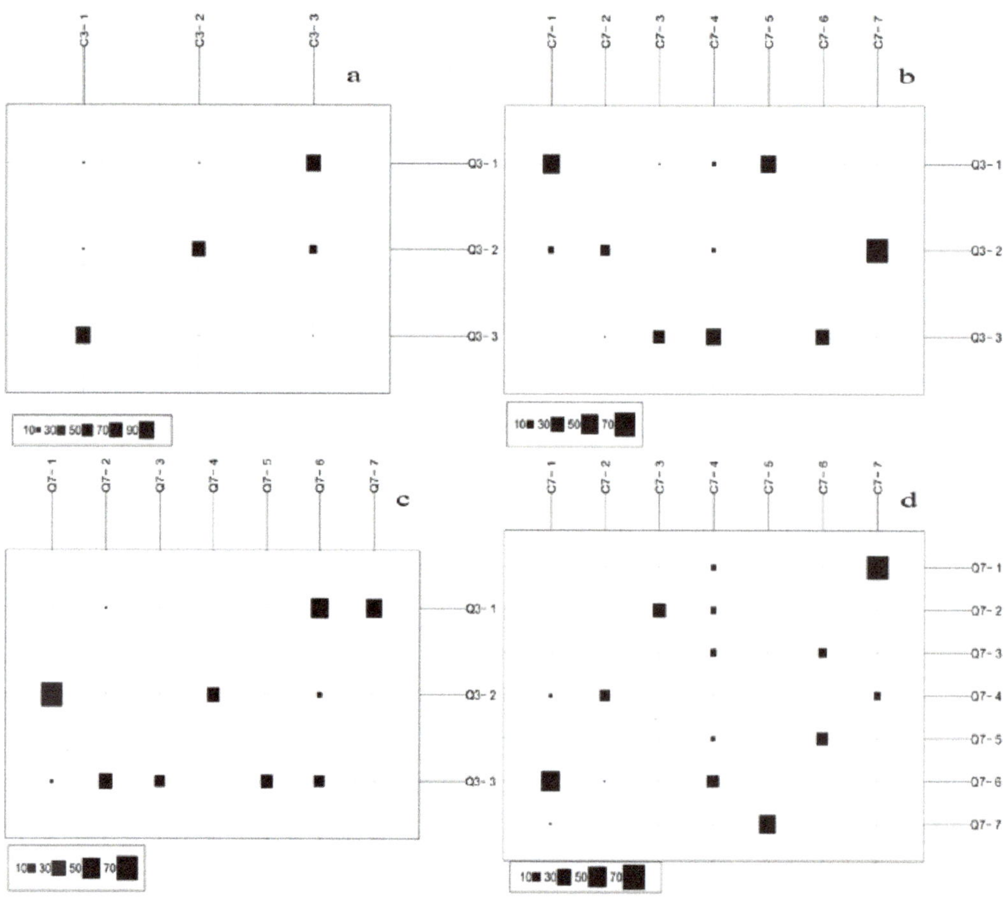

S3 Figure: Comparison of accession grouping of different statistical methods. (a) STRUCTURE k = 3 and Awclust k = 3. (b) STRUCTURE k = 3 and adegenet k = 7. (c) STRUCTURE k = 3 and k = 7. (d) STRUCTURE k = 7 and adegenet k = 7.

S1 Table: Origin of 272 indica accessions of the association panel

IRGC No.	Origin	IRGC No.	Origin	IRGC No.	Origin	IRGC No.	Origin
121159	China	120988	Philippines	121857	Senegal	122239	Mali
117793	China	122112	Philippines	122297	Senegal	121771	Mali
121137	China	122093	Philippines	121849	Senegal	121799	Mali
120970	China	122156	Philippines	121853	Senegal	121827	Mali
117280	China	121752	Philippines	122289	Senegal	121821	Mali
121156	China	121705	Philippines	122290	Senegal	122140	Mali
120862	China	120986	Philippines	121847	Senegal	121738	Mali
120909	China	121136	Madagascar	121852	Senegal	121642	Viet Nam
120880	China	121151	Madagascar	121848	Senegal	121043	Viet Nam
117636	China	117527	Madagascar	121858	Senegal	117574	Viet Nam
117681	China	121050	Madagascar	121850	Senegal	117622	Viet Nam
121157	China	121840	Madagascar	121851	Senegal	121104	Viet Nam
121063	China	121074	Madagascar	121855	Senegal	117682	Viet Nam
121040	China	121049	Madagascar	122292	Senegal	117478	Viet Nam
121092	China	121124	Madagascar	122291	Senegal	121100	Bhutan
120929	China	121836	Madagascar	122285	Senegal	121576	Bhutan
120914	China	120859	Madagascar	122284	Senegal	121099	Bhutan
121093	China	121073	Madagascar	121846	Senegal	121161	Bhutan
121110	China	121785	Madagascar	122288	Senegal	121015	Bhutan
120925	China	121823	Madagascar	122232	Senegal	121701	Bhutan
121164	China	122159	Madagascar	121024	Sri Lanka	117841	Taiwan
121176	China	121144	Madagascar	120892	Sri Lanka	117784	Taiwan
121028	China	117470	Madagascar	121079	Sri Lanka	117826	Taiwan
117745	China	121140	Madagascar	121011	Sri Lanka	117849	Taiwan
117840	China	121811	Madagascar	121094	Sri Lanka	117848	Taiwan
120912	China	121990	Madagascar	121153	Sri Lanka	117907	Taiwan
121042	China	121963	Madagascar	117525	Sri Lanka	121235	Thailand
120946	China	121142	Madagascar	121010	Sri Lanka	122179	Thailand
120861	China	121834	Madagascar	120893	Sri Lanka	121019	Thailand
117271	China	121717	Madagascar	121152	Sri Lanka	117561	Thailand
120972	China	121052	Madagascar	121131	Sri Lanka	121020	Thailand
117912	China	121839	Madagascar	121070	Sri Lanka	117587	Thailand
120947	China	121829	Madagascar	121098	Sri Lanka	122025	Colombia
117277	China	121830	Madagascar	121158	Sri Lanka	121147	Colombia
121001	China	121141	Madagascar	121059	Sri Lanka	117829	Colombia
121111	China	121812	Madagascar	121077	Sri Lanka	120863	Colombia
121163	China	121102	India	121127	Sri Lanka	122011	Colombia
121044	China	120867	India	121065	Sri Lanka	121016	Myanmar
121162	China	122206	India	121041	Sri Lanka	121080	Myanmar
122088	Philippines	120998	India	117567	Sri Lanka	121045	Myanmar
122099	Philippines	121774	India	120948	Sri Lanka	120904	Myanmar
122089	Philippines	121881	India	121083	Sri Lanka	120883	Myanmar
121835	Philippines	120882	India	120856	Sri Lanka	120949	Brazil
121591	Philippines	121130	India	120958	Sri Lanka	117463	Brazil
121753	Philippines	122181	India	117460	Bangladesh	117520	Brazil
120984	Philippines	121725	India	121034	Bangladesh	121053	Brazil
122272	Philippines	117915	India	117623	Bangladesh	121663	Nepal
121748	Philippines	117501	India	120879	Bangladesh	121225	Nepal
122091	Philippines	117500	India	121101	Bangladesh	121091	Nepal
122098	Philippines	121066	India	120913	Bangladesh	120959	Nepal
120983	Philippines	121118	India	121035	Bangladesh	121085	Ecuador
121087	Philippines	121064	India	121095	Bangladesh	120979	Ecuador
122236	Philippines	117564	India	117447	Bangladesh	121105	Iran
121760	Philippines	121172	India	121237	Bangladesh	120855	Iran
121023	Philippines	117521	India	121106	Bangladesh	117276	Korea Rep
122000	Philippines	117533	India	121146	Bangladesh	117531	Korea Rep
120981	Philippines	121139	India	120895	Bangladesh	120964	Laos
117757	Philippines	117563	India	121122	Bangladesh	120952	Laos
117758	Philippines	117691	India	120924	Bangladesh	120994	Cuba
121750	Philippines	121708	India	122298	Indonesia	120860	Egypt
117659	Philippines	122255	India	122029	Indonesia	120977	Guatemala
120982	Philippines	117454	India	120878	Indonesia	121169	Guinea
117268	Philippines	121689	India	121965	Indonesia	121062	Kenya
122094	Philippines	117598	India	117559	Indonesia	121420	Malaysia
121751	Philippines	120885	Senegal	117880	Indonesia	121051	Mauritius
120987	Philippines	122287	Senegal	117443	Indonesia	120939	Upper volta
122090	Philippines	121854	Senegal	121889	Indonesia	120916	Venezuela
121089	Philippines	121856	Senegal	122076	Mali	121165	Zambia

S2 Table: Distributions of markers on chromosomes

Chr	Marker No.	Size (Mb)	Average Distance (Kb)
Chr1	2,361	43.2	18.3
Chr2	1,965	35.9	18.3
Chr3	1,763	36.3	20.6
Chr4	1,782	35.4	19.9
Chr5	1,171	29.9	25.6
Chr6	1,589	31.0	19.5
Chr7	1,431	29.6	20.7
Chr8	1,388	28.3	20.4
Chr9	1,178	22.8	19.4
Chr10	891	23.1	25.9
Chr11	1,771	28.8	16.3
Chr12	1,534	27.4	17.9
Total	18,824	371.7	20.2

S3 Table: Basic characteristics of LD pattern in the whole panel and three subpopulations.

Population[a]	Maximum r^2	Basal r^2	Decay distance (kb)	R^{2b}
Whole	0.65	0.11	150	0.87
Q3-1	0.76	0.16	110	0.93
Q3-2	0.88	0.16	240	0.88
Q3-3	0.71	0.19	80	0.95

[a] Subpopulations were inferred by STRUCTURE *k*=3.

[b] R^2 was the determination coefficient of the power law model.

S4 Table: Comparison of 16 association models based the mean squared difference (MSD) between the observed and expected p values for 40 traits (10)-environment (4) combinations.

Env_Trait	naive	Q3	Q7	C7	Kp	Ks	Kv	Q3Kp	Q3Ks	Q3Kv	Q7Kp	Q7Ks	Q7Kv	C7Kp	C7Ks	C7Kv
HZ_BRR	0.0009	0.0012	**0.0003**	0.0007	0.0013	0.0016	0.0010	0.0009	0.0006	0.0010	0.0010	0.0019	0.0008	0.0009	0.0004	0.0008
HZ_DEC	0.0325	0.0219	0.0086	0.0157	0.0003	0.0003	0.0007	0.0003	0.0003	0.0008	0.0007	**0.0002**	0.0018	**0.0002**	**0.0002**	0.0010
HZ_GL	0.3075	0.0284	0.0144	0.0156	0.0046	0.0046	0.0022	0.0024	0.0038	0.0015	0.0016	0.0017	**0.0012**	0.0015	0.0014	0.0018
HZ_GLWR	0.8830	0.0702	0.0105	0.0214	0.0026	0.0026	0.0011	0.0013	0.0013	0.0005	0.0013	0.001	0.0008	0.0005	0.0012	**0.0003**
HZ_GT	0.1789	0.0106	0.0034	0.0076	0.0018	0.0018	**0.0012**	0.0019	0.0015	0.0031	0.0016	**0.0012**	0.0027	0.0021	0.0018	0.0028
HZ_GW	0.7308	0.0987	0.0100	0.0173	0.0012	0.0012	0.0005	0.0018	0.0018	0.0007	0.0003	0.0010	0.0004	0.0003	0.0004	**0.0001**
HZ_HMRR	0.0014	**0.0011**	**0.0011**	**0.0011**	0.0020	0.0063	0.0062	0.0013	0.0047	0.0063	0.0021	0.0041	0.0067	0.002	0.0052	0.0063
HZ_TGW	0.0053	0.0031	0.0037	0.0022	0.0018	0.0008	0.0047	0.0007	0.0007	**0.0003**	0.0012	0.0007	0.0048	0.0007	0.0007	0.0006
HZ_MRR	0.1554	0.1186	0.1321	0.1298	0.0094	0.0060	0.0111	0.0076	0.0073	0.0099	0.0052	0.0117	0.0081	0.0060	0.0121	**0.0027**
HZ_PGWC	0.0259	0.0286	0.0044	0.0100	**0.0001**	**0.0001**	0.0002	0.0002	0.0002	**0.0001**	0.0004	0.0002	**0.0001**	0.0002	0.0003	0.0003
JZ_BRR	0.0871	0.0888	0.0904	0.1131	0.0004	0.0002	0.0788	0.0004	**0.0001**	0.0789	0.0003	**0.0001**	0.0764	0.0005	0.0003	0.0900
JZ_DEC	0.9647	0.0050	0.0019	**0.0007**	0.0022	0.0019	0.0031	0.0019	0.0009	0.0044	0.0013	0.0008	0.0035	0.0018	0.0011	0.0049
JZ_GL	0.7024	0.0056	0.0053	0.007	0.0013	0.0013	**0.0002**	**0.0002**	**0.0002**	0.0005	0.0003	0.0004	**0.0002**	0.0004	0.0004	0.0005
JZ_GLWR	1.3391	0.0596	0.0058	0.0168	0.0037	0.0037	0.0007	0.0020	0.0020	**0.0002**	0.0005	0.0005	0.0016	0.0008	0.0010	0.0012
JZ_GT	0.3433	0.0036	0.0009	0.0007	0.0017	0.0017	0.0056	0.0009	0.0010	0.0066	**0.0001**	**0.0001**	0.0062	**0.0001**	0.0003	0.0052
JZ_GW	1.2121	0.1174	0.0072	0.0133	0.0047	0.0047	**0.0007**	0.0042	0.0042	0.0028	0.0015	0.0033	0.0012	0.0014	0.0017	0.0029
JZ_HMRR	0.0622	0.0218	0.0174	0.015	0.0007	0.0005	0.0031	0.0007	**0.0004**	0.0126	0.0015	0.0008	0.0068	0.0011	0.0007	0.0041
JZ_TGW	0.0008	0.0009	0.0004	0.0016	0.0015	0.0011	0.0051	0.0009	0.0005	0.0047	0.0009	0.0004	0.0042	0.0007	**0.0002**	0.0034
JZ_MRR	0.0232	0.0181	0.0270	0.0269	0.0016	0.0014	0.0093	0.0016	0.0013	0.0057	0.0017	0.0016	0.0071	0.0015	**0.0012**	0.0116
JZ_PGWC	0.7402	0.0185	0.0139	0.0213	0.0010	0.0007	0.0276	0.0025	**0.0002**	0.0222	0.0020	0.0026	0.0194	0.0013	0.0011	0.0229
SY_BRR	**0.0006**	0.0055	0.0079	0.0057	0.0066	0.0018	0.0027	0.0154	0.0073	0.0191	0.0189	0.0106	0.0225	0.0154	0.0085	0.0502
SY_DEC	0.1313	0.0247	0.0196	0.0296	0.0007	0.0007	0.0124	0.0007	0.0008	0.0103	**0.0001**	**0.0001**	0.0094	0.0008	0.0008	0.0117
SY_GL	0.6315	0.0646	0.0365	0.0617	0.0090	0.0090	0.0140	0.0061	0.0061	0.0112	0.0040	0.0040	**0.0023**	0.0060	0.0060	0.0148
SY_GLWR	1.3241	0.1881	0.0242	0.1120	0.0074	0.0074	0.0061	0.0043	0.0043	0.0014	0.0061	0.0079	**0.0010**	0.0039	0.0039	0.0069
SY_GT	0.0746	0.0097	0.0050	0.0062	0.0008	0.0009	0.0051	0.0006	0.0007	0.0046	0.0002	**0.0001**	0.0057	0.0006	0.0005	0.0040
SY_GW	0.9470	0.1487	0.0033	0.0402	0.0017	0.0017	**0.0004**	0.0014	0.0014	0.0004	0.0028	0.0252	0.0009	0.0021	0.0011	0.0007
SY_HMRR	0.0084	0.0025	0.0028	0.0017	0.0006	0.0006	0.0157	**0.0005**	0.001	0.0115	0.0013	0.0024	0.0009	0.0009	0.0011	0.0125
SY_TGW	0.0075	0.0072	0.0072	0.0072	0.0017	0.0014	0.0023	0.0011	0.0011	**0.0007**	0.0010	0.0010	0.0018	0.0016	0.0013	0.0019
SY_MRR	0.0037	0.0019	**0.0013**	0.0023	0.0108	0.022	0.0081	0.0074	0.0173	0.0059	0.006	0.0167	0.0324	0.0081	0.0196	0.0357
SY_PGWC	0.2034	0.0486	0.0514	0.0712	0.0004	0.0004	0.0122	0.0003	0.0004	0.0107	**0.0002**	**0.0002**	0.0100	**0.0002**	**0.0002**	0.0121
SZ_BRR	0.1388	0.0092	0.0033	0.0009	0.0004	0.0006	0.0046	0.0004	0.0034	0.0008	0.0006	0.0007	0.0006	0.0004	**0.0001**	0.0006
SZ_DEC	0.0443	0.0090	0.0063	0.0093	0.0015	0.002	0.0012	0.0009	0.0015	0.0008	0.0004	**0.0002**	0.0008	0.0015	0.0021	0.001
SZ_GL	0.3291	0.0392	0.0135	0.0209	0.0041	0.0041	0.0033	0.0022	0.0022	0.0028	0.0018	0.0014	0.0015	**0.0012**	**0.0012**	0.0026
SZ_GLWR	1.1181	0.1524	0.0116	0.0570	0.0039	0.0039	0.0044	0.0021	0.0021	0.0028	0.0027	0.0021	0.0038	0.0019	**0.0009**	0.0020
SZ_GT	0.2610	0.0294	0.0063	0.0109	0.004	0.0040	0.0042	0.0041	0.0038	0.0042	0.0035	0.0035	0.0039	**0.0034**	**0.0034**	0.0039

ACKNOWLEDGMENTS

This research was granted by Global Rice Science Partnership (GRiSP), National 863 High Technology Research and Development Program from the Ministry of Science and Technology of China (2014AA10A601), Shenzhen Peacock Plan, the Natural Science Foundation of China (31461143014, 31261140369), and Hubei Collaborative Innovation Center for Grain Industry.

AUTHOR CONTRIBUTIONS

Conceived and designed the experiments: JX GY. Performed the experiments: XQ ZY. Analyzed the data: YP GY. Contributed reagents/materials/analysis tools: MD DX ZL. Wrote the paper: YP GY.

REFERENCES

1. Unnevehr L, Duff B, Juliano BO. Consumer Demand for Rice Grain Quality: Terminal Report of IDRC Projects, National Grain Quality (Asia), and International Grain Quality Economics (Asia): Int. Rice Res. Inst.; 1992.
2. Tan YF, Xing YZ, Li JX, Yu SB, Xu CG, Zhang Q. Genetic bases of appearance quality of rice grains in Shanyou 63, an elite rice hybrid. TAG Theoretical and applied genetics Theoretische und angewandte Genetik. 2000;101(5–6):823–9. doi: 10.1007/s001220051549.
3. Cheng FM, Zhong LJ, Wang F, Zhang GP. Differences in cooking and eating properties between chalky and translucent parts in rice grains. Food Chemistry. 2005;90(1–2):39–46.
4. Zhao X, Zhou L, Ponce K, Ye G. The Usefulness of Known Genes/Qtls for Grain Quality Traits in an Indica Population of Diverse Breeding Lines Tested using Association Analysis. Rice. 2015;8(1). doi: 10.1186/s12284-015-0064-3.
5. Yu S, Yang C, Fan Y, Zhuang J, Li X. Genetic dissection of a thousand-grain weight quantitative trait locus on rice chromosome 1. Chinese Science Bulletin. 2008;53(15):2326–32. doi: 10.1007/s11434-008-0281-x.
6. Singh R, Singh A, Sharma T, Singh A, Singh N. Fine mapping of grain length QTLs on chromosomes 1 and 7 in Basmati rice (*Oryza sativa* L.). Journal of Plant Biochemistry and Biotechnology. 2012;21(2):157–66. doi: 10.1007/s13562-011-0080-3.
7. Zhang W, Sun P, He Q, Shu F, Wang J, Deng H. Fine mapping of *GS2*, a dominant gene for big grain rice. The Crop Journal. 2013;1(2):160–5. doi: 10.1016/j.cj.2013.10.003.
8. Guo L, Ma L, Jiang H, Zeng D, Hu J, Wu L, et al. Genetic Analysis and Fine Mapping of Two Genes for Grain Shape and Weight in Rice. Journal of integrative plant biology. 2009;51(1):45–51. doi: 10.1111/j.1744-7909.2008.00793.x. pmid:19166493
9. Kato T, Segami S, Toriyama M, Kono I, Ando T, Yano M, et al. Detection of QTLs for grain length from large grain rice (Oryza sativa L.). Breeding Science. 2011;61(3):269–74. doi: 10.1270/jsbbs.61.269.
10. Zhou L, Chen L, Jiang L, Zhang W, Liu L, Liu X, et al. Fine mapping of the grain chalkiness QTL *qPGWC-7* in rice (*Oryza sativa* L.). TAG Theoretical and applied genetics Theoretische und angewandte Genetik. 2009;118(3):581–90. doi: 10.1007/s00122-008-0922-0.
11. Bai X, Luo L, Yan W, Kovi MR, Zhan W, Xing Y. Genetic dissection of rice grain shape using a recombinant inbred line population derived from two contrasting parents and fine mapping a pleiotropic quantitative trait locus *qGL7*. BMC genetics. 2010;11:16. doi: 10.1186/1471-2156-11-16 pmid:20184774; PubMed Central PMCID: PMC2846863.
12. Xie X, Song MH, Jin F, Ahn SN, Suh JP, Hwang HG, et al. Fine mapping of a grain weight quantitative trait locus on rice chromosome 8 using near-isogenic lines derived from a cross between *Oryza sativa* and *Oryza rufipogon*. TAG Theoretical and applied genetics Theoretische und angewandte Genetik. 2006;113(5):885–94. doi: 10.1007/s00122-006-0348-5 pmid:16850315.
13. Xie X, Jin F, Song MH, Suh JP, Hwang HG, Kim YG, et al. Fine mapping of a yield-enhancing QTL cluster associated with transgressive variation in an *Oryza sativa* x *O.rufipogon* cross. TAG Theoretical and applied genetics Theoretische und angewandte Genetik. 2008;116(5):613–22. doi: 10.1007/s00122-007-0695-x pmid:18092146.
14. Oh J-M, Balkunde S, Yang P, Yoon D-B, Ahn S-N. Fine mapping of grain weight QTL,*tgw11* using near isogenic lines from a cross between *Oryza sativa* and *O.grandiglumis*. Genes & Genomics. 2011;33(3):259–65. doi: 10.1007/s13258-011-0038-9.
15. Song XJ, Huang W, Shi M, Zhu MZ, Lin HX. A QTL for rice grain width and weight encodes a previously unknown RING-type E3 ubiquitin ligase. Nature genetics. 2007;39(5):623–30. doi: 10.1038/ng2014 pmid:17417637.
16. Mao H, Sun S, Yao J, Wang C, Yu S, Xu C, et al. Linking differential domain functions of the GS3 protein to natural variation of grain size in rice. Proceedings of the National Academy of Sciences of the United States of America. 2010;107(45):19579–84. doi: 10.1073/pnas.1014419107 pmid:20974950; PubMed Central PMCID: PMC2984220.
17. Zhang X, Wang J, Huang J, Lan H, Wang C, Yin C, et al. Rare allele of *OsPPKL1*associated with grain length causes extra-large grain and a significant yield increase in rice. Proceedings of the National Academy of Sciences of the United States of America. 2012;109(52):21534–9. doi: 10.1073/pnas.1219776110. pmid:23236132
18. Shomura A, Izawa T, Ebana K, Ebitani T, Kanegae H, Konishi S, et al. Deletion in a gene associated with grain size increased yields during rice domestication. Nature genetics. 2008;40(8):1023–8. doi: 10.1038/ng.169 pmid:18604208.
19. Li Y, Fan C, Xing Y, Jiang Y, Luo L, Sun L, et al. Natural variation in *GS5* plays an important role in regulating grain size and yield in rice. Nature genetics. 2011;43(12):1266–9. doi: 10.1038/ng.977 pmid:22019783.
20. Li Y, Fan C, Xing Y, Yun P, Luo L, Yan B, et al. *Chalk5* encodes a vacuolar H^+-translocating pyrophosphatase influencing grain chalkiness in rice. Nature genetics. 2014;46(4):398–404. doi: 10.1038/ng.2923.http://www.nature.com/ng/journal/v46/n4/abs/ng.2923.html#supplementary-information. pmid:24633159
21. Ishimaru K, Hirotsu N, Madoka Y, Murakami N, Hara N, Onodera H, et al. Loss of function of the IAA-glucose hydrolase gene *TGW6* enhances rice grain weight and increases yield. Nature genetics. 2013;45(6):707–11. doi: 10.1038/ng.2612 pmid:23583977.

22. Song XJ, Kuroha T, Ayano M, Furuta T, Nagai K, Komeda N, et al. Rare allele of a previously unidentified histone H14 acetyltransferase enhandes grain weitht, yield, and plalnt biomass in rice. Proceedings of the National Academy of Sciences of the United States of America. 2014;112:76–81. doi: 10.1073/pnas.1421127112. pmid:25535376

23. Abe Y, Mieda K, Ando T, Kono I, Yano M, Kitano H, et al. The *SMALL AND ROUND SEED1* (*SRS1/DEP2*) gene is involved in the regulation of seed size in rice. Genes Genet Syst. 2010;85:327–39. pmid:21317545 doi: 10.1266/ggs.85.327

24. Wang Y, Xiong G, Hu J, Jiang L, Yu H, Xu J, et al. Copy number variation at the *GL7*locus contributes to grain size diversity in rice. Nature genetics. 2015;47(8):944–8. doi: 10.1038/ng.3346.http://www.nature.com/ng/journal/v47/n8/abs/ng.3346.html#supplementary-information. pmid:26147619

25. Wang S, Li S, Liu Q, Wu K, Zhang J, Wang S, et al. The *OsSPL16-GW7* regulatory module determines grain shape and simultaneously improves rice yield and grain quality. Nature genetics. 2015;47(8):949–54. doi: 10.1038/ng.3352 pmid:26147620.

26. Wang S, Wu K, Yuan Q, Liu X, Liu Z, Lin X, et al. Control of grain size, shape and quality by *OsSPL16* in rice. Nature genetics. 2012;44(8):950–4.http://www.nature.com/ng/journal/v44/n8/abs/ng.2327.html#supplementary-information. doi: 10.1038/ng.2327. pmid:22729225

27. Qi P, Lin Y-S, Song X-J, Shen J-B, Huang W, Shan J-X, et al. The novel quantitative trait locus *GL3.1* controls rice grain size and yield by regulating Cyclin-T1;3. Cell research. 2012;22(12):1666–80.http://www.nature.com/cr/journal/v22/n12/suppinfo/cr2012151s1.html. doi: 10.1038/cr.2012.151. pmid:23147796

28. Guo LB, Ye GY. Use of major quantitative trait loci to improve grain yield of rice. Rice Science. 2014;21(2):65–82. doi: 10.1016/S1672-6308(13)60174-2.

29. Zhang P, Liu X, Tong H, Lu Y, Li J. Association mapping for important agronomic traits in core collection of rice (*Oryza sativa* L.) with SSR markers. PloS one. 2014;9(10):e111508. doi: 10.1371/journal.pone.0111508 pmid:25360796; PubMed Central PMCID: PMC4216065.

30. Jiangbo Z. Association analysis of important agronomic traits in japonica rice germplasm. African Journal of Biotechnology. 2012;11(12). doi: 10.5897/ajb11.1912.

31. Borba TCdO, Brondani RPV, Breseghello F, Coelho ASG, Mendonça JA, Rangel PHN, et al. Association mapping for yield and grain quality traits in rice (*Oryza sativa* L.). Genetics and Molecular Biology. 2010;33(3):515–24. doi: 10.1590/S1415-47572010005000065. pmid:21637426

32. Agrama HA, Eizenga GC, Yan W. Association mapping of yield and its components in rice cultivars. Molecular Breeding. 2007;19(4):341–56. doi: 10.1007/s11032-006-9066-6.

33. Huang X, Wei X, Sang T, Zhao Q, Feng Q, Zhao Y, et al. Genome-wide association studies of 14 agronomic traits in rice landraces. Nature genetics. 2010;42(11):961–7. doi: 10.1038/ng.695 pmid:20972439.

34. Huang X, Zhao Y, Wei X, Li C, Wang A, Zhao Q, et al. Genome-wide association study of flowering time and grain yield traits in a worldwide collection of rice germplasm. Nature genetics. 2012;44(1):32–9. doi: 10.1038/ng.1018 pmid:22138690.

35. Zhao K, Tung CW, Eizenga GC, Wright MH, Ali ML, Price AH, et al. Genome-wide association mapping reveals a rich genetic architecture of complex traits in Oryza sativa. Nature communications. 2011;2:467. doi: 10.1038/ncomms1467 pmid:21915109; PubMed Central PMCID: PMC3195253.

36. Begum H, Spindel JE, Lalusin A, Borromeo T, Gregorio G, Hernandez J, et al. Genome-wide association mapping for yield and other agronomic traits in an elite breeding population of tropical rice (*Oryza sativa*). PloS one. 2015;10(3):e0119873. doi: 10.1371/journal.pone.0119873 pmid:25785447; PubMed Central PMCID: PMC4364887.

37. Yang W, Guo Z, Huang C, Duan L, Chen G, Jiang N, et al. Combining high-throughput phenotyping and genome-wide association studies to reveal natural genetic variation in rice. Nature communications. 2014;5:5087. doi: 10.1038/ncomms6087 pmid:25295980; PubMed Central PMCID: PMC4214417.

38. R Core Team. R: A language and environment for statistical computing. R Foundation for Statistical Computing, Vienna, Austria. 2015.

39. Jr FEH. Hmisc: Harrell Miscellaneous. R package version 3.16–0. 2015.

40. Risterucci AM, Grivet L, N'Goran JAK, Pieretti I, Flament MH, Lanaud C. A high-density linkage map of *Theobroma cacao* L. TAG Theoretical and applied genetics Theoretische und angewandte Genetik. 2000;101(5–6):948–55. doi: 10.1007/s001220051566.

41. Courtois B, Audebert A, Dardou A, Roques S, Ghneim-Herrera T, Droc G, et al. Genome-wide association mapping of root traits in a japonica rice panel. PloS one. 2013;8(11):e78037. doi: 10.1371/journal.pone.0078037 pmid:24223758; PubMed Central PMCID: PMC3818351.

42. Kawahara Y, dlB M., Hamilton JP, Kanamori H, McCombie WR, Ouyang S, et al. Improvement of the *Oryza sativa* Nipponbare reference genome using next generation sequence and optical map data. Rice. 2013;6:1–10. doi: 10.1186/1939-8433-6-4

43. Langmead B, Trapnell C, Pop M, Salzberg SL. Ultrafast and memory-efficient alignment of short DNA sequences to the human genome. Genome biology. 2009;10(3):R25. doi: 10.1186/gb-2009-10-3-r25 pmid:19261174; PubMed Central PMCID: PMC2690996.

44. Browning SR, Browning BL. Rapid and accurate haplotype phasing and missing-data inference for whole-genome association studies by use of localized haplotype clustering. American journal of human genetics. 2007;81(5):1084–97. doi: 10.1086/521987 pmid:17924348; PubMed Central PMCID: PMC2265661.

45. Pritchard JK, Stephens M, Donnelly P. Inference of Population Structure Using Multilocus Genotype Data. Genetics. 2000;155(2):945–59. pmid:10835412
46. Evanno G, Regnaut S, Goudet J. Detecting the number of clusters of individuals using the software STRUCTURE: a simulation study. Molecular ecology. 2005;14(8):2611–20. doi: 10.1111/j.1365-294X.2005.02553.x pmid:15969739.
47. Gao X, Starmer J. Human population structure detection via multilocus genotype clustering. BMC genetics. 2007;8:34. doi: 10.1186/1471-2156-8-34 pmid:17592628; PubMed Central PMCID: PMC1934381.
48. Jombart T, Ahmed I. *adegenet 1.3–1*: new tools for the analysis of genome-wide SNP data. 2011. doi: 10.1093/bioinformatics/btr521
49. Bradbury PJ, Zhang Z, Kroon DE, Casstevens TM, Ramdoss Y, Buckler ES. TASSEL: software for association mapping of complex traits in diverse samples. Bioinformatics. 2007;23(19):2633–5. doi: 10.1093/bioinformatics/btm308 pmid:17586829.
50. VanRaden PM. Efficient methods to compute genomic predictions. Journal of dairy science. 2008;91(11):4414–23. doi: 10.3168/jds.2007-0980 pmid:18946147.
51. Lipka AE, Tian F, Wang Q, Peiffer J, Li M, Bradbury PJ, et al. GAPIT: genome association and prediction integrated tool. Bioinformatics. 2012;28(18):2397–9. doi: 10.1093/bioinformatics/bts444 pmid:22796960.
52. Phung NT, Mai CD, Mournet P, Frouin J, Droc G, Ta NK, et al. Characterization of a panel of Vietnamese rice varieties using DArT and SNP markers for association mapping purposes. BMC plant biology. 2014;14:371. doi: 10.1186/s12870-014-0371-7 pmid:25524444; PubMed Central PMCID: PMC4279583.
53. Yu J, Pressoir G, Briggs WH, Vroh Bi I, Yamasaki M, Doebley JF, et al. A unified mixed-model method for association mapping that accounts for multiple levels of relatedness. Nature genetics. 2006;38(2):203–8. doi: 10.1038/ng1702 pmid:16380716.
54. Zhang Z, Ersoz E, Lai CQ, Todhunter RJ, Tiwari HK, Gore MA, et al. Mixed linear model approach adapted for genome-wide association studies. Nature genetics. 2010;42(4):355–60. doi: 10.1038/ng.546 pmid:20208535; PubMed Central PMCID: PMC2931336.
55. Stich B, Mohring J, Piepho HP, Heckenberger M, Buckler ES, Melchinger AE. Comparison of mixed-model approaches for association mapping. Genetics. 2008;178(3):1745–54. doi: 10.1534/genetics.107.079707 pmid:18245847; PubMed Central PMCID: PMC2278052.
56. Huang R, Jiang L, Zheng J, Wang T, Wang H, Huang Y, et al. Genetic bases of rice grain shape: so many genes, so little known. Trends in plant science. 2013;18(4):218–26. doi: 10.1016/j.tplants.2012.11.001 pmid:23218902.
57. Xu F, Sun X, Chen Y, Huang Y, Tong C, Bao J. Rapid identification of major QTLs associated with rice grain weight and their utilization. PloS one. 2015;10(3):e0122206. doi: 10.1371/journal.pone.0122206 pmid:25815721; PubMed Central PMCID: PMC4376791.
58. Zheng TQ, Xu JL, Li ZK, Zhai HQ, Wan JM. Genomic regions associated with milling quality and grain shape identified in a set of random introgression lines of rice (Oryza sativa L.). Plant Breeding. 2007;126(2):158–63. doi: 10.1111/j.1439-0523.2007.01357.x.
59. McNally KL, Childs KL, Bohnert R, Davidson RM, Zhao K, Ulat VJ, et al. Genomewide SNP variation reveals relationships among landraces and modern varieties of rice. Proceedings of the National Academy of Sciences of the United States of America. 2009;106(30):12273–8. doi: 10.1073/pnas.0900992106 pmid:19597147; PubMed Central PMCID: PMC2718348.
60. Mather KA, Caicedo AL, Polato NR, Olsen KM, McCouch S, Purugganan MD. The extent of linkage disequilibrium in rice (*Oryza sativa* L.). Genetics. 2007;177(4):2223–32. doi: 10.1534/genetics.107.079616 pmid:17947413; PubMed Central PMCID: PMC2219496.
61. Pasam RK, Sharma R, Malosetti M, van Eeuwijk FA, Haseneyer G, Kilian B, et al. Genome-wide association studies for agronomical traits in a world wide spring barley collection. BMC plant biology. 2012;12:16. doi: 10.1186/1471-2229-12-16 pmid:22284310; PubMed Central PMCID: PMC3349577.
62. Jin L, Lu Y, Xiao P, Sun M, Corke H, Bao J. Genetic diversity and population structure of a diverse set of rice germplasm for association mapping. TAG Theoretical and applied genetics Theoretische und angewandte Genetik. 2010;121(3):475–87. doi: 10.1007/s00122-010-1324-7 pmid:20364375.
63. lint-Garcia SA, Thornsberry JM, Buckler ESt. Structure of linkage disequilibrium in plants. Annual review of plant biology. 2003;54:357–74. doi: 10.1146/annurev.arplant.54.031902.134907 pmid:14502995.
64. Schwarz G. Estimating the Dimension of a Model. 1978:461–4. doi: 10.1214/aos/1176344136.
65. Yan S, Zou G, Li S, Wang H, Liu H, Zhai G, et al. Seed size is determined by the combinations of the genes controlling different seed characteristics in rice. TAG Theoretical and applied genetics Theoretische und angewandte Genetik. 2011;123(7):1173–81. doi: 10.1007/s00122-011-1657-x pmid:21805338.

CHAPTER 9

COMPLETE CHLOROPLAST GENOME SEQUENCE AND ANNOTATION OF THE TROPICAL JAPONICA GROUP OF ASIAN CULTIVATED RICE (ORYZA SATIVA L.)

Shuo Wang and Li-zhi Gao

GENOME ANNOUNCEMENT

Rice (*Oryza sativa* L.), as an important staple crop of the family *Poaceae*, is distributed widely across diverse tropical-to-temperate regions of both hemispheres and provides the vast majority of daily caloric intake for half the world's population. It is also well known for its great genetic diversity within species (1, 2), which can be categorized into five distinct varietal groups: *indica*, *aus/boro*, *aromatic* (basmati/sadri), temperate*japonica*, and tropical *japonica* (alias *javanica*) (3,–6). Along with the arrival of genome sequencing era, the*indica* and temperate *japonica* rice became some of the first few crop species having their nuclear, chloroplast, and mitochondrial genomes completely sequenced (7,–12), which consolidated their valuable status as model systems for other grass species.

The tropical *japonica* rice (*O. sativa* subsp. tropical *japonica*) was most likely domesticated in northern parts of Southeast Asia or South China, and then introduced southward to Southeast Asia and from there to West Africa and Latin American countries (13). Here we first announce its complete chloroplast genome sequence and functional annotations. The plant was collected from Indonesia and deposited as a germplasm accession of the International Rice Collection (IRGC 66630) at the International Rice Research Institute (IRRI), which was then prepared for sequencing as an accession of the 3,000 Rice Genomes Project (3K RGP [14]). Genomic DNA was extracted by a modified cetyltrimethylammonium bromide (CTAB) method at IRRI and shipped to BGI-Shenzhen for the construction of Illumina index libraries. After quality control, at least 3 μg of genomic DNA was randomly fragmented by sonication and size fractionated by electrophoresis. DNA fragments of approximately 500 bp in length were then purified, labeled with 6-bp nucleotide multiplex identifiers, and followed by pooling prior to library construction for next-generation sequencing (NGS). Each library was sequenced in six or more lanes on the Illumina HiSeq 2000 platform to generate 90-bp paired-end reads. The reads were subsequently extracted based on the corresponding unique nucleotide multiplex identifiers and filtered by deleting those with adapter contamination or containing >50% low-quality bases (quality value, ≤5) (14).

Approximately 1.5 Gb of Illumina paired-end total DNA sequencing data (GigaScience database [15]) were filtered with NCBI-blast version 2.2.31+ (ftp://ftp.ncbi.nih.gov/blast/) to obtain chloroplast DNA reads. The filtered chloroplast DNA reads were then subjected to SOAP*denovo*2 (16), ABySS version 1.9.0 (17), and SPAdes version 3.1.0 (18) for several runs of assembly. The final assembly resulted in a complete circular genome sequence with a length of 134,536 bp and G+C content of 39.7%. Annotation was performed with Dual Organellar GenoMe Annotator (DOGMA [19]) using default parameters to predict protein-coding genes, tRNA genes, and ribosomal RNA (rRNA) genes. For genes with low sequence identity, manual annotation was performed to determine the positions of start and stop codons depending on the translated amino acid sequence using the chloroplast/bacterial genetic code.

Citation: Wang S, Gao L-Z. 2016. Complete chloroplast genome sequence and annotation of the Tropical japonica group of Asian cultivated rice (Oryza sativa L.). Genome Announc 4(1):e01703-15. doi:10.1128/genomeA.01703-15.

Nucleotide sequence accession numbers.

The source accession for this DNA sample at IRRI is IRGC 66630 (http://iris.irri.org/germplasm2/id/365929), and the accession number for the genetic stock attached to the DNA sample is IRGC 126310 (http://iris.irri.org/germplasm2/id/3059145). The complete chloroplast genome sequence with all genes annotated has been submitted to GenBank under the accession numberKT289404.

REFERENCES

1. Li ZK, Rutger JN 2000. Geographic distribution and multilocus organization of isozyme variation of rice (*Oryza sativa* L.). Theor Appl Genet 101:379–387. doi:.10.1007/s001220051494
2. Yu SB, Xu WJ, Vijayakumar CHM, Ali J, Fu BY, Xu JL, Marghirang R, Domingo J, Jiang YZ, Aquino C, Virmani SS, Li ZK 2003. Molecular diversity and multilocus organization of the parental lines used in the International Rice Molecular Breeding Program. Theor Appl Genet 108:131–140. doi:.10.1007/s00122-003-1400-3
3. Vaughan DA. 1994. The wild relatives of rice: a genetic resources handbook, p 66 International Rice Research Institute, Manila, Philippines.
4. Izawa T. 2008. The process of rice domestication: a new model based on recent data. Rice 1:127–134. doi:.10.1007/s12284-008-9014-7
5. Kovach MJ, Sweeney MT, McCouch SR 2007. New insights into the history of rice domestication.Trends Genet 23:578–587. doi:.10.1016/j.tig.2007.08.012
6. Garris AJ, Tai TH, Coburn J, Kresovich S, McCouch S 2005. Genetic structure and diversity in *Oryza sativa* L. Genetics 169:1631–1638. doi:.10.1534/genetics.104.035642
7. Yu J, Hu S, Wang J, Wong GK, Li S, Liu B, Deng Y, Dai L, Zhou Y, Zhang X, Cao M, Liu J, Sun J, Tang J, Chen Y, Huang X, Lin W, Ye C, Tong W, Cong L, Geng J, Han Y, Li L, Li W, Hu G, Huang X, Li W, Li J, Liu Z, Li L, Liu J, Qi Q, Liu J, Li L, Li T, Wang X, Lu H, Wu T, Zhu M, Ni P, Han H, Dong W, Ren X, Feng X, Cui P, Li X, Wang H, Xu X, Zhai W, Xu Z, et al. 2002. A draft sequence of the rice genome (*Oryza sativa* L. ssp. *indica*). Science 296:79–92. doi:.10.1126/science.1068037
8. Goff SA, Ricke D, Lan TH, Presting G, Wang R, Dunn M, Glazebrook J, Sessions A, Oeller P, Varma H, Hadley D, Hutchison D, Martin C, Katagiri F, Lange BM, Moughamer T, Xia Y, Budworth P, Zhong J, Miguel T, Paszkowski U, Zhang S, Colbert M, Sun WL, Chen L, Cooper B, Park S, Wood TC, Mao L, Quail P, Wing R, Dean R, Yu Y, Zharkikh A, Shen R, Sahasrabudhe S, Thomas A, Cannings R, Gutin A, Pruss D, Reid J, Tavtigian S, Mitchell J, Eldredge G, Scholl T, Miller RM, Bhatnagar S, Adey N, Rubano T, Tusneem N, et al. 2002. A draft sequence of the rice genome (*Oryza sativa* L. ssp. japonica). Science296:92–100. doi:.10.1126/science.1068275
9. Tang J, Xia H, Cao M, Zhang X, Zeng W, Hu S, Tong W, Wang J, Wang J, Yu J, Yang H, Zhu L 2004. A comparison of rice chloroplast genomes. Plant Physiol 135:412–420. doi:.10.1104/pp.103.031245
10. Tian X, Zheng J, Hu S, Yu J 2006. The rice mitochondrial genomes and their variations. Plant Physiol140:401–410. doi:.10.1104/pp.105.070060
11. Shimda H, Sugiuro M 1991. Fine structural features of the chloroplast genome: comparison of the sequenced chloroplast genomes. Nucleic Acids Res 19:983–995. doi:.10.1093/nar/19.5.983
12. Notsu Y, Masood S, Nishikawa T, Kubo N, Akiduki G, Nakazono M, Hirai A, Kadowaki K 2002. The complete sequence of the rice (*Oryza sativa* L.) mitochondrial genome: frequent DNA sequence acquisition and loss during the evolution of flowering plants. Mol Genet Genomics 268:434–445. doi:.10.1007/s00438-002-0767-1
13. Khush GS. 1997. Origin, dispersal, cultivation and variation of rice p 25–34. *In* Sasaki T, Moore G (ed) Oryza: From molecule to plant. Springer International Publishing, The Netherlands.
14. The 3,000 Rice Genomes Project 2014. The Rice 3,000 Genome Project. GigaScience 3:e7. doi:.10.1186/2047-217X-3-7
15. The 3,000 Rice Genomes Project 2014. The Rice 3,000 Genome Project. GigaScience Database.http://dx.doi.org/10.5524/200001.
16. Luo R, Liu B, Xie Y, Li Z, Huang W, Yuan J, He G, Chen Y, Pan Q, Liu Y, Tang J, Wu G, Zhang H, Shi Y, Liu Y, Yu C, Wang B, Lu Y, Han C, Cheung DW, Yiu SM, Peng S, Xiaoqian Z, Liu G, Liao X, Li Y, Yang H, Wang J, Lam TW, Wang J 2012. SOAP*denovo*2: an empirically improved memory-efficient short-read *de novo* assembler. GigaScience 1:e18. doi:.10.1186/2047-217X-1-18
17. Simpson JT, Wong K, Jackman SD, Schein JE, Jones SJM, Birol I 2009. ABySS: a parallel assembler for short read sequence data. Genome Res 19:1117–1123. doi:.10.1101/gr.089532.108
18. Bankevich A, Nurk S, Antipov D, Gurevich AA, Dvorkin M, Kulikov AS, Lesin VM, Nikolenko SI, Pham S, Prjibelski AD, Pyshkin AV, Sirotkin AV, Vyahhi N, Tesler G, Alekseyev MA, Pevzner PA 2012.SPAdes: a new genome assembly algorithm and its applications to single-cell sequencing. J Comput Biol19:455–477. doi:.10.1089/cmb.2012.0021
19. Wyman SK, Jansen RK, Boore JL 2004. Automatic annotation of organellar genomes with DOGMA. Bioinformatics 20:3252–3255. doi:.10.1093/bioinformatics/bth352

CHAPTER 10

GENOME-WIDE DISSECTION OF THE MICRORNA EXPRESSION PROFILE IN RICE EMBRYO DURING EARLY STAGES OF SEED GERMINATION

Dongli He, Qiong Wang, Kun Wang, and Pingfang Yang

INTRODUCTION

Rice is not only the staple food for over half of the world population, but also an ideal plant model for monocot seed germination as it has an annotated genome and a well-developed germination system. A series of physiological and biochemical events were involved in seed germination, during which embryo cells transit from a quiescent state to a metabolically active state. In rice, classical studies have defined seed germination as a sequential process following a triphasic model based on water uptake, namely, rapid water uptake in the first 20 hours after imbibition (HAI) (phase I), lag phase for metabolism reactivation (phase II, 20–48 HAI), and radicle emergence for seedling establishment (phase III, after 48 HAI) [1,2]. Albeit no significant change has been observed in phenotypes of the 24 HAI seed, the oxygen uptake was found to be enhanced in parallel with seed wet weight increase [3]. During this period, the pro-mitochondria are converted to the typical cristae-rich mitochondrial structures following recovery of the aerobic respiration for seedling establishment [2,4]. Previous transcriptome study has shown that a significant proportion of transcripts changed in the first 24 HAI, with most changes being observed between 3 and 12 HAI (5,396 up-regulated and 4,935 down-regulated transcripts) [5]. In addition, a phosphorproteome study also suggests that the first 12 HAI is a potentially important signal transduction phase for the initial rice seed germination [6]. Therefore, the first 24 HAI greatly determines the recovery of metabolism activity in rice seed during the germination progress.

Plant microRNAs (miRNAs), generated from stem-loop regions of long primary transcripts by a Dicer-like (DCL) enzyme, are a class of non-coding sRNAs that are abundant and best-characterized with 20-24-nucleotide (nt) length [7]. MicroRNAs can guide the RNA-induced silencing complex (RISC) to cleave mRNA or remodel chromatin through fully or partly complementary base pairing, resulting in mRNA slicing and degradation, translational repression, or chromatin modifications [8,9]. MicroRNAs have been shown to play important roles in plant growth and development, phyto-hormone homeostasis, and stress response. The ancient miRNAs often functionally target the evolutionarily conserved genes. For example, miR167 is involved in ovule and anther development by repressing Auxin Responsive Factor 6 (*ARF6*) [10]; Arabidopsis seeds expressing the miR160-resistant form of *ARF10* are hypersensitive to abscisic acid (ABA) in a dose-dependent manner [11]. Additionally, miR164 and miR319 are important for organ morphogenesis by targeting NAM, ATAF1/2 and CUC2 domain-containing proteins (*NAC*) and TEOSINTE BRANCHED1/CYCLOIDEA/PROLIFERATING (*TCP*) transcription factor families, respectively [12,13]. Some new miRNA-target pairs have been confirmed experimentally, i.e., miR402 targets DEMETER-LIKE protein3 (*DML3*), a putative DNA glycosylase involved in DNA demethylation. Overexpression of miR402 and deficiency of *dml3* accelerates seed germination under stress conditions [14]. As a key regulator, miRNA has been extensively studied in rice for its role in root development, seed maturation, and pollen development [15–18]. The recently released rice dataset of miRBase (V21.0) has documented 332 miRNA families. These

Citation: He D, Wang Q, Wang K, Yang P (2015) Genome-Wide Dissection of the MicroRNA Expression Profile in Rice Embryo during Early Stages of Seed Germination. PLoS ONE 10(12): e0145424. doi:10.1371/journal.pone.0145424.

identified miRNAs are important for understanding small RNA (sRNA)-mediated gene regulations in rice. However, to date, limited knowledge about miRNA regulation has been obtained in the early stage of rice seed germination.

Rice seed is composed of a dominant starchy endosperm and a genetically vigorous embryo. Proteomic analysis of the dissected endosperms has shown that the biological processes operating in the endosperm are heavily regulated by the embryo during seed germination [19]. Therefore, embryo is an important tissue in seed germination control. In this study, we sequenced the sRNA populations from embryos of 0, 12, and 24 HAI rice seeds using next-generation deep sequencing technology. A series of miRNAs were identified, including both known and novel miRNAs. We also predicted the potential targets for the miRNAs. Real-time RT-PCR and 5'-RACE assay were performed to confirm some deep sequencing and target prediction results. This study provides the unique composition and expression profiles of miRNAs and their potential regulations in the embryo at the early stages of rice seed germination.

MATERIALS AND METHODS

Plant Material and Seed Germination

The de-hulled rice (*Oryza sativa* Nipponbare) seeds were washed three times with distilled water, then placed in 9-cm-diameter plates and imbibed in distilled water at 30°C in a dark biochemical incubator. Fifty seeds were placed in one plate and set up as a biological replicate. Four replicates were prepared for each time-point sample. After 0 h, 12 h, and 24 h of imbibition, embryos were dissected manually and collected for RNA extraction, respectively. For phytohormone treatment, seeds were incubated with 200 μM GA_3 and 50 mM ABA (Sigma, St. Louis, MO, USA) for 24 h, after which the embryos were collected.

Small RNA Libraries Construction and Sequencing

Total RNA was extracted with TRIzol Reagent (Invitrogen, Carlsbad, CA, USA) according to the manufacturer's instructions. Equal amounts (about 25 μg) of total RNA from four replicates of each time-point sample were mixed together for sRNA libraries construction. The sRNA libraries were constructed following the methods described by Lu et al [20]. Briefly, the total RNA was fractionated by 15% denaturing polyacrylamide gel (8 M urea) electrophoresis, and sRNAs in the range of 18–30 nt were excised and purified with a Spin-X cellulose acetate filter (2 mL, 0.45 um; Thermo Fisher, Waltham, MA, USA). After dephosphorylation and sequential ligation of 5- and 3- Solexa adaptors, the sRNAs were reverse-transcribed using superscript™II reverse transcription kit (Invitrogen), and then amplified by PCR with phusion High-Fidelity PCR kit (Finnzymes, Espoo, Finland) to produce sRNA sequencing libraries. Next generation sequencing was performed on an Illumina platform (Nextomics Science and Technology Limited, Wuhan, China). All the sequence data have been deposited as a series with the accession number GSE73657 at NCBI' GEO database (http://www.ncbi.nlm.nih.gov/geo/query/acc.cgi?acc=GSE73657).

Deep Sequencing and Processing

The raw sequencing data were filtered with FastQC (http://www.bioinformatics.babraham.ac.uk/projects/fastqc/) to delete low quality reads, adapters, contamination, and false sequences (such as reads with poly A/T/C/G). After filtering, sRNAs with size ranging from 18 to 30 nt were collected and mapped to the rice genome (MSU Rice Annotation Release 7.0) using Bowtie software, and then the sequences matching known rice rRNA, tRNA, snRNA, and snoRNA in Rfam11 (https://www.sanger.ac.uk/resources/databases/rfam.html) were discarded. Subsequently, the unique sRNAs were aligned with the data in miRBase 21 (www.mirbase.org) to search for the conserved miRNAs.

De novo prediction of novel miRNA was performed by miRDeep2. Genomic sequences with 300 bp length, surrounding the sRNAs (150 nt upstream and 150 nt downstream), were extracted to assess the potential secondary structures of pre-miRNA. The candidates were first filtered with default criteria, and then with the stringent criterion of minimal folding free energy index (MFEI) > 0.85 (MFEI = [(MFE/length of the sRNA sequence) × 100] / (G+C) %) [21].

Target Prediction of Identified miRNAs

The prediction of putative targets for identified miRNAs was performed with Mireap, PsRNATarget and degradome analyses based on the rice genome (MSU 7.0) [22–24]. For the Mireap software analysis, the following modified parameters were used: 1) no more than four mismatches between miRNA and the target, the G-U base pair was counted as 0.5 mismatches; 2) no more than two adjacent mismatches in the miRNA/target duplex; 3) no adjacent mismatches in positions 2–12 (5' of miRNA); 4) perfect matches in positions 10–11; 5) no more than 2.5 mismatches in positions 1–12; 6) MFE between miRNA and target gene not less than 75% of the MFE for a perfect matching. PsRNATarget prediction software (http://plantgrn.noble.org/psRNATarget) was used with default parameters. Degradome analysis was performed based on GSE18251, GSE17398, and GSE19050 dataset as previously described with $p < 0.05$ [15,22]. WEGO software (http://wego.genomics.org.cn/) was used for plotting gene ontology annotation results, and Agrigo software (http://bioinfo.cau.edu.cn/agriGO) was used to enrich the GO terms.

Expression Analysis of miRNAs, Target Genes, and Host Genes

The miRNA expression profiles in multiple tissues were obtained from the published dataset [16]. To compare the miRNA expression in multiple tissues, the Log_2TPM (Transcripts per million = number of actual reads/total number of clean reads × 1,000,000) value was presented. MeV4.9 was used for hierachical clustering (HCL) and principal component analysis (PCA). The expression patterns of target genes were extracted from the published microarray dataset (GSE43780) [5]. To analyze the expression relevance between a miRNA and its corresponding target gene(s), the TPM value of miRNA or the genechip robust multi-array average (GCRMA) value of the target gene was divided by its maximum to reach abundance between 0 and 1.

Stem-Loop Reverse Transcription Polymerase Chain Reaction (RT-PCR)

RNA reverse transcription was performed with specific stem-loop primers as previously described [25] (S1 Table). The stem-loop RT-PCR reactions contained 25 ng of RNA samples, 50 nM stem-loop RT primer, 1 × RT buffer, 0.25 mM of each dNTP, 5 U/μL SuperScript II reverse transcriptase (Invitrogen, USA) and 0.25 U/μL RNase inhibitor. The 10 μL reactions were incubated in a 96-well plate for 30 min at 16°C, 30 min at 42°C and 5 min at 85°C, and then held at 4°C in a Bio-Rad professional thermocycler (Bio-Rad, Hercules, California, USA). The real-time PCR amplification program consisted of initial denaturation at 95°C for 3 min, followed by 40 cycles of 95°C for 10 s, 58°C for 15 s, and 72°C for 15 s. Three biological replicates were performed on the CFX96 Real-time PCR system (Bio-Rad, California, USA). The melting curves (65°C to 95°C) were analyzed to check the specificity of PCR products. PCR efficiency was calculated by LinRegPCR (http://www.hartfaalcentrum.nl/; subject: LinRegPCR) considering an ideal value range (1.8–2.05) with correlation $R^2 > 0.995$ [26]. The relative miRNA expression was calculated using the following formula:

$$\text{Ration} = E_{\text{target}}^{\text{target}\Delta\text{Ct(Control}-\text{sample)}} / E_{\text{ref.}}^{\text{ref.}\Delta\text{Ct(Control}-\text{sample)}}$$

In which E_{target} represents PCR efficiency of measured miRNA/gene, E_{ref} represents PCR efficiency of reference gene 18S RNA, and ΔCt represents the difference in threshold cycles between the control and sample.

RNA Ligase-Mediated Rapid-Amplification of 5′-cDNA Ends (5′-RACE)

To investigate the cleavage sites occurring in the targeted mRNAs, 5'-RACE was performed using the FirstChoice® RLM-RACE kit (Ambion, USA) according to the manufacturer's instructions. Briefly, high quality mRNAs were directly ligated to the 5-RACE RNA adapter (44 nt), and the ligation reaction was then reverse-transcribed into cDNA with random primers. Two 5-RACE gene-specific outer and inner primers were used for nested PCR. The PCR products with the anticipated size were cloned and sequenced.

RESULTS

Deep Sequencing of Small RNAs

Previous study has shown that large-scale rearrangement of transcripts mediated by RNA synthesis and degradation occurs in the first 24 h of rice seed germination [5]. Using an experimental system similar to that utilized in the previous study [5], the miRNA composition and dynamic expression patterns were investigated in the early stages of rice seed germination. About 200 embryos from 0, 12, and 24 HAI rice seeds were collected for total RNA extraction and sRNA libraries construction, respectively (Fig 1). RNA concentration and integrity were detected using Agilent 2100 Bioanalyzer and Agilent RNA 6000 Nano kit. About 100 μg of high-quality RNA from each sample was used for sRNAs isolation and sequencing. In total, 29230963, 34302280, and 33901503 raw reads with high quality (Q30 > 94%) were obtained for 0, 12, and 24 HAI sRNA libraries, respectively. After removal of low quality sequences and adapter contaminants, 19203513, 22332607, and 17359816 reads for each library were perfectly mapped to the reference rice genome (MSU 7.0) with size ranging from 18 to 30 nt (Table 1). Consistent with the representative size range of DCLs cleavage products [27], most of these sRNAs were 21 or 24 nt in length. The 24-nt sRNAs occurred in a particularly high abundance, accounting for 25.3%, 34.7%, and 37.5% in the three libraries, respectively (Fig 2).

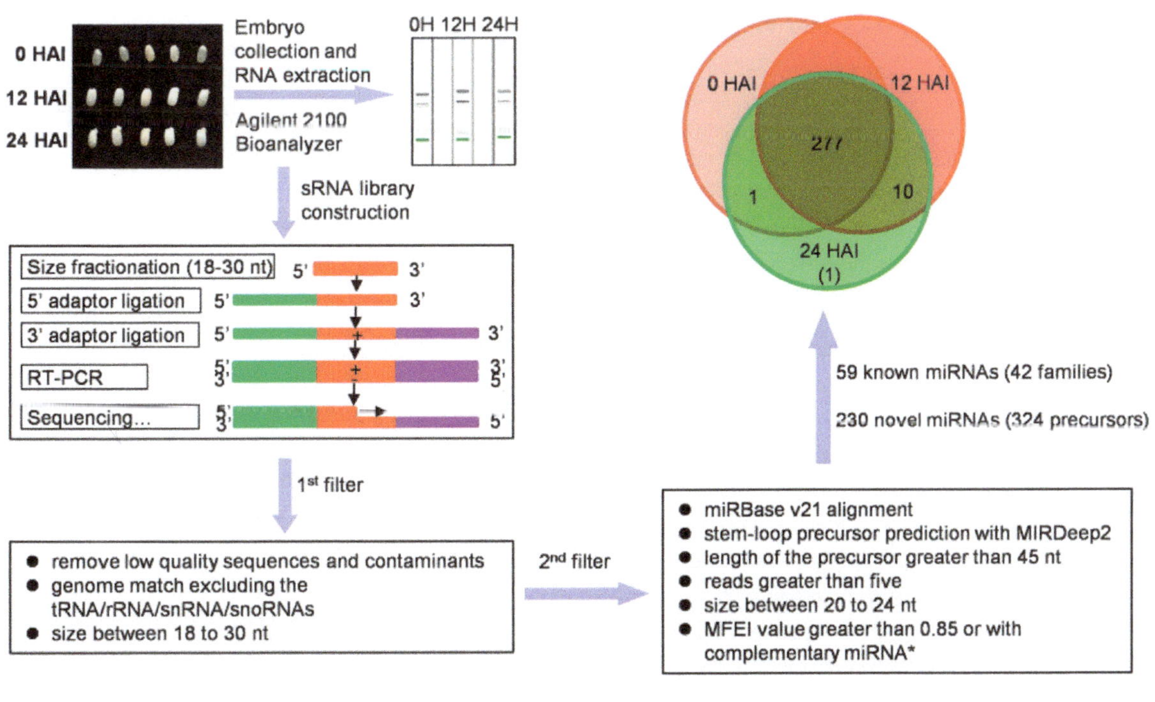

Figure 1: Schematic representation of the miRNA analysis in rice embryos at the early stages of seed germination.

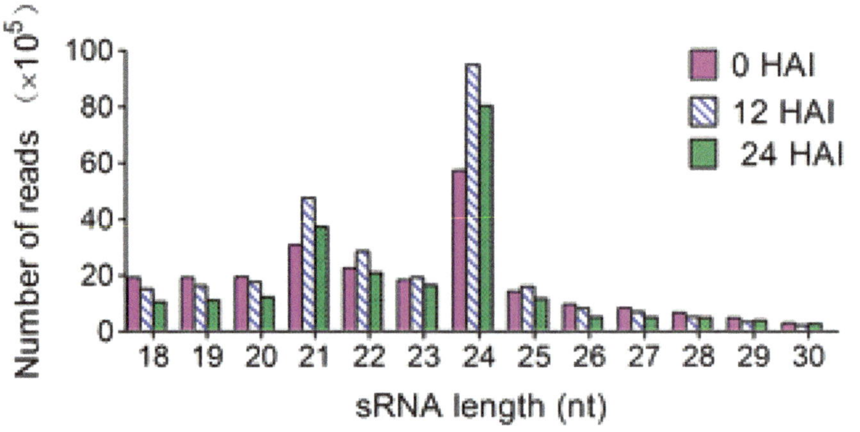

Figure 2: Size distribution of the sRNAs in libraries from 0, 12, and 24 HAI rice seed embryos.

Table 1: Statistics for sRNA sequencing from three libraries

Category	0 HAI		12 HAI		24 HAI	
	Reads[b]	Unique[b]	Reads[b]	Unique[b]	Reads[b]	Unique[b]
high quality raw Reads[a]	29230963 (Q30 = 94.74%)	-	34302280 (Q30 = 95.29%)	-	33901503 (Q30 = 94.58%)	-
sequences of 18–30 nt	22265577(100%)	6596317 (100%)	27075645(100%)	10712075 (100%)	21236500(100%)	9420781 (100%)
mapping to genome	19203513(86.3%)	4856925 (73.6%)	22332607(82.5%)	7892704 (73.7%)	17359816(81.8%)	7115379 (75.5%)
CDS matched	600910(2.7%)	211949(3.2%)	1570839(5.8%)	338298(3.2%)	1322306(6.2%)	243265(2.6%)
Non-coding RNA						
total	17642144(100%)	-[c]	19201655(100%)	-[c]	14737124(100%)	-[c]
known miRNA	400047(2.3%)	1724(0.4%)	873690(4.6%)	2506(2.9%)	781907(5.3%)	2314(29.6%)
rRNA	2289143(13.0%)	24748(1.1%)	1388914(7.2%)	23099(1.7%)	747877(5.1%)	19579(2.6%)
tRNA	277627(1.6%)	8121(2.9%)	274039(1.4%)	7349(2.7%)	421275(2.9%)	6878(1.6%)
snoRNA	40642(0.2%)	8790(21.6%)	44927(0.2%)	6001(13.4%)	28636(0.2%)	4755(16.6%)
snRNA	5956(0.34%)	1679(28.2%)	6005(0.32%)	1390(23.1%)	3623(0.25%)	1079(29.8%)
others	14628729(82.9%)	4345729 (29.7%)	16614080(86.5%)	7138503 (43.0%)	12753806(86.5%)	6489573 (50.9%)

[a]: Q30 is the percentage of base with correct recognition rate over than 99.9%.
[b]: the percentage to the 100% reads.
[c]: the percentage to the sequences of corresponding 'reads'.

doi:10.1371/journal.pone.0145424.t001

The composition of the genome-matched sRNAs of rice seed was complex, which included protein-coding RNA fragments (3.13–7.62%) and other non-coding transcripts. The non-coding sRNAs contained various categories, including known miRNAs (2.09–4.5%), rRNAs (4.31–11.9%), tRNAs (1.45–2.43%), snoRNAs (0.16–0.21%), snRNAs (0.02–0.03%) and unannotated sRNAs, (80.9–84.31%). The unannotated sRNAs were further used to predict the novel miRNAs.

Identification of Known and Novel miRNAs in Dry and Imbibed Rice Seeds

To scan the known miRNAs, the unique sRNAs were aligned to the rice miRNA datasets in miRBase 21 (http://www.mirbase.org/), allowing up to two mismatches. In the three libraries, 59 known rice miRNAs homologs from 83 precursors corresponding to 42 miRNA families were matched. Of which, 14 families had two or more miRNA members, and the other 28 families contained only one miRNA member. In this study, 17 conserved families were detected in the imbibed rice seeds, most of them were found in relatively high abundance. In particular, miR319, miR168, miR156, miR166, and miR159 were mapped with more than 10,000 reads. However, some conserved miRNAs, i.e., miR164, mi395, and miR393, were lowly expressed in rice seeds. Nineteen families are known rice-specific miRNAs, and most of them had only one member and were in relatively low abundance. A homolog of ptc-miR6478, previously reported only in *Populus trichocarpa*, was also identified in this study. Moreover, we detected a new precursor with stringent criteria for osa-miR1862 family (named osa-MIR1862d-2). These results show that our sequencing depth was sufficient to reflect the expression profiles of miRNAs during seed imbibition.

For novel miRNAs identification, the 300-bp flanking sequences of unannotated sRNAs were initially extracted, and then subjected to MIRDeep2 to scan potential miRNA precursors with classical stem-loop structures using the default criteria. A total of 382 novel miRNA candidates with length of 18–26 nt processed from 481 putative precursors were obtained from the three libraries. MFEI has been identified to be useful in evaluating genuine miRNAs [21], which depends on the MFE, length and G+C content of the potential precursors. These criteria were used in our investigation of miRNAs in rice seeds. Low MFE is an important characteristic of miRNA precursors. In this study, the miRNA precursor's MFE values varied from -180.6 to 5.2 kcal·mol^{-1} according to Mfold 3.2. In animal, the size of miRNA precursor ranges typically from 70 to 80 nt, but it is more variable in plant.[28]. In rice seeds, the size of miRNA precursors varied from 37 to 268 nt, with most being 51–110 nt. In addition, the high G+C content can make the pre-miRNA secondary structure more stable and thus difficult to be processed into mature miRNA by the RISC complex [21]. Here, the G+C content ranged from 14.03% to 88%. With all these factors in consideration, we calculated the MFEI value of all identified miRNA precursors as previously described [21]. Previous research has shown that more than 90% of known plant pre-miRNAs have an MFEI value greater than 0.85 [21]. In this study, the MFEI values ranged from 0.20 to 3.75, with an average of 1.26. To minimize the false positive miRNAs, we further set the novel miRNA criteria as follows: 1) the mature miRNA length ranged from 20 to 24 nt; 2) precursor had a classical fold-back structure and the length was no less than 45 nt; 3) the number of reads was greater than five in at least one library; 4) MFEI value was greater than 0.85 or the complementary miRNA* could be found. Finally, 230 novel miRNAs processed from 324 precursors were obtained (Fig 1). Of these novel miRNAs, 218 were shared by all three libraries. The novel miRNA was denoted with OsmiR plus a serial number, such as OsmiR-1. Different miRNA precursors that produced the same mature sequence were noted with -1, -2, and -3.

In plants, most of the canonical miRNAs (~21 nt) are in high abundance [29]. Similarly, in this study, the 21-nt miRNAs occupied an overwhelming majority of the reads. Moreover, a great number of long miRNAs (24 nt) with low abundance existed in the early stages of rice seed germination. Specifically, of the 230 novel miRNAs, 176 were 24-nt. The miRNAs function through interacting with AGO proteins, which bind to the miRNA with particular 5' end nucleotide. AGO1 mainly recruits miRNAs that begin with a 5'-U/T; AGO2 prefers to bind sRNAs with 5' -A [30,31]. The first base distribution of all 289 identified miRNAs showed that the first base of 24-nt miRNAs were tended to be A (66.2%), suggesting these miRNAs function with AGO2; while that of the shorter miRNAs were inclined to be U/T (69.3%) (Fig 3C). Genome mapping revealed that most known and novel miRNAs were located in the intergenic regions, and fewer in the intragenic regions (i.e., exon, intron). In addition, the miRNAs were located randomly on the 5p or 3p arm of their precursors (Fig 3D and 3E).

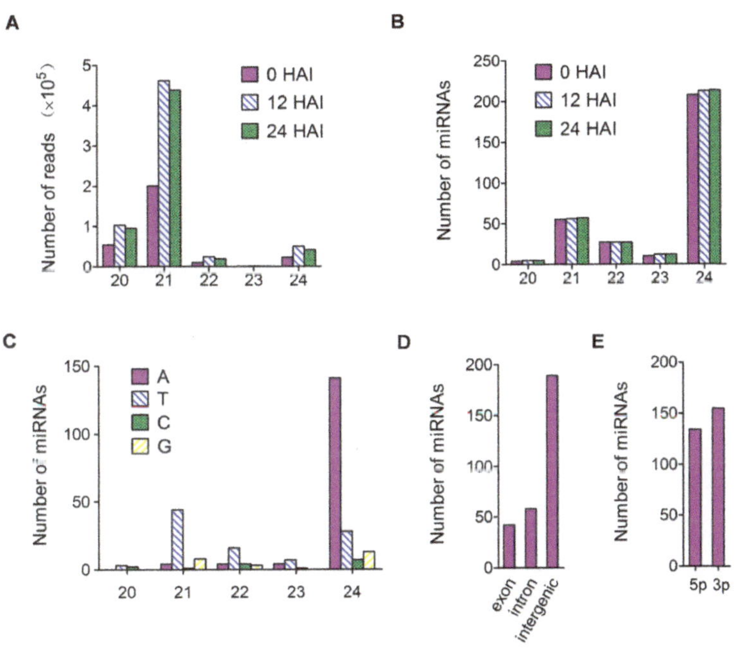

Figure 3: Characteristics of all identified miRNAs in libraries from 0, 12 and 24 HAI rice seed embryos. Distribution of sequencing reads (A), number of miRNAs (B) and the first base of miRNAs (C). (D) Genome and (E) precursor arm locations of all identified miRNAs.

To validate the deep sequencing results, 35 (17 known and 18 novel) miRNAs with sequence reads higher than 100 were randomly selected for stem-loop RT-PCR analysis. All the 35 amplified miRNAs produced a single PCR band with the length of about 60 bp on the gel (The amplified PCR product included the 20–24 bp miRNA sequence, part of the 44 bp stem-loop primer sequence (that is 34 bp) and several primer balance bases). The PCR products were cloned and sequenced. All the tested miRNAs sequences were verified to be quite correct to that of the sequencing, demonstrating a high reliability of the deep sequencing results.

The Expression Patterns of Known and Novel miRNAs During Rice Seed Imbibition

To investigate the tissue expression pattern of the miRNAs, the expression data of known miRNAs in bicellular pollen (BCP), leaf, root and callus of rice were obtained from the previous report [16]. HCL analysis and PCA were performed with Log_2TPM value of known miRNAs from BCP, leaf, root, callus, and embryos of 0, 12, and 24 HAI rice seeds. Results indicated that 7 samples were significantly clustered into two clades, and the seed embryo samples were separated from the others (Fig 4).

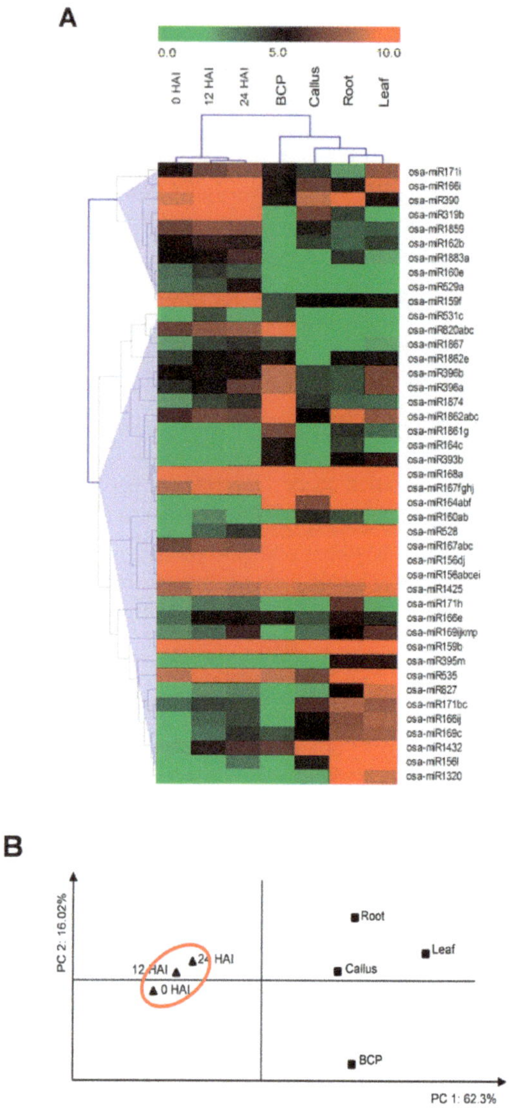

Figure 4: Analyses of tissue-specific expression pattern of known miRNAs. Hierarchical clustering analyses **(A)** and principal component analysis **(B)** of known miRNAs in various tissues. Clustering of known miRNAs based on Log_2TPM. The bar represents the scale of miRNAs expression levels of miRNAs.

In total, 286499, 636508, and 592581 miRNA reads were obtained from 0, 12, and 24 HAI libraries, respectively. Although a significantly smaller amount of miRNAs were mapped in the dry seeds (0 HAI) than that in the imbibed seeds (12 HAI and 24 HAI), the amount of genome-mapped sRNAs reads was substantial (Table 1), which implied the existence of a large number of other categories of sRNAs, i.e., siRNAs, in dry seeds. HCL analysis using the LOG_2TPM value of miRNAs showed that 0 HAI was separated from the other two time points. In order to directly plot the miRNA expression patterns in germinating rice seed, we normalized the TPM values through dividing each miRNA TPM by its corresponding maximum and reaching the abundance between 0 and 1. After normalization, one miRNA that was detected in at least one library with the normalized value lower than 0.5 (ratio >2 fold) was regarded as differentially expressed. A total of 178 differentially expressed miRNAs were obtained in germination initiation stage, most (128 miRNAs) of which were enriched in the 24 HAI. Only four miRNAs were enriched in the dry seeds, especially for ptc-miR6478 and a novel miRNA OsmiR-201.

To validate the miRNAs expression patterns obtained by deep sequencing, the 35 selected miRNAs were further verified by stem-loop real-time RT-PCR. The results showed that 27 miRNAs (about 77%) had expression patterns consistent to those in initial sequencing (Fig 5). However, some inconsistent results were also observed, such as OsmiR-220 and OsmiR-26, which were identified to be up-regulated remarkably with imbibition by sequencing but down-regulated by RT-PCR analysis, reflecting the dynamic change of the miRNAs expressions.

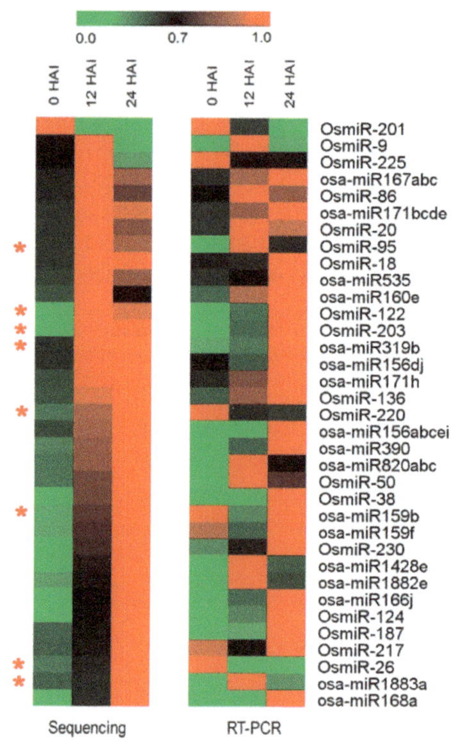

Figure5: Heatmaps of sequencing data and stem-loop real-time RT-PCR. The bar represents the scale of the expression levels of miRNAs. The stars indicate inconsistent results between sequencing and RT-PCR.

The intragenic miRNA, especially for the intronic miRNA, may be derived from a common precursor transcript and display coincidental expression pattern with the host gene [32,33]. We also investigated the expression relationship between miRNAs and the host genes inferred from the transcriptome dataset (GSE43780) [5]. However, neither the exonic nor intronic miRNAs had significant co-expression with their host genes, suggesting that these miRNAs were controlled by different promoters with their host genes.

Target Prediction and Function Analysis of Identified miRNAs

The target genes of all identified miRNAs were predicted using three classical methods: Mireap, degradome and psRNATarget analysis. After searching the rice genome dataset MSU7.0, 132, 163, 1011 target genes were obtained for 72, 133, and

242 miRNAs by the three methods, which were assembled into 207, 217, and 1508 miRNA-target pairs, respectively. Finally, a total of 1197 target genes for 259 miRNAs were found, which constituted 1784 miRNA-target pairs. Unexpectedly, only 25 pairs (21 genes targeted by 15 known miRNAs and one novel miRNA) were shared by all three prediction methods, and 123 (82 genes targeted by 26 known miRNAs and 24 novel miRNAs) pairs were shared by at least two methods (Fig 6A). Of the 21 common targets, most were conserved transcription factors for ancient miRNAs, including SQUAMOSA PROMOTER BINDING PROTEIN-LIKE (*SPLs*) protein coding genes targeted by miR156/miR529, *ARFs* targeted by miR160, *TCPs* targeted by miR319, and *MYBs* targeted by miR159.

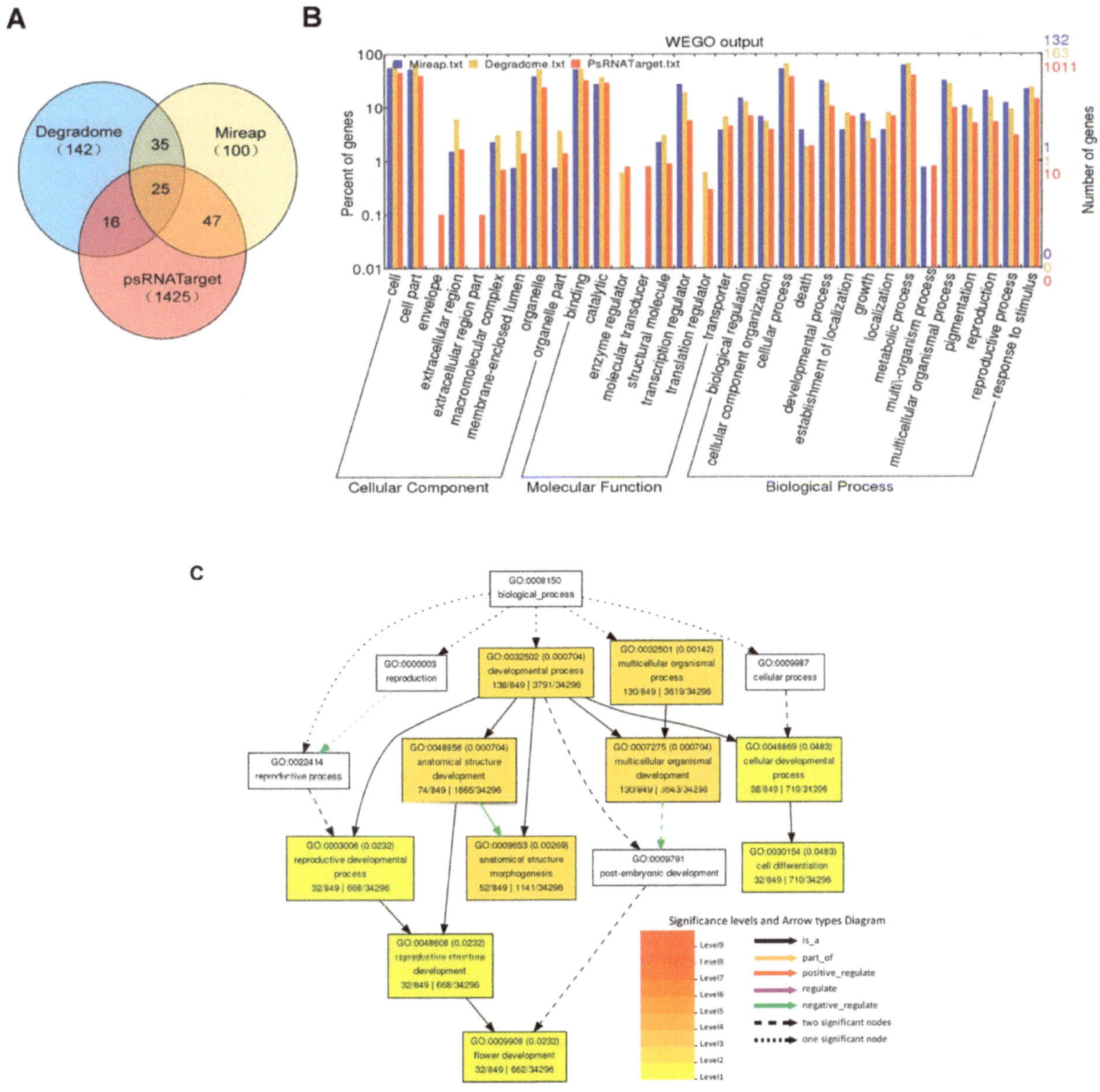

Figure 6: Gene ontology (GO) analysis of miRNAs' target genes. (A) Venn diagram shows the overlapping relationship of the target prediction results among the three methods. (B) Categorization of miRNAs target genes with WEGO (http://wego.genomics.org.cn/cgi-bin/wego/index.pl). (C) GO enrichment of target genes using Agrigo (http://bioinfo.cau.edu.cn/agriGO/). The color scale shows the P-value cutoff levels for each biological process. The more statistically significant it is, the darker or redder the color is.

All the putative target genes were categorized through gene ontology (GO) analysis. The results showed that these genes were involved in various molecular functions, i.e., binding, catalytic activity, and transcription regulation, and played roles in many biological processes, i.e., metabolic and cellular processes and signaling (Fig 6B). The Singular Enrichment Analysis (SEA) results of Agrigo software indicated that target genes related to the anatomical structure development ($p = 2*10^{-6}$) and multicellular organismal development ($P = 5.7*10^{-6}$) were significantly enriched in the biological process category (Fig 6C). In contrast, no GO term was enriched for the categories of cellular component and molecular function.

Furthermore, the expression profiles of target genes were extracted from the transcriptome dataset (GSE43780) [5]. In general, the mRNA expression profiles were negatively correlated with their corresponding miRNAs. Of the 178 differentially expressed miRNAs, 28 (16 known and 12 novel) miRNAs with sequencing reads over 100 were significantly expressed opposite to their targets. These targets included 39 genes encoding proteins for major and minor metabolism, protein degradation and post-translational modification, transportation, and transcriptional regulation, which participated in cellular, metabolic, growth, developmental, and other biological processes.

Verification of miRNAs-Induced Cleavage by 5′-RACE

To validate the authentic cleavage event of the known or novel miRNAs in the germinating rice seed, some inversely expressed miRNA-target pairs were selected to verify the cleavage site through 5-RACE.

GAMYB is well documented as the gibberellin (GA) signaling regulator in cereal aleurone cells [34]. The transcriptome data of germinating rice seeds indicated that both *OsGAMYBL1*(*LOC_Os06g40330*) and *OsGAMYB* (*LOC_Os01g59660*) were remarkably down-regulated in the first 24 HAI. Our RT-PCR results showed that the mRNA level of *OsGAMYBL1* was constant while the level of *OsGAMYB* decreased in the first 24 h of imbibition (Fig 7A). ABA and GA are two vital phytohormones in regulating seed germination. We detected ABA and GA responses of the selected miRNAs and their targets in this study. RT-PCR results showed that both miR159 and *GAMYB*s were insensitive to ABA and GA in rice embryos. RNA ligase-mediated 5-RACE indicated that *OsGAMYBL1* and *OsGAMYB* produced PCR products with canonical cleavage sites of miR159 (Fig 7B). Unexpectedly, we also found some fragments mapped downstream of the canonical cleavage sites, which reflected the existence of RNA degradation after cleavage, and hence the post-transcriptional modification.

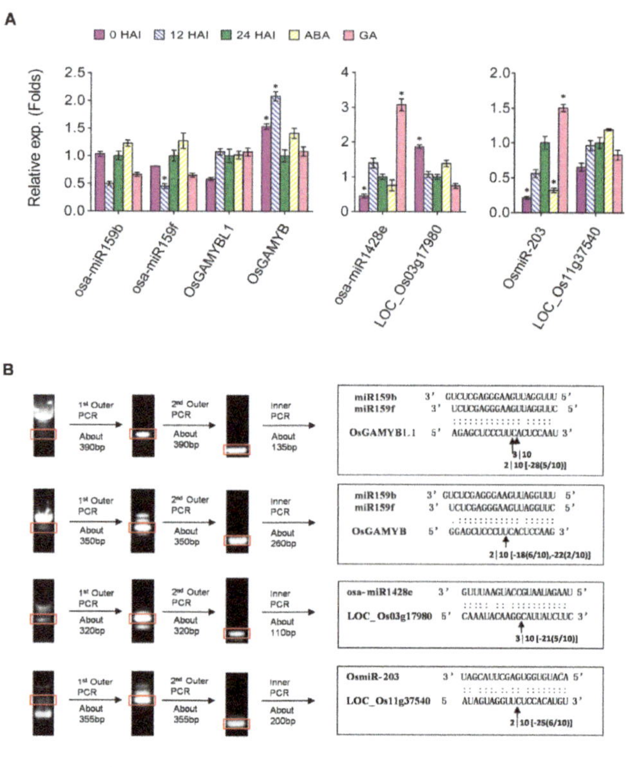

Figure7: Real-time RT-PCR and 5'-RACE analysis. (A) Real-time RT-PCR analysis of the relative expression of osa-miR159b, osa-miR159f, osa-miR1428e and OsmiR203 and their corresponding targets in embryos of 0, 12, and 24 HAI and ABA- or GA- treated seeds. The data represent the mean values ± SD of three replicates. The rice 18S rRNA fragment was amplified as internal control. Asterisks (* $p < 0.01$, Student's t-test) represent significant differences from the expression level in 24 HAI seed embryos. (B) Validation of target mRNA cleavage sites by RNA ligase mediated 5□ -RACE. The rectangles on the gel show the bands of nested PCR products. The arrows indicate the cleavage sites, the numbers under the arrows show the frequency of RACE clones sequenced, and the numbers in the brackets show the position and frequency of cleavage sites downstream of the canonical sites.

Real-time RT-PCR results indicated that rice special miRNA osa-miR1428e-3p was increased during seed imbibition and induced greatly by GA. Inversely, its target *OSK3* (a carbon catabolite depressing protein kinase encoded by *LOC_Os03g17980*) was decreased during seed imbibition and insensitive to GA and ABA. OsmiR-203 was induced by GA as well, and its target *LOC_Os11g37540* (encoding a protein phosphatase 2C family protein) displayed consistent expression under all detected conditions (Fig 7A). The 5'-RACE also detected the canonical cleavage sites on the *OSK3* and *LOC_Os11g37540* for osa-miR1428e-3p and OsmiR-203, while most clones carried fragments matched to the downstream sequences, suggesting that mRNA fragments induced by miRNA-mediated cleavage were easily degraded from the 5 terminal (Fig 7B).

DISCUSSION

Specific miRNA Profiles and Expression Patterns in Dry and Imbibed Rice Seeds

The rice dataset of the miRBase (version 21) has recorded 688 precursors for mature miRNAs from 332 families. However, we identified only 83 precursors processed for 59 known miRNAs belonging to 42 families in rice seeds; this ratio is relatively lower than that in leaf, root and pollen [16]. Moreover, the HCL analysis and PCA indicated that the known miRNA expression profiles in the embryo of rice seeds were significantly separated from that of the leaf, root, callus and developing pollen (Fig 4). Particularly, some known miRNAs, i.e., osa-miR1876 and osa-miR5150, were abundantly expressed in rice seed but were not detected in other tissues, and osa-miR159f, osa-miR166l, and osa-miR319b were predominantly expressed in the seed, which suggested the pivotal regulatory roles of these miRNAs in seed germination. The size distribution of miRNAs has been revealed to be tissue-specific in Arabidopsis and rice [16,35]. Most of the conserved miRNAs are short sRNAs with size ranging from 20 to 22 nt [7]. The 24-nt miRNAs are involved in DNA methylation or histone modifications leading to transcriptional gene silencing and chromatin remodeling [36,37]. In imbibed seeds, many 24-nt miRNAs with low abundance and fluctuating expression patterns were identified, and most of them were novel. From the above analysis, we concluded that the unique miRNAs composition and expression profiles existed in dry and imbibed rice seeds.

The composition of miRNAs in germinating rice seed was relatively stable as 277 of 289 identified miRNAs were shared by three libraries (Fig 1), whereas their expression pattern varied greatly. Altogether, 178 differentially expressed miRNAs were identified, and only few miRNAs had higher

TPM values in dry seeds, such as ptc-miR6478 and OsmiR-201. However, the miRNAs expression patterns between 12 and 24 HAI were similar. HCL analysis showed that the expression profile of miRNAs in 0 HAI seed embryos was separated from that of the other two time points. The drastic turnover of the miRNAs profile from dry seeds to imbibed seeds suggested that miRNA-mediated regulatory metabolisms were rapidly activated after imbibition.

Roles of miRNAs in Initiation of Rice Seed Germination

Since protein translation from the stored long-lived mRNAs is indispensable for seed germination, the miRNA-mediated post-transcription regulation can play key roles in the mRNA selective translation and protein turnover during this process [38,39]. To comprehensively explore the possible roles of identified miRNAs during seed imbibition, two widely accepted softwares, Mireap and PsRNATarget, were used to predict the miRNA targets. According to the miRNA-target sequence complementation, it is relatively easy to predict the plant miRNA targets with the reference genomic or transcriptomic sequence data [23]. However, the computational prediction is challenged by a high level of false positives that exist in miRNA-target pairing [40]. Degradome sequencing is a recently established, efficient strategy to identify miRNA targets on a large scale [22,41]. In this study, degradome analysis was also used for targets prediction based on the published datasets. In total, 1197 target genes involved in various functions for 259 miRNAs were obtained by three prediction methods. However, the remaining 30 miRNAs were failed to find potential targets. These miRNAs might regulate gene expression through non-stringent complementary paring, such as translational repression [42,43], or guide the non-coding locus cleavage and produce phasi/tasiRNA [8]. GO enrichment results indicated that target genes involved in anatomical structure development and multicellular organismal development, i.e., wall-associated protein kinases (*OsWAKs*), expansin precursors, and *OsSPLs*, were over-represented. This is consistent with the development process of germination initiation, during which many cell structures need to be repaired or synthesized *de novo*. Unexpectedly, only 25 miRNA-target pairs overlapped in the three methods of analysis. The ultra-low overlap among these target prediction methods reflected a difference in the parameter setting or prediction principle. This is also strongly suggestive of the predicted targets which need to be confirmed experimentally since the result varied with different prediction methods.

Transcription factors such as *TCPs*, *ARFs* and *SPLs* have been confirmed to be easily targeted by conserved miRNAs. By repressing *TCPs*, miR319 played a central role in coordinating multiple miRNAs (i.e., miR396 and miR164) and phytohormones (including auxin, ABA and GA) pathways to control lateral organ development [44,45]. Auxin alone is not generally considered as a typical seed germination hormone, but it was revealed to interact with ABA through *ARF10*targeted by miR160 and to be involved in seed germination and post-germination control [11]. The auxin signal affects rice root development through regulation of miR156-SPLs [46]. SPL proteins have also been revealed to restrain the transition from post-germination to autotrophic growth and affect seedling establishment [31,47]. In this study, osa-miR319 and osa-miR156abcdejl showed high abundance, but miR160abe was relatively low-expressed in all three libraries. Since the complex biological process of seed germination is regulated by a great interaction network of phytohormones, these conserved miRNAs are devoted to the germination initiation and the following seedling establishment through targeting the related signaling regulation factors. However, until now, the mechanisms of miRNAs-mediated phytohormone crosstalk involving in seed germination are largely unknown.

The first 24 HAI is critical for initiation of rice seed germination, during which RNA transcription, metabolism resumption and cellular repair are highly activated [19,38,48,49]. According to the published transcriptome data, 28 miRNAs were found to be expressed remarkably opposite to their targeted genes, including conserved miRNAs, i.e., miR169ijk targeted nuclear factors, miR390 targeted ATP binding proteins, and miR166e targeted SGT1 protein, which are known to participate in plant development and disease resistance [50]. Novel miRNA-target pairs provide useful information to explore the new functions of miRNAs in rice seed germination. For example, OsmiR-129 sharply increased during seed imbibition, it might reduce the phytic acid synthesis in rice seeds by targeting a multi-drug resistance-associated ABC transporter and release micronutrients and organic phosphorus for seed germination [51]. Notably, among those inversely expressed miRNA-target pairs, several miRNAs were found to target genes with unknown function, i.e., OsmiR-100 (LOC_Os07g07030), OsmiR-217 (LOC_Os07g08669), OsmiR-232 (LOC_Os05g06910). It is worth in-depth investigation on the roles of those miRNA-target pairs in the seed germination regulation.

The expression level of miRNAs are not only inversely but sometimes also positively correlated with their target genes for feedback regulation or other interference factors [16]. In this study, many differentially expressed miRNAs were identified to be positively correlated with their target genes, such as OsmiR-201. OsmiR-201 is highly expressed in dry seeds but the expression decreases rapidly during imbibition. Three potential targets of OsmiR-201, LOC_Os05g50660 (encoding a PX domain containing protein), LOC_Os01g69940 (encoding an F-box domain containing protein), and LOC_Os01g62800 (encoding a methyltransferase) decrease during seed imbibition. Exploring whether the feedback regulations exist in these kinds of miRNA-target pairs will help to understand the miRNA-mediated regulations from a new angle.

MicroRNA Mediates the Crosstalk between ABA and GA in Controlling Rice Seed Germination

ABA is well documented as a seed dormancy maintainer that functions antagonistically to GA. MicroRNA159 has been reported extensively to be involved in the crosstalk between ABA and GA through cleavage of *GAMYB*-like genes in leaves, flowers, seed maturation and germination process [34,52,53]. In Arabidopsis seeds, exogenous ABA results in the accumulation of miR159 that participates in the regulation of seed germination by targeting MYB33 and MYB101, two positive regulators of ABA response [53]. This negative feedback might contribute to the developmental switch from seed dormancy to germination in Arabidopsis [54]. In rice, expression of miR159 was detected in various organs except in seeds. However, its target genes, *OsGAMYBL1* and *OsGAMYB*, which regulate almost all GA-regulated genes in aleuronic cells, are highly expressed in the seed aleurone layer and promote storage hydrolysis during seed germination [34]. The expression level of miR159 is found to be very low in the embryo of non-dormant maize seeds [55]. Unexpectedly, in this study, we identified abundant miR159, especially miR159f, in the embryos of germinating rice seeds. In addition, we observed that miR519 was insensitive to ABA or GA. Thus, the plentiful miR159 in rice embryo cannot be induced by ABA as it is in Arabidopsis. RACE analysis verified that *OsGAMYBL1* and *OsGAMYB* were both cleaved by miR159 at the canonical positions. Meanwhile, *OsGAMYB* (the ortholog of Arabidopsis MYB33) was detected to be down-regulated by RT-PCR and microarray analysis in the embryos of germinating rice seeds. It is of interest to reveal the distinctive expression patterns of osa-miR159 and the role of osa-miR159 mediated crosstalk between ABA and GA in the embryo of germinating rice seed in future studies.

Two other targets potentially involved in the ABA signaling were also verified by 5-RACE. One is *OSK3*, targeted by rice known miRNA osa-miR1428e-3p. *OSK3* has been identified as an important gene for starch accumulation in the early stages of rice endosperm development [56]. During seed germination, most proteins related to rice endosperm development are rapidly degraded [19,57]. Similarly, *OSK3* was detected to decrease drastically by RT-PCR in seed imbibition. *OSK3* belongs to the sucrose non-fermenting related kinase 1 (*SnRK1*) family, which is well documented as a central component of the regulatory response to glucose starvation and participates in ABA signaling pathways [58,59]. However, RT-PCR results showed that*OSK3* was insensitive to ABA or GA in the embryo of germinating rice seed, suggesting that the function of *OSK3* might be regulated by ABA at the post-translational level. In contrast, osa-miR1428e-3p was induced significantly by GA, implying that GA displayed antagonistic role to ABA through up-regulating the inhibitor of the positive regulator in ABA signaling. Gene*LOC_Os11g37540* encoding a PP2C domain containing protein was predicted to be targeted by a novel miRNA OsmiR-203. Rice has 90 PP2C proteins, and most of them participate in ABA signaling by interacting with SnRKs and suppressing SnRKs activation through dephosphorylation of Ser/Thr residues in the activation loop [58,60]. However, the RT-PCR results indicated that *LOC_Os11g37540* was insensitive to ABA or GA, and OsmiR-203 was slightly increased by GA but inhibited by ABA. These observations reflect the complex crosstalk between ABA and GA mediated by miRNAs in controlling rice seed germination.

ACKNOWLEDGMENTS

We appreciate Rebecca Njeri Damaris's help in English editing.

AUTHOR CONTRIBUTIONS

Conceived and designed the experiments: PY DH. Performed the experiments: DH QW. Analyzed the data: DH KW. Contributed reagents/materials/analysis tools: DH QW. Wrote the paper: DH PY.

REFERENCES

1. Yang PF, Li XJ, Wang XQ, Chen H, Chen F, Shen SH. Proteomic analysis of rice (*Oryza sativa*) seeds during germination. Proteomics 2007; 7: 3358–3368. pmid:17849412 doi: 10.1002/pmic.200700207
2. Howell KA, Millar AH, Whelan J. Ordered assembly of mitochondria during rice germination begins with promitochondrial structures rich in components of the protein import apparatus. Plant Mol Biol. 2006; 60: 201–223. pmid:16429260 doi: 10.1007/s11103-005-3688-7
3. Howell KA, Cheng K, Murcha MW, Jenkin LE, Millar AH, Whelan J. Oxygen initiation of respiration and mitochondrial biogenesis in rice. J Biolog Chem. 2007; 282: 15619–15631. doi: 10.1074/jbc.m609866200
4. He DL, Han C, Yao JL, Shen SH, Yang PF. Constructing the metabolic and regulatory pathways in germinating rice seeds through proteomic approach. Proteomics. 2011; 11: 2693–2713. doi: 10.1002/pmic.201000598. pmid:21630451
5. Howell KA, Narsai R, Carroll A, Ivanova A, Lohse M, Usadel B, et al. Mapping metabolic and transcript temporal switches during germination in rice highlights specific transcription factors and the role of RNA instability in the germination process. Plant Physiol. 2009; 149: 961–980. doi: 10.1104/pp.108.129874. pmid:19074628
6. Han C, Yang PF, Sakata K, Komatsu S. Quantitative proteomics reveals the role of protein phosphorylation in rice embryos during early stages of germination. J Proteome Res. 2014; 13: 1766–1782. doi: 10.1021/pr401295c. pmid:24460219
7. Voinnet O. Origin, biogenesis, and activity of plant microRNAs. Cell. 2009; 136: 669–687. doi: 10.1016/j.cell.2009.01.046. pmid:19239888
8. Eckardt NA. The plant cell reviews aspects of microRNA and PhasiRNA regulatory function. Plant Cell. 2013; 25: 2382–2391. doi: 10.1105/tpc.113.250713. pmid:23897925
9. Hannon GJ. RNA interference. Nature. 2002; 418: 244–251. pmid:12110901 doi: 10.1038/418244a
10. Wu MF, Tian Q, Reed JW. Arabidopsis microRNA167 controls patterns of ARF6 and ARF8 expression, and regulates both female and male reproduction. Development. 2006; 133: 4211–4218. pmid:17021043 doi: 10.1242/dev.02602
11. Liu PP, Montgomery TA, Fahlgren N, Kasschau KD, Nonogaki H, Carrington JC. Repression of AUXIN RESPONSE FACTOR10 by microRNA160 is critical for seed germination and post-germination stages. Plant J. 2007; 1 52: 133–146. pmid:17672844 doi: 10.1111/j.1365-313x.2007.03218.x
12. Palatnik JF, Wollmann H, Schommer C, Schwab R, Boisbouvier J, Rodriguez R, et al. Sequence and expression differences underlie functional specialization of Arabidopsis microRNAs miR159 and miR319. Dev. Cell. 2007; 13: 115–125. pmid:17609114 doi: 10.1016/j.devcel.2007.04.012
13. Koyama T, Mitsuda N, Seki M, Shinozaki K, Ohme-

Takagi M. TCP Transcription Factors Regulate the Activities of ASYMMETRIC LEAVES1 and miR164, as Well as the Auxin Response, during Differentiation of Leaves in Arabidopsis. Plant Cell. 2010; 22: 3574–3588. doi: 10.1105/tpc.110.075598. pmid:21119060

14. Kim JY, Kwak KJ, Jung HJ, Lee HJ, Kang H. MicroRNA402 affects seed germination of Arabidopsis thaliana under stress conditions via targeting DEMETER-LIKE Protein3 mRNA. Plant Cell Physiol. 2010; 51: 1079–1083. doi: 10.1093/pcp/pcq072. pmid:20460498

15. Ma XX, Shao CG, Wang HZ, Jin YF, Meng YJ. Construction of small RNA-mediated gene regulatory networks in the roots of rice (Oryza sativa). BMC Genomics. 2013; 14: 1–13. doi: 10.1186/1471-2164-14-510

16. Wei LQ, Yan LF, Wang T. Deep sequencing on genome-wide scale reveals the unique composition and expression patterns of microRNAs in developing pollen of Oryza sativa. Genome Biol. 2011; 12: 1–14. doi: 10.1186/gb-2011-12-6-r53

17. Xue LJ, Zhang JJ, Xue HW. Characterization and expression profiles of miRNAs in rice seeds. Nucleic Acids Res. 2009; 37: 916–930. doi: 10.1093/nar/gkn998. pmid:19103661

18. Watanabe KA, Ringler P, Gu LK, Shen QXJ. RNA-sequencing reveals previously unannotated protein- and microRNA-coding genes expressed in aleurone cells of rice seeds. Genomics. 2014; 103: 122–134. doi: 10.1016/j.ygeno.2013.10.007. pmid:24200500

19. Han C, He D, Li M, Yang P. In-depth proteomic analysis of rice embryo reveals its important roles in seed germination. Plant Cell Physiol. 2014; 55: 1826–1847. doi: 10.1093/pcp/pcu114. pmid:25231964

20. L C, Meyers BC, Green PJ. Construction of small RNA cDNA libraries for deep sequencingMethods. 2007; 43: 110–117. doi: 10.1016/j.ymeth.2007.05.002

21. Zhang BH, Pan XP, Cox SB, Cobb GP, Anderson TA. Evidence that miRNAs are different from other RNAs. Cell Molr Life Sci. 2006; 63: 246–254. doi: 10.1007/s00018-005-5467-7

22. Wang HJ, Zhang PJ, Chen WJ, Jie D, Dan F, Jia YH, et al. Characterization and Identification of Novel Serum Micrornas in Sepsis Patients with Different Outcomes. Shock. 2013; 39: 480–487. doi: 10.1097/SHK.0b013e3182940cb8. pmid:23612084

23. Dai XB, Zhao PX. psRNATarget: a plant small RNA target analysis server. Nucleic Acids Res. 2011; 39: 155–159. doi: 10.1093/nar/gkr319

24. Addo-Quaye C, Miller W, Axtell MJ. CleaveLand: a pipeline for using degradome data to find cleaved small RNA targets. Bioinformatics. 2009; 25: 130–131. doi: 10.1093/bioinformatics/btn604. pmid:19017659

25. Chen CF, Ridzon DA, Broomer AJ, Zhou ZH, Lee DH, Nguyen JT, et al. Real-time quantification of microRNAs by stem-loop RT-PCR. Nucleic Acids Res. 2005; 33: 1–14. doi: 10.1093/nar/gni178

26. Ramakers C, Ruijter JM, Deprez RH, Moorman AF. Assumption-free analysis of quantitative real-time polymerase chain reaction (PCR) data. Neurosci Lett. 2003; 339: 62–66. pmid:12618301 doi: 10.1016/s0304-3940(02)01423-4

27. Henderson IR, Zhang X, Lu C, Johnson L, Meyers BC, Green PJ, et al. Dissecting Arabidopsis thaliana DICER function in small RNA processing, gene silencing and DNA methylation patterning. Nat Genet. 2006; 38: 721–725. doi: 10.1038/ng1804

28. Zhang BH, Pan XP, Wang QL, Cobb GP, Anderson TA. Computational identification of microRNAs and their targets. Comput Biol Chem. 2006; 30: 395–407. pmid:17123865 doi: 10.1016/j.compbiolchem.2006.08.006

29. Wu L, Zhou HY, Zhang QQ, Zhang JG, Ni FR, Liu C, et al. DNA methylation mediated by a MicroRNA pathway. Mol Cell. 2010; 38: 465–475. doi: 10.1016/j.molcel.2010.03.008. pmid:20381393

30. Mi SJ, Cai T, Hu YG, Chen Y, Hodges E, Ni FR, et al. Sorting of small RNAs into Arabidopsis argonaute complexes is directed by the 5 ‘ terminal nucleotide. Cell. 2008; 133: 116–127. doi: 10.1016/j.cell.2008.02.034. pmid:18342361

31. Wu L, Zhang QQ, Zhou HY, Ni FR, Wu XY, Qi YJ. Rice microRNA effector complexes and targets. Plant Cell. 2009; 21: 3421–3435. doi: 10.1105/tpc.109.070938. pmid:19903869

32. Baskerville S, Bartel DP. Microarray profiling of microRNAs reveals frequent coexpression with neighboring miRNAs and host genes. RNA. 2005; 11: 241–247. pmid:15701730 doi: 10.1261/rna.7240905

33. Yang GD, Yan K, Wu BJ, Wang YH, Gao YX, Zheng CC. Genomewide analysis of intronic microRNAs in rice and Arabidopsis. J Genetics. 2012; 91: 313–324. doi: 10.1007/s12041-012-0199-6

34. Tsuji H, Aya K, Ueguchi-Tanaka M, Shimada Y, Nakazono M, Watanabe R, et al. GAMYB controls different sets of genes and is differentially regulated by microRNA in aleurone cells and anthers. Plant J. 2006; 47: 427–444. pmid:16792694 doi: 10.1111/j.1365-313x.2006.02795.x

35. Slotkin RK, Vaughn M, Borges F, Tanurdzic M, Becker JD, Feijo JA, et al. Epigenetic reprogramming and small RNA silencing of transposable elements in pollen. Cell. 2009; 136: 461–472. doi: 10.1016/j.cell.2008.12.038. pmid:19203581

36. Wang XF, Elling AA, Li XY, Li N, Peng ZY, He GM, et al. Genome-wide and organ-specific landscapes of epigenetic modifications and their relationships to mRNA and small RNA transcriptomes in maize. Plant Cell. 2009; 21: 1053–1069. doi: 10.1105/tpc.109.065714. pmid:19376930

37. Groszmann M, Greaves IK, Albertyn ZI, Scofield GN, Peacock WJ, Dennis ES. Changes in 24-nt siRNA levels in Arabidopsis hybrids suggest an epigenetic contribution to hybrid vigor. Proc Natl Acad Sci. 2011; 108: 2617–2622. doi: 10.1073/pnas.1019217108. pmid:21266545

38. Galland M, Huguet R, Arc E, Cueff G, Job D, Rajjou L. Dynamic proteomics emphasizes the importance of selective mRNA translation and protein turnover during

arabidopsis seed germination. Mol & Cell Proteomics. 2014; 13: 252–268. doi: 10.1074/mcp.m113.032227

39. Gallandt M, Rajjou L. Regulation of mRNA translation controls seed germination and is critical for seedling vigor. Frontiers Plant Sci. 2015; 6: 1–13. doi: 10.3389/fpls.2015.00284

40. Jones-Rhoades MW, Bartel DP. Computational identification of plant MicroRNAs and their targets, including a stress-induced miRNA. Mol Cell. 2004; 14: 787–799. pmid:15200956 doi: 10.1016/j.molcel.2004.05.027

41. Nobuta K, Venu RC, Lu C, Belo A, Vemaraju K, Kulkarni K, et al. (2007) An expression atlas of rice mRNAs and small RNAs. Nature Biotech. 25: 473–477. doi: 10.1038/nbt1291

42. Braun JE, Huntzinger E, Izaurralde E. A molecular link between miRISCs and deadenylases provides new insight into the mechanism of gene silencing by microRNAs. CSH Perspect Biol. 2012; 4:1–8. doi: 10.1101/cshperspect.a012328

43. Iwakawa H, Tomari Y. Molecular Insights into microRNA-mediated translational repression in plants. Mol Cell. 2013; 52: 591–601. doi: 10.1016/j.molcel.2013.10.033. pmid:24267452

44. Rodriguez RE, Mecchia MA, Debernardi JM, Schommer C, Weigel D, Palatnik JF. Control of cell proliferation in Arabidopsis thaliana by microRNA miR396. Development. 2010; 137: 103–112. doi: 10.1242/dev.043067. pmid:20023165

45. Curaba J, Singh MB, Bhalla PL. miRNAs in the crosstalk between phytohormone signalling pathways. J Exp Bot. 2014; 65: 1425–1438. doi: 10.1093/jxb/eru002. pmid:24523503

46. Xie KB, Shen JQ, Hou X, Yao JL, Li XH, Xiao JH, et al. Gradual increase of miR156 regulates temporal expression changes of numerous genes during leaf development in rice. Plant Physiol. 2012; 158: 1382–1394. doi: 10.1104/pp.111.190488. pmid:22271747

47. .Klein J, Saedler H, Huijser P. A new family of DNA binding proteins includes putative transcriptional regulators of the Antirrhinum majus floral meristem identity gene SQUAMOSA. Mol Gen Genet. 1996; 250: 7–16. pmid:8569690 doi: 10.1007/bf02191820

48. He D, Yang P. Proteomics of rice seed germination. Front Plant Sci. 2013; 4: 246–256. doi: 10.3389/fpls.2013.00246. pmid:23847647

49. Holdsworth MJ, Finch-Savage WE, Grappin P, Job D. Post-genomics dissection of seed dormancy and germination. Trends Plant Sci. 2008; 13: 7–13. pmid:18160329 doi: 10.1016/j.tplants.2007.11.002

50. Azevedo C, Betsuyaku S, Peart J, Takahashi A, Noel L, Sadanandom A, et al. Role of SGT1 in resistance protein accumulation in plant immunity. Embo J. 2006; 25: 2007–2016. pmid:16619029 doi: 10.1038/sj.emboj.7601084

51. Xu XH, Zhao HJ, Liu QL, Frank T, Engel KH, An GH, et al. Mutations of the multi-drug resistance-associated protein ABC transporter gene 5 result in reduction of phytic acid in rice seeds. Theor and Appl Genet. 2009; 119: 75–83. doi: 10.1007/s00122-009-1018-1

52. Millar AA, Gubler F. The Arabidopsis GAMYB-like genes, MYB33 and MYB65, are MicroRNA-regulated genes that redundantly facilitate anther development. Plant Cell. 2005; 17: 705–721. pmid:15722475 doi: 10.1105/tpc.104.027920

53. Reyes JL, Chua NH. ABA induction of miR159 controls transcript levels of two MYB factors during Arabidopsis seed germination. Plant J. 2007; 49: 592–606. pmid:17217461 doi: 10.1111/j.1365-313x.2006.02980.x

54. Nonogaki H. MicroRNA gene regulation cascades during early stages of plant development. Plant Cell Physiol. 2010; 51: 1840–1846. doi: 10.1093/pcp/pcq154. pmid:20937608

55. Wang LW, Liu HH, Li DT, Chen HB. Identification and characterization of maize microRNAs involved in the very early stage of seed germination. BMC Genomics. 2011; 12: 1–13. doi: 10.1186/1471-2164-12-154

56. Kanegae H, Miyoshi K, Hirose T, Tsuchimoto S, Mori M, Nagato Y, et al. Expressions of rice sucrose non-fermenting-1 related protein kinase 1 genes are differently regulated during the caryopsis development. Plant Physiol Biochem. 2005; 43: 669–679. pmid:16087344 doi: 10.1016/j.plaphy.2005.06.004

57. He DL, Han C, Yang PF. Gene Expression profile changes in germinating rice. J Integr Plant Biol. 2011; 53: 835–844. doi: 10.1111/j.1744-7909.2011.01074.x. pmid:21910826

58. Raghavendra AS, Gonugunta VK, Christmann A, Grill E. ABA perception and signalling. Trends Plant Sci. 2010; 15: 395–401. doi: 10.1016/j.tplants.2010.04.006. pmid:20493758

59. Jossier M, Bouly JP, Meimoun P, Arjmand A, Lessard P, Hawley S, et al. SnRK1 (SNF1-related kinase 1) has a central role in sugar and ABA signalling in Arabidopsis thaliana. Plant J. 2009; 59: 316–328. doi: 10.1111/j.1365-313X.2009.03871.x. pmid:19302419

60. Singh A, Giri J, Kapoor S, Tyagi AK, Pandey GK. Protein phosphatase complement in rice: genome-wide identification and transcriptional analysis under abiotic stress conditions and reproductive development. BMC Genomics. 2010; 11: 435: 1–8. doi: 10.1186/1471-2164-11-435. pmid:20637108

CHAPTER 11

IDENTIFICATION OF A CANDIDATE GENE FOR PANICLE LENGTH IN RICE (ORYZA SATIVA L.) VIA ASSOCIATION AND LINKAGE ANALYSIS

Erbao Liu, Yang Liu, Guocan Wu, Siyuan Zeng, Thu G. Tran Thi, Lijun Liang, Yinfeng Liang, Zhiyao Dong, Dong She, Hui Wang, Imdad U. Zaid, and Delin Hong

Keywords: Association analysis, gene identification, map-based cloning, panicle length, quantitative trait locus, rice

INTRODUCTION

Rice (*Oryza sativa* L.) is an important staple food that feeds approximately 50% of the world's population. At present, rice is grown globally on approximately 160 million hectares annually and the average yield is 4.4 tons per hectare (GRiSP, 2013). Higher productivity is needed to meet the demands of the rapidly increasing population (Yuan, 1997; Khush, 1999). Panicle length is one aspect of panicle architecture and is usually measured as a yield-related trait. Panicle length, together with spikelet number and density, seed setting rate and grain plumpness, determines the grain number per panicle; hence, yield increases in rice. Studies focusing on traits that are components of grain yield and quality, such as grain number, panicle number and grain weight, have revealed a few genes associated with these traits, such as *GS3, GS5, GW2, GW5, GW8, Gn1, GL3*, and *GIF1* (Ashikari et al., 2005; Fan et al., 2006; Song et al., 2007; Wang et al., 2008, 2012;Weng et al., 2008; Li Y. et al., 2011; Zhang et al., 2012). Panicle length QTL were commonly co-identified with heading date, and some genes were cloned in recent years, such as *Hd1, EHD4, Hd6, Ghd7, DTH7*, and *DTH8* (Yano et al., 2000; Takahashi et al., 2001; Xue et al., 2008; Wei et al., 2010; Gao et al., 2013, 2014). However, panicle length has received relatively less attention. There are two subspecies, i.e., indica and japonica, in *Oryza sativa* L. In general, subspecies indica has longer panicle length and looser spikelet density than those of japonica. Even within subspecies there is considerable genetic variation of panicle length (range, 12–40 cm; genetic coefficient of variation, 10%; Jambhulkar and Bose, 2014; Zuo et al., 2014). Panicle length is inherited in a quantitative manner and controlled by both major and minor QTLs (Liu et al., 2011). To date, at least 253 QTLs for panicle length have been detected distributed on 12 chromosomes (Xiao et al., 1998; Hittalmani et al., 2002, 2003; Xing et al., 2002; Kobayashi et al., 2003; Thomson et al., 2003; Ashikari et al., 2005; Lee et al., 2005; Mei et al., 2005; Cho et al., 2007; Liu et al., 2011;Marathi et al., 2012; Yao et al., 2015; Zhang et al., 2015). However, only a few genes have been cloned and applied in rice plant architecture breeding. The gene *short panicle1* (*SP1*) encodes a putative polypeptide transporter protein (PTR) family regulating the activity of the spike meristem, resulting in the short-panicle phenotype. SP1 contains a conserved PTR2 domain consisting of 12 transmembrane domains (Li et al., 2009). The *dense and erect panicle1* (*DEP1*) locus is a gain-of-function mutation that results in the truncation of a phosphatidylethanolamine-binding protein-like domain protein. This allele enhances meristematic activity, resulting in a reduced length of the inflorescence internode, an increased number of grains per panicle and a consequent increase in grain yield (Huang et al., 2009). The gene *DEP2* encodes a plant-specific protein without any known functional domain is essential for determining panicle outgrowth and elongation (Li et al., 2010). And the gene *DEP3*, predicted to encode a patatin-like phospholipase A2, plays an important role in high grain yield (Qiao et al., 2011). *LARGER PANICLE* (*LP*) encodes a Kelch repeat-containing F-box protein and leads to an increase in spikelets and branches. *LP* expression is enriched in the branch primordial region, and *LP* may be involved in

Citation: Liu E, Liu Y, Wu G, Zeng S, Tran Thi TG, Liang L, Liang Y, Dong Z, She D, Wang H, Zaid IU and Hong D (2016) Identification of a Candidate Gene for Panicle Length in Rice (Oryza sativa L.) Via Association and Linkage Analysis. Front. Plant Sci. 7:596. doi: 10.3389/fpls.2016.00596.

modulating cytokinin levels in plant tissues (Li M. et al., 2011). Linkage mapping using cross-populations is the traditional method for QTL identification due to its high power and simple genetic background (Huang and Han, 2014). However, linkage mapping exploits only those loci with the strongest influence, hindering the detection of phenotypes occurring at lower frequencies in samples of natural populations. As a new approach to the use of natural populations for QTL analysis, genome-wide association (GWA) mapping has been widely applied for the mining of useful alleles in many species (Breseghello and Sorrells, 2006; Agrama et al., 2007; Kump et al., 2011; Morris et al., 2013; Zhang et al., 2014; Liu et al., 2015) due to its greater power to identify variants with weak effects compared with linkage studies (Risch and Merikangas, 1996). However, few studies have verified the loci and candidate genes underlying panicle length via association and linkage mapping. In this study, three populations, including a population of 254 recombinant inbred lines (RILs) derived from a cross between the cultivars Xiushui 79 (short panicle) and C-bao (long panicle), 540 rice accessions covering a wide geographical expanse (from 12.3°N to 47.4°N) and a single segment segregating F_2 population, were used to identify QTLs and candidate genes associated with panicle length.

MATERIALS AND METHODS

Plant Materials, Cultivation, and Measurements

Three mapping populations were used in the present study. A population consisting of 540 rice accessions from the geographical regions of East and Southeast Asia were used for association mapping, including 121 accessions from Vietnam (17°N–23°N), 400 from China and 11 from Japan (20°N–54°N). A population consisting of 254 RILs derived from a cross between two japonica rice cultivars, Xiushui79 (female parent, panicle length 15.53 cm) and C-bao (male parent, panicle length 26.63 cm) were used for primary QTL analysis. A single segment segregating F_2 population derived by crossing a single segment substitution line with Xiushui79 was used to confirm and finely map the major QTL controlling panicle length. All plants were grown in a paddy field at Jiangpu Experimental Station, Nanjing Agricultural University, Nanjing, China (31°56′N, 119°4′E). The seeds of natural and RILs populations were sown in the seedling nursery on 15 May in 2011 and 2012, and the seedlings were transplanted on 15 June in 2011 and 2012, planting one seedling per hill, with three replicates. The seeds of single segment segregating F_2 population were sown in the seedling nursery on 12 May in 2014, and the seedlings were transplanted on 15 June in 2014. Each plot consisted of five rows with eight hills per row, and the hill spacing was 17 × 20 cm. Panicle length was measured as the length from the panicle neck to the panicle tip of the main panicle. The average of three replicates was subjected to association mapping and linkage analysis.

Genotype Analysis

DNA was extracted from fresh leaves of individuals of the 540 rice accessions, 254 RILs and single segment segregating F_2 population using the method reported by Monna et al. (2002). A total of 262 SSR markers selected from the rice maps (Temnykh et al., 2000; McCouch et al., 2002) were used to genotype 540 rice accessions. PCR amplification was conducted in a 10-μL reaction mixture containing 1 μL of 20 ng μL^{-1} template DNA, 0.6 μL of 25 mmol L^{-1} $MgCl_2$, 0.7 μL of 2 pmol μL^{-1}forward primers, 0.7 μL of 2 pmol μL^{-1} reverse primers, 0.2 μL of 2.5 mmol L^{-1} dNTP, 1 μL of 10 × PCR buffer, 0.1 μL of 5 U μL^{-1} rTaq DNA polymerase (TaKaRa, Japan) and 5.7 μL of ddH_2O. DNA amplification was performed using a PTC-100 Peltier Thermal Cycler (MJ Research Inc., USA). The PCR program included denaturation at 95°C for 5 min, followed by 35 cycles of 95°C for 30 s, 55°C for 30 s, and 72°C for 30 s, and a final extension step at 72°C for 5 min. The PCR products were separated through electrophoresis on 8% non-denaturing polyacrylamide gels at a voltage of 180 V for approximately 100 min and then visualized via silver staining (Creste et al., 2001).

Population Structure and Linkage Disequilibrium

The population structure of 540 rice accessions was analyzed with STRUCTURE 2.2 (Falush et al., 2003). Twenty independent runs were performed for each k (from 2 to 10) using a burn-in length of 50,000, a run length of 100,000 and a model for the admixture and independent allele frequency. The mean log-likelihood value over 20 runs at each K value was used to identify the true number of populations (K), which was determined when the mean log-likelihood value reached the highest value for the model parameter K. The genetic distance was calculated from the 262 molecular markers using Nei's distance (Nei et al., 1983), and phylogenetic reconstruction was based on the Neighbour-joining method implemented in PowerMarker version 3.25 (Liu and Muse, 2005). A principal component analysis (PCA) was performed using the pcaMethods package of R.2.11.1 (Stacklies et al., 2007) to examine the population structure.The decay of LD (with distance in cM) between the SSR loci within the same chromosome was evaluated using 1000 permutations and calculated using TASSEL 2.1 software (Bradbury et al., 2007).

Association Mapping

The general linear model (GLM) and mixed linear model (MLM) in TASSEL 2.1 software were used for association mapping. The population structure (Q) was included as a covariate in the GLM to test for marker-trait associations and the matrices Q and K were used as covariates in the MLM analysis (Dang et al., 2014). The K matrix (kinship matrix) was obtained from the results of the relatedness analysis using SPAGeDi (Hardy and Vekemans, 2002). The allelic effects were estimated compared with the "null allele" (non-amplified alleles) for each locus (Flavio et al., 2006). The formula that was used to calculate the average positive (negative) allelic effects (AAE) within a locus was:

$$AAE = \sum {}^{a_c}/_{n_c}$$

where a_c represents the phenotypic value of the cth allele with a positive (negative) effect, and n_c represents the number of alleles with positive (negative) effects within the locus.

Genetic Linkage Map and QTL Analysis

A total of 254 RILs derived from a cross between the cultivars Xiushui 79 and C-bao were used to conduct QTL analysis of panicle length using 91 SSR markers. The linkage map was constructed with MapMaker3.0/EXP version 3.0 (Lander et al., 1987). QTL analysis was performed using the inclusive composite interval mapping method in IciMapping, version 3.3 (http://www.isbreeding.net/), based on a stepwise linear regression model (Wang, 2009).

Map-Based Cloning of *LP1*

Primers were designed around *LP1* on chromosome 9 of rice using indels identified between the indica cultivar 93-11 and the japonica cultivar Nipponbare using the software Primer Premier 5.0. Twelve of the 82 selected SSR primers in the preliminary mapping region showed polymorphisms between Xiushui79 and C-bao, which were available from the Rice Annotation Project Database (RAP-DB, http://rapdb.dna.affrc.go.jp/). These primers were used to genotype and identify the recombinants of the single-locus segregating F_2 population, which included a total of 8650 individuals. The number of recombinants was identified based on a combination of genotype and phenotype data.

Expression Analysis of Candidate Genes

Total RNA was extracted from various tissues of Xiushui79 and NIL-*LP1* plants using the Ultrapure RNA Kit (CoWin Biotech Co., China). Genomic DNA contamination was removed by treatment with RNase-free DNase I. First-strand cDNA was synthesized using 6 μg of RNA and 4 μg of reverse transcriptase mix (PrimeScript RT Master Mix Perfect Real Time; TaKaRa Bio, Inc.) in a volume of 20 μL. Real-time quantitative RT-PCR was performed in a total volume of 20 μL containing 2 μL of template cDNA, 10 μL of qPCR SYBR Green Master Mix (Vazyme), 0.4 μL of forward and reverse gene-specific primers, 0.4 μL of ROX reference dye1, and 6.8 μL of ddH_2O. Gene expression was normalized against 18S rRNA as an internal control. The amplification reaction was performed in a 96-well thermocycler (Roche Applied Science LightCycler 480) using the AceQ qPCR Kit (Vazyme). The cycling program consisted of 5 min at 95°C, followed by 40 cycles of amplification (95°C for 10 s and 60°C for 30 s).. The relative gene expression of the target gene was calculated using the following equation:

$$Exp = 2^{-\Delta Ct}$$

where $\Delta Ct = Ct_{targetgene} - Ct_{18SrRNA}$ (Livak and Schmittgen, 2001).

Sequencing Analysis of *LP1*

Gene-specific PCR primers were designed to amplify the full-length DNA sequence of *LP1* from Xiushui79 and NIL-*LP1* . The PCR product was gel purified and sequenced by GenScript Corporation Ltd., Nanjing, China. Multiple sequence alignments were performed using DNAMAN.

Phylogenetic Analysis of *LP1*

The amino acid sequences of *LP1* homologs were obtained from NCBI (http://www.ncbi.nlm.nih.gov/). Then, the protein sequences were aligned using CLUSTAL W in MEGA5.0, and the result was used to construct a bootstrap N-J phylogenetic tree. A total of 1000 replicates were conducted to determine the statistical support for each node.

Sequence Polymorphisms of *LP1* in Rice Accessions

A total of 103 rice accessions showing abundant diversity in panicle length were selected to sequence a 5.4-kb genomic DNA fragment of *LP1*. Association analysis of the SNPs and indel markers in the 5.4-kb region of the 103 accessions with panicle length was conducted.

RESULTS

Phenotypic Variation of Panicle Length in Natural and Ril Populations

Panicle length differed significantly among the 540 accessions in both 2011 and 2012, with CVs of 19.45 and 19.56%, respectively. A high broad-sense heritability value was observed in both years. We selected ten varieties to represent the diversity of panicle length among the 540 rice accessions, including the shortest-panicle accession, Longjing25 (11.91 cm), and the longest-panicle accession, Haonuopie (39.98 cm) (Figure 1A). The frequency distribution of panicle length in the 540 accessions recorded in 2011 and 2012 is presented in Figure 1B. The means of panicle length over 254 RILs were 21.1 and 20.9 cm in 2011 and 2012, with CVs of 15.25% and 15.57%, respectively. RIL242 had the smallest panicle length (14.0 cm in 2011 and 14.9 cm in 2012), and RIL31 had the largest panicle length (29.0 cm in 2011 and 27.1 cm in 2012).

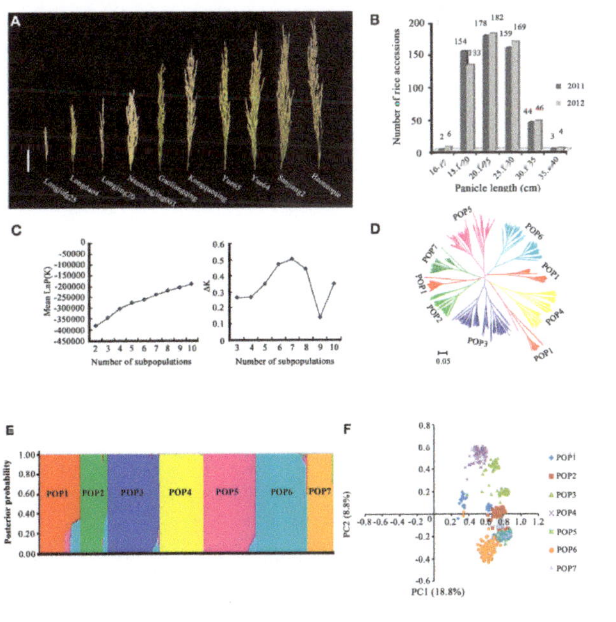

Figure 1: Population structure analysis of 540 rice accessions. (A) Panicle length of 10 rice accessions selected from 540 accessions to represent the diversity of panicle length in the population. Bar = 5 cm. (B) Frequency distribution of the panicle length of the 540 accessions in 2011 and

2012. (C) Changes in mean LnP(K) and ΔK. (D) Neighbor-joining tree for the 540 accessions generated using Nei's genetic distance. (E) Posterior probability of the 540 accessions belonged to seven subpopulations. (F) Principal components analysis (PCoA) for 540 accessions and reference cultivars genotyped with 262 molecular markers.

GWA Mapping and Linkage Analysis for Panicle Length

STRUCTURE analysis using 262 simple sequence repeat (SSR) markers revealed that the log-likelihood increased as the model parameter K increased; thus, the statistic ΔK was used to determine a suitable value for K. The ΔK value was much higher for the model parameter $K = 7$ than for other values of K (Figure 1C), indicating that the population used in this study was a mixed population consisting of seven subpopulations (Figure 1E). A neighbor-joining tree among the 540 rice accessions was constructed based on Nei›s (1983) genetic distance (Nei et al., 1983) (Figure1D), and the results were consistent with the results of the structure analysis. The results of the PCA were essentially consistent with those from STRUCTURE. The two major PCs from the PCA explained 27.6% (18.8 and 8.8%) of the total variance (Figure 1F).

Regression analysis between the D′ value and the genetic distance of syntenic marker pairs revealed that the genomes of the seven subpopulations fit the equation$y = b\ln x + c$. The minimum distances of LD decay were 60.2, 13.0, 85.4, 70.8, 29.8, 72.9, and 61.8 cM for the seven subpopulations.

Total fifteen SSR markers associated with panicle length were detected by both GLM and MLM analysis. GLM analysis of marker-trait associations revealed 10 SSR markers associated with panicle length ($p < 0.05$), located on chromosomes 2, 3, 5, 6, 8, 9, and 10, in both years. PVE ranged from 10.32% to 32.63%. RM5475 on Chr3 explained the maximum percentage of phenotypic variation: 28.38% in 2011 and 32.63% in 2012 (Table 1). RM3600 on Chr9, residing on site of 1.46 Mb, exhibited the second highest PVE values, of 22.19% in 2011 and 22.17% in 2012 (Table 1). MLM analysis revealed seven SSR markers associated with panicle length ($p < 0.05$), located on chromosomes 2, 6, 7, 9, and 11, in both years. PVE ranged from 2.09 to 25.82%. RM295 on Chr7 exhibited the highest PVE: 24.52% in 2011 and 25.82% in 2012. RM3600 on Chr9 explained 16.97% PVE in 2011 and 19.69% PVE in 2012. Two SSR markers, RM3600 on Chr9 and RM3688 on Chr2, were both detected by GLM and MLM methods (Table 1). In this study, the alleles with positive effects were considered elite alleles for panicle length. The allele RM3600-130 bp exhibited the largest phenotypic effect (8.69 cm) on panicle length, and the typical carrier accession was Yue 33.

GWA mapping revealed that the SSR marker RM3600 was associated with PL and was located in the region between RM5652 and RM410 of the QTL *LP1* detected through linkage analysis in the RILs (Tables 1, 2). The locus explained more than 20% of the variation of panicle length in both the linkage and GWA mapping analyses, indicating that *LP1* is a major QTL controlling panicle length in rice.

Table 1: Marker-trait associations with *P*-values of less than 0.05 by GLM and MLM analysis, the proportion of phenotypic variance explained (PVE) and the average positive (negative) allele effects (AAE) in 2011 and 2012.

No.	Marker names	Chr.	Start position (bp)	2011		2012		AAE		Analysis methods
				P-value	PVE (%)	*P*-value	PVE (%)	Positive	Negative	
1	RM3688	2	22,400,777	0.0500	19.28	0.0251	22.59	+2.65	−1.88	GLM
				0.0384	2.57	0.0480	2.09			MLM
2	RM213	2	34,658,315	0.0274	17.67	0.0028	27.27	+5.64	−2.06	GLM
3	RM266	2	35,431,738	0.0347	2.95	0.0377	2.77	+3.26	−1.38	MLM
4	RM535	2	35,784,312	0.0271	12.21	0.0454	10.32	+4.00	−3.08	GLM
5	RM5475	3	30,577,227	0.0372	28.38	0.0157	32.63	+2.37	−3.83	GLM
6	RM480	5	27,376,099	0.0323	11.50	0.0194	13.55	+3.44	−2.35	GLM
7	RM314	6	4,845,258	0.0222	18.60	0.0229	18.46	+2.42	−1.65	GLM
8	RM276	6	6,231,149	0.0303	21.70	0.0104	26.57	+3.60	−3.04	GLM
9	RM5389	6	8,276,087	0.0324	10.33	0.0067	9.34	+2.98	−1.24	MLM
10	RM6811	6	28,600,179	0.0270	4.51	0.0398	3.96	+3.66	−1.15	MLM
11	RM295	7	414,668	0.0482	24.51	0.0123	25.82	+2.88	−1.34	MLM
12	RM264	8	27,926,632	0.0424	10.40	0.0223	13.00	+2.29	−3.11	GLM
13	RM3600	9	17,054,142	0.0109	22.19	0.0110	22.17	+6.16	−2.25	GLM
				0.0086	16.97	0.0042	19.69			MLM
14	RM184	10	16,430,316	0.0161	14.30	0.0198	13.48	+1.62	−2.98	GLM
15	RM286	11	384,837	0.0335	10.33	0.0380	8.05	+1.98	−2.04	MLM

GLM, general linear model; MLM, mixed linear model.

A total of four additive QTLs were detected for panicle length on chromosomes 4, 6, and 9 in the RILs, including SSR marker RM551 (0 cM) on Chr4 and regions between RM3288 and RM349 (7.0 cM) on Chr4, between RM5314 and RM454 (0.9 cM) on Chr6, and between RM5652 and RM410 (7.0 cM) on Chr9 (Table 2). Each QTL explained 1.2–34.9% of the observed phenotypic variation (Table 2), with *LP1* explaining 34.9% of the phenotypic variation, representing the highest value among the four QTLs.

Table 2: Estimated position, flanking markers, additive effects and percentage of phenotypic variation explained (R^2) for the QTLs for panicle length in RILs population.

QTL	Marker interval	Distance (cM)	Position (bp)	LOD	*A* effect	*P*-value	R^2 (%)
PL4-1	RM551	0.0	(4)168,711	12.8	−0.56	$<1\times10^{-4}$	5.6
PL4-2	RM3288-RM349	7.0	(4)27,516,234–32,718,571	9.1	−0.59	$<1\times10^{-4}$	3.6
PL6	RM5314-RM454	0.9	(6)23,381,619-24,460,234	22.1	−0.81	$<1\times10^{-4}$	1.2
LP1	RM5652-RM410	7.0	(9)14,731,280–17,589,271	115.8	−1.35	$<1\times10^{-4}$	34.9

Phenotypic Performance of Panicle Length in a Fine-Mapping Population

A NIL plant containing C-bao allele at the *LP1* in genetic background of Xiushui79 isolated from BC_4F_2 generation (Figures 2A,B,D) was employed to develop a single segment segregating F_2population to confirm and finely map the major QTL *LP1*. The panicle length of NIL-*LP1*homozygous plant was significantly longer than that of Xiushui79 (i.e., 21.90 ±2.70 cm vs. 15.83 ± 1.90 cm) in 2014, which was consistent with the measurements obtained in 2013 (Table 3, Figure2C). By contrast, the F_1progeny of NIL-*LP1*/Xiushui79 exhibited a panicle length that was intermediate to those of the two parents (19.26 ± 1.21 cm) in 2014, indicating a semi-dominant pattern of expression of the allele for long panicle length. Among the 1980 plants randomly selected from the single-segment segregating F_2population, the phenotypic separation ratio was fitted to 1:2:1 [460 Xiushui79 homozygote type: 1017 heterozygote type: 503 NIL-*LP1* homozygote type, χ2(1:2:1) χ(1:2:1)2 = 3.22 < χ2(0.05,2)χ(0.05,2)2 = 5.99], suggesting that panicle length is controlled by a single-gene locus. The other agronomic traits of NIL-*LP1* were evaluated under field conditions. Compared with Xiushui79, NIL-*LP1* exhibited an increase in yield of 13.73% resulting from increases of 41.02% in panicle length, 10.90% in the number of filled grains per panicle, 2.69% in grain width, and 1.05% in the thousand filled-grain weight (Table 3). No significant differences were observed between Xiushui79 and NIL-*LP1* regarding days to heading, the panicle number per plant, seed setting rates, grain length, and grain thickness (Table 3). Therefore, the increase in grain yield observed in NIL-*LP1* was mainly due to the improvement of panicle length.

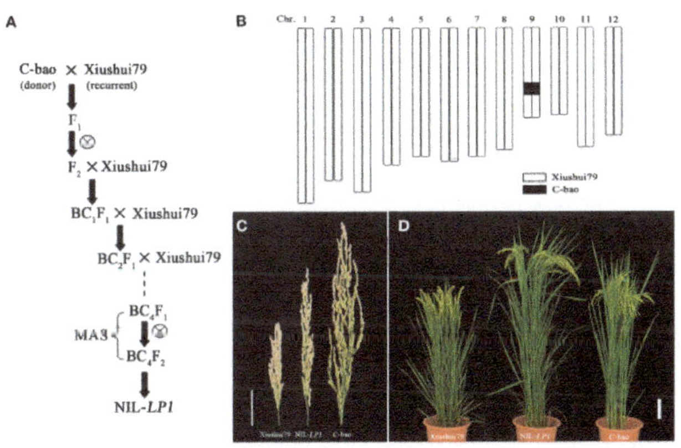

Figure 2: Genotypic and phenotypic performance of the parents. (A) Workflow of the construction of one NIL plant (BC_4F_2), NIL-*LP1*. (B) Graphical genotype of NIL-*LP1*. The black bar indicates the fragment from C-bao, and the remainder was derived from Xiushui79. (C) Panicle morphology of Xiushui79, NIL-*LP1* and C-bao. Bar = 5 cm. (D)Xiushui79, NIL-*LP1*, and C-bao plants. Bar = 10 cm.

Table 3: Comparison of panicle length and other agronomic traits between Xiushui79 and NIL-*LP1*.

Traits	Xiushui79	NIL-*LP1*	*P*-value
PL (cm)	15.83 ± 1.90	21.90 ± 2.70	5.18×10^{-6}
HD(d)	103.0 ± 0.3	104 ± 0.3	0.37
PH (cm)	80.33 ± 3.45	102.33 ± 4.20	3.09×10^{-4}
PP	15.3 ± 1.2	17.0 ± 1.6	0.28
NSP	110.0 ± 15.5	126.7 ± 23.8	1.57×10^{-2}
NFGP	107.3 ± 12.6	119.0 ± 17.5	3.22×10^{-2}
SS (%)	97.55 ± 0.21	94.01 ± 0.33	0.11
TGW (g)	28.413 ± 1.041	28.712 ± 1.261	1.53×10^{-2}
GL (mm)	7.51 ± 0.11	7.52 ± 0.13	0.63
GW (mm)	3.35 ± 0.23	3.44 ± 0.21	3.53×10^{-2}
GT (mm)	2.39 ± 0.15	2.40 ± 0.19	0.84
YPP (kg)	2.04 ± 0.51	2.32 ± 0.66	4.26×10^{-3}

All trait data are presented as the mean ± SD. Each P-value for each trait was obtained from a t-test between Xiushui79 and NIL-LP1. PL, panicle length; HD, days to heading (summer 2015, Nanjing, Jiangsu, China); PH, plant height; PP, panicle number per plant; NSP, number of spikelets per panicle; NFGP, number of filled grains per panicle; SS, seed setting rate; TGW, thousand filled-grain weight; GL, grain length; GW, grain width; GT, grain thickness; YPP, yield per plot.

Map-Based Cloning of *LP1*

Further QTL analysis was conducted in 176 individuals from the NIL-*LP1*/Xiushui79 F_2 population. The *LP1* QTL was delimited to a 1.60-cM region between RM3600 and RM410 with an LOD peak of 101.67. A total of 8650 F_2 plants were subjected to marker analysis by scanning RM3600 and RM410 (confidence interval markers of *LP1* in preliminary mapping). Analysis of RM3600 identified 25 recombination events between the marker and *LP1* on one side, whereas an analysis of RM410 detected 156 recombination events between the marker and *LP1* on the other side. The insertion/deletion (indel) marker L04 revealed two recombinants, whereas RM24496, RM24489, and RM7289 revealed 84, 24, and 3 recombinants, respectively, on the other side (Figure 3B). Therefore, *LP1* was mapped to a 90-kb DNA region between L04 and RM7289 (Figure 3C). This region contains six predicted ORFs according to the Nipponbare genome sequence (annotated by Rice Genome Annotation Project,http://rice.plantbiology.msu.edu/index.shtml), including *LOC_Os09g28300, LOC_Os09g28310, LOC_Os09g28340, LOC_Os09g28354, LOC_Os09g28370*, and *LOC_Os09g28390* (Figure 3C).

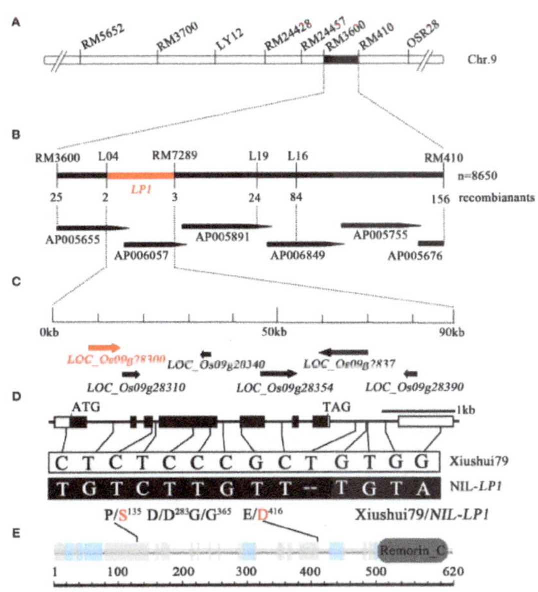

Figure 3: Map-based cloning of *LP1*. (A) Primary mapping of *LP1*. *LP1* was mapped between the SSR markers RM3600 and RM410. (B) High-resolution mapping of *LP1*. *LP1*was delimited to

a 90-kb region between the markers L04 and RM7289 in the BAC clones AP005655 and AP006057, using a total of 8650 plants from the NIL-*LP1*/Xiushui79 F_2population. (C) Predicted ORFs based on the Nipponbare genome sequence in the Rice Genome Annotation Project database (http://rice.plantbiology.msu.edu/index.shtml). The horizontal arrows indicate the six predicted ORFs. (D) Gene structure of*LOC_Os09g28300*. The empty boxes refer to the 5′ and 3′ UTRs; the black boxes, to exons; and the lines between the boxes, to introns. The fourteen SNPs in *LP1* are indicated by solid lines. (E) The Remorin_C domain predicted in the protein encoded by *LOC_Os09g28300*. The solid lines indicate the positions of two amino acid transitions.

Real-time quantitative RT-PCR was performed to further analyze the candidate genes by detecting differences in expression between the parents. Only *LOC_Os09g28300*, *LOC_Os09g28340*, and*LOC_Os09g28370* were expressed. Among these three genes, *LOC_Os09g28340* and*LOC_Os09g28370* exhibited no differential expression in the roots, stems, flag leaves and young panicles between Xiushui79 and NIL-*LP1* (Figures 4B,C,D). The expression of *LOC_Os09g28300*was significantly up-regulated (6.08-fold) in the young panicles of Xiushui79 compared with NIL-*LP1* (Figures 4A,D).

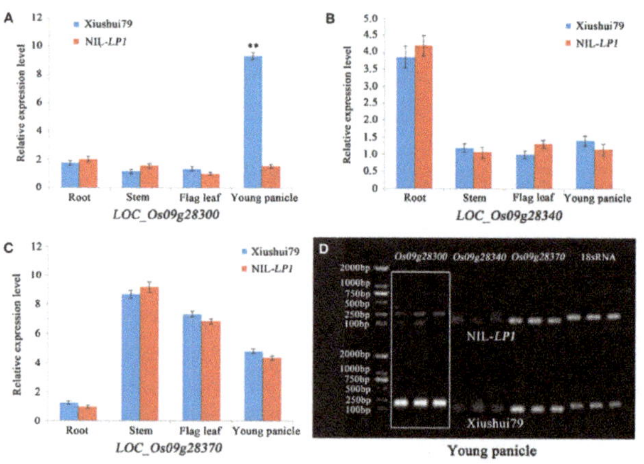

Figure 4: Analysis of the expression of the three expressed candidate genes via real-time quantitative RT-PCR. (A) Relative expression of *LOC_Os09g28300* in the roots, stems, flag leaves and young panicles. (B) Relative expression of *LOC_Os09g28340*in the roots, stems, flag leaves and young panicles. (C) Relative expression of*LOC_Os09g28370* in the roots, stems, flag leaves and young panicles. (D) Real-time quantitative RT-PCR results for the three candidate genes in the young panicles. The 18S rRNA gene was used as a control. ** indicates significance at the $\alpha = 0.01$ probability level.

Sequence comparison of *LOC_Os09g28300* between Xiushui79 and NIL-*LP1* revealed 14 SNPs, four of which were located in exons and 10 in introns. SNP1, a single nucleotide transition from C (Xiushui79) to T (NIL-*LP1*), was identified in the third exon of *LOC_Os09g28300*, and SNP2, a single nucleotide transversion from G (Xiushui79) to T (NIL-*LP1*), was found in the fifth exon (Figure 3D). These two polymorphisms cause amino acid residue changes from proline to serine (P/S^{135}) and glutamic acid to aspartic acid (E/D^{416}), respectively (Figure 3D). SNP3 (C → T, in the 4th exon, D/D^{283}) and SNP4 (C → T, in the 4th exon, G/G^{365}) do not cause amino acid substitutions (Figure 3D).

LP1 Encodes a Remorin_C-containing Protein

LOC_Os09g28300 encodes a C-terminal region domain-containing protein (Remorin_C). These proteins are plant-specific plasma membrane-associated proteins of unknown function (Figure 3E). Phylogenetic analysis revealed the presence of Remorin_C in many species, including rice, maize, millet, *Bachypodium distachyon*, wheat, and *Arabidopsis* (Figure 5A). Among the identified homologs, the maize homolog exhibits the highest homology with rice, sharing 67.4% amino acid identity with LP1. In addition, LP1 shares 54.2% amino acid identity with Os08g36760, which is encoded on chromosome 8 in rice. The two polymorphisms that lead to amino acid substitutions, P/S^{135} and E/D^{416}, are located at conserved positions among the homologs (Figure 5B). At the E/D^{416} locus, only Xiushui79 exhibits glutamic acid, whereas NIL-*LP1*and the other species exhibit aspartic acid. These differences in conserved amino acids may affect LP1 function.

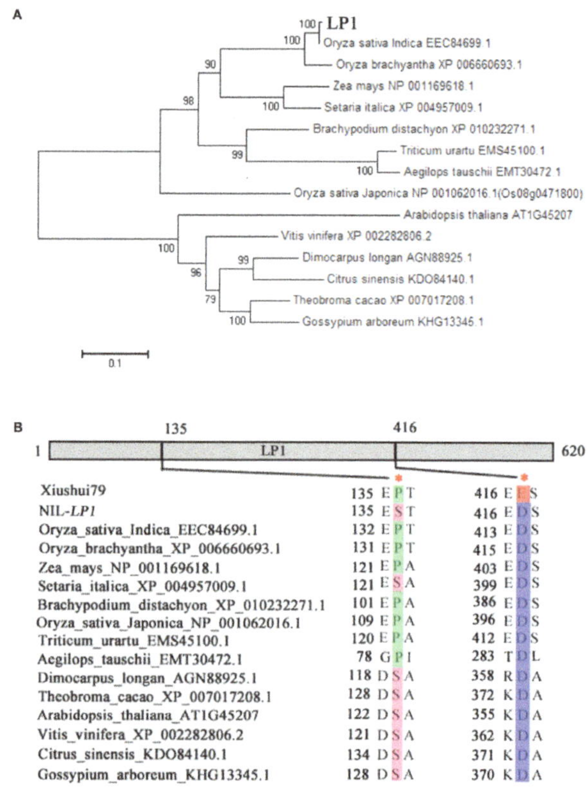

Figure 5: Phylogenetic analysis of the *LP1* **protein. (A)** Phylogenetic tree of the*LP1*-like proteins in higher plants. The amino acid sequences of 15 *LP1* homologs were obtained from NCBI (http://www.ncbi.nlm.nih.gov/) and used to construct a bootstrap N-J phylogenetic tree. A total of 1000 replicates were performed to determine the statistical support for each node. *LP1* is denoted in bold. (B) Diagram indicating the two amino acid substitutions in *LP1* at positions conserved across LP1-like proteins.

Sequence Polymorphisms of *LP1* in Rice Accessions

A total of 103 rice accessions with abundant diversity in panicle length selected from the 540 accessions were used to sequence a 5.4-kb genomic DNA fragment encompassing the mutation sites leading to amino acid substitutions in *LP1*. A total of 22 SNPs were identified. Based on the sequencing results, seven accessions carrying the same SNP1 as C-bao were identified, whereas polymorphisms in SNP2, SNP3, and SNP4 were widely distributed in both japonica and indica accessions. Association analysis of the SNPs and indel markers in the 5.4-

kb region with panicle length in the two parents and 103 rice accessions revealed that SNP1 was significantly associated with panicle length at $\alpha = 0.01$ probability level, whereas other polymorphic sites made no significant contributions to panicle length. These results indicate that SNP1 identical with Xiushui 79 is a major functional mutation for short panicles.

DISCUSSION

Panicle length is one of the most important traits for rice yields and ideal plant breeding. Previous studies have demonstrated that panicle length is a typical quantitative trait controlled by multiple genes and may be significantly influenced by environmental conditions (Yao et al., 2015). Therefore, the number and effect of QTLs for panicle length detected via linkage mapping may differ among segregating populations, such as F_2 populations, RILs or backcrossed inbred lines (Huang and Han, 2014), and these differences may underlie the identification of approximately 253 QTLs for panicle length distributed on 12 chromosomes (Xiao et al., 1998; Hittalmani et al., 2002,2003; Xing et al., 2002; Kobayashi et al., 2003; Thomson et al., 2003; Ashikari et al., 2005; Lee et al., 2005; Mei et al., 2005; Cho et al., 2007; Liu et al., 2011; Marathi et al., 2012; Yao et al., 2015;Zhang et al., 2015). GWA mapping, a new method that fully exploits ancient recombination events to identify the genetic loci underlying traits, is becoming a powerful tool for detecting natural variation underlying complex traits in crops (Rafalski, 2010). The results obtained using these two types of mapping methods may enable mutual authentication. Consequently, both association mapping using a natural population and linkage mapping using RILs were conducted in this study. The QTL *LP1* was detected through both association mapping and linkage analysis and was identified as a major QTL associated with high phenotypic variation for panicle length.

A natural population of 540 rice accessions with abundant phenotypic and genotypic diversity was employed to perform a succession study of panicle length in rice. Population structure and linkage disequilibrium are the basis for association analysis (Pritchard and Rosenberg, 1999; Flint-Garcia et al., 2003). Determining population structure can prevent false-positive results regarding associations between phenotypes and genotypes in association mapping due to the linkage disequilibrium in natural populations (Pritchard and Rosenberg, 1999). The seven subpopulations (POP1–POP7) grouped by structure were essentially consistent with geographic regions. For example, the accessions from Vietnam were basically classified as POP6, and the accessions from northeastern China were mostly classified as POP1. The LD decay distances among the seven subpopulations identified in this study ranged from 13 to 80 cM, which was longer than has been observed in previous studies (Agrama et al., 2007; Jin et al., 2010). These results indicate that the 540 rice accessions underwent artificial hybridization and selection.

By comparing the mapping results of population-based association mapping with those of the linkage mapping of the families, 10 marker-panicle length associations were detected in both 2011 and 2012, whereas only four related QTLs were identified, indicating that GWA mapping has greater power than linkage analyses to identify variants with weak effects. *LP1* was detected via the two approaches due to the strong effect of this QTL on panicle length in rice. Therefore, *LP1* was selected as a major QTL for further studies, including map-based cloning and candidate gene identification.

NIL development is a useful method for confirming and evaluating the genetic effects of a QTL and provides useful materials for population development during fine mapping of QTLs (Ding et al., 2011). This method has been successfully applied for fine mapping of QTLs in most crop species (Benson et al., 2015; Jang et al., 2015; Wang et al., 2015; Zheng et al., 2015). In the present study, we developed NILs in a backcrossing program using two parents, Xiushui79 (the receipt parent) and C-bao (the donor parent). The NILs exhibited an additive effect similar to that observed in the QTL analysis. Introgression of the *LP1* allele into the Xiushui79 background produced the longest panicles among the examined allele combinations, which will be valuable for breeding applications. NIL-*LP1* plants will serve as an important parent for the generation of a single-locus segregating F_2population for fine mapping.

In the present study, the chromosome segment containing *LP1* was delimited to a 90-kb region for the first time. Within the fine-mapping region, we identified six annotated genes:*LOC_Os09g28300, LOC_Os09g28310, LOC_Os09g28340, LOC_Os09g28354, LOC_Os09g28370*, and *LOC_Os09g28390*. Among the six candidate genes, *LOC_Os09g28300*was validated as the gene locus controlling panicle length. *LP1* is predicted to be a major regulator of panicle length in rice. Furthermore, real-time quantitative RT-PCR analysis revealed significant differences in the expression of *LOC_Os09g28300* in young panicles between Xiushui79 and NIL-*LP1*. The expression of *LOC_Os09g28300* was significantly higher in Xiushui79 than in NIL-*LP1*, indicating that the *LP1* allele of Xiushui79 suppresses panicle development in rice. Map-based cloning revealed that *LP1* encodes a Remorin_C-containing protein that may be closely related to plant development. The two SNP polymorphisms resulting in amino acid substitutions likely affect the function of the LP1 protein.

Sequence polymorphism analysis of *LP1* indicated that SNP1 is a functional mutation leading to short panicles. Most varieties with short panicles exhibit the same amino acid as Xiushui79 at the P/S^{135} locus, indicating that a change from proline to serine may relieve the suppression of short panicles. Consequently, the C-bao allele at *LP1* locus can be used to improve varieties with short panicles. As a major QTL, the NIL-*LP1* allele has beneficial effects on panicle length, the number of filled grains per panicle, the thousand filled-grain weight and grain width, and *LP1* should be selected in rice breeding.

Compared the QTLs detected in this study for panicle length with other studies, we found that*DEP1* (Huang et al., 2009), mapped to the interval between the marker RM3700 and RM7424 on Chr9, was within the region of QTL *LP1* detected in this study (Figure 6). However, the results of fine mapping show that the physical distance between *LP1* and *DEP1* is about 770 kb, which indicated that *LP1* was a novel gene within the region of QTL *LP1* (Figures 3, 6).Although other genes associated with the panicle phenotype have been cloned, the present study is the first to observe the involvement of *LP1*. The results of this study provide an opportunity for further functional analysis of *LP1* to elucidate the mechanism of rice panicle improvement. *LP1* can be used to create new varieties to improve rice yields. Compared the marker-PL associations with previous studies, we found 11 of the 15 SSR markers detected in this study were novel, and the other four SSR markers were located near to the chromosome regions harboring panicle and yield related QTLs or genes which have been reported. Among them, RM5472 on chromosome3 was near to the chromosome regions covering *Hd6* of heading-related gene shown in Figure 6 (Takahashi et al., 2001). Marker RM480 on Chr4 associated with panicle length was near to chromosome region harboring the gene *GS5* (Li Y. et al., 2011) (Figure 6). The marker RM6811 was near to the region of*DEP3* mapped on chromosome 6 (Qiao et al., 2011) (Figure 6). RM264 on chromosome 8 was near to the region containing *GW8* (Wang et al., 2012) (Figure 6). Utilization of all of the elite alleles that were detected in this study may improve the panicle length of parents of F_1 hybrid rice via pyramiding breeding.

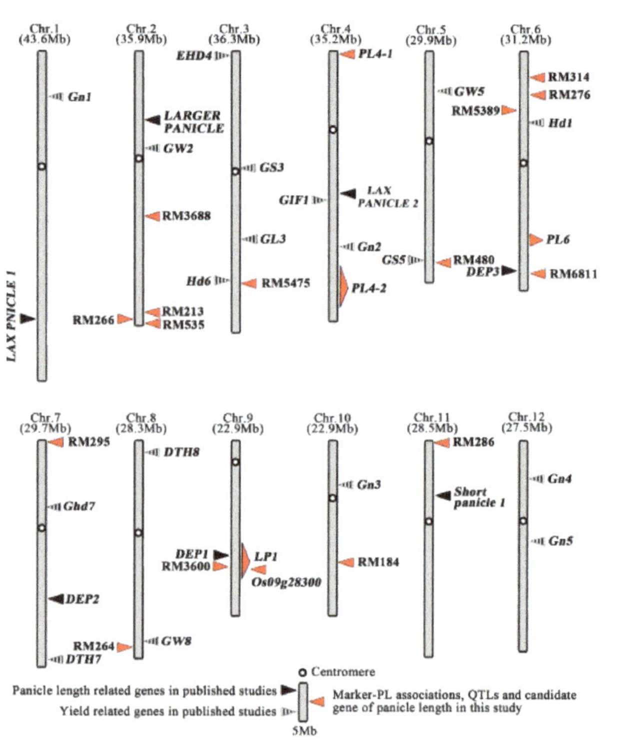

Figure 6: Distribution of marker-PL and QTLs or genes for panicle length and panicle length related on rice chromosome.

AUTHOR CONTRIBUTIONS

DH designed the research; EL, YL, GW, SZ, and TT carried out the field experiment; EL, YL, LL, YFL, ZD, DS, HW and IZ carried out the molecular experiment; EL analyzed data; and EL wrote the manuscript; DH revised the manuscript.

CONFLICT OF INTEREST STATEMENT

The authors declare that the research was conducted in the absence of any commercial or financial relationships that could be construed as a potential conflict of interest.

The reviewer DG and handling Editor declared their shared affiliation, and the handling Editor states that the process nevertheless met the standards of a fair and objective review.

REFERENCES

1. Agrama, H. A., Eizenga, G. C., and Yan, W. (2007). Association mapping of yield and its components in rice cultivars. *Mol Breeding.* 19, 341–356. doi: 10.1007/s11032-006-9066-6
2. Ashikari, M., Sakakibara, H., Lin, S., Yamamoto, T., Takashi, T., Nishimura, A., et al. (2005). Cytokinin oxidase regulates rice grain production. *Science* 309, 741–745. doi: 10.1126/science.1113373
3. Benson, J. M., Poland, J. A., Benson, B. M., Stromberg, E. L., and Nelson, R. J. (2015). Resistance to gray leaf spot of maize: genetic architecture and mechanisms elucidated through nested association mapping and near-isogenic line analysis. *PLOS Genet.* 11:e1005045. doi: 10.1371/journal.pgen.1005045
4. Bradbury, P. J., Zhang, Z., Kroon, D. E., Casstevens, T. M., Ramdoss, Y., and Buckler, E. S. (2007). TASSEL: software for association mapping of complex traits in diverse samples. *Bioinformatics* 23, 2633–2635. doi: 10.1093/bioinformatics/btm308
5. Breseghello, F., and Sorrells, M. E. (2006). Association mapping of kernel size and milling quality in wheat (Triticum aestivum L.) cultivars. *Genetics* 172, 1165–1177. doi: 10.1534/genetics.105.044586
6. Cho, Y., Kang, H., Lee, J., Lee, Y., Lim, S., Gauch, H., et al. (2007). Identification of quantitative trait loci in rice for yield, yield components, and agronomic traits across years and locations. *Crop Sci.* 47, 2403–2417. doi: 10.2135/cropsci2006.08.0509
7. Creste, S., Neto, A. T., and Figueira, A. (2001). Detection of single sequence repeat polymorphisms in denaturing polyacrylamide sequencing gels by silver staining. *Plant Mol Bio Rep.* 19, 299–306. doi: 10.1007/BF02772828
8. Dang, X., Tran Thi, T. G., Dong, G., Wang, H., Edzesi, W. M., and Hong, D. (2014). Genetic diversity and association mapping of seed vigor in rice (Oryza sativa L.). *Planta* 239, 1309–1319. doi: 10.1007/s00425-014-2060-z
9. Ding, X., Li, X., and Xiong, L. (2011). Evaluation of near-isogenic lines for drought resistance QTL and fine mapping of a locus affecting flag leaf width, spikelet number, and root volume in rice. *Theor. Appl. Genet.* 123, 815–826. doi: 10.1007/s00122-011-1629-1
10. Falush, D., Stephens, M., and Pritchard, J. K. (2003). Inference of population structure using multilocus genotype data: linked loci and correlated allele frequencies. *Genetics* 164, 1567–1587.
11. Fan, C., Xing, Y., Mao, H., Lu, T., Han, B., Xu, C., et al. (2006). GS3, a major QTL for grain length and weight and minor QTL for grain width and thickness in rice, encodes a putative transmembrane protein. *Theor Appl Genet.* 112, 1164–1171. doi: 10.1007/s00122-006-0218-1
12. Flavio, B., Mark, E., and Sorrells. (2006). Association mapping of kernel size and milling quality in wheat (Triticum aestivum L.) cultivars. *Genetics* 172, 1165–1177. doi: 10.1534/genetics.105.044586
13. Flint-Garcia, S. A., Thornsberry, J. M., and Buckler,

E. S. (2003). Structure of linkage disequilibrium in plants. *Annu Rev Plant Biol.* 54, 357–374. doi: 10.1146/annurev.arplant.54.031902.134907

14. Gao, H., Jin, M. N., Zheng, X. M., Chen, J., Yuan, D. Y., Xin, Y. Y., et al. (2014). Days to heading 7, a major quantitative locus determining photoperiod sensitivity and regional adaptation in rice. *Proc. Natl. Acad. Sci. U.S.A.* 111, 16337–16342. doi: 10.1073/pnas.1418204111

15. Gao, H., Zheng, X. M., Fei, G. L., Chen, J., Jin, M. N., Ren, Y. L., et al. (2013). Ehd4 encodes a novel and oryza-genus-specific regulator of photoperiodic flowering in rice. *PLoS Genet.* 9:e1003281. doi: 10.1371/journal.pgen.1003281

16. GRiSP (Global Rice Science Partnership). (2013). *Rice Almanac, 4th Edn.* Los Baños: International Rice Research Institute. 272–275.

17. Hardy, O., and Vekemans, X. (2002). SPAGeDi: a versatile computer program to analyse spatial genetic structure at the individual or population levels. *Mol. Ecol.* 2, 618–620. doi: 10.1046/j.1471-8286.2002.00305.x

18. Hittalmani, S., Huang, N., Courtois, B., Venuprasad, R., Shashidhar, H. E., Zhuang, J. Y., et al. (2003). Identification of QTL for growth- and grain yield-related traits in rice across nine locations of Asia. *Theor Appl Genet.* 107, 679–690. doi: 10.1007/s00122-003-1269-1

19. Hittalmani, S., Shashidhar, H. E., Bagali, P. G., Huang, N., Sidhu, J. S., Singh, V. P., et al. (2002). Molecular mapping of quantitative trait loci for plant growth, yield and yield related traits across three diverse locations in a doubled haploid rice population. *Euphytica* 125, 207–214. doi: 10.1023/A:1015890125247

20. Huang, X., and Han, B. (2014). Natural variations and genome-wide association studies in crop plants. *Annu Rev Plant Biol.* 65, 531 551. doi: 10.1146/annurev-arplant-050213-035715

21. Huang, X., Qian, Q., Liu, Z., Sun, H., He, S., Luo, D., et al. (2009). Natural variation at the DEP1 locus enhances grain yield in rice. *Nat Genet.* 41, 494–497. doi: 10.1038/ng.352

22. Jambhulkar, N. N., and Bose, L. K. (2014). Genetic variability and association of yield attributing traits with grain yield in upland rice. *Genetika* 46, 831–838. doi: 10.2298/GENSR1403831J

23. Jang, S. J., Sato, M., Sato, K., Jitsuyama, Y., Fujino, K., Mori, H., et al. (2015). A single-nucleotide polymorphism in an endo-1,4-β-glucanase gene controls seed coat permeability in soybean. *PLoS ONE* 10:e0128527. doi: 10.1371/journal.pone.0128527

24. Jin, L., Lu, Y., Xiao, P., Sun, M., Corke, H., and Bao, J. (2010). Genetic diversity and population structure of a diverse set of rice germplasm for association mapping. *Theor Appl Genet.* 121, 475–487. doi: 10.1007/s00122-010-1324-7

25. Khush, G. S. (1999). Green revolution: preparing for the 21st century. *Genome* 42, 646–655. doi: 10.1139/g99-044

26. Kobayashi, S., Fukuta, Y., Sato, T., Osaki, M., and Khush, G. S. (2003). Molecular marker dissection of rice (Oryza sativa L.) plant architecture under temperate and tropical climates. *Theor Appl Genet.* 107, 1350–1356. doi: 10.1007/s00122-003-1388-8

27. Kump, K. L., Bradbury, P. J., Wisser, R. J., Buckler, E. S., Belcher, A. R., Oropeza-Rosas, M. A., et al. (2011). Genome-wide association study of quantitative resistance to southern leaf blight in the maize nested association mapping population. *Nat Genet.* 43, 163–168. doi: 10.1038/ng.747

28. Lander, E. S., Green, P., Abrahamson, J., Barlow, A., Daly, M. J., Lincoln, S. E., et al. (1987). MAPMAKER: an interactive computer package for constructing primary genetic linkage maps of experimental and natural populations. *Genomics* 1, 174–181. doi: 10.1016/0888-7543(87)90010-3

29. Lee, S. J., Oh, C. S., Suh, J. P., McCouch, S. R., and Ahn, S. N. (2005). Identification of QTLs for domestication-related and agronomic traits in an Oryza sativa x O. rufipogon BC_1F_7 population. *Plant Breeding* 124, 209–219. doi: 10.1111/j.1439-0523.2005.01092.x

30. Li, F., Liu, W., Tang, J., Chen, J., Tong, H., Hu, B., et al. (2010). Rice DENSE AND ERECT PANICLE 2 is essential for determining panicle outgrowth and elongation. *Cell Res* 20, 838–849. doi: 10.1038/cr.2010.69

31. Li, M., Tang, D., Wang, K., Wu, X., Lu, L., Yu, H., et al. (2011). Mutations in the F-box gene LARGER PANICLE improve the panicle architecture and enhance the grain yield in rice. *Plant Biotechnol J.* 9, 1002–1013. doi: 10.1111/j.1467-7652.2011.00610.x

32. Li, S., Qian, Q., Fu, Z., Zeng, D., Meng, X., Kyozuka, J., et al. (2009). Short panicle1 encodes a putative PTR family transporter and determines rice panicle size. *Plant J.* 58, 592–605. doi: 10.1111/j.1365-313X.2009.03799.x

33. Li, Y., Fan, C., Xing, Y., Jiang, Y., Luo, L., Sun, L., et al. (2011). Natural variation in GS5 plays an important role in regulating grain size and yield in rice. *Nat Genet.* 43, 1266–1270. doi: 10.1038/ng.977

34. Liu, E., Liu, X., Zeng, S., Zhao, K., Zhu, C., Liu, Y., et al. (2015). Time-course association mapping of the grain-filling rate in rice (Oryza sativa L.). *PLoS ONE* 10:e0119959. doi: 10.1371/journal.pone.0119959

35. Liu, K., and Muse, S. V. (2005). PowerMarker: integrated analysis environment for genetic marker data. *Bioinformatics* 21, 2128–2129. doi: 10.1093/bioinformatics/bti282

36. Liu, T., Li, L., Zhang, Y., Xu, C., Li, X., and Xing, Y. (2011). Comparison of quantitative trait loci for rice yield, panicle length and spikelet density across three connected populations. *J Genet.* 90, 377–382. doi: 10.1007/s12041-011-0083-9

37. Livak, K. J., and Schmittgen, T. D. (2001). Analysis of relative gene expression data using realtime quantitative PCR and the $2^{(-\Delta\Delta C(T))}$ method. *Methods* 25, 402–408. doi: 10.1006/meth.2001.1262

38. Marathi, B., Guleria, S., Mohapatra, T., Parsad, R., Mariappan, N., Kurungara, V. K., et al. (2012). QTL analysis of novel genomic regions associated with

yield and yield related traits in new plant type based recombinant inbred lines of rice (Oryza sativa L.). *BMC Plant Biol.* 12:137. doi: 10.1186/1471-2229-12-137

39. McCouch, S. R., Teytelman, L., Xu, Y., Lobos, K. B., Clare, K., Walton, M., et al. (2002). Development and mapping of 2240 new SSR markers for rice (Oryza sativa L.). *DNA Res.* 9, 199–207. doi: 10.1093/dnares/9.6.199
40. Mei, H. W., Li, Z. K., Shu, Q. Y., Guo, L. B., Wang, Y. P., Yu, X. Q., et al. (2005). Gene actions of QTLs affecting several agronomic traits resolved in a recombinant inbred rice population and two backcross populations. *Theor Appl Genet.* 110, 649–659. doi: 10.1007/s00122-004-1890-7
41. Monna, L., Lin, X., Kojima, S., Sasaki, T., and Yano, M. (2002). Genetic dissection of a genomic region for a quantitative trait locus, Hd3, into two loci, Hd3a and Hd3b, controlling heading date in rice. *Theor Appl Genet.* 104, 772–778. doi: 10.1007/s00122-001-0813-0
42. Morris, G. P., Ramu, P., Deshpande, S. P., Hash, C. T., Shah, T., Upadhyaya, H. D., et al. (2013). Population genomic and genome-wide association studies of agroclimatic traits in sorghum. *Proc. Natl. Acad. Sci. U.S.A.* 110, 453–458. doi: 10.1073/pnas.1215985110
43. Nei, M., Tajima, F., and Tateno, Y. (1983). Accuracy of estimated phylogenetic trees from molecular data. II. Gene frequency data. *J. Mol. Evol.* 19, 153–170. doi: 10.1007/BF02300753
44. Pritchard, J. K., and Rosenberg, N. A. (1999). Use of unlinked genetic markers to detect population stratification in association studies. *Am. J. Hum. Genet.* 65, 220–228. doi: 10.1086/302449
45. Qiao, Y., Piao, R., Shi, J., Lee, S. I., Jiang, W., Kim, B. K., et al. (2011). Fine mapping and candidate gene analysis of dense and erect panicle 3, DEP3, which confers high grain yield in rice (*Oryza sativa* L.). *Theor Appl Genet.* 122, 1439–1449. doi: 10.1007/s00122-011-1543-6
46. Rafalski, J. A. (2010). Association genetics in crop improvement. *Curr Opin Plant Biol.* 13, 174–180. doi: 10.1016/j.pbi.2009.12.004
47. Risch, N., and Merikangas, K. (1996). The future of genetic studies of complex human diseases. *Science* 273, 1516–1517. doi: 10.1126/science.273.5281.1516
48. Song, X. J., Huang, W., Shi, M., Zhu, M. Z., and Lin, H. X. (2007). A QTL for rice grain width and weight encodes a previously unknown RING-type E3 ubiquitin ligase. *Nat Genet.* 39, 623–629. doi: 10.1038/ng2014
49. Stacklies, W., Redestig, H., Scholz, M., Walther, D., and Selbig, J. (2007). pcaMethods a bioconductor package providing PCA methods for incomplete data. *Bioinformatics* 23, 1164–1167. doi: 10.1093/bioinformatics/btm069
50. Takahashi, Y., Shomura, A., Sasaki, T., and Yano, M. (2001). Hd6, a rice quantitative trait locus involved in photoperiod sensitivity, encodes the alpha subunit of protein kinase CK2. *Proc. Natl. Acad. Sci. U.S.A.* 98, 7922–7927. doi: 10.1073/pnas.111136798
51. Temnykh, S., Park, W. D., Ayres, N., Cartinhour, S., Hauck, N., Lipovich, L., et al. (2000). Mapping and genome organization of microsatellite sequences in rice (Oryza sativa L.). *Theor Appl Genet.* 100, 697–712. doi: 10.1007/s001220051342
52. Thomson, M. J., Tai, T. H., McClung, A. M., Lai, X. H., Hinga, M. E., Lobos, K. B., et al. (2003). Mapping quantitative trait loci for yield, yield components and morphological traits in an advanced backcross population between Oryza rufipogon and the Oryza sativa cultivar Jefferson. *Theor Appl Genet.* 107, 479–493. doi: 10.1007/s00122-003-1270-8
53. Wang, E. T., Wang, J. J., Zhu, X. D., Hao, W., Wang, L., Li, Q., et al. (2008). Control of rice grain-filling and yield by a gene with a potential signature of domestication. *Nat Genet.* 40, 1370–1374. doi: 10.1038/ng.220
54. Wang, J. (2009). Inclusive composite interval mapping of quantitative trait genes. *Acta Agron Sin.* 35, 239–245. doi: 10.3724/SP.J.1006.2009.00239
55. Wang, S., Wu, K., Yuan, Q., Liu, X., Liu, Z., Lin, X., et al. (2012). Control of grain size, shape and quality by OsSPL16 in rice. *Nat Genet.* 44, 950–955. doi: 10.1038/ng.2327
56. Wang, Y., Liu, X., Ji, X., Zhang, L., Liu, Y., Lv, X., et al. (2015). Identification and validation of a major QTL controlling the presence/absence of leaf lobes in Brassica rapa L. *Euphytica* 205, 761–777. doi: 10.1007/s10681-015-1403-6
57. Wei, X. J., Xu, J. F., Guo, H. N., Jiang, L., Chen, S. H., Yu, C. Y., et al. (2010). DTH8 suppresses flowering in rice, influencing plant height and yield potential simultaneously1[W][OA]. *Plant Physiol.* 153, 1747–1758. doi: 10.1104/pp.110.156943
58. Weng, J., Gu, S., Wan, X., Gao, H., Guo, T., Su, N., et al. (2008). Isolation and initial characterization of GW5, a major QTL associated with rice grain width and weight. *Cell Res.* 18, 1199–1209. doi: 10.1038/cr.2008.307
59. Xiao, J., Li, J., Grandillo, S., Ahn, S. N., Yuan, L., Tanksley, S. D., et al. (1998). Identification of trait-improving quantitative trait loci alleles from a wild rice relative, Oryza rufipogon. *Genetics* 150, 899–909
60. Xing, Z., Tan, F., Hua, P., Sun, L., Xu, G., and Zhang, Q. (2002). Characterization of the main effects, epistatic effects and their environmental interactions of QTLs on the genetic basis of yield traits in rice. *Theor Appl Genet.* 105, 248–257. doi: 10.1007/s00122-002-0952-y
61. Xue, W. Y., Xing, Y. Z., Weng, X. Y., Zhao, Y., Tang, W. J., Wang, L., et al. (2008). Natural variation in Ghd7 is an important regulator of heading date and yield potential in rice. *Nat Genet.* 40, 761–767. doi: 10.1038/ng.143
62. Yano, M., Katayose, Y., Ashikari, M., Yamanouchi, U., Monna, L., Fuse, T., et al. (2000). Hd1, a major photoperiod sensitivity quantitative trait locus in rice, is closely related to the arabidopsis flowering time gene CONSTANS. *Plant Cell.* 12, 2473–2483. doi: 10.1105/tpc.12.12.2473
63. Yao, X. -Y., Li, Q., Liu, J., Jiang, S. -K., Yang, S. -L., Wang, J. -Y., et al. (2015). Dissection of QTLs for plant height and panicle length traits in rice under different

environment. *Sci. Agr. Sin.* 48, 407–414. doi: 10.3864/j.issn.0578-1752.2015.03.01

64. Yuan, L. P. (1997). Hybrid rice breeding for super high yield. *Hybrid Rice* 12, 1–6.

65. Zhang, D., Song, H., Cheng, H., Hao, D., Wang, H., Kan, G., et al. (2014). The acid phosphatase-encoding gene GmACP1 contributes to soybean tolerance to low-phosphorus stress. *PLoS Genet.* 10:e1004061. doi: 10.1371/journal.pgen.1004061

66. Zhang, L., Wang, J., Wang, J., Wang, L., Ma, B., Zeng, L., et al. (2015). Quantitative trait locus analysis and fine mapping of the qPL6 locus for panicle length in rice. *Theor. Appl. Genet.* 128, 1151–1161. doi: 10.1007/s00122-015-2496-y

67. Zhang, X., Wang, J., Huang, J., Lan, H., Wang, C., Yin, C., et al. (2012). Rare allele of OsPPKL1 associated with grain length causes extra-large grain and a significant yield increase in rice. *Proc. Natl. Acad. Sci. U.S.A.* 109, 21534–21539. doi: 10.1073/pnas.1219776110

68. Zheng, Z., Ma, J., Stiller, J., Zhao, Q., Feng, Q., Choulet, F., et al. (2015). Fine mapping of a large-effect QTL conferring fusarium crown rot resistance on the long arm of chromosome 3B in hexaploid wheat. *BMC Genomics* 16:850. doi: 10.1186/s12864-015-2105-0

69. Zuo, S., Kang, H., Li, Q., Chen, Z., Zhang, Y., Liu, W., et al. (2014). Genome-wide association analysis on genes controlling panicle traits of varieties from international rice core collection bank and its breeding utilization. *Chin. J. Rice Sci.* 28, 649–658. doi: 10.3969/j.issn.1001-7216.2014.06.011

CHAPTER 12

POPULATION STRUCTURE, DIVERSITY AND TRAIT ASSOCIATION ANALYSIS IN RICE (ORYZA SATIVA L.) GERMPLASM FOR EARLY SEEDLING VIGOR (ESV) USING TRAIT LINKED SSR MARKERS

Annamalai Anandan, Mahender Anumalla, Sharat Kumar Pradhan, and Jauhar Ali

INTRODUCTION

Rice (*Oryza sativa* L.) is widely cultivated under irrigated and rainfed conditions, with cultivated varieties and landraces that specifically adapt to these situations. In India, more than 50% of rice areas under rainfed conditions are cultivated as direct seeded rice (DSR). The existing rainfed upland varieties are of short duration type, drought tolerant and low yielding and possess early vigor trait. Direct seeding is also a common practice in low land rice of India and weed is the major problem in those situations. Early seedling vigor (ESV) trait of rice will help in competing with weed and this trait is needed in lowland rice. While, irrigated rice varieties are high yielding and relatively poorer in early vigor as compared to rainfed varieties. In rice growing areas of South and South East Asia, ESV trait has been exploited in rainfed upland cultivar development but not to the same extent in rainfed low land and irrigated situations, mainly due to the narrow genetic base of selecting attributes primarily for higher grain yields. Attempts were made to introgress ESV from upland landraces to high yielding cultivars; however, significant progress could not be achieved due to complex nature of the trait [1]. Further, only a small fraction of available landraces has been used in practical breeding. Traditional native landraces are widely grown under rainfed lowland conditions by resource poor farmers in India due to lack of suitable improved varieties for such harsh conditions. Most of these landraces are less productive, but they possess weed-smothering ability by early vigor and excellent growth adaptation features to direct seeded conditions. Due to unpredictable rainfall patterns, direct seeded rice has become regular practice under rainfed situation. Conversely, due to labor scarcity and unavailability of timely irrigation water, the farmers in the irrigated ecosystems are also adapting to dry direct seeding techniques. However, currently available modern rice varietal architecture with semi-dwarf stature and reduced seedling vigor is not amenable to dry direct seeded conditions [1]. It is clear that, ESV is relatively restricted to upland varieties as compared to transplanted rice varieties. Therefore, for DSR ESV is an important trait to achieve higher biomass, smothering weed ability with higher grain yields [2].

For better understanding of ESV trait, it is important to know the genetic relationship among a large set of cultivated genotypes by using DNA markers efficiently. Among the different DNA markers, simple sequence repeats (SSR) are most commonly used, as they are hyper variable, co-dominant, robust, multi-allelic in nature, chromosome specific and greatly facilitated linkage map construction and use in plant breeding. SSR markers are widely employed to assess genetic diversity and genetic structure in rice and several other crops. However,

Citation: Anandan A, Anumalla M, Pradhan SK, Ali J (2016) Population Structure, Diversity and Trait Association Analysis in Rice (Oryza sativa L.) Germplasm for Early Seedling Vigor (ESV) Using Trait Linked SSR Markers. PLoS ONE 11(3): e0152406. doi:10.1371/journal.pone.0152406.

limited information is available on classification of ESV in rice cultivars using DNA markers. Molecular markers-based genetic maps facilitate applied genetics and breeding program [3].

ESV and its component traits being complex quantitative nature, their involvement with several physiological processes and influences of G x E interactions makes these traits to achieve slower progress for genetic enhancement. Therefore, integration of marker tools into breeding schemes provides an opportunity to handle such challenges and facilitates marker-assisted selection. Continuous variation could be observed for ESV within a segregating population and frequency of favorable alleles at many loci could also be achieved by selection. QTL mapping based on segregating population helps to identify loci for quantitative and qualitative traits with limited number of segregating alleles at any locus and low resolution typically in the range of 10-30cM [4]. To overcome such situations, association mapping based on the linkage disequilibrium is popularly adapted to achieve a higher resolution and targets multiple alleles at individual loci by exploiting historical recombination events available in germplasm and by identifying association between marker and phenotype of larger number of traits. Till date, very few researches on association studies in rice for ESV have been done. Therefore, the present study examined to explore; i) the characteristic differences for ESV of 629 eco-geographical rice genotypes, ii) genetic diversity of the selected genotypes using molecular markers, and iii) genomic regions associated with ESV related traits.

MATERIALS AND METHODS

Experiment 1

The total of 629 rice (*Oryza sativa*. L) genotypes (Fig 1) were representative of breeding lines and landraces, comprising of 25 tropical *japonica*, 57*indica* landraces, 127 breeding lines, and 427 ARC (Assam Rice Collection) accessions cited in S1 Table. The seeds were obtained from Gene bank of National Rice Research Institute (NRRI), Cuttack, India. The experiment was conducted at NRRI during 2013 dry season (experiment 1) in upland situation (20°.48'N, 85°.86'E). The experimental site soil had 34.4% sand, 29.1% silt, 32.4% clay with bulk density of 1.46 Mg/m^3. Seeds of each of the genotype were direct sown in 4 m long row with 15 cm apart and 20 cm between rows in augumented block design. Pre-emergence herbicide Pendimethalin was sprayed at the recommended rate within 24 hr of irrigation and maintained weed-free through the growing period by hand weeding. Crop was raised with recommended fertilizers doses of 80 kg nitrogen ha^{-1}, 40 kg phosphorus ha^{-1}, and 40 kg potash ha^{-1}. The field was maintained under non-saturated aerobic condition with supplemental surface irrigation.

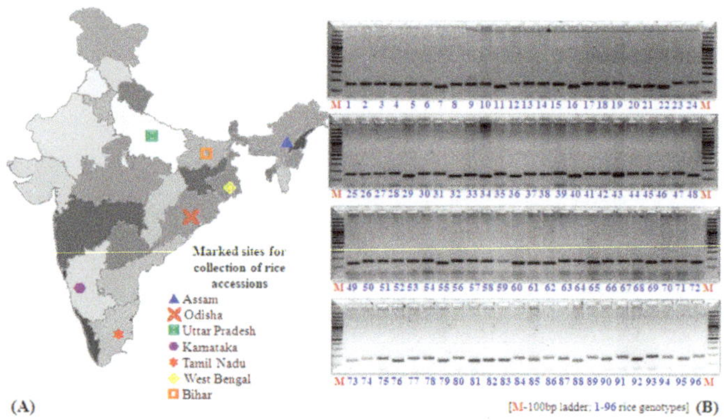

Figure 1: (A) Diagrammatic representation of collection of rice germplasm from different states of India; (B) Molecular profiling of mini-core 96 rice accessions with polymorphic microsatellite markers.

In order to study the variability among genotypes for ESV, several vegetative traits were observed. Fourteen days after sowing (DAS), six rice plants in the middle row from each genotype were selected to study the shoot length (cm), number of leaves, root length (cm) and shoot dry weight (DW) (g). The same set of traits was observed on 28 and 56 DAS with number of tillers per plant and leaves per tiller. Growth rate of seedlings were measured to assess ESV by vigor index (VI) as suggested by Maguire [5].

$$VI = \frac{\text{Rate of germination (\%)}}{\text{Seedling dry weight (mg)}}$$

While, germination rate (GR) was calculated as $GR = \frac{X_1}{Y_1} + \frac{(X_2 - X_1)}{Y_2} + \dots + \frac{(X_n - X_{n-1})}{Y_n}$ Where, X_n is the number of germinated seeds on the nth day and Y_n is the number of days from the first experimental day.

Absolute growth rate (AGR) was calculated as $AGR = \frac{(h_2 - h_1)}{(t_2 - t_1)} \text{cm day}^{-1}$ Where, h_1 and h_2 are the plant height at times t_1 and t_2 respectively.

Relative growth rate (RGR) was determined by measuring plant dry weight periodically during growth and is represented as mg g^{-1}day^{-1}. $RGR = \frac{(\log_e W_2 - \log_e W_1)}{(t_2 - t_1)}$ Where, W_1 and W_2 are the plant dry weights at times t_1 and t_2 respectively.

Crop growth rate (CGR) measured the plant dry weight of a particular ground area at a regular interval of time divided by land area and represented as g m^{-2} day^{-1}. $CGR = \frac{(W_2 - W_1)}{P(t_2 - t_1)}$ Where w_1 and w_2 are the plant dry weights at times t_1 and t_2 respectively, P = spacing (m^2).

An initial descriptive statistics, including mean, standard deviation, minimum and maximum values, coefficient of variation (%), skewness, kurtosis, and distribution pattern was performed by box plot technique. PCA was performed to estimate Euclidean distance between two genotypes in the multivariate space and adaptation of genotypes to specific environment [6–7]. These analyses were performed using the Windostat 7.5 software (Indostat Services, Hyderabad, India.). Discriminate function analysis was applied to find the best variable(s) that can discriminate high and low seedling vigor genotypes. For this, the genotypes were divided arbitrarily into three groups based on the shoot length and dry weight (Tables 1 and 2).

Table 1: Arbitrary classification of seedling vigor on 14th **and 28th days after sowing.**

S. no	Shoot length (cm)	Categories	Shoot weight(g)	Categories
1	0–15	Short	<0.04	Low
2	16–30	Intermediate	0.05–0.1	Intermediate
3	>30	Tall	>0.1	High

doi:10.1371/journal.pone.0152406.t001

Table 2: Arbitrary classification of seedling vigor on 56th DAS.

S. NO	Shoot length (cm)	Categories	Shoot weight(g)	Categories
1	0–30	Short	<2.9	Low
2	31-55-	Intermediate	3-6.00	Intermediate
3	>55	Tall	>6.1	High

doi:10.1371/journal.pone.0152406.t002

Based on the broad trend observed in the PCA and phenotypic data. Forward stepwise discriminate function analysis (DFA) was carried out to understand the combination of variables, which could best explain the grouping in STATISTICA ver.10 (Stat soft, Inc, Tulsa, ok, USA). In DFA, a larger eigen value indicates that the discriminant function is more useful in distinguishing between the groups.

Experiment 2

Ninety-six genotypes were selected representing high, medium and low vigor genotypes with almost equal proportions from each group were selected for the panel from the results of the experiment 1. Groups were defined by following the scale presented in Table 1. The selected genotypes were raised in earthen pots (20 cm height and 15 cm diameter) containing up-land soil and farmyard manure in 3:1 ratio in three replications during 2014 dry season. Three plants were maintained per pot. The same set of 96 genotypes was raised on upland situation in augumented design with similar agronomic practices in four-meter row length as followed in experiment 1. Five plants from each genotype were selected for measurement of shoot length (cm), root length (cm), leaf length (cm), leaf width (cm), number of leaves, shoot dry weight (g), root dry weight (g), and VI on 14th and 28th DAS. By using these variables, RGR, AGR and CGR were derived to measure growth rate of seedlings as mentioned in experiment 1. PCA based clustering was performed to find out the pattern of phenotypic diversity and variables contributions to diversity using the Windostat 7.5 software.

DNA isolation and selection of SSR markers.

The experimental material comprised of selected 96 genotypes that were direct sown in the pot and grown in net house at NRRI, Cuttack. After 20 days of sowing, young leaves were harvested and genomic DNA was isolated from the bulked leaf sample following CTAB method [8]. The quantity was estimated by spectrophotometer and agarose gel electrophoresis using lambda DNA of known concentration. The isolated DNA samples were diluted in $T_{10}E_1$ buffer (10 mM Tris-HCl, 1 mM EDTA, pH 8.0) to obtain the final concentration of 15 ng/µl for amplification. A *priori* information was used to select ESV linked 52 SSR markers from the reports of earlier mapping population studies and the distribution of markers covered all the chromosomes to illustrate the diversity [1].

PCR amplification and electrophoresis.

The PCR amplification was executed with 20 µl of PCR mixture contained 30 ng of genomic DNA, 1x PCR buffer, 200 µM dNTP mix, 4 picomoles of each forward and reverse primers, 2 mM of $MgCl_2$, and 1U of Taq DNA polymerase (MBI, Fermentas, USA). Template DNA initially denatured at 94°C for 5 min followed by 36 cycles of denaturation at 94°C for 1 min, annealing at 55–67°C (varies upon primers) for 1 min and extension at 72°C for 5 min followed by final extension at 72°C for 10 min. The amplification products were separated on 2.5% agarose gel containing ethidium bromide using 1x TBE buffer. DNA fragments were visualized under UV transillumination using Alpha Innotech gel documentation system (Flour Chem™ 5500, Alpha Innotech, USA).

Data analysis.

Amplicons were scored according to their product size for each genotype and primer combination. Number of alleles (N), major allele frequency (A), observed heterozygosity (H_o), expected heterozygosity (He), and PIC, for each SSR locus were determined by using Power marker 3.25[9]. A PIC value of each marker was determined as suggested by Botstein et al. [10].

Further, the allelic data were subjected to estimation of genetic distances among genotypes using simple matching coefficients by bootstrapping 10,000 times and they were clustered using neighbor joining method [11]. PCoA was performed to highlight the resolving power of the ordination and the first two components were used to represent the genotypes in the graphical form. PCoA and dissimilarity matrix were performed by using DARwin software version 5.0 [12]. Further, Analysis of molecular variance (AMOVA) was performed to describe variance components among individuals and the population differentiation among the seven assumed sub populations using GeneAlEx 6.41 program [13] with 1000 permutations. Genetic differentiation among the assumed sub population was analyzed using Nei's gene diversity statistics [14] using POPGENE program version 1.31 [15]. To identify the genetic structure of given population and assign individuals to populations, the software STRUCTURE version 2.3.3 [16] were used. To derive the optimal number of groups (K), STRUCTURE was run with K varying from 1 to 10, with five runs for each K value. To determine the true value of K, *ad hoc*statistic ΔK was followed [17]. Parameters were set to 1,00,000 burn-in periods and 10,000 Markov Chain Monte

Carlo (MCMC) replications after burn-in with an admixture and allele frequencies correlated model.

Association analysis.

To find the genetic relatedness between phenotypic performances of the rice accessions and SSR makers, two statistical models were used; i) General linear model (GLM) and ii) Mixed linear model (MLM) in TASSEL version 5.0[18]. Significantly associated markers with traits were identified on the traits of their R^2 and p-value.

RESULTS

Experiment 1

Variation in ESV related morphological traits on 14, 28and 56 days after sowing.

In the present investigation, distribution pattern of ESV traits of 629 genotypes were presented in S1A–S1H Fig. Among the ESV traits studied on 14th, 28th and 56th days after sowing (DAS), number of leaves and germination rate were negatively skewed (S <-1) on 14 DAS (S1 Fig), while dry weight (DW), vigor index (VI) and crop growth rate (CGR) were positively skewed (S>1) on 56 DAS. Normal distributions were observed for all other traits. Visible trait differences were observed among the genotypes studied. Among the ESV traits studied at different stages, vigor index on 56 DAS showed the largest relative difference with 40.75 fold followed by DW on 56 DAS (37.40 fold) and CGR between 28 and 56 DAS (37.40 fold). The higher level of difference could provide an opportunity to select genotypes with ESV. Inter-quartile range (IQR) is a measure of variability. Among all the traits studied, lower IQR was observed for DW on 14 DAS and RGR between 28 and 56 DAS followed by VI on 14 DAS. Germination rate and number of leaves/tiller on 56 DAS have significant number of outliers. Among the traits studied on 14 DAS, shoot length varied from 9.4to 26.6cm as measured in 'Sharvesh'(9.40 cm) and 'ARC 10797' (26.58 cm)respectively with mean of 16.75 cm and 'ARC 10797' exhibited 1.58 times higher than the mean value. Coefficients of variation (CV) indicating variability among the genotypes were high for VI on 56 DAS (57.93%) followed by CGR between 28th and 56th(52.14%), DW on 56 DAS (51.44%) and VI on 28 DAS (42.23%). Higher CV suggests that, the selected material in this experiment exhibited higher variability.

Grouping pattern of genotypes and association between variables.

Principal component analysis (PCA) was performed to identify association between traits, responsible traits for ESV and grouping pattern of rice genotypes on the basis of the traits. The first two PCs accounted for about 71.47% of the total variability with >1.0 eigen value. The biplot in Fig 2 shows a strong relationship between the DW and VI (r = >0.900), dry weight and shoot length (r = >0.400) and VI and CGR (r = >0.800) irrespective of different days of observation (14, 28 and 56 DAS). Cultivar-by-trait biplot indicated that the traits were grouped into two, based on days of observation. Traits observed on 14 and 28 DAS, were grouped together and separated from traits observed on 56 DAS (Fig 2). Genotypes with early vigor and growth on 14 and 28 DAS were grouped together in quadrant 2, whereas genotypes with high biomass on 56 DAS were grouped in quadrant 3. The discriminate function is statistically significant with high canonical correlation (r = 0.8398). Further, it discriminated high and low seedling vigor genotypes and found that shoot length on 28 DAS showed the lowest partial lambda, the highest F-remove value and standardized coefficient, followed by shoot weight on 28 DAS.

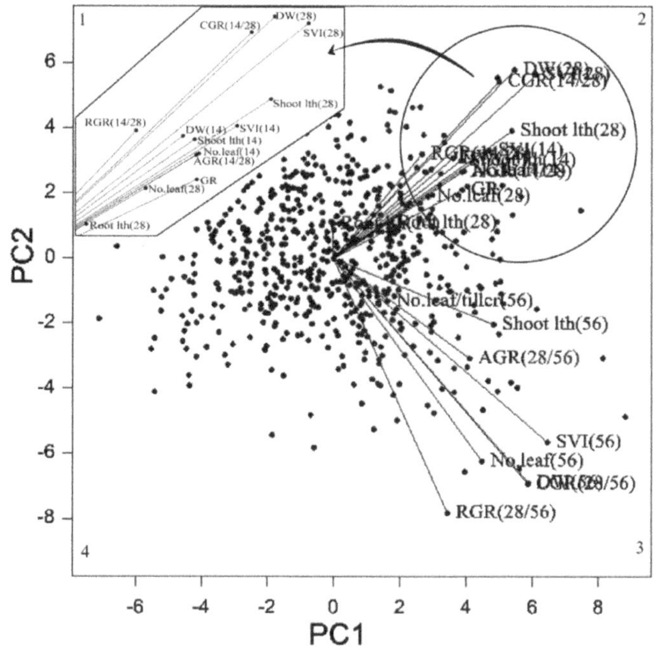

Figure 2: Spatial distribution of 629 rice genotypes for early seedling vigor on 14, 28 and 56 days after sowing for the first two principal components.

Experiment 2

Traits and grouping pattern of selected genotypes.

The selected 96 genotypes were raised under net house and field condition during dry season. PCA was executed for all the 96 genotypes under both situations to identify trends among low and high early seedling vigor genotypes and the traits responsible for source of the variability (Fig 3A and 3B). The PC_1 accounted for about 60.71% and PC_2 accounted for 21.22% of the total variance (totaling 81.93%) under net house condition. From Fig 3A, it is inferred that the longest vector loading such as AGR, leaf length and shoot length on 28 DAS are the major discriminators. However, under field condition RGR, leaf length and shoot length on 14 DAS are the major discriminators (Fig 3B). In PCA under field condition, first and second component explained 75.08% and 15.88% (totaling 90.96%) of the variance respectively. The PCA under both conditions, groups the genotypes into two groups early vigor genotypes on right side and non-early vigor genotypes on left side.

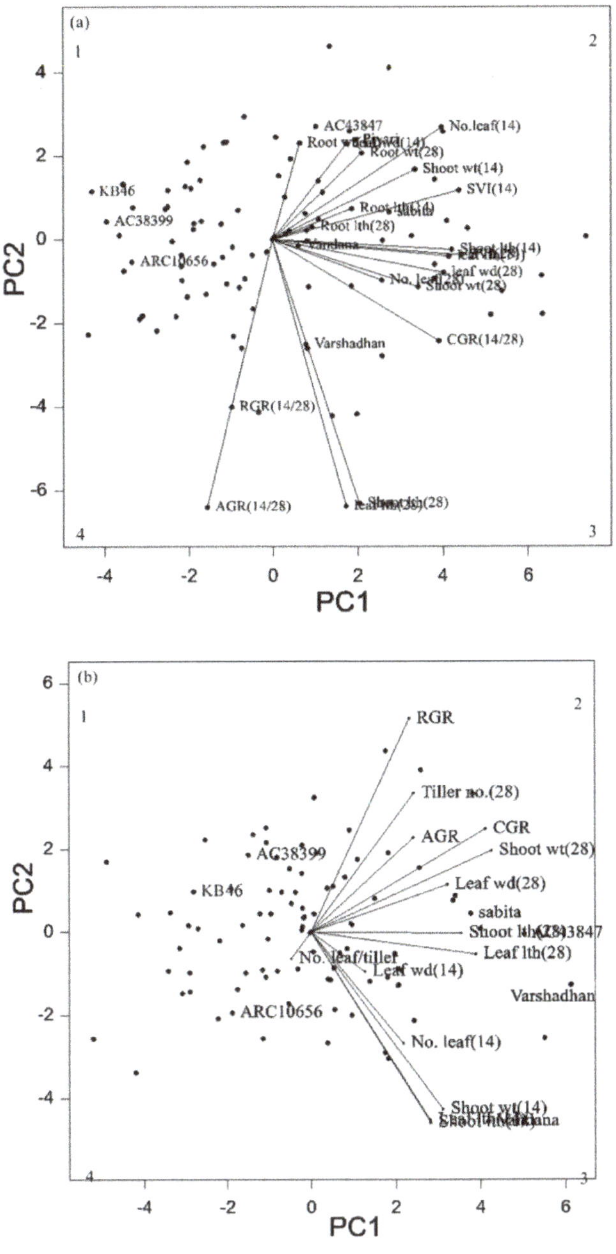

Figure 3: (a, b). Spatial distribution of selected 96 rice genotypes for early seedling vigor on 14 and 28 days after sowing for the first two principal components. (a) Net house and (b) Field conditions.

SSR marker segregation.

In the present experiment, 96 rice genotypes include 85 landraces and 11 modern/breeding lines representing six major rice growing states of India. Initially the experiment was begun with 52 SSR markers to assess the genetic relatedness. Among them, 39 markers generated polymorphic bands (Table 3). The number of alleles amplified per marker varied from 2 to 7 with an average of 3.28 per locus in 96 genotypes. PIC values ranged from 0.04 (RM3839) to 0.37 (RM148) with a mean of 0.24. Observed heterozygosity (H_o) ranged from 0.04(RM3839) to 0.97 (RM148) with an average heterozygosity across all 39 loci was 0.42. None of the loci exhibited heterozygosity as zero. The expected heterozygosity (gene diversity) (H_e) ranged from 0.04 to 0.50 with an average of 0.30. Major allele frequency was also calculated for all 39 loci, which ranged from 0.52 to 0.98 with an average of 0.79 (Table 4).

Table 3: List of early seedling vigor (ESV) traits associated polymorphic microsatellite markers in rice (1).

S.No	Marker	Traits associated	Chr	Amplicon Size (bp)	Tm	Reference
1	RM13	SL,RL,DW,GR,SDW	5	141	55	[19]
2	RM125	GR, GP and GI	7	127	55	[20]
3	RM26	TDW,SDW,RDW,VI,RARL,SL,GR	5	112	55	[21–24]
4	RM87	SDW	5	151	55	[19,21]
5	RM3839	SDW	4	218	50	[25]
6	RM161	RL,DW,SEV,FV,GR	5	187	61	[26]
7	RM218	SL,FV	3	148	55	[27]
8	RM228	SS,SL,FW,WS	10	154	55	[26–27]
9	RM9	GR,SDW,SFW	1	136	55	[20]
10	RM6091	GR,GI	11	126	50	[20]
11	RM7075	GR	1	155	50	[20]
12	RM258	RDW,	10	148	55	[21]
13	RM3428	GP,GI,NL,SS	11	156	55	[20]
14	RM334	SDW,	5	182	55	[19, 21]
15	RM148	GR,SL,SEV,FW	3	129	55	[19, 26, 27]
16	RM250	SDW,	2	153	55	[28]
17	RM221	SEV,	2	192	55	[26]
18	RM264	GR,SL,RL	8	178	55	[19]
19	RM230	GR,SL,RL	8	257	55	[19]
20	RM253	SL, GR,	6	141	55	[19, 23]
21	RM85	GR,SL,RL,SEV,FW	3	107	55	[19, 26]
22	RM340	SL,GP	6	163	55	[19, 20]
23	RM249	SV, FV,	5	121	55	[19]
24	RM225	SV,GR, SDW, SL	6	140	55	[19, 29]
25	RM263	SL,SDW	2	199	55	[22]
26	RM30	GR, SDW, SL	6	105	55	[29]
27	RM223	SW,GR,SL,SEV	8	165	55	[19, 26]
28	RM1339	SL	1	144	55	[30]
29	RM16	SL,SDW,RL,GR	3	181	55	[19]
30	RM252	SEV,GP,GR,RA	5	216	55	[19, 21, 26]
31	RM336	GR, SDW, SL	7	154	55	[29]
32	RM8085	SL,	1	126	55	[30]
33	RM19	GR, SDW, SL	12	226	55	[29]
34	RM6	SDW	2	163	55	[22]
35	RM168	SL,DW,GR,RL,SEV,FW	3	116	55	[19, 26]
36	RM224	GR, GR, SDW, SL	11	157	55	[29, 31]
37	RM7389	SL	3	111	55	[30]
38	RM21	GR, SDW, SL	11	157	55	[29]
39	RM341	GR,	2	172	55	[23]

SDW-Shoot dry weight; SL-Shoot length; RDW-Root dry weight: RL-Root length; TFW-Total fresh weight; TDW-Total dry weight; RA-Root activity; GP-Germination percentage; GR-Germination rate; GI-Germination index; FGC-First germination count; EGC-Ending germination count; ME-Mesocotyl elongation; PL-Plumule length; COL-Coleorhiza length; CL-Coleoptile length; VI-Vigor index; RSC-Reducing sugar content; SW-Seed weight, SE-Seedling establishment, SFW-Seedling fresh weight, SS-Seed size, SEV-Seedling early vigor, LA-Leaf area, NL-Number of leafs; SR-Seed reserve utilization efficiency, WS-Weight of mobilized seed reserve, FV-Field vigor, SV-Seed vigor.

doi:10.1371/journal.pone.0152406.t003

Table 4: List of microsatellite markers used for genotyping rice accessions for early seedling vigor traits along with their genetic diversity parameters.

S. No	Marker	Chromosome	Molecular Weight	Gene diversity	Heterozygosity	PIC	Allele frequency	Total allele	Unique allele	Rare allele
1	RM125	7	127	0.10	0.10	0.09	0.95	3	0	1
2	RM13	5	141	0.14	0.15	0.13	0.93	3	0	0
3	RM1339	1	144	0.30	0.36	0.25	0.82	5	0	2
4	RM148	3	129	0.50	0.97	0.37	0.52	3	0	0
5	RM16	3	181	0.30	0.38	0.26	0.81	6	0	0
6	RM161	5	187	0.30	0.36	0.25	0.82	3	0	0
7	RM168	3	116	0.27	0.32	0.23	0.84	2	0	0
8	RM19	12	226	0.47	0.75	0.36	0.63	6	0	0
9	RM21	11	157	0.38	0.50	0.30	0.75	3	0	0
10	RM218	3	148	0.46	0.71	0.35	0.65	3	0	0
11	RM221	2	192	0.49	0.86	0.37	0.57	3	0	0
12	RM223	8	165	0.11	0.11	0.10	0.94	3	0	1
13	RM224	11	157	0.44	0.66	0.34	0.67	3	0	1
14	RM225	6	140	0.40	0.56	0.32	0.72	2	0	0
15	RM228	10	154	0.12	0.13	0.11	0.94	2	0	0
16	RM230	8	257	0.12	0.13	0.11	0.94	3	0	0
17	RM249	5	121	0.42	0.60	0.33	0.70	2	0	0
18	RM250	2	153	0.11	0.11	0.10	0.94	3	0	0
19	RM252	5	216	0.44	0.67	0.35	0.67	4	0	0
20	RM253	6	141	0.14	0.16	0.13	0.92	2	0	0
21	RM258	10	148	0.39	0.53	0.31	0.73	5	0	0
22	RM26	5	112	0.45	0.68	0.35	0.66	4	1	1
23	RM263	2	199	0.35	0.45	0.29	0.78	3	0	0
24	RM264	8	178	0.45	0.70	0.35	0.65	3	0	0
25	RM30	6	105	0.47	0.76	0.36	0.62	2	0	0
26	RM334	5	182	0.38	0.51	0.31	0.74	4	0	1
27	RM336	7	154	0.48	0.82	0.37	0.59	2	0	0
28	RM340	6	163	0.13	0.14	0.12	0.93	5	0	1
29	RM341	2	172	0.46	0.73	0.36	0.64	4	0	0
30	RM3428	11	156	0.13	0.14	0.12	0.93	2	0	0
31	RM3839	4	218	0.04	0.04	0.04	0.98	5	0	1
32	RM6	2	163	0.45	0.68	0.35	0.66	3	0	0
33	RM6091	11	126	0.28	0.34	0.24	0.83	2	0	0
34	RM7075	1	155	0.16	0.18	0.15	0.91	4	0	0
35	RM7389	3	111	0.29	0.35	0.25	0.82	2	0	0
36	RM8085	1	126	0.14	0.16	0.13	0.92	2	0	0
37	RM85	3	107	0.25	0.29	0.22	0.85	2	0	0
38	RM87	5	151	0.11	0.11	0.10	0.94	3	0	0
39	RM9	1	136	0.23	0.27	0.21	0.86	7	0	0
	Mean			0.30	0.42	0.24	0.79	3.28	0.03	0.23

doi:10.1371/journal.pone.0152406.t004

Genetic relatedness by cluster and principal coordinate analysis (PCoA).

Cluster analysis was carried out to assess genetic distance and the dissimilarity matrix-using neighbor joining method. In the Unrooted tree genotypes were grouped into three major clusters (Fig 4). Cluster 1 has 59 genotypes, they sub grouped into cluster 1a and cluster 1b with 43 and 16 genotypes respectively. Similarly, cluster 2 was sub grouped into cluster 2a that contains 18 genotypes and cluster 2b with 10 genotypes. Cluster 3 was the smallest containing 9 genotypes.

On the basis of geographical location, no significant grouping was observed. However, sub group 1b and 2b and cluster 3 having genotypes of landraces alone. Further, sub group 1a and 2a clustered genotypes of both improved and landraces. The genotypes of tropical *japonica* were grouped into two sub groups as 2a and 2b.

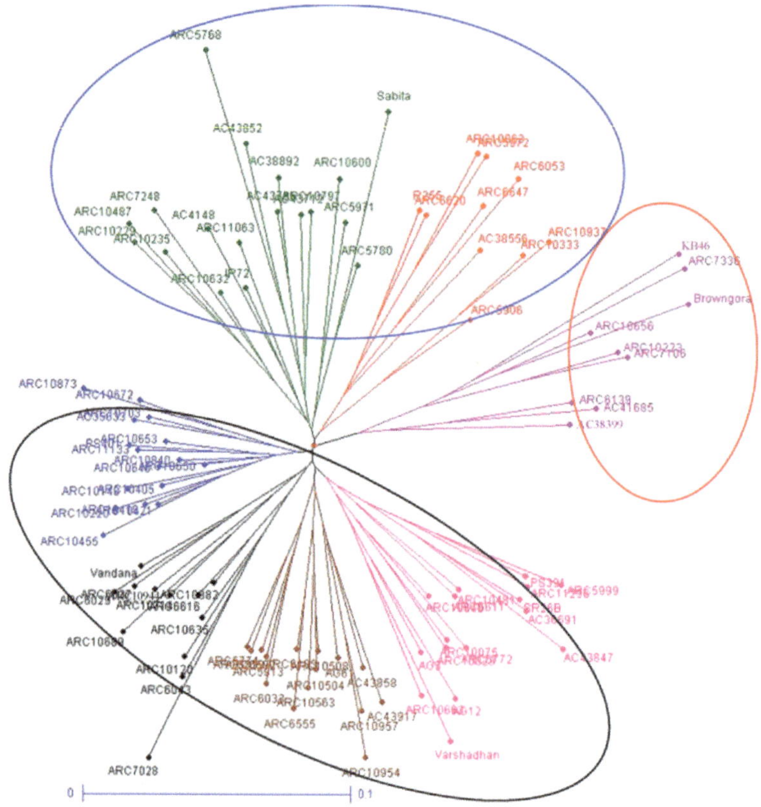

Figure 4: Unrooted neighbor joining tree constructed for 96 rice genotypes based on trait linked SSRs.

PCoA with SSR markers were used to determine the genetic relatedness among the genotypes (Fig 5). The first three axis of differentiation explained 19.98% of the total variation. In PCoA, genotypes were presented in colors corresponding to the clusters observed in unrooted tree. Similar to clusters analysis, PCoA also did not group the genotypes based on the location-specific and intermixing of colors across the coordinates were observed. However, improved genotypes were inclined to stay on the quadrant I and IV; tropical *japonica* AC 38892 and AC 38556 were on the opposite quadrant, indicating their dissimilarity.

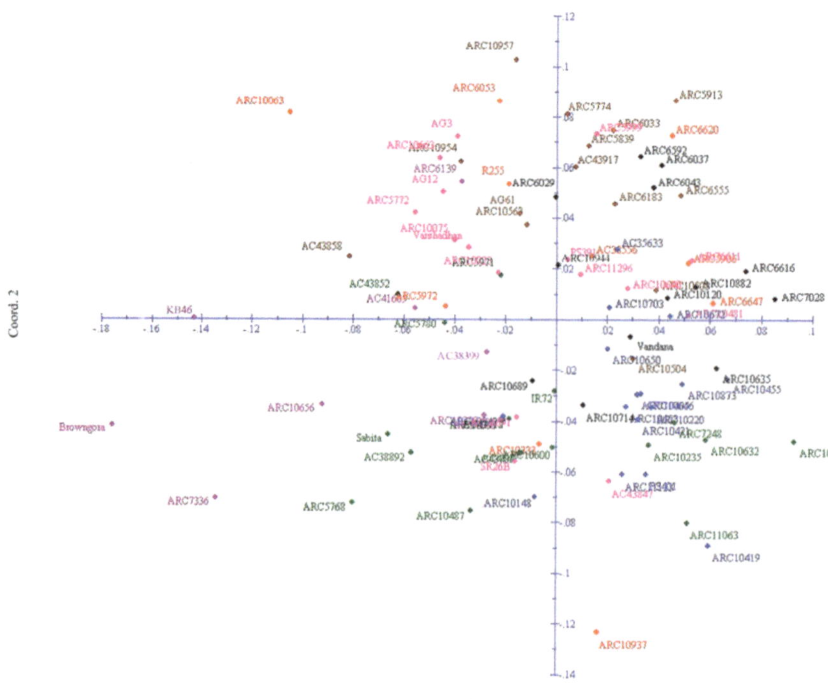

Figure 5: Principal Coordinate Analysis (PCoA) of rice genotypes based on trait linked SSRs. (genotypes represented in colors corresponding to the cluster observed in unrooted tree).

Analysis of molecular variation (AMOVA) and Nei's genetic distance.

The assumed populations were subjected to AMOVA. AMOVA revealed that maximum variation was among individuals with greater variance (99%), while sub-populations were grouped together, the variance was only 1% and the F_{st} (0.016) was non-significant. The inbreeding co-efficient F_{IS} (-0.419) and F_{IT} (-0.396) were found to be 1.000. It indicated that individuals from different populations are genetically more closely related than within population. Nei's genetic distance was estimated between assumed seven populations. The pairwise Nei's genetic distance ranged from 0.0024 (lowest) between population of pop 6 and pop 7 to 0.0316 (highest) between population of pop 3 and pop 4.

Population structure.

STRUCTURE software was used to determine relationship among the genotypes studied and distinguished 96 genotypes into two populations with ΔK value of 31 (Fig 6). In population 1, 46 genotypes and rest of the 50 genotypes were grouped into population 2. Two tropical *japonica*, 11 improved lines, 26 ARC lines and seven landraces were grouped into one population with proportion of 47.91%. While, population 2 has the membership proportion of 52.08% consisted of 44 ARC lines, five landraces and one improved genotype IR72. By structure analysis, pure or admixture genotypes were categorized. The genotype with score >0.80 was considered as pure and <0.80 as admixture. Among the 96 genotypes, 27 genotypes were pure and 19 were admix in population 1, while 34 were pure and 16 were admix in population 2. In total, 61 were pure and 35 were identified as admixture and the same can be visualized in graphical representation (Fig 7). The fixation index (F_{st}) values of two population ranged between 0.0218 (population 1) and 0.3397 (population 2), while allele frequency divergence between two population was 0.057.

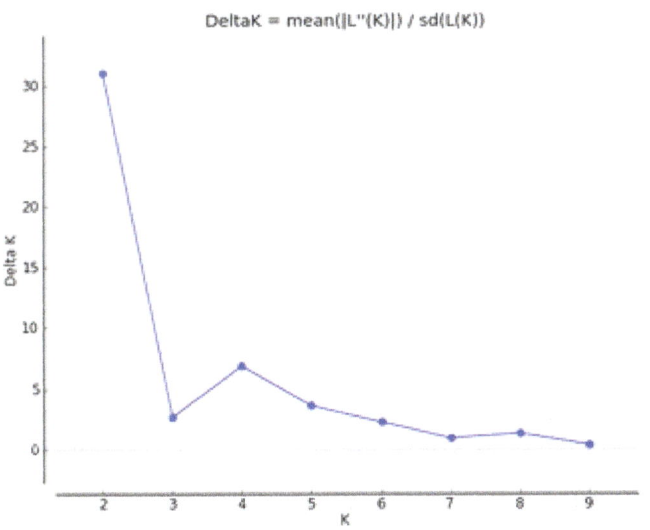

Figure 6: Estimation of population using LnP(D) derived delta K for determining optimum number of subpopulations.

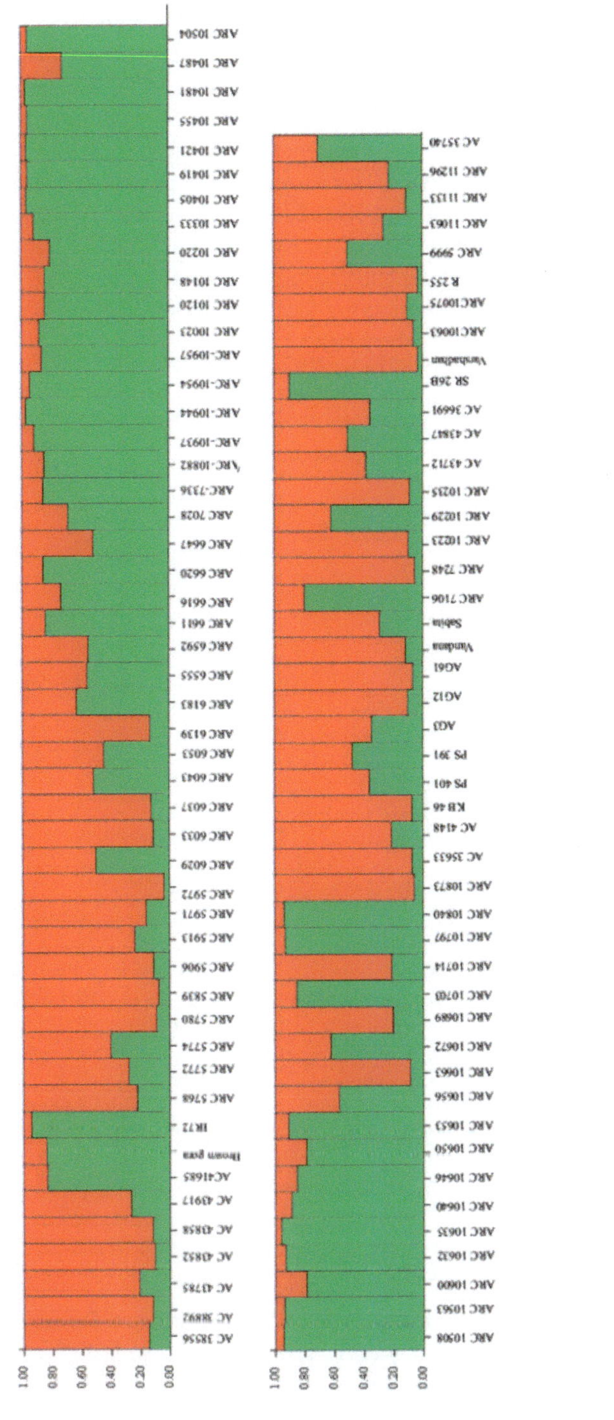

Figure 7: Distribution pattern of 96 rice genotypes based on SSRs within sub-populations.

Association of SSRs with early seedling vigor traits.

The results of association analysis using GLM and MLM (Q+K) are presented in the Table 5. The structured association mapping revealed 16 and 10 marker-trait association for 12 and 11 ESV related parameters were observed by GLM and MLM approach respectively. GLM analysis of marker-trait association among the population revealed markers RM13 and RM334 with leaf length observed on 14 DAS located on chromosome 1 and 5 respectively. By accounting population structure and relative kinship effects (Q+K model), RM13 marker associated leaf length on 14 DAS with 7.3% of phenotypic variability. GLM and MLM analysis together they detected RM223 strongly associated with leaf length on 28 DAS with 10.3% and 8.4% phenotypic variability respectively. There were two common markers associated with leaf width on 14 DAS distributed over chromosome 8 and 7, of which RM230 on chromosome 8 had described 5.05% of the phenotypic variability. For root length on 14 DAS, two QTLs were detected among the population by both analyses, and RM3839 (chromosome 4) was associated with root length followed by RM230 (chromosome 8). RM249 and RM6091 on chromosome 4 and 8 had described 4.8% and 3.8% of the phenotypic variability respectively for root length on 28 DAS by GLM analysis. Followed by MLM, RM249 exhibited association with root length on 28 DAS with 5.04% of phenotypic variability. Three SSR markers (RM13, RM3348 and RM16) related to shoot length on 14 DAS were detected by GLM. Among them RM13 and RM16 accounting 6.05% and 5.14% phenotypic variability for shoot length detected by MLM. None of the markers were found to be associated with shoot length on 28 DAS and dry shoot weight on 14 DAS by GLM and MLM analyses. Marker RM341 was found to be associated with vigor index QTLs as detected by GLM and MLM with >4.5% phenotypic variability.

Table 5: Association of marker alleles with phenotypic traits of early seedling vigor in rice under direct seeded situation using GLM and MLM (Q+K) model.

Trait	Marker	Chromosome	GLM Model			MLM (Q+K) model		
			F ratio	p-value	R^2	*F* ratio	p-value	R^2
Leaf length (14)	RM13	1	6.65	0.01	0.07	6.82	0.01	0.07
	RM334	5	6.24	0.01	0.06	-	-	-
Leaf length (28)	RM223	8	11.58	0.00	0.10	8.85	0.00	0.08
	RM6	2	4.53	0.04	0.04	-	-	-
Leaf width (14)	RM230	8	5.20	0.02	0.05	4.97	0.03	0.05
	RM125	7	4.17	0.04	0.04	4.03	0.05	0.04
	RM218	3	3.91	0.05	0.04	-	-	-
No. of leaf (14)	RM224	11	5.90	0.02	0.06	-	-	-
	RM21	11	4.41	0.04	0.05	-	-	-
	RM13	5	4.11	0.05	0.04	-	-	-
Root length (14)	RM3839	4	4.60	0.03	0.04	5.10	0.03	0.05
	RM230	8	4.33	0.04	0.04	4.78	0.03	0.05
Root length (28)	RM249	5	5.35	0.02	0.05	4.95	0.03	0.05
	RM6091	11	4.19	0.04	0.04	-	-	-
Root weight (14)	RM7075	1	5.21	0.02	0.05	-	-	-
	RM249	5	4.82	0.03	0.05	4.74	0.03	0.05
Root weight (28)	RM250	2	4.06	0.05	0.04	3.92	0.05	0.04
Shoot length (14)	RM13	1	5.43	0.02	0.05	4.75	0.03	0.05
	RM334	5	5.26	0.02	0.05	-	-	-
	RM16	3	3.99	0.05	0.04	5.60	0.02	0.06
Shoot weight (28)	RM341	2	5.50	0.02	0.05	5.26	0.02	0.05
	RM224	11	4.10	0.05	0.04	4.23	0.04	0.04
Vigor index (14)	RM341	2	4.37	0.04	0.04	5.10	0.03	0.05
Vigor index (28)	RM341	2	4.35	0.04	0.05	4.85	0.03	0.05

doi:10.1371/journal.pone.0152406.t005

DISCUSSION

Genetic variability is the key determinant for any breeding program. The sporadic reports on germplasm evaluation of landraces and genetics of ESV traits in rice encouraged us to take up this study. Traditional native landraces those grown under rainfed are competitive to weeds. By introduction of modern high yielding with low ESV semi dwarf varieties, weed competitive landraces are almost extinct and disappearing in farmers' field. Therefore, the present experiment was executed to identify the suitable genotypes with high ESV and to find the loci responsible for those traits of the gene by association mapping. We began with a panel of 629 rice accessions (S1 Table). Landraces included in this study were from seven states of India*viz*., Assam, Odisha, Uttar Pradesh, Karnataka, Tamil Nadu, West Bengal and Bihar. The early seedling vigor trait distribution pattern of the panel of genotype studied on 14th, 28th and 56^{th}days after sowing (DAS), exhibited that germination rate, number of leaves on 14 DAS, DW on 56 DAS, VI on 56 DAS and CGR distribution was asymmetrical. VI and DW on 56 DAS and CGR between 28 and 56 DAS exhibited maximum variability and portrayed that the population was heterogeneous.

Spatial distribution of genotypes and variation of ESV traits

In this study, association between dry weight and shoot length or vigor index suggests that accumulation of dry weight is an important trait contributing for ESV. Cultivar-by-trait biplot analysis grouped the vectors of the variables of 14 and 28 DAS into a separate group (highly correlated). Similarly, variables of 56 DAS had formed another group (highly correlated). The reason might have been due to higher differences in the trait variability observed before 28 and 56 DAS. Further, it clarifies that the observation taken on 14 and 28 DAS was the reliable estimator to study ESV in rice than observed on 56 DAS. The position and perpendicular projection of genotypic points onto variable vectors grouped early vigor genotypes on 14 and 28 DAS in quadrant 2 and these genotypes were moderately vigorous on 56 DAS than genotypes captured in quadrant 1 and 4. Genotypes captured in quadrant 3, exhibited vigorous growth on 56 DAS with moderate early vigor than genotypes grouped in quadrant 1 and 4. The PCA analysis of selected 96 genotypes under net house and field condition classified early vigor genotypes on right of the biplot and non-early vigorous genotypes on left side of the plot. This is evident from the genotypes like 'Sabita' (a known early vigor genotype) [32] positioned on right side with other early vigor genotypes *viz*., Varshadhan, Vandana, AC4387 and Pyari. In contrary, genotypes KB46, AC 38399 and ARC 10656 moved on the left side of the plot, where they exhibited reduced vigor index and dry weight.

Diversity estimation, genetic relationships and grouping of genotypes

The genetic architecture of diverse rice accessions can be assessed by model based and distance based clustering approach using the SSRs or SNPs marker system [33–37]. This is the first such study where ESV trait linked SSR markers were utilized to assess genetic diversity and population structure in 96 germplasm lines of Indian rice collections. We are confident enough that, comprehensive analysis of trait linked SSR markers in such a diverse collection will be helpful for breeders to design breeding strategies for direct seeded situation with ESV.

The 39 SSRs generated a total of 128 alleles with an average of 3.28 alleles per locus (Table 4) and PIC value of 0.24. Comparison of our findings with previous studies in rice showed mixed results. Singh et al (2013) also observed a lower number of alleles (2.22) per locus and an average PIC value of 0.25 in 375 Indian rice collections. Similarly, Shah et al. [38] reported lower genetic diversity with an average of 2.75 allele per locus and an average PIC value of 0.38 from 40 Pakistan rice accessions. On the other hand, Behera et al. [39] also reported 4.69 alleles per locus with an average PIC value of 0.81 among the 36 landraces of having different therapeutic values from India. In the present context, reason for reduced number of alleles may be the use of trait linked SSR markers and the exclusion of monomorphic and spurious bands from analysis [35]. Several factors were reported to improve average PIC values of SSRs such as inclusion of diverse lines, size of collection, breeding behaviors of the species, genotypic method and location of markers in the genome [35]. Lower PIC value observed in the present study might be due to inclusion of large number of ARC accessions. One unique and nine rare alleles were observed in the present study. Unique allele can be considered for diagnostic of a particular genotype and increase in number of rare alleles contributes for overall genetic diversity. The proportion low or high frequency allele also contributes to overall genetic diversity. In our case of study, high frequency allele has 58.7% contribution over low frequency allele.

Genetic relatedness among the accessions was assessed on the basis of their clustering pattern in the unrooted tree. The unrooted tree grouped varieties into three major clusters with varying number of genotypes per cluster. The clustering pattern of genotypes obtained in the present study was found to be trait based grouping and no geographical isolation clustering has been observed. Inclusion of large number of ARC collections might be the reason for lack of geographical isolation. Accessions of subgroup 1b possessing 11 ESV traits measured on 14 and 28 DAS were grouped together with higher crop growth rate (CGR). The genotypes were grouped together under the subgroup1b were tall and possessed high vigor index (VI) as preferred for ESV. While, cluster 3 positioned opposite to subgroup 1b grouped the genotypes of least vigor type and exhibited inferiority for 11 ESV traits including vigor index. Singh et al. [35] also reported similar grouping pattern in Indian rice collections for biotic stress. The grouping pattern of the PCoA corresponded well with the pattern of Neighbor Joining Tree. Genotypes of subgroup 1b (deep blue color) tends to stay in quadrant 3, whereas, corresponding genotypes of cluster 3 (purple color) moved apart to the left extreme of PCoA plot. However, few exceptions were observed in position of genotypes compared to cluster analysis due to the complexity of the variations could not explained in two dimensional graphical representation, as the first two PCs explained the total variation less than 80% [40].

Source of variation, population structure and marker-trait association

AMOVA study indicated higher partitioning of variation within individuals and lower portion existed among populations. Estimates of the fixation indices revealed that F_{st} indicates little divergence existing between subgroups of the population. F_{IT} is the overall inbreeding coefficient of an individual within the total population. Negative value of F_{IT} indicates higher proportion of heterozygote genotypes on the level of sub populations and all evaluated individuals as well. Negative F_{IS} value also suggests that there is an excess of heterozygotes in these loci (out breeding) compared with Hardy Weinberg Equilibrium (HWE) expectations and reflects the genetic variability of general population. The reason for heterozygotes might be inclusion of landraces in the present study and these genotypes might have heterozygous allele for the studied trait.

The model based approach by structure analysis revealed two subpopulations. The genetic structure of populations has been previously reported in rice to vary from 2 to 8 subpopulations [33, 37, 41–44]. Courtois et al. [33] and Das et al. [42] have detected two and four subgroups in their study population of 425 and 91 accessions of rice landraces respectively. Threshold level to identify accessions belonging to a specific subpopulation varies among different scientific groups of 60–80% [33, 45, 46]. In the present study, 35 landrace accessions were categorized as admixtures with the stringent threshold of 80%. Based on criterion of maximum membership probabilities, 96 accessions were grouped into two populations of 46 and 50 genotypes each. This finding is in accordance with the previous studies where they obtained two distinct subgroups in their core collection of rice [33, 47]. In the current rice collection panel, the use of trait linked SSR markers might also be the reason for the formation of less number of groups. Besides, grouping of the genotypes into two were noticed based on the traits of ESV observed on 14thand 28th DAS. Between the two sub-populations, second sub-population exhibited higher shoot length, root length, number of leaves, leaf length, dry weight and VI on both DAS than the sub-population-1. However, genotypes of admixtures were observed to have superior in early vigor trait than pure lines. Since, this study includes large set of landraces, which might have diverse ancestry still possessing heterozygosity for the trait studied in the present experiment.

The application of association mapping for ESV will be helpful in deciding the molecular markers to be used in stacking the multiple QTLs responsible for this trait. The significant SSR markers of GLM explained on an average of 5.0% and MLM demonstrated 5.2% of phenotypic variation. Maximum 10.3% and 8.4% phenotypic variation was explained by RM223 for leaf length on 28 DAS by GLM and MLM respectively. Similarly, Q-Q plot exhibits the significant association markers for leaf length on 28 DAS (Fig 8). A QTL for leaf length on 14 DAS detected in the present study on chromosome 1 (RM13) was observed in both GLM and MLM models.

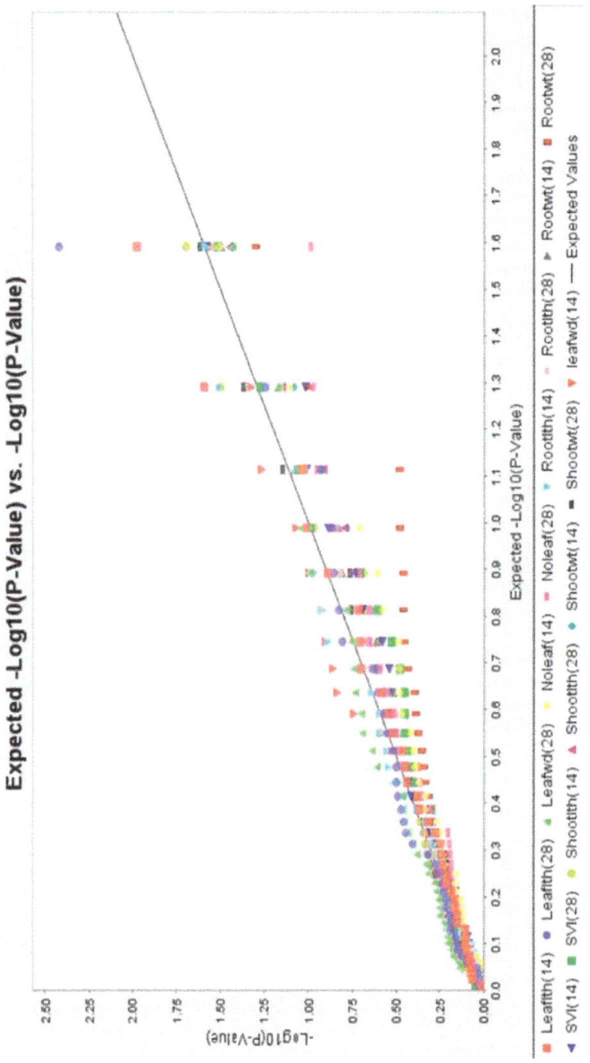

Figure 8: Quantile-Quantile (QQ) plot and distribution of SSR-trait association.

SSR genomic region RM230 and RM125 mapped on chromosome 8 and 7 were significantly associated with QTL controlling leaf width variation on 14 DAS. Increasing leaf width by utilizing favorable allele at this locus would likely to be useful for increasing photosynthetic area of seedling and weed smothering ability. On the other hand, the same marker (RM230) allele was significantly associated with root length on 14 DAS. Previously, Zhang et al. [19] reported this marker on chromosome 8 as contributing to root length andRM3839 was also observed to be significantly associated with root length on 14 DAS by both GLM and MLM approach. QTLs*qRL-1* and *qRL-2* of root length were reported on chromosome 1 and 2 respectively from RIL population of Xiushui 79/C-Bao [48]. The root dry weight QTL was detected on chromosome 5 (by linked marker allele RM249) on 14 DAS coincided with QTL of root length on 28 DAS. For root dry weight on 28 DAS, significant association of RM250 was found by both GLM and MLM on chromosome 2. Further, a QTL *qRDW10-1*was identified for root dry weight on chromosome of 10 with LOD score of 5.6 from RIL population of Zhenshan 97/Ming hui63 [21].

Two significant QTLs (chromosome 1 and 3) were identified for shoot length on 14 DAS. Marker alleles RM13 on chromosome 1 and RM16 on chromosome 3 were positively associated with shoot length on 14 DAS. Moreover, previous report of Zhang et al. [19] supports the association of genomic region RM13 with QTLs for shoot length. Hence, selection of these alleles can be expected to result in genetic improvement for this trait to obtain vigorous growth in the early stage of seedling under direct seeded condition. Cairns et al. [28] observed a QTL *qSL-2*for shoot length on chromosome 2 from back cross population of Vandana/Moroberekan. In our study, allele RM224 located on chromosome 11 was significantly associated with shoot dry weight on 28 DAS and another tightly linked marker allele (RM341) showed a stronger association with the same trait and detected by both GLM and MLM on chromosome 2. However, Zhou et al. [27] and Cui et al. [21] observed QTL *qFV-5-1* and *qDW5*for shoot dry weight on chromosome 5 from Lemont/Teqing and IR64/Azucena mapping population respectively. Interestingly, RM341 coincided with QTLs for vigor index on chromosome 2 at 14 and 28 DAS and showed a stronger association detected by both GLM and MLM. Conversely Diwan et al. [23] observed QTL *qVI* on chromosome 5 with LOD score of 4.4. Therefore, marker-assisted selection in favor of RM341 could increase the shoot length and vigor index and might also contribute favorably to grain yield and biomass under direct seeded conditions.

CONCLUSION

Screening large number of accessions in the present study for early seedling vigor brings out valuable information. We conclude from this study that the vigor index should be studied on 14th and 28th DAS. The genotypes utilized for studying diversity pattern by employing trait-linked markers have categorized early vigor and non-early vigor genotypes. The results of distance based approach and principal coordinate analysis were in accordance with model based structure analysis. Further, the study has identified significantly associated markers and markers with pleiotropic effects indicating that association mapping served as an effective tool. The putative QTLs, identified in this study, need to be validated before they could be applied for marker assisted breeding in rice. Genotypes identified from this study possessing favorable alleles can be utilized for improving early seedling vigor after functionally characterizing them.

AUTHOR CONTRIBUTIONS

Conceived and designed the experiments: AA SKP JA. Performed the experiments: AA SKP. Analyzed the data: AA MA. Contributed reagents/materials/analysis tools: AA MA SKP. Wrote the paper: AA JA.

REFERENCES

1. Mahender A, Anandan A, Pradhan SK. Early seedling vigor, an imperative trait for direct seeded rice: an overview on physio-morphological parameters and molecular markers. Planta. 2015;241: 1027–1050. doi: 10.1007/s00425-015-2273-9. pmid:25805338
2. Joshi E, Dinesh K, Lal B, Nepalia V, Gautam P, Vyas AK. Management of direct seeded rice for enhanced resource—use efficiency. Pl Know J. 2013;2(3): 119–134.
3. Kannan B, Senapathy S, Gajaraj A, Raj B, Chandra S, Muthiah A, et al. Association Analysis of SSR Markers with Phenology, Grain, and Stover-Yield Related Traits in Pearl Millet (*Pennisetum glaucum* (L.) R. Br.) The Scientific World Journal; vol. 2014; Article ID 562327. doi: 10.1155/2014/562327
4. Flint-Garcia SA, Thuillet AC, Yu J, Pressoir G, Romero SM, Mitchell SE, et al. Maize association population: a high-resolution platform for quantitative trait locus dissection. Plant J. 2005; 44(6): 1054–64. pmid:16359397 doi: 10.1111/j.1365-313x.2005.02591.x
5. Maguire JD. Speed of germination—aid in selection and evaluation for seedling emergence and vigor. Crop Sci. 1962;2: 176–177. doi: 10.2135/cropsci1962.0011183x000200020033x
6. Anandan A, Rajiv G, Ramarao A, Prakash M. Internode elongation pattern and differential response of rice genotypes to varying levels of flood water. Funct Plant Biol. 2012;39: 137–145. doi: 10.1071/fp11184
7. Anandan A, Pradhan SK, Das SK, Behera L, Sangeetha G. Differential responses of rice genotypes and physiological mechanism under prolonged deepwater flooding. Field Crops Res 2015;172: 153–163. doi: 10.1016/j.fcr.2014.11.007
8. Murray MG, Thompson WF. Rapid isolation of high molecular weight plant DNA. Nucleic Acids Res. 1980;8: 4321–4325. pmid:7433111 doi: 10.1093/nar/8.19.4321
9. Liu K, Muse SV. Power Marker: Integrated analysis environment for genetic marker data; Bioinformatics. 2005;21: 2128–2129. pmid:15705655 doi: 10.1093/bioinformatics/bti282
10. Botstein D, White RL, Skolnick M, Davis RW. Construction of a genetic linkage map in man using restriction fragment length polymorphisms. Amer J Hum Genet. 1980;32: 314–331. pmid:6247908
11. Felsenstien J. Confidence limits on phylogenies: an approach using the bootstrap. Evolution. 1985; 39: 783–91. doi: 10.2307/2408678
12. Perrier X, Jacquemoud-Collet JP. DARwin software. 2006. Available:http://darwin.cirad.fr/darwin.
13. Peakall ROD, Smouse PE. GENALEX 6: genetic analysis in Excel. Population genetic software for teaching and research. Mol Ecol Notes. 2006;6: 288–295. doi: 10.1111/j.1471-8286.2005.01155.x
14. Nei M. Analysis of gene diversity in subdivided populations. Proc Natl Acad Sci USA. 1973; 70: 3321–3323. pmid:4519626 doi: 10.1073/pnas.70.12.3321
15. Yeh FC, Yang RC, Boyle T. POPGENE, version 1.31. Microsoft windows based freeware for population genetic analysis. University of Alberta, Canada. 1999.
16. Pritchard JK, Stephens M, Donnelly P. Inference of population structure using multilocus genotype data. Genetics. 2000;155: 945–959. pmid:10835412
17. Evanno G, Regnaut S, Goudet J. Detecting the number of clusters of individuals using the software STRUCTURE: A simulation study. Mol Ecol. 2005; 14(8): 2611–2620. pmid:15969739 doi: 10.1111/j.1365-294x.2005.02553.x
18. Bradbury PJ, Zhang Z, Kroon DE, Casstevens TM, Ramdoss Y. Buckler ES. TASSEL: Software for association mapping of complex traits in diverse samples. Bioinformatics 2007;23: 2633–2635. pmid:17586829 doi: 10.1093/bioinformatics/btm308
19. Zhang ZH, Qu XS, Wan S, Chen LH, Zhu YG. Comparison of QTL controlling seedling vigor under different temperature conditions using recombinant inbred lines in rice (*Oryza sativa*). Ann Bot. 2005;95(3): 423–429. pmid:15574482 doi: 10.1093/aob/mci039
20. Wang Z, Wang J, Bao Y, Wang F, Zhang H. Quantitative trait loci analysis for rice seed vigor during the germination stage. J Zhejiang Univ-Sci B (Biomed and Biotechnol). 2010; 11(12):958–964. doi: 10.1631/jzus.b1000238
21. Cui KH, Peng SB, Xing YZ, Xu CG, Yu SB, Zhang Q. Molecular dissection of seedling-vigor and associated physiological traits in rice. Theor Appl Genet. 2002;105: 745–753. pmid:12582488 doi: 10.1007/s00122-002-0908-2
22. Han L, Qiao Y, Zhang S, Zhang Y, Cao G, Kim J, et al. Identification of quantitative trait loci for cold response of seedling vigor traits in rice. J Genet Genomics. 2007;34:239–246. pmid:17498621 doi: 10.1016/s1673-8527(07)60025-3
23. Diwan J, Channbyregowda M, Shenoy V, Salimath P, Bhat R. Molecular mapping of early vigor related QTLs in rice. Res J Biol. 2013;1: 24–30.
24. Xie L, Tan Z, Zhou Y, Xu R, Feng L, Xing Y, et al. Identification and fine mapping of quantitative trait loci for seed vigor in germination and seedling establishment in rice. J Integr Plant Biol. 2014;56:749–759. doi: 10.1111/jipb.12190. pmid:24571491
25. Cheng XX, Cheng JP, Huang X, Lai YY, Wang L, Du WL, et al. Dynamic quantitative trait loci analysis of seed reserve utilization during three germination stages in rice. PLoS One. 2013;8(11):e80002. doi: 10.1371/journal.pone.0080002. pmid:24244592
26. Lu XL, Niu AL, Cai HY, Zhao Y, Liu JW, Zhu YG, et al. Genetic dissection of seedling and early vigor in a recombinant inbred line population of rice. Plant Sci. 2007;172:212–220. doi: 10.1016/j.plantsci.2006.08.012
27. Zhou L, Wang JK, Yi Q, Wang YZ, Zhu YG, Zhang ZH. Quantitative trait loci for seedling vigor in rice under field conditions. Field Crops Res. 2006;100: 294–301. doi: 10.1016/j.fcr.2006.08.003
28. Cairns JE, Namuco OS, Torres R, Simborio FA, Courtois B, Aquino GA, et al. Investigating early vigor in upland rice (*Oryza sativa* L.): Part II. Identification of QTLs

controlling early vigor under greenhouse and field conditions. Field Crops Res. 2009;113: 207–217. doi: 10.1016/j.fcr.2009.05.007

29. Huang Z, Yu T, Su L, Yu SB, Zhang ZH, Zhu YG. Identification of chromosome regions associated with seedling vigor in rice. Acta Genetica Sinica. 2004;31:596–603. pmid:15490878

30. Abe A, Takagi H, Fujibe T, Aya K, Kojima M, Sakakibara H, et al. OsGA20ox1, a candidate gene for a major QTL controlling seedling vigor in rice. Theor Appl Genet. 2012;125(4):647–657. doi: 10.1007/s00122-012-1857-z. pmid:22481119

31. Jiang L, Liu S, Hou M, Tang J, Chen L, Zhai H, et al. Analysis of QTLs for seed low temperature germinability and anoxia germinability in rice (Oryza sativa L.). Field Crops Res: 2006; 98:68–75 doi: 10.1016/j.fcr.2005.12.015

32. Namuco OS, Cairns JE, Johnson DE. Investigating early vigor in upland rice (*Oryza sativa* L.): Part I. Seedling growth and grain yield in competition with weeds. Field Crops Res. 2009;113(3): 197–206. doi: 10.1016/j.fcr.2009.05.008

33. Courtois B, Frouin J, Greco R, Bruschi G, Droc G, Hamelin C, et al. Genetic diversity and population structure in a European collection of rice. Crop Sci. 2012;52(4): 1663–1675 doi: 10.2135/cropsci2011.11.0588

34. Choudhury B, Khan ML, Dayanandan S. Genetic structure and diversity of indigenous rice (*Oryza sativa*) varieties in the Eastern Himalayan region of Northeast India. Springerplus. 2013 Dec; 2(1):228. doi: 10.1186/2193-1801-2-228. pmid:23741655

35. Singh N, Choudhury DR, Singh AK, Kumar S, Srinivasan K, Tyagi RK, et al. Comparison of SSR and SNP Markers in Estimation of Genetic Diversity and Population Structure of Indian Rice Varieties. PLoS ONE. 2013;8 (12): e84136. doi: 10.1371/journal.pone.0084136. pmid:24367635

36. Sow M, Ndjiondjop MN, Sido A, Mariac C, Laing M, Bezançon G. Genetic diversity, population structure and differentiation of rice species from Niger and their potential for rice genetic resources conservation and enhancement. Genet Resour Crop Evol. 2014;61(1): 199–213. doi: 10.1007/s10722-013-0026-9

37. Nachimuthu VV, Muthurajan R, Duraialaguraja S, Sivakami R, Pandian BA, Ponniah G, et al. Analysis of population Structure and Genetic Diversity in Rice Germplasm Using SSR Markers: An Initiative Towards Association Mapping of Agronomic Traits in Oryza Sativa. Rice. 2015;8: 30. doi: 10.1186/s12284-015-0062-5. pmid:26407693

38. Shah SM, Naveed SA, Arif M. Genetic diversity in Basmati and non- Basmati rice varieties based on microsatellite markers. Pak J Bot.2013;45: 423–431

39. Behera L, Patra BC, Sahu RK, Nanda A, Sahu SC, Patnaik A, et al. Assessment of genetic diversity in medicinal rice using microsatellite markers. Aust J Crop Sci. 2012;6(9): 1369–1376.

40. Patel ML, Reddi MV, Rona BS, Reddy BJ. Genetic divergence in safflower (*Carthamus tinctorius* L.) Indian J Genet. 1989;49(1): 113–118.

41. Zhao K, Tung CW, Eizenga GC, Wright MH, Ali ML, Price AH, et al. Genome-wide association mapping reveals a rich genetic architecture of complex traits in *Oryza sativa*. Nat Commun. 2011;2: 467. doi: 10.1038/ncomms1467. pmid:21915109

42. Das B, Sengupta S, Parida SK, Roy B, Ghosh M, Prasad M, et al. Genetic diversity and population structure of rice landraces from Eastern and North Eastern States of India. BMC Genet. 2013;14(1): 71. doi: 10.1186/1471-2156-14-71

43. Zhang D, Zhang H, Wei X, Qi Y, Wang M, Sun J, et al. Genetic structure and diversity of*Oryza sativa* L. in Guizhou, China. Chinese Sci Bull. 2007; 52(3): 343–351. doi: 10.1007/s11434-007-0063-x

44. Jin L, Lu Y, Xiao P, Sun M, Corke H, Bao J. Genetic diversity and population structure of a diverse set of rice germplasm for association mapping. Theor Appl Genet. 2010; 121(3): 475–487. doi: 10.1007/s00122-010-1324-7. pmid:20364375

45. Salgotra RK, Gupta BB, Bhat JA, Sharma S. Genetic Diversity and Population Structure of Basmati Rice (*Oryza sativa* L.) Germplasm Collected from North Western Himalayas Using Trait Linked SSR Markers. PLoS ONE. 2015;10(7): e0131858. doi: 10.1371/journal.pone.0131858

46. Liakat Ali M, McClung AM, Jia MH, Kimball JA, McCouch SR, Georgia CE. A rice diversity panel evaluated for genetic and agro-morphological diversity between subpopulations and its geographic distribution. Crop Sci. 2011; 51(5): 2021–2035 doi: 10.2135/cropsci2010.11.0641

47. Zhang P, Li J, Li X, Liu X, Zhao X, Lu Y. Population Structure and Genetic Diversity in a Rice Core Collection (*Oryza sativa* L.) Investigated with SSR Markers. PLoS ONE. 2011;(12): e27565. doi: 10.1371/journal.pone.0027565. pmid:22164211

48. Yang W, Yuan G, Lin HD. Discovery of elite alleles for seed vigor traits in two populations of japonica rice in Taihu lake region. Acta Agron Sin. 2010;36(5): 754–763. doi: 10.1016/s1875-2780(09)60048-1

CHAPTER 13 THE RICE MINUTE-LIKE1 (RML1) GENE, ENCODING A RIBOSOMAL LARGE SUBUNIT PROTEIN L3B, REGULATES LEAF MORPHOLOGY AND PLANT ARCHITECTURE IN RICE

Ming Zheng, Yihua Wang, Xi Liu, Juan Sun, Yunlong Wang, Yang Xu, Jia Lv , Wuhua Long, Xiaopin Zhu, Xiuping Guo, Ling Jiang, Chunming Wang, and Jianmin Wan

Keywords: *Minute* mutants, *Oryza sativa*, plant development, plant growth, ribosomal proteins, ribosome biogenesis, rice, vascular patterning

INTRODUCTION

Ribosomes, machines for protein synthesis, are composed of small (40S) and large (60S) subunits in eukaryotes. The small subunit consists of ~33 ribosomal proteins (designated RPS) and 18S rRNA, whereas the large subunit consists of ~47 ribosomal proteins (designated RPL) and 25–28S, 5.8S, and 5S rRNAs (McIntosh and Bonham-Smith, 2006). The genes for most ribosomal proteins (RPs) appear to be evolutionarily conserved among species, such as those of the Archaea, Bacteria, and Eukarya (Lecompte *et al.*, 2002; Mears *et al.*, 2002). In yeast (*Saccharomyces cerevisiae*), two-thirds of the RPs are duplicated (Deutschbauer *et al.*, 2005; Komili *et al.*, 2007; Dean *et al.*, 2008). Most RP families in mammalian species are composed of single-expressed genes and multiple pseudo-copies (Dudov and Perry, 1984). In contrast to mammals, Arabidopsis (*Arabidopsis thaliana*) RPs are involved in development and consist of two to seven copies, making them quite difficult to characterize in detail (Barakat *et al.*, 2001; Blanc and Wolfe, 2004; Thomas *et al.*, 2006).

Citation: Ming Zheng, Yihua Wang, Xi Liu, Juan Sun, Yunlong Wang, Yang Xu, Jia Lv, Wuhua Long, Xiaopin Zhu, Xiuping Guo, Ling Jiang, Chunming Wang, and Jianmin Wan, "The Rice Minute-Like1 (RML1) gene, encoding a ribosomal large subunit protein L3B, regulates leaf morphology and plant architecture in rice," J. Exp. Bot. (2016) 67 (11): 3457-3469 doi:10.1093/jxb/erw167.

Previous studies have shown that mutations in ribosomal proteins generally cause deleterious effects on growth and development of an organism (Horiguchi *et al.*, 2012). RP mutations in *Drosophila melanogaster* cause *minute* phenotypes that include delayed larval development, pleiotropic morphological aberrations, smaller body size, and recessive embryo lethality (Kongsuwa *et al.*, 1985; Lambertsson, 1998; Marygold *et al.*, 2007). Growth defects of most *minute* mutants are attributed to decreased levels of ribosomes, which might perturb the translation of specific targets or result in a reduced capacity for global protein synthesis in *Drosophila* development (Marygold *et al.*, 2007).

Mutations of RPs of both small and large subunits in Arabidopsis also exhibit multiple abnormalities, including vascular pattern defects, embryo lethality, retarded root growth, late flowering, or reduced plant size (Byrne, 2009). Some of these (*rps6*, *rps11*, *rpl2*, *rpl8*, *rpl23*, *rpl19*, and*rpl40*) are related to embryo-defective abnormalities (Tzafrir *et al.*, 2004;Meinke *et al.*, 2008). Other RP mutations affect leaf development, such as*pfl1* (*s18a*), *pfl2* (*s13b*), *ae5* (*l28a*), o*l17* (*l5b*), *pgy1* (*l10a*), *pgy2* (*l9*), *pgy3*(*l5a*), and *rpl4a/d* (Van Lijsebettens *et al.*, 1994; Ito *et al.*, 2000; Nishimura*et al.*, 2005; Pinon *et al.*, 2008; Yao *et al.*, 2008; Fujikura *et al.*, 2009). The capacity of protein synthesis in these mutants might be sufficiently decreased to retard cell division or might involve genes that affect auxin distribution in the developing leaves (Scarpella *et al.*, 2006; Byrne, 2009).*PGY* genes are thought to be involved in ribosome-mediated translational regulation of genes in the

HD-ZIPIII-KANADI pathway (Pinon *et al.*, 2008;Yao *et al.*, 2008). Similar defects were found to be present in the recently described *short valve1* (*stv1*)/*rpl24* mutant, with cotyledon and leaf vascular patterning defects (Nishimura *et al.*, 2005). In addition, this mutant displayed variable apical–basal gynoecium patterning defects. Recent studies of *stv1*/*rpl24*, *rpl4d*, *rpl5a*, and elongation factor *eif3h*mutants have found that they are involved in the auxin-signaling pathway through an uORF-dependent mechanism by perturbing translation reinitiation of *AUXIN RESPONSE FACTOR* (*ARF*) transcripts, such as *ETTIN*(*ETT*)/*ARF3* and *MONOPTEROS* (*MP*)/*ARF5* (Nishimura *et al.*, 2005; Zhou *et al.*, 2010; Horiguchi *et al.*, 2012; Rosado *et al.*, 2012).

RIBOSOMAL PROTEIN L3 (RPL3) is a highly conserved protein across yeast, animals and plants. In yeast, RPL3 is an essential and indispensable component for the formation of a peptidyltransferase centre (PTC) (Schulze and Nierhaus, 1982). Depletion of RPL3 *in vivo* arrests early assembly of the 60S ribosomal subunits and impairs nucleocytoplasmic export of pre-60S ribosomal particles. Additionally, RPL3-depleted cells are arrested in the G1 phase (Rosado *et al.*, 2007). Several studies have shown that mutant forms of RPL3 have altered ribosome structures, reduced ribosomal peptidyltransferase activity, and decreased rates of cell growth and protein synthesis (Petrov *et al.*, 2004; Meskauskas and Dinman, 2007). Mutations in the 'W finger' of RPL3 also affect the structure of 25S rRNA and maturation of pre-40S (Meskauskas and Dinman, 2007; Garcia-Gomez *et al.*, 2014). Mutation of RPL3 in *E. coli* also increases resistance to the peptidyltransferase inhibitor tiamulin by alteration of the binding site for the drug (Bosling *et al.*, 2003; Klitgaard *et al.*, 2015). In Arabidopsis, T-DNA insertion of *RPL3A* results in embryo lethality (Tzafrir *et al.*, 2004). In *Nicotiana tabacum*, *RPL3* has been shown to positively regulate cell division, and silencing of *RPL3* led to retarded development, inhibition of lateral root growth, and a decrease in accumulation of pre-rRNA (Popescu and Tumer, 2004). These studies suggest that the RPL3 also has a regulatory role in plant development.

Despite the characterization of several ribosomal genes that function in yeast and Arabidopsis, mutations of such genes have not yet been identified in rice. In this study, we characterize a *rice minute-like1* mutant (*rml1*), which displays retarded growth, as evidenced by reduced plant height, narrow leaves, inhibited lateral root growth, reduced seed size, and delayed flowering. We show that the *RML1* encodes ribosome large subunit protein L3B (RPL3B). In addition, there is a second copy, *RPL3A*, in the rice genome. We demonstrate that *RPL3A* and *RPL3B* have different expression profiles and functions. We suggest that ribosome aberrancy or polysome insufficiency might be responsible for the aberrant growth and development in the *rml1* mutant. This work enhances our understanding of the regulatory roles of ribosomal genes in plant development.

MATERIALS AND METHODS

Plant materials and growth conditions

The *rml1* mutant was obtained from a ^{60}Co-irradiated population of *Oryza sativa indica* rice cv. 93-11. Genetic analysis showed that the mutant phenotype was controlled by one recessive gene. Plants were grown in a paddy field at Nanjing Agricultural University, China.

Histological analysis of leaves and stems

To analyse leaf vasculature, *rml1* and wild-type flag leaves at the mature stage were fixed in FAA (formalin–acetic acid–alcohol) solution, and samples were treated as described by Zheng *et al.* (2015). Embedded tissues were sectioned at 8 μm thickness and stained with 0.05% Toluidine Blue. Whole-mount clearing of *rml1* and wild-type flag leaves was performed as described by Qi *et al.* (2008). Images were observed with a Nikon ECLIPSE80i light microscope. For analysis of cell morphology of the stem, internode I of wild-type and *rml1* were sectioned by a vibratome at 100 μm and stained with a mixture of Calcofluor White (Sigma) and 10% KOH. Microscopic examinations were made under UV light.

Map-based cloning and complementation test for *RML1*

To identify and map the *RML1* gene, *rml1* was crossed with 02428, a*japonica* cultivar. Ten individuals showing the recessive mutant phenotype were identified in the *rml1*/02428 F_2 population for preliminary mapping. A further 1200 F_2 plants with the mutant phenotype were used for fine mapping with SSR/InDel markers designed by comparison of the genomic sequences of Nipponbare (*japonica*) and cv. 93-11.

For complementation of the *rml1* mutation, the wild-type *RML1* genomic DNA sequence was cloned into the binary vector pCAMBIA1300 under its native promoter to generate the binary vector *pRML1-gRML1*. Due to ongoing difficulties in performing transformation of the *rml1* mutant allele in the *indica* background, the allele was back-crossed to the *japonica* cv. Dian Jing You (DJY) to derive a line named *r-3*. Plasmid *pRML1-gRML1* was introduced into calli of *r-3* by Agrobacterium-mediated transformation as described previously (Hiei and Komari, *et al.*, 2008). To determine whether the *RPL3A* transcript can rescue the mutant phenotype, the full-length*RPL3A* cDNA was cloned into the binary vector under control of the *RPL3B*promoter. Plasmid *pRPL3B-cRPL3A* was also transformed into *r-3* calli.

RNA extraction and real-time RT-PCR analysis

Total RNA was extracted using a RNA Prep Pure Plant Kit (TIANGEN, Beijing) and cDNA was synthesized with Oligo (dT) 18 or a random primer, and reverse-transcribed using PrimeScript Reverse Transcriptase (TaKaRa Bio Inc., Dalian). Real-time RT-PCR was performed using a SYBRGreen Mix Kit (Bio-Rad, Hercules, CA) on an ABI 7500 real-time PCR system with three biological replicates. The rice *ubiquitin* gene *LOC_Os03g13170* was used as an endogenous control. The $2^{-\Delta\Delta CT}$ method was adopted to analyse relative gene expression (Livak and Schmittgen, 2001).

Subcellular localization of RML1 protein

The cDNA of *RML1* was amplified from the wild-type and fused with green fluorescent protein (GFP) to generate a pCAMBIA1305 vector. The fusion protein was transiently expressed in epidermal cells of *Nicotiana benthamiana* leaves. GFP alone was used as the control, and FIB2-mCherry was used as a nuclear marker (Degenhardt and Bonham-Smith, 2008*a*). After 18 or 48h transformation, GFP signals were observed with a confocal laser scanning microscope (Carl Zeiss LSM780).

RML1pro:GUS reporter gene construction and analysis

A 1.8-kb fragment upstream of the *RML1* translation start site was cloned into binary vector pCAMBIA1381Z to fuse with the *GUS* reporter gene and transformed into the Nipponbare cultivar by the *Agrobacterium*-mediated method. Hygromycin-resistant calli were regenerated and eight positive homozygous lines were obtained. GUS staining was performed as described by Jefferson *et al.* (1987).

Ribosome profile analysis

Ribosomes were isolated from rice leaves as described previously (Mustroph *et al.*, 2009; Rivera *et al.*, 2015) with minor modifications. Tissue (20g) from 1-week-old seedlings was pulverized in liquid nitrogen and homogenized in 50ml of plant extraction buffer [50mM Tris-HCl (pH 7.5), 400mM KCl, 30mM $MgCl_2$, 5mM dithiothreitol, 50mg ml^{-1}cycloheximide, 50mg ml^{-1} chloramphenicol, 1% Triton X-100]. Cell debris was removed by centrifugation at 1076 *g* for 7min at 4 °C. A 0.1 volume of 20% Triton X-100 was added to the supernatant and centrifuged at 17210*g* for 20min at 4 °C. The supernatant was then loaded onto a 13-ml sucrose cushion [20mM Tris-HCl (pH 7.6), 5mM $MgCl_2$, 50mM NH_4Cl, 60% (w/w) sucrose] and centrifuged at 117000 *g* for 18h at 4 °C. Pellets were re-suspended in 200 µl of cold re-suspension buffer [200mM Tris-HCl (pH 9.0), 200mM KCl, 25mM EGTA, 35mM $MgCl_2$, 5mM DTT, 50mg ml^{-1}cycloheximide, and 50mg ml^{-1} chloramphenicol] and incubated at 4 °C for 1h. The ribosomes were layered onto an 11-ml 5–55% linear sucrose gradient and centrifuged at 170000 *g* for 2.5h at 4 °C, and the absorbance of each fraction was measured at 254nm with a fraction collection system (Biocomp). Proteins in each fraction were precipitated in two volumes of ice-chilled ethanol at 4 °C for about 12h.

Western blotting

Total proteins were extracted from wild-type and *rml1* leaves with buffer B [50mM Tris-HCl, pH 7.5, 150mM NaCl, 0.5% Triton X-100, and protease inhibitor cocktail Complete Mini tablets pH 8.0 (Roche)] (Willige *et al.*, 2011), and boiled in 1× SDS loading buffer at 95 °C for 5min. Proteins were separated in 8–12% SDS-PAGE gradient gels and transferred to polyvinylidene difluoride (PVDF) membranes (0.45 µm, Bio-Rad). Western blots were detected with anti-OsRPL3 (Abmart), anti-hsRPS14 (Millipore), anti-α-tubulin (Sigma) antibodies, and related secondary antibodies, respectively. An ECL reagent (Bio-Rad) was used for imaging.

RESULTS

Multiple developmental defects in the *rml1* mutant

We screened a ^{60}Co-irradiated population of *indica* rice variety 93-11 in a search for leaf morphology and plant architectural variants. We identified a*minute* mutant that displayed retarded plant growth and development, and named it *rice minute-like 1* (*rml1*). The *rml1* mutant showed delayed seed germination and inhibition of root growth. We performed a time-course comparison of primary root lengths of wild-type and *rml1* mutant over days 1–7 following seed germination And found that the wild-type grew faster than *rml1* (Fig. 1C). Wild-type seedlings at the 5th leaf stage in the field were already forming new tillers, whereas the *rml1* mutant was not. At the mature stage, the *rml1* plants had reduced size compared to the wild-type, including dwarfing, narrow leaves, and short panicles (Fig. 1E–G;Table 1). The mutant was ~40 d later flowering than the wild-type (Fig. 1E). Thus, the *rml1* mutant clearly displayed retarded and delayed plant growth relative to its wild-type.

Table 1: Morphological traits of wild-type and *rml1* plants

	Tillers per plant	Plant height (cm)	Main panicle length (cm)	Primary branch number per main panicle	Number of spike-lets per panicle
Wild-type	9.9±2.6	116.1±4.3	25.0±2.0	11.9±1.1	218.7±13.0
rml1	7.4±1.6*	84.0±5.7**	20.3±1.2**	8.2±0.9**	88.0±11.8**

Error bars indicate ±SD (*n*=10). Student's *t*-test was used for statistical analysis (*, $P<0.05$; **, $P<0.01$).

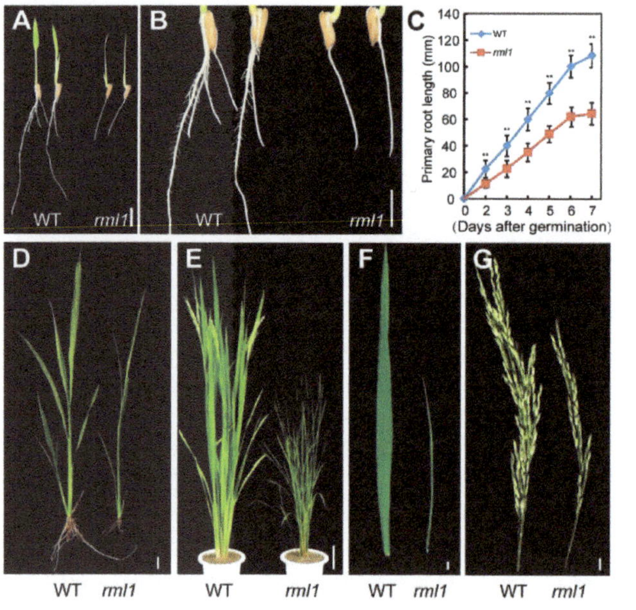

Figure 1: Gross morphology analyses of the wild-type (WT) and*rml1* mutant. (A) Root and shoot phenotypes of 2-d-old wild-type and *rml1*mutant plants. (B) Magnification of wild-type and *rml1* mutant roots in (A). (C) Comparison of primary root growth after germination between wild-type and *rml1* mutants (n=20). (D) Phenotypes of the wild-type and the *rml1* mutant at the five-leaf stage when grown in the field. (E) Morphologies of the wild-type and *rml1*mutant plants at heading. (F) Flag leaves of wild-type and *rml1*plants at the heading stage. (G) Main panicles of the wild-type and *rml1* mutant plants. Scale bars: 10mm (A, B, D, F, G); 10cm (E). Student's t-test was used for statistical analysis (*, P<0.05; **, P<0.01).

Abnormal vascular numbers and defects in *rml1* leaves

One of the obvious phenotypes of *rml1* mutant was leaf blade morphology. Leaf blade width in the *rml1* mutant was consistently narrower than in the wild-type from the 5th leaf stage to maturity (Fig. 2A,B). Some leaf blades showed irregular edges and twisted basal midribs. The lengths and widths of mature flag leaves of *rml1* were about 56% and 17% of the wild-type, respectively (Fig. 2L), and sometimes the flag leaves formed a curled structure. Histological observations showed that the number of large veins (LVs) in the leaf blades was significantly reduced in *rml1* flag leaves. Moreover, the number of small veins (SVs) between adjacent LVs was also reduced to less than one-half of the wild-type.

We performed transverse sections of mature flag leaf blades in order to characterize the arrangements of vascular bundles in detail. The midrib of the wild-type leaf consisted of several air cavities and vascular bundles (VBs) (Fig. 2C). In contrast, development of midrib in *rml1* mutant was suppressed and sometimes irregular (Fig. 2G, H). The leaf blade of the wild-type had ~13 LVs, with five to six SVs between adjacent LVs. In contrast, the number of LVs in *rml1* was greatly reduced to about half of that in the wild-type. Remarkably, the number of SVs between adjacent LVs was reduced to only one or two. *rml1* plants also had abnormal vascular patterning in which the sizes of xylem (X) and bulliform cells (BCs) were also reduced (Fig. 2E, F, J, K). At the edge of the leaf blades, the morphology of VBs was asymmetric. Other tissues had no noticeable differences except that the size of mesophyll cells in the mutant was slightly smaller than those of the wild-type. These results indicated that *RML1* controls vascular number and patterning during leaf development.

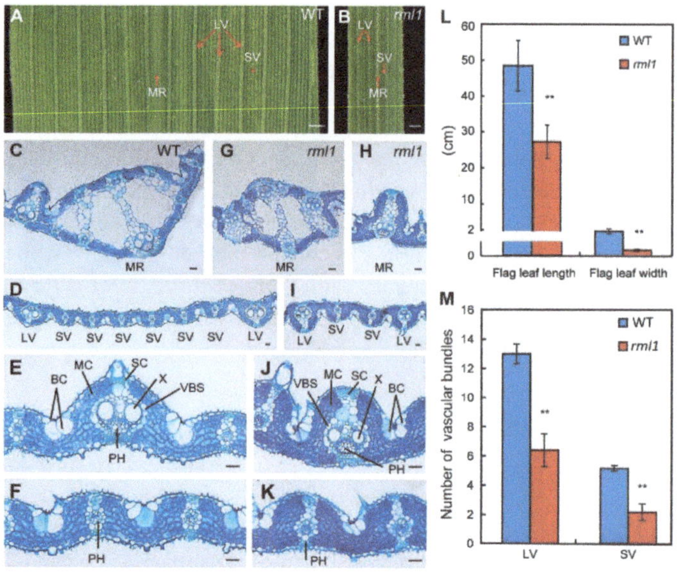

Figure 2: Patterning of vascular bundles in leaves of wild-type (WT) and *rml1* plants. (A, B) Enlarged views of adaxial surfaces of mature flag leaf blades in wild-type and *rml1* plants. (C–K) Transverse sections of the middle part of flag leaf blades in the wild-type and*rml1* plants stained with toluidine blue. (C–F) Magnifications of midrib (C), large vascular bundles (E), and small vascular bundles (F) in the wild-type. (G–K) Magnifications of midrib (G,H) and vascular bundles (I–K) in the *rml1* mutants. (L) Length and width of wild-type and *rml1* flag leaves at maturity (n=15). (M) Numbers of large and small vascular bundles (between the two adjacent LVs) in wild-type and *rml1* flag leaves at maturity (n=10). Abbreviations: BC, bulliform cells; LV, large vascular bundle; MC, mesophyll cell; MR, midrib; PH, phloem; SC, sclerenchymatous tissue or cell; SV, small vascular bundle; VBS, vascular bundle sheath; X, xylem. Scale bars: 1mm (A, B); 50 μm (C–K). Student's t-test was used for statistical analysis (*,P<0.05; **, P<0.01).

rml1 displays semi-dwarfness

In addition to narrow leaves, *rml1* plants exhibited a semi-dwarf stature. The height of *rml1* plants at maturity was about 72% of wild-type, and every internode, including that of the panicle, of *rml1* was significantly shorter (Fig. 3A; Table 1). Examination of cross-sections revealed that the size and number of vascular bundles in the first internode of *rml1* were reduced (Fig. 3B). The cross-sectional width did not differ, but the number of parenchyma cells (PCs) was significantly increased (Fig. 3B, C). In addition, longitudinal sections of the first internode showed that the size of PCs was reduced, but the number was higher in the *rml1* mutant (Fig. 3B, D). These data revealed that the *rml1* mutant exhibited reduced cell size in conjunction with increased cell number, but overall the result was a smaller plant.

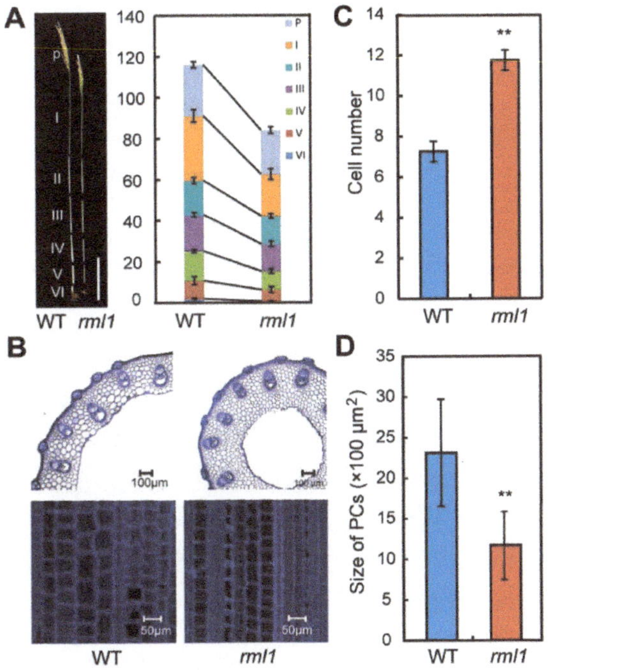

Figure 3: Morphological comparison of wild-type and *rml1* plant height. (A) Phenotypic characterization of internodes of wild-type and*rml1* plants at maturity. P, panicle. I to VI indicate corresponding internodes from top to bottom. (B) Transverse (top) and longitudinal (bottom) sections of the first internode (I) of wild-type and *rml1* plants at maturity. (C) Comparison of parenchyma cell numbers in the transverse sections of the first internode of wild-type and *rml1* plants. (D) Comparison of parenchyma cell (PC) size in longitudinal sections of the first internode of wild-type and *rml1* plants. Student's *t*-test was used for statistical analysis (*, $P<0.05$; **, $P<0.01$). Scale bar in (A) is 20cm.

A mutation in a the *Ribosomal Protein L3* gene causes the *minute-like rml1*phenotypes

To isolate the *rml1* gene, ten plants with the recessive *rml1* phenotype were selected from the F_2 progeny of a cross between *rml1* and the*japonica* cultivar 02428. The mutant gene was located in a 16.5-cM interval on the short arm of chromosome 11. We further mapped the *rml1*locus to a 24-kb region between markers sn-16 and sn-9 on the BAC clone OSJNBb0073K23 using a further 1200 F_2 plants with the recessive phenotype. Six ORFs in this region were predicted by the Rice GAAS database. Sequence analysis revealed that the third ORF (*LOC_Os11g06750*) in *rml1* had a four-base deletion in the last exon, leading to a frame shift and presumably forming a protein 4kDa larger than the wild-type. To further confirm whether *ORF3* was the *RML1*allele, a 10.3-kb genomic DNA fragment including the native promoter was transformed into calli derived from a homozygous line of *r-3*backcrossed to the *japonica* cv. Dian Jing You (DJY). Of 21 T_0 plants generated, 13 independent positive transgenic plants phenocopied the wild-type (Fig. 4C). Therefore, we concluded that the mutation of *ORF3*was responsible for *rml1*. The *ORF3* encodes a putative ortholog of yeast Ribosomal L3 protein, RPL3B, and consists of 389 amino acids. Another highly homologous protein (RPL3A) with almost the same molecular size also occurs in rice. We next used a specific antibody against OsRPL3 to confirm that the mutation caused the redundant protein. Only one thick band was detected in the wild-type, whereas *rml1* had two bands (Fig. 4D). Given the molecular size of the RPL3 proteins, we believe that the upper band in *rml1* is the mutant rpl3b protein and the lower band is the RPL3A protein.

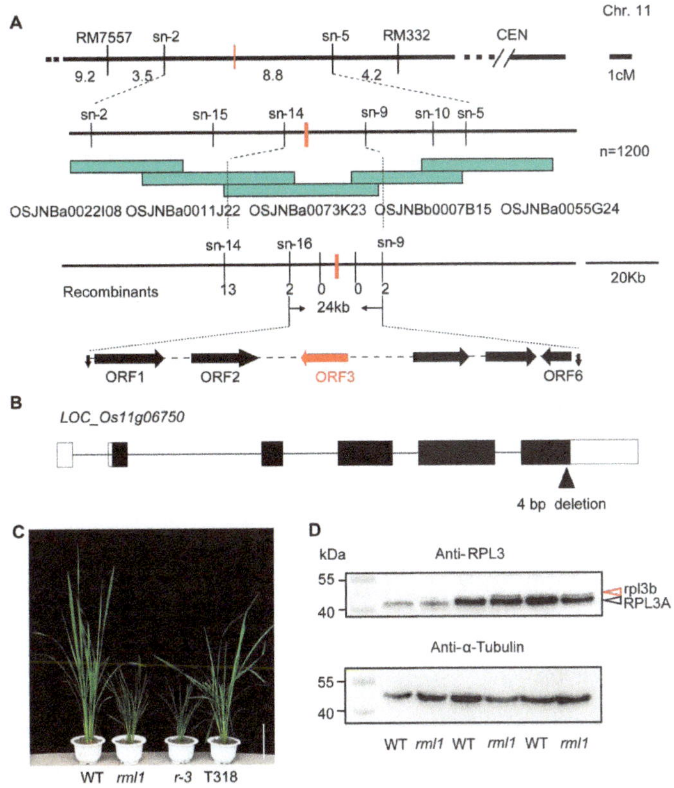

Figure 4: Map-based cloning and characterization of the *RML1*gene. (A) Fine mapping of the *RML1* gene on chromosome 11. The *RML1*locus was narrowed to a 24-kb region containing six predicted ORFs. (B) Schematic of the *RML1* gene and mutations in the *rml1*mutant. *rml1* has a 4-bp deletion in the last exon. Lines, open boxes, and black boxes indicate introns, non-transcribed regions, and exons, respectively. (C) Gross morphologies of wild-type, *rml1*, *r-3*(recessive genotype from BC_2F_3), and T318 transgenic lines (complemented with a 10.3-kb genomic fragment of *ORF3*) plants. (D) Immunoblot analysis of the OsRPL3 protein in wild-type and *rml1* seedlings. A gradient experiment was conducted with samples loaded in gels (2, 5 and 8 μl). Anti-α-Tubulin antibody was used as a loading control. Black and red arrowheads indicate RPL3A and rpl3b proteins, respectively. Scale bar in (C) is 20cm.

Expression pattern and subcellular localization of RML1

Expression analysis revealed that *RML1* was expressed in all organs tested, including roots, leaves, leaf sheaths, stems, and spikelets (Fig. 5A). The expression levels of *RPL3B* were higher in leaves, sheaths, and spikelets than in roots and stems. To further confirm the expression profile of*RML1*, a -*glucuronidase* (*GUS*) gene driven by the *RML1* promoter (approximately 1.8kb upstream of the translation start site) was transformed into calli of cv. Nipponbare. GUS activity was detected in young roots, seedlings, young leaf blades, leaf sheaths, stems, and panicles, and this was consistent with results from real time-PCR (Fig. 5B). The ubiquitous expression of *RML1* suggested that it might have a pleiotropic role in rice plant development.

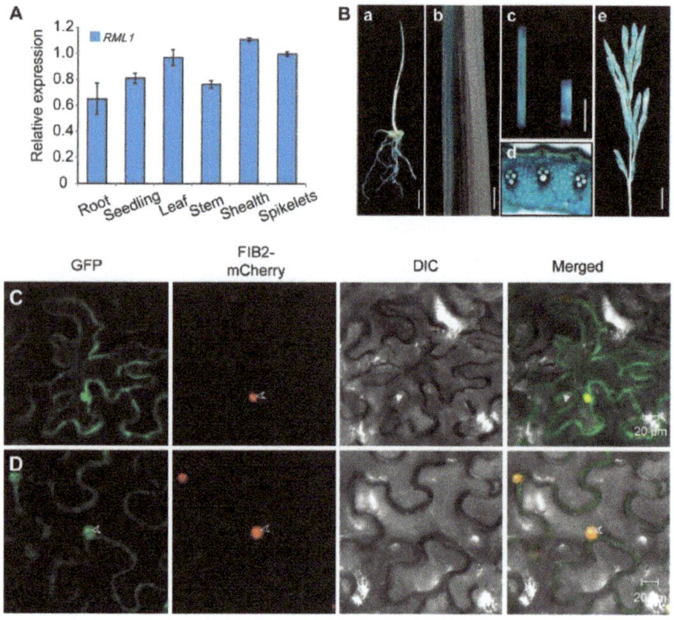

Figure 5: Expression pattern of *RML1*and the subcellular location of RML1. (A) Real-time PCR data showing that *RML1* is expressed in various tissues. (B) GUS staining of various tissues in the *pRML1*: *GUS*transgenic lines: root (a); leaf (b); stem and sheath (c); stem cross- section (d); spikelet (e). (C) Subcellular localization of the GFP protein in *Nicotiana benthamiana* epidermal cells. Free GFP signals were located in the cytoplasm. (D) RML1-GFP fusion protein was localized to the nucleoplasm and cytoplasm and concentrated in the nucleolus. FIB2-mCherry was specifically localized to the nucleolus as a marker. Arrowheads indicate the nucleolus. Scale bars: 10mm (B-a, c); 1mm (B-b); 100 μm (B-d); 20 μm (C, D). *Ubiquitin* (*UBQ*) was used as an internal control in the real-time PCR analyses. Error bars indicate ±SD (*n*=3).

BLASTP (http://blast.ncbi.nlm.nih.gov/Blast.cgi) found that RML1 was predicted to be a Ribosomal L3 superfamily protein. During the process of ribosome biogenesis, ribosomal proteins are synthesized in the cytoplasm and then imported into the nucleolus to participate in ribosome subunit assembly, and ribosomes that have translation ability are present in the cytoplasm (Byrne, 2009). To determine the subcellular localization of RML1, GFP was fused to the C-terminus of RML1 and the fusion gene was transiently expressed in leaf epidermal cells of *Nicotiana benthamiana*. RML1-GFP protein was localized to the nucleoplasm and cytoplasm, and concentrated in the nucleolus (Fig. 5C, D). Thus, RML1 is mainly localized in the cytosol as well as in the nucleus.

RPL3A and *RPL3B* are differentially expressed

The two RPL3 family members, *RPL3A* and *RPL3B*, in rice share 89.9% identity at the transcript level and 98.7% identity at the amino acid level (Fig. 6A). Real-time-PCR showed that *RPL3B* is more abundant than *RPL3A*in all tissues (Fig. 6B). Previous studies have shown that genetic defects in individual ribosomal components cause deleterious developmental effects in a gene dosage-dependent manner (Ito *et al.*, 2000; Rosado *et al.*, 2010). To determine whether *RPL3A* provides dosage compensation for*rpl3b*, we analysed their expression levels. In the *rml1* mutant, the expression of *RPL3B* was greatly decreased, but expression of *RPL3A* was unchanged (Fig. 6C), suggesting that *RPL3A* cannot compensate for the mutation of *RPL3B*. In addition, we fused the cDNA of *RPL3A* to the promoter of *RPL3B* and transformed this plasmid into the calli of the mutant. Transgenic lines T317 rescued the mutant phenotypes, despite a slight delay in plant growth compared to T318 lines (*RPL3B* promoter:*RPL3B* genomic DNA). The small difference in complementation efficiency might result from lower expression of *RPL3A* under control of the 3B promoter in T317 compared to *RPL3B* in T318 lines. Thus, the different expression levels of *RPL3A* and *RPL3B* could be responsible for the distinct functional difference in plant development.

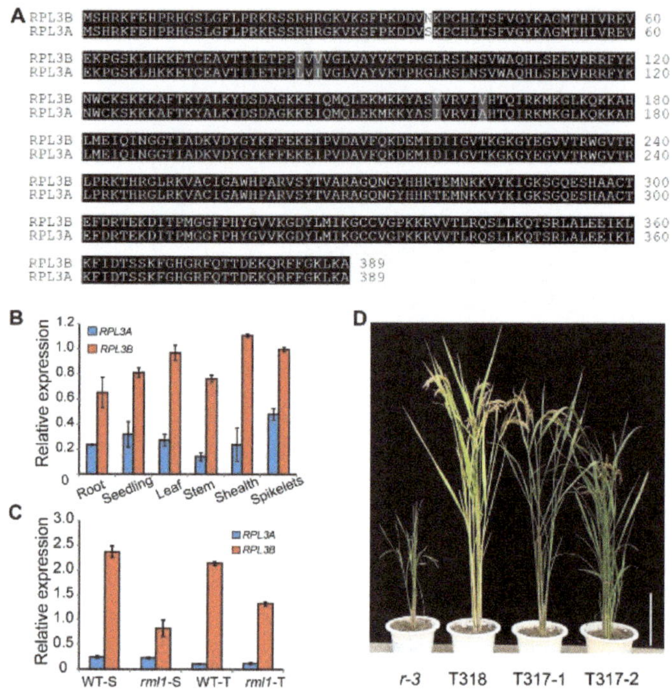

Figure 6: Function comparisons between RPL3A and RPL3B. (A) Clustal alignment of the two rice RPL3 amino acid sequences (RPL3A and RPL3B). Identical and similar residues are shaded black and grey, respectively; the difference is highlighted with no shading. (B) Expression of *RPL3B* was higher than *RPL3A* in all tissues analysed. (C) Results from real-time PCR assay showing that expression of the *RPL3B*gene is reduced in the *rml1* mutant, whereas that of *RPL3A* is unchanged relative to the wild-type. (D) Genetic complementation of *rml1* with *pRML1:gRML1* (T318) and*pRML1:cRPL3A* (T317) constructs. *Ubiquitin* (*UBQ*) was used as an internal control in real-time PCR. Error bars indicate ±SD (*n* 3). Abbreviations: S, seedling stage; T, tilling stage. Scale bar in (D) is 20cm.

The *rml1* mutant has altered ribosomal structure

RPL3 is one of only two proteins capable of initiating *in vitro* assembly of*E. coli* large ribosomal subunits (Nowotny and Nierhaus, 1982), and it is one of the few essential proteins for peptidyltransferase activity (Schulze and Nierhaus, 1982). We hypothesized that the mutant rpl3b ribosomal protein might have undergone a structural change in the *rml1* mutant. In*E. coli*, mutation of *RPL3* increases its tiamulin resistance by alteration of the drug-binding site at the peptidyl transferase centre, probably by its effect on the rRNA structure responsible for the tiamulin resistance (Bosling *et al.*, 2003; Klitgaard *et al.*, 2015). We therefore analysed the responses of wild-type and *rml1* plants treated with a concentration series of tiamulin by measuring the lengths of primary roots, and found that the*rml1* mutant displayed mild resistance to tiamulin compared to the wild-type (Fig. 7). Treatment with other antibiotics that target different ribosomal locations in prokaryotes resulted in no obvious changes between the wild-type and *rml1* . These results suggest that mutation of *RPL3B* probably causes a structural alteration in the peptidyltransferase centre.

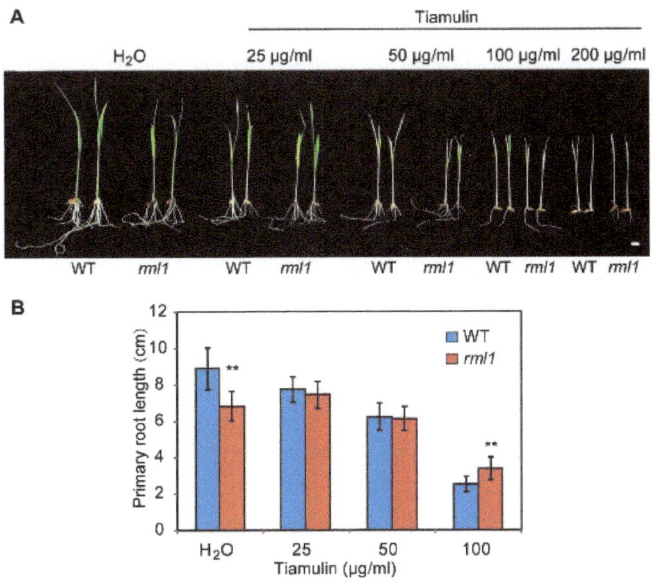

Figure 7: The *rml1* mutant shows slight resistance to tiamulin antibiotic. (A) Wild-type and*rml1* mutant plants treated with different concentrations of tiamulin antibiotic. Wild-type and *rml1* seeds were surface-sterilized and germinated on agar plates with or without the antibiotic. (B) Lengths of primary roots were measured after 7 d. Error bars indicate ±SD (n=10). Student's t-test was used for statistical analysis (*, P<0.05; **, P<0.01).

Defective ribosome biogenesis alters the ribosome profile in *rml1*

To investigate a potential role of RPL3B in ribosomal biogenesis, we performed ribosomal profiling in wild-type and *rml1* plants. Total ribosome particles were isolated from cell extracts and centrifuged on a sucrose cushion overnight. The re-suspended polysomes were fractionated by sucrose density gradient ultracentrifugation and measured at 254nm using a UV detector. The *rml1* mutant showed a clear deficit of the free 60S ribosomal subunit fraction compared to the wild-type, while the accumulation of 40S small subunits was slightly increased (Fig. 8A). The profile for 80S subunits (monosomes) was similar to the wild-type. Significantly reduced accumulation of polysomes in *rml1* plants indicated that the mutant might be mildly defective in translation initiation or have repressed global protein translation activity (Fig. 8A; Garcia-Gomez *et al.*, 2014). The protein compositions of the gradient fractions were subsequently determined by protein gel blotting and the ribosomal small protein 14 (HsRPS14) was used as a marker to monitor sedimentation of 60S and 40S ribosomes. In *rml1* plants, the rpl3b protein also occurred in 60S and 80S subunits and polysomes, suggesting that the mutant rpl3b protein also participated in ribosome biogenesis. Both normal and mutant types of RPL3B also bound 25S rRNA*in vitro* .

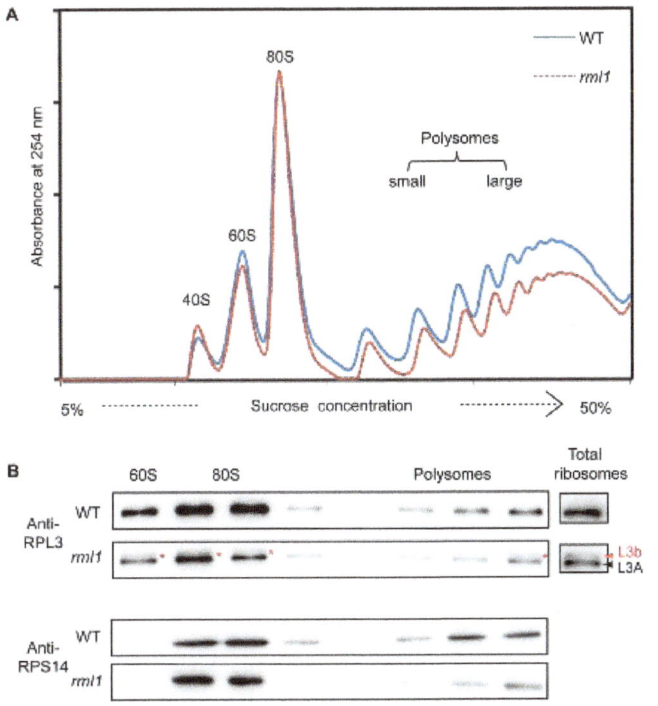

Figure 8: Absorbance profiles of ribosomes at 254nm and immunoblot analysis of the OsRPL3 protein in ribosomal fractions. (A) Ribosome extracts from wild-type and*rml1*seedlings were fractionated on a 5–50% sucrose density gradient. Absorbamce peaks at 254nm representing free 40S and 60S subunits, 80S monosomes, and polysomes are indicated. (B) Fractions from the gradient shown in (A) were collected and subjected to immunoblotting with the indicated antibodies. HsRPS14 was used as the marker for the 40S subunit. Red and black arrowheads indicate rpl3b and RPL3A, respectively. Red asterisks indicate rpl3b. All assays were performed at least three times.

Real-time PCR analyses to detect whether the mutation of RPL3B resulted in defects in pre-rRNA processing or rRNA accumulation showed that 35S pre-RNA, 25S rRNA, and 18S rRNA were slightly increased in the *rml1*mutant. These phenotypes have also been observed in a yeast *rpl3* [I282T] mutant that showed a similar polysome profile to *rml1* (Garcia-Gomez *et al.*, 2014). Our results indicate that depletion of RPL3B in the *rml1* mutant causes defective biogenesis of the 60S ribosome large subunit and polysomes as well as defects in pre-rRNA processing.

DISCUSSION

RPL3 family members are involved in rice plant growth and development

Multiple copies of the ribosomal proteins (RPs) in yeast have both redundant and non-redundant roles in function (Deutschbauer *et al.*, 2005; Komili *et al.*, 2007; Dean *et al.*, 2008). Most RP families in Arabidopsis consist of multiple members, ranging from two to seven (Barakat *et al.*, 2001), but some copies are non-expressed pseudogenes (Rosado *et al.*, 2010). RP genes in rice also have multiple copies (Shi *et al.*, 2014), but it remains unclear whether or how the duplicated genes function in plant development. In Arabidopsis, some members of the RP family with different functions have been associated with different expression levels or patterns. *RPL23aA* has a higher expression level than*RPL23aB*, and knock-down of *RPL23aA* leads to various developmental defects, whereas knock-down of the *rpl23aB* mutant does not (Degenhardt and Bonham-Smith, 2008*a*, *b*). In addition, paralogs of *RPL5* and *RPL10*family members also have distinct roles caused by their differential expression patterns (Fujikura *et al.*, 2009; Falcone *et al.*, 2010, 2013).

In this study we report for the first time in rice that RPs have functions in plant growth and development. We identified a *rice minute-like* mutant caused by mutation of *RPL3B*. In rice, *RPL3B* has an expressed paralog,*RPL3A*, and the coding regions as well as amino acid sequences are highly similar. We found that *RPL3A* is expressed at a lower level than *RPL3B*during all periods of plant development; however, the expression level of*RPL3A* in *rml1* plants is unchanged compared to the wild-type (Fig. 6). According to the ribosome heterogeneity model (Horiguchi *et al.*, 2012), the differential expression levels of the two *RPL3* genes might be responsible for their distinct functions in rice plant development, and it was observed that *RPL3A* provides no dosage compensation for the *RPL3B*mutation. Conversely, *RPL3B* is essential for translation of certain specific transcripts during plant development, a function that could not be supplemented by *RPL3A*. The relationship between the two isoforms in rice needs to be further clarified with mutants of *RPL3A*.

Mutation of *RPL3B* had a deleterious effect on polysome synthesis

Mutation or depletion of RPL3p has been shown to have effects on ribosome structure and function in yeast (Petrov *et al.*, 2004; Meskauskas and Dinman, 2007; Rosado *et al.*, 2007). In this study, the rpl3b protein had a deletion at its C-terminus and formed a protein ~4kDa larger than the wild-type. Ribosomal profile analysis showed that levels of both the 60S subunit and polysomes were decreased. In addition, immunoblot analysis of the fractions indicated that the mutant rpl3b protein was present in 60S, 80S, and polysome particles, although the band was weak (Fig. 8). This indicates that a few rpl3b proteins are assembled into 60S subunits as well. The *rml1* mutant also displayed mild resistance to tiamulin compared to the wild-type (Fig. 7). An EMSA assay proved that the mutant rpl3b protein can also bind the sarcin/ricin RNA domain of 25S rRNA *in vitro* in a similar manner to the wild-type. These results suggest that the peptidyltransferase centre was slightly changed.

Reduction in the polysome/monosome ratio in yeast is associated with translation initiation defects (Deplazes *et al.*, 2009; Garcia-Gomez *et al.*, 2014; Visweswaraiah *et al.*, 2015). Our polysome profile assays found that polysomes were significantly decreased in the *rml1* mutant, while monosomes were unchanged (Fig. 8). Notably, the polysome/monsome ratio in the *rml1* mutant was reduced relative to the wild-type. One explanation would be that the 60S particles are not well stabilized and a few aberrant 60S r-particles containing the mutated form of RPL3B were lost after assembly, leading to a decrease in the level of 60S and, in turn, to a small excess of 40S subunits. Considering the latest model for ribosome function of Horiguchi *et al.* (2012) together with our results, we deduce that ribosome insufficiency and aberrancy, or defective translation initiation, caused developmental abnormalities in the *rml1* mutant.

A deficit in ribosomal biogenesis alters auxin-related responses

Our study demonstrated that deletion of *RPL3B* disturbs the vascular patterns in leaves and stems of rice, including reduced numbers of large veins and small veins, and altered midrib morphology and vascular bundle size. In addition, some leaves had no proper blade at their lower sections. Previous studies have reported that polar auxin transport plays an essential role in formation of the vascular system in Arabidopsis (Scarpella *et al.*, 2006, 2010). We therefore hypothesize that the abnormal vascular patterns in the rice *rml1* mutant might be associated with auxin distribution.

Active auxin in plants is synthesized in areas that are associated with rapidly dividing and growing tissues. The shoot apical meristems and young leaves are considered to be the primary sites of auxin synthesis, and auxin moves from apical to basal regions (basipetally) in fulfilling its essential role in plant development (Ljung *et al.*, 2001). In Arabidopsis, polar localization on the plasma membrane protein PIN1 has been proposed as the auxin efflux carrier (Palme and Galweiler, 1999). Recent reports have shown that auxin displays a key role in both initiation and elaboration of final morphology of both leaves and vascular networks (Scarpella *et al.*, 2010). Some rice mutations that display vascular defects have also been associated with auxin transport (Qi *et al.*, 2008; Cho *et al.*, 2013; Fujita *et al.*, 2013).

We found that the free contents of IAA in *rml1* were higher than in the wild-type, whereas auxin signal transport and transduction genes, such as *OsPINs* and *auxin response factors* (*ARFs*), were significantly down regulated. Moreover, expression of the *NAL1* gene that is related to leaf vascular and plant architecture development was decreased in the *rml1* mutant. Reduction in polar auxin transport capacity in the *nal1* mutant has been shown to affect vein patterning (Qi *et al.*, 2008;Fujita *et al.*, 2013). Antisense knock-down of *OsARF1* (*ARF23*) also caused dwarf plants with small curled leaves, defects in reproductive development, and late flowering (Waller *et al.*, 2002), similar phenotypic effects to the *rml1* mutant. We therefore deduce that auxin transport or transduction defects may lead to the abnormal leaf morphology and plant architecture phenotypes displayed in the *rml1* mutant. Alternatively, defects in ribosome biogenesis may disrupt translation of mRNAs that are necessary to guide auxin distribution or transduction in developing leaves.

Individual mutations in RPs (*rpl24*, *rpl4d*, and *rpl5a*) cause specific auxin-related phenotypes in Arabidopsis, and these RPs regulate the translation of the auxin response factors (*ARF3/5*) via translation of the upstream opening reading frames at the *ARF*'s 5′ leader sequence (Nishimura *et al.*, 2005; Horiguchi *et al.*, 2012; Rosado *et al.*, 2012). However, functional analyses of rice *ARF* genes have been reported only for *OsARF1* (*ARF23*),*ARF12*, *ARF16*, *ARF19*, *ARF24*, and *ARF25* where most mutants have defects in root growth and nutritional uptake (Waller *et al.*, 2002; Li *et al.*, 2014; Shen *et al.*, 2015; Zhang *et al.*, 2015). Further studies are needed to clarify the relationship between ribosomal proteins and *ARF*s, and to demonstrate whether a similar mechanism occurs in ribosomal protein-mediated translation of *ARF*s in rice.

ACKNOWLEDGMENTS

This research was supported by grants from the 863 Program of China (2014AA10A604-4, 2014AA10A603-15), the National Science and Technology Supporting Program (2013BAD01B02-16), the Jiangsu Science and Technology Development Program (BE2014394, BE2015363), and the PAPD Program, and also supported by the Key Laboratory of Biology, Genetics and Breeding of Japonica Rice in Mid Lower Yangtze River, Ministry of Agriculture, P.R. China, and the Jiangsu Collaborative Innovation Center for Modern Crop Production. We thank Dr Xiangning Jiang (College of Life Sciences and Biotechnology, Beijing Forestry University) for assistance in analysing endogenous IAA contents.

REFERENCES

1. Barakat A Szick-Miranda K Chang IF Guyot R Blanc G Cooke R Delseny M Bailey-Serres J . 2001. The organization of cytoplasmic ribosomal protein genes in the Arabidopsis genome. Plant Physiology 127, 398–415.
2. Blanc G Wolfe KH . 2004. Functional divergence of duplicated genes formed by polyploidy during Arabidopsis evolution. The Plant Cell 16, 1679–1691.
3. Bosling J Poulsen SM Vester B Long KS . 2003. Resistance to the peptidyl transferase inhibitor tiamulin caused by mutation of ribosomal protein l3. Antimicrobial Agents and Chemotherapy 47, 2892–2896.
4. Byrne ME . 2009. A role for the ribosome in development. Trends in Plant Science 14, 512–519.
5. Cho SH Yoo SC Zhang H Pandeya D Koh HJ Hwang JY Kim GT Paek NC . 2013. The rice narrow leaf2 and narrow leaf3 loci encode WUSCHEL-related homeobox 3A (OsWOX3A) and function in leaf, spikelet, tiller and lateral root development. New Phytologist 198, 1071–1084.
6. Dean EJ Davis JC Davis RW Petrov DA . 2008. Pervasive and persistent redundancy among duplicated genes in yeast. PLoS Genetics 4, e1000113.
7. Degenhardt RF Bonham-Smith PC . 2008 a. Transcript profiling demonstrates absence of dosage compensation in Arabidopsis following loss of a single RPL23a paralog. Planta 228, 627–640.
8. Degenhardt RF Bonham-Smith PC . 2008 b. Arabidopsis ribosomal proteins RPL23aA and RPL23aB are differentially targeted to the nucleolus and are disparately required for normal development. Plant Physiology 147, 128–142.
9. Deplazes A Mockli N Luke B Auerbach D Peter M . 2009. Yeast Uri1p promotes translation initiation and may provide a link to cotranslational quality control. Embo Journal 28, 1429–1441.
10. Deutschbauer AM Jaramillo DF Proctor M Kumm J Hillenmeyer ME Davis RW Nislow C Giaever G . 2005. Mechanisms of haploinsufficiency revealed by genome-wide profiling in yeast. Genetics 169, 1915–1925.
11. Dudov KP Perry RP . 1984. The gene family encoding the mouse ribosomal protein L32 contains a uniquely expressed intron-containing gene and an unmutated processed gene. Cell 37, 457–468.
12. Falcone FM Casadevall R Luciani MD Pezza A Casati P . 2013. New evidence for differential roles of l10 ribosomal proteins from Arabidopsis . Plant Physiology 163, 378–391.
13. Falcone FM Pezza A Biarc J Burlingame AL Casati P . 2010. Plant L10 ribosomal proteins have different roles during development and translation under ultraviolet-B stress. Plant Physiology 153, 1878–1894.
14. Fujikura U Horiguchi G Ponce MR Micol JL Tsukaya H . 2009. Coordination of cell proliferation and cell expansion mediated by ribosome-related processes in the leaves of Arabidopsis thaliana . The Plant Journal 59, 499–508.
15. Fujita D Trijatmiko KR Tagle AG et al. 2013. NAL1 allele from a rice landrace greatly increases yield in modern indica cultivars. Proceedings of the National Academy of Sciences, USA 110, 20431–20436.
16. Garcia-Gomez JJ Fernandez-Pevida A Lebaron S Rosado IV Tollervey D Kressler D de la Cruz J . 2014. Final pre-40S maturation depends on the functional integrity of the 60S subunit ribosomal protein L3. PLoS Genetics 10, e1004205.
17. Hiei Y Komari T . 2008. Agrobacterium-mediated transformation of rice using immature embryos or calli induced from mature seed. Nature Protocols 3, 824–834.
18. Horiguchi G Van Lijsebettens M Candela H Micol JL Tsukaya H . 2012. Ribosomes and translation in plant developmental control. Plant Science 191–192, 24–34.
19. Ito T Kim GT Shinozaki K . 2000. Disruption of an Arabidopsis cytoplasmic ribosomal protein S13-homologous gene by transposon-mediated mutagenesis causes aberrant growth and development. The Plant Journal 22, 257–264.
20. Jefferson RA Kavanagh TA Bevan MW . 1987. GUS fusions: beta-glucuronidase as a sensitive and versatile gene fusion marker in higher plants. Embo Journal 6, 3901–3907.
21. Klitgaard RN Ntokou E Norgaard K Biltoft D Hansen LH Traedholm NM Kongsted J Vester B . 2015. Mutations in the bacterial ribosomal protein l3 and their association with antibiotic resistance. Antimicrobial Agents and Chemotherapy 59, 3518–3528.
22. Komili S Farny NG Roth FP Silver PA . 2007. Functional specificity among ribosomal proteins regulates gene expression. Cell 131, 557–571.
23. Kongsuwa K Yu Q Vincent A Frisardi M Rosbash M Lengyel J Merriam J . 1985. A Drosophila minute gene encodes a ribsomal protein. Nature 317, 555–558.
24. Lambertsson A . 1998. The minute genes in Drosophila and their molecular functions. Advances In Genetics Incorporating Molecular Genetic Medicine 38, 69–134.
25. Lecompte O Ripp R Thierry J Moras D Poch O . 2002. Comparative analysis of ribosomal proteins in complete genomes: an example of reductive evolution at the domain scale. Nucleic Acids Research 30, 5382–5390.
26. Li G Liang W Zhang X Ren H Hu J Bennett MJ Zhang D . 2014. Rice actin-binding protein RMD is a key link in

the auxin-actin regulatory loop that controls cell growth. Proceedings of the National Academy of Sciences, USA 111, 10377–82.

27. Livak KJ Schmittgen TD . 2001. Analysis of relative gene expression data using real-time quantitative PCR and the 2 (-Delta Delta C(T)) method. Methods 25, 402–408.

28. Ljung K Bhalerao RP Sandberg G . 2001. Sites and homeostatic control of auxin biosynthesis in Arabidopsis during vegetative growth. The Plant Journal 28, 465–474.

29. Marygold SJ Roote J Reuter G et al. 2007. The ribosomal protein genes and minute loci of Drosophila melanogaster . Genome Biology 8, R216.

30. McIntosh KB Bonham-Smith PC . 2006. Ribosomal protein gene regulation: what about plants? Canadian Journal of Botany 84, 342–362.

31. Mears JA Cannone JJ Stagg SM Gutell RR Agrawal RK Harvey SC . 2002. Modeling a minimal ribosome based on comparative sequence analysis. Journal of Molecular Biology 321, 215–234.

32. Meinke D Muralla R Sweeney C Dickerman A . 2008. Identifying essential genes in Arabidopsis thaliana . Trends in Plant Science 13, 483–491.

33. Meskauskas A Dinman JD . 2007. Ribosomal protein L3: gatekeeper to the A site. Molecular Cell 25, 877–888.

34. Mustroph A Juntawong P Bailey-Serres J . 2009. Isolation of plant polysomal mRNA by differential centrifugation and ribosome immunopurification methods. Methods in Molecular Biology 553, 109–126.

35. Nishimura T Wada T Yamamoto KT Okada K . 2005. The Arabidopsis STV1 protein, responsible for translation reinitiation, is required for auxin-mediated gynoecium patterning. The Plant Cell 17, 2940–2953.

36. Nowotny V Nierhaus KH . 1982. Initiator proteins for the assembly of the 50S subunit from Escherichia coli ribosomes. Proceedings of the National Academy of Sciences, USA 79, 7238–7242.

37. Palme K Galweiler L . 1999. PIN-pointing the molecular basis of auxin transport. Current Opinion In Plant Biology 2, 375–381.

38. Petrov A Meskauskas A Dinman JD . 2004. Ribosomal protein L3: influence on ribosome structure and function. RNA Biology 1, 59–65.

39. Pinon V Etchells JP Rossignol P Collier SA Arroyo JM Martienssen RA Byrne ME . 2008. Three PIGGYBACK genes that specifically influence leaf patterning encode ribosomal proteins. Development 135, 1315–1324.

40. Popescu SC Tumer NE . 2004. Silencing of ribosomal protein L3 genes in N. tabacum reveals coordinate expression and significant alterations in plant growth, development and ribosome biogenesis. The Plant Journal 39, 29–44.

41. Qi J Qian Q Bu Q et al. 2008. Mutation of the rice Narrow leaf1 gene, which encodes a novel protein, affects vein patterning and polar auxin transport. Plant Physiology 147, 1947–1959.

42. Rivera MC Maguire B Lake JA . 2015. Isolation of ribosomes and polysomes. Cold Spring Harbor Protocols 2015, 293–299.

43. Rosado A Li R van de Ven W Hsu E Raikhel NV . 2012. Arabidopsis ribosomal proteins control developmental programs through translational regulation of auxin response factors. Proceedings of the National Academy of Sciences, USA 109, 19537–19544.

44. Rosado A Sohn EJ Drakakaki G Pan S Swidergal A Xiong Y Kang BH Bressan RA Raikhel NV . 2010. Auxin-mediated ribosomal biogenesis regulates vacuolar trafficking in Arabidopsis. The Plant Cell 22, 143–158.

45. Rosado IV Kressler D de la Cruz J . 2007. Functional analysis of Saccharomyces cerevisiae ribosomal protein Rpl3p in ribosome synthesis. Nucleic Acids Research 35, 4203–4213.

46. Scarpella E Barkoulas M Tsiantis M . 2010. Control of leaf and vein development by auxin. Cold Spring Harbor Perspectives in Biology 2, a1511.

47. Scarpella E Marcos D Friml J Berleth T . 2006. Control of leaf vascular patterning by polar auxin transport. Genes Development 20, 1015–1027.

48. Schulze H Nierhaus KH . 1982. Minimal set of ribosomal components for reconstitution of the peptidyltransferase activity. Embo Journal 1, 609–613.

49. Shen C Yue R Sun T Zhang L Yang Y Wang H . 2015. OsARF16, a transcription factor regulating auxin redistribution, is required for iron deficiency response in rice (Oryza sativa L.). Plant Science 231, 148–58.

50. Shi Y Liu X Li R Gao Y Xu Z Zhang B Zhou Y . 2014. Retention of OsNMD3 in the cytoplasm disturbs protein synthesis efficiency and affects plant development in rice. Journal of Experimental Botany 65, 3055–3069.

51. Thomas BC Pedersen B Freeling M . 2006. Following tetraploidy in an Arabidopsis ancestor, genes were removed preferentially from one homeolog leaving clusters enriched in dose-sensitive genes. Genome Research 16, 934–946.

52. Tzafrir I Pena-Muralla R Dickerman A et al. 2004. Identification of genes required for embryo development in Arabidopsis. Plant Physiology 135, 1206–1220.

53. Uchiumi T Sato N Wada A Hachimori A . 1999. Interaction of the sarcin/ricin domain of 23 S ribosomal RNA with proteins L3 and L6. Journal of Biological Chemistry 274, 681–686.

54. Van Lijsebettens M Vanderhaeghen R De Block M Bauw G Villarroel R Van Montagu M . 1994. An S18 ribosomal protein gene copy at the Arabidopsis PFL locus affects plant development by its specific expression in meristems. Embo Journal 13, 3378–3388.

55. Visweswaraiah J Pittman Y Dever TE Hinnebusch AG . 2015. The beta-hairpin of 40S exit channel protein Rps5/uS7 promotes efficient and accurate translation initiation in vivo. eLife 4, e07939.

56. Waller F Furuya M Nick P . 2002. OsARF1, an auxin response factor from rice, is auxin-regulated and classifies as a primary auxin responsive gene. Plant Molecular Biology 50, 415–425.

57. Willige BC Isono E Richter R Zourelidou M

Schwechheimer C . 2011. Gibberellin regulates PIN-FORMED abundance and is required for auxin transport-dependent growth and development in Arabidopsis thaliana . The Plant Cell 23, 2184–2195.

58. Yao Y Ling Q Wang H Huang H . 2008. Ribosomal proteins promote leaf adaxial identity. Development 135, 1325–1334.

59. Zhang S Wang S Xu Y Yu C Shen C Qian Q Geisler M Jiang D Qi Y . 2015. The auxin response factor, OsARF19, controls rice leaf angles through positively regulating OsGH3-5 and OsBRI1. Plant Cell and Environment 38, 638–54.

60. Zheng M Wang Y Wang Y et al. 2015. DEFORMED FLORAL ORGAN1 (DFO1) regulates floral organ identity by epigenetically repressing the expression of OsMADS58 in rice (Oryza sativa). New Phytologist 206, 1476–1490.

61. Zhou F Roy B von Arnim AG . 2010. Translation reinitiation and development are compromised in similar ways by mutations in translation initiation factor eIF3h and the ribosomal protein RPL24. BMC Plant Biology 10, 193.

CHAPTER 14

THE RICE TCM5 GENE ENCODING A NOVEL DEG PROTEASE PROTEIN IS ESSENTIAL FOR CHLOROPLAST DEVELOPMENT UNDER HIGH TEMPERATURES

Kailun Zheng, Jian Zhao, Dongzhi Lin, Jiaying Chen, Jianlong Xu, Hua Zhou, Sheng Teng, and Yanjun Dong

Keywords: Chloroplast development, Deg protease, Photosystem, Rice

BACKGROUND

Chloroplast is a semi-autonomous organelle and carries out photosynthesis: the capture of light energy and its conversion into chemical energy. It is known that chloroplast development consists of a series of complex events related to chloroplast differentiation and can be divided into three steps which are coordinately regulated by plastid and nucleus genes (Jarvis and López-Juez 2013; Kusumi et al. 2011; Kusumi and Iba, 2014). The first step involves the activation of plastid replication and plastid DNA synthesis. The second step is the chloroplast 'build-up', characterized by the establishment of chloroplast genetic system. At this step nuclear-encoded plastid RNA polymerase (NEP) preferentially transcribes plastid genes encoding plastid gene expression machineries (Hajdukiewicz et al. 1997), and the transcription/translation activities in the chloroplast are dramatically increased. At third step, the plastid and nuclear genes encoding photosynthetic apparatus are expressed at very high levels. Plastid genes for photosynthetic apparatus are mainly transcribed by plastid-encoded RNA polymerase (PEP) (Santis-Maciossek et al. 1999). All expressions of these genes lead to the synthesis and assembly of photosynthetic machineries. In spite of these, the mechanisms of the major genes in higher plants remain largely unknown (Pfalz and Pfannschmidt 2013).

The family of Deg proteases (for degradation of periplasmic proteins) (Strauch and Beckwith 1988), also named as HtrA proteases (for high temperature requirement A) (Lipinska et al. 1988), is one important group among proteolytic enzymes. The Deg proteases, consisting of three representative proteins (DegP, DegQ and DegS), were initially found in *Escherichia coli* and well-studied in *Escherichia coli* and mammals. Most Deg members contain one to four PDZ protein-protein interaction domains (Clausen et al. 2002). A first *E.coli* DegP was identified based on the fact that it is required for the survival at high temperatures (Lipinska et al. 1988, 1990; Strauch and Beckwith 1988). A second *E.coli* protease, DegS, acts in a stress signaling cascade sensing misfolded proteins in the periplasm and transducing the signal to the cytoplasm (Walsh et al. 2003). A third *E.coli* protease, DegQ, is a periplasmic protease but also shares some functional features with DegP (Kolmar et al. 1996). However, relatively little is known about members of this family in plants (Kato and Sakamoto, 2010). To date, although it was predicated to contain 16 genes encoding putative Deg proteases in *Arabidopsis thaliana* (Huesgen et al. 2005), 15 in *Oryza sativa* (Tripathi and Sowdhamini 2006) and 20 in *Populus trichocarpa* (Garcia-Lorenzo et al. 2006), but barely a few Deg proteases from Arabidopsis have been well-studied.

Citation: Zheng K, Zhao J, Lin D, et al. The Rice TCM5 Gene Encoding a Novel Deg Protease Protein is Essential for Chloroplast Development under High Temperatures. Rice. 2016;9:13. doi:10.1186/s12284-016-0086-5.

Six AtDeg proteases are located in chloroplasts (Itzhaki et al. 1998), one in peroxisomes (Schuhmann et al. 2008), one in mitochondria (for review, see Schuhmann et al. 2012), and one in the nucleus (Pendle et al. 2005). AtDeg3 and AtDeg10 were reported to locate in both chloroplast and mitochondrion, while AtDeg7 is located in both nucleus and mitochondrion (Tanz et al. 2014). Additionally, the Deg proteases were reported to chloroplast-located be involved in the degradation of damaged photosynthetic proteins, especially the photosystem II (PSII) reaction center D1 protein (Kato et al. 2012). For example, the chloroplast-targeted AtDeg1 acts as a chaperone, assisting in the assembly of PSII dimers and supercomplexes (Sun et al.2010b). The AtDeg15 is a processing enzyme, cleaving the N-terminal peroxisomal targeting signal 2 (Helm et al.2007; Schuhmann et al. 2008). Recently, it was precisely predicated to contain 10 Deg proteases and 6 DegP-like proteases (Schuhmann et al. 2012) in *Oryza sativa*. However, to our knowledge, any *deg* rice mutants have not been reported yet, still less the function.

Here we report a detailed phenotypic analysis of the rice *deg* mutant, *tcm5*, which exhibited the albino phenotype before the 3-leaf stage at 32 °C, whereas normal green at 20 °C, and the cloning of *TCM5*, encoding a Deg protease protein. Additionally, the transcripts of certain genes for chlorophyll biosynthesis, photosynthesis and chloroplast development were severely affected in *tcm5* mutants at high temperatures. Our work implicates that rice*TCM5* plays an important role in chloroplast development and maintenance of PSII function under high temperatures.

RESULTS

Characterization of the *tcm5* Mutant

The leaves of *tcm5* mutants displayed an albino phenotype (Fig. 1a) before the 3-leaf-stage, and could not survive after the 4–5 leaf stage at 32 °C (Fig. 1b), but they turned yellowish green at 28 °C. Interestingly, the *tcm5* mutants showed no obvious phenotypic differences from the wild type (WT) plants at 20 °C. Similarly, at 32 °C, the Chl a, Chl b and carotenoid (Car) contents in *tcm5* mutants at the 3-leaf stage were drastically lower than those in WT plants (Fig. 1e), however, the pigment levels were similar to WT plants at 20 °C (Fig. 1f). In addition, during the whole growth periods, leaf Chl contents (SPAD values) in *tcm5* plants always showed lower levels than those in WT plants, moreover, the difference ranges become smaller when climate temperatures gradually decreased from summer to autumn in 2010 (Shanghai, China). These results indicated that the mutant trait was high-temperature sensitive PSII complexes are located in the chloroplast. PSII activity of the 3-leaf-stage *tcm5* and WT seedlings at 20 and 32 °C was surveyed by measuring the ratio of variable fluorescence to maximum fluorescence as follows: $Fv/Fm=(Fm-Fo)/Fm$, where *Fo* and *Fm* are minimum and maximum chlorophyll a fluorescence of dark-adapted leaves, respectively. *Fv/Fm* reflects the maximum potential capacity of the PSII photochemical reactions (Krause and Weis1991). At 32 °C, the *Fv/Fm* was 0.787 ± 0.03 in WT plants and 0.365 ± 0.02 in *tcm5* mutants, indicating that the photochemical efficiency of PSII was significantly suppressed in the *tcm5* mutants. In addition, *PSII* represents the actual photosynthesis efficiency. At 32 °C, the *PS*II value was 0.624 ± 0.04 in WT plants and 0.361 ± 0.04 in *tcm5*mutants, showing that the photosynthesis efficiency was significantly reduced in *tcm5* mutants. By contrast, at 20 °C, all respective values in the *tcm5* mutants were similar to WT levels. Thus, the*tcm5* mutation severely inhibited the PSII activity under high temperatures.

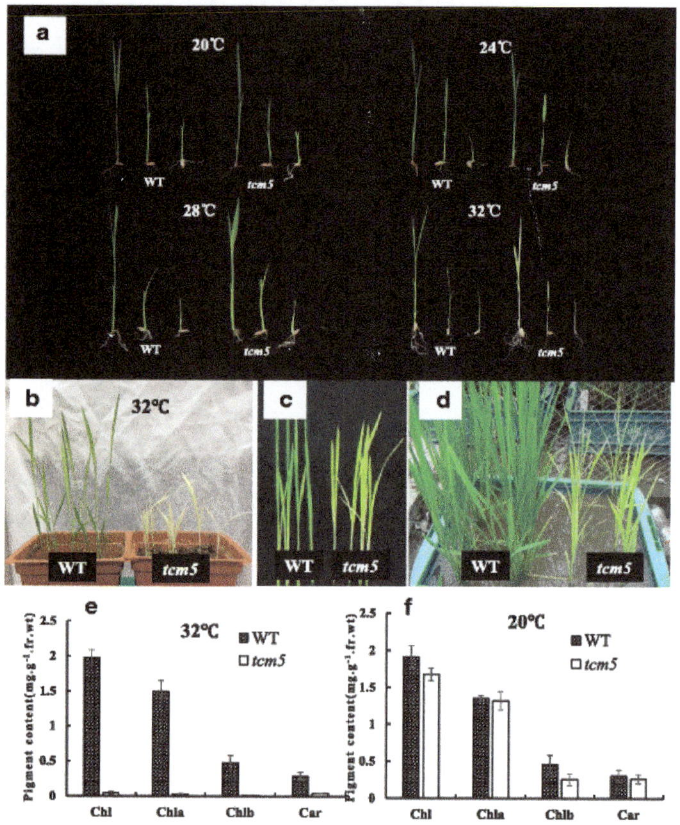

Figure 1: Characterization of the *tcm5* mutants; **a**, 1-, 2-, 3-leaf stage seedlings of wild type (WT) (*left*) and *tcm5* mutants (*right*) grown at 20, 24, 28 and 32 °C, respectively; **b**, 4-leaf-stage plants of wild type (*left*) and *tcm5* mutants (*right*) grown at 32 °C; **c**,**d**, indicate 25-day-old (**c**) and 50-day-old (**d**) plants grown at rice paddy field, respectively (2010, Shanghai, China); **e**, **f**, indicate the pigment contents in the 3rd leaves of WT and mutants grown at 32 and 20 °C, respectively. *Chl*total chlorophyll, *Chla* chlorophyll a, *Chlb* chlorophyll b, *Car* carotenoid, *WT* wide type

We next examined whether the lack of photosynthetic pigments in *tcm5* mutants was accompanied by ultrastructural changes in the chloroplasts. At 32 and 20 °C, we observed well-developed chloroplasts, with dense and well-structured grana lamella stacks, in WT plants, (Fig. 2a,c). By contrast, at 32 °C we did not observe any well-developed chloroplasts and well-structured grana lamella stacks (Fig. 2b) in *tcm5* mutants. However, there was no obvious difference between *tcm5* and WT plants at 20 °C (Fig. 2d). These results indicated that the *tcm5* mutation causes abnormal chloroplast development at high temperatures.

We also surveyed some agronomic characters of the *tcm5* and WT plants and found that, under field condition, the grain and panicle number per plant was significantly decreased, showing that the mutation inhibited the plant growth in *tcm5* mutants.

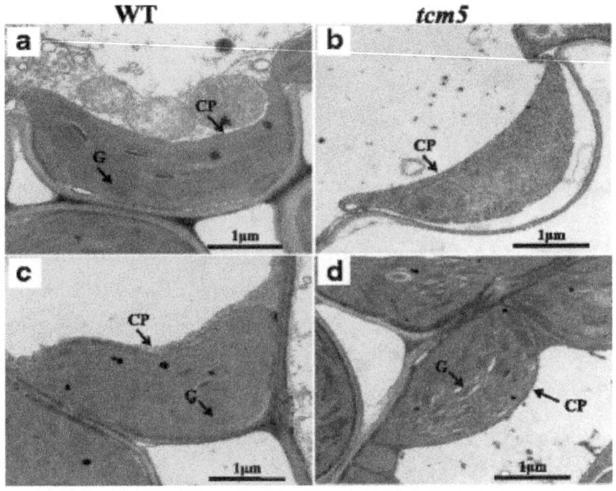

Figure 2: Transmission electron microscopic images of chloroplasts in WT and *tcm5* mutant; **a**, **b**, Chloroplast structure in WT(**a**) and *tcm5* (**b**) cell at 32 °C; **c**, **d**, Chloroplast structure in WT(**c**) and *tcm5*(**d**) cell at 20 °C. *CP, chloroplast; G, grana lamella stacks*

Map-based Cloning of *TCM5*

We used map-based cloning to identify the *TCM5* locus. To this end, we isolated 96 individuals with the albino phenotype from the segregating 412 F_2 plants derived from the cross of Pei'ai 64S and *tcm5* mutant. The segregation ratio (3:1) of green to albino plants showed that the mutant phenotype is controlled by a single recessive gene (*tcm5*). First, by genotyping the 96 mutant individuals, we localized *tcm5*between RM2474 and RM18692 on chromosome 5 (Fig. 3a). To fine-map *TCM5* gene, we developed new InDel and CAPS markers between RM2474 and RM18692. The *TCM5* locus was finally narrowed to an 18.9-kb between P1 and P3 markers using 6400 F_2 mutant individuals (Fig. 3b). The target region was predicated to contain two candidate genes (*LOC_Os05g34450*, *LOC_Os05g34460*) (Fig. 3c). We sequenced both candidate genes and only found a single nucleotide mutation (T→G) at position 194 bp from the ATG start codon in*LOC_Os05g34460*, which resulted in an amino acid change from Val to Gly (Fig. 3d).

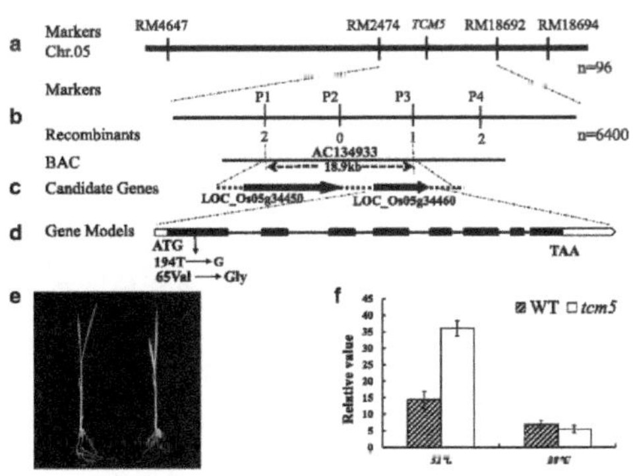

Figure 3: Genetic analysis and cloning of the *TCM5* gene; **a**, The *TCM5* gene was initially located between the RM2474 and RM18692 SSR markers on chromosome 5 using 96 F_2 mutant individuals; **b**, *TCM5* was narrowed to a 18.9 kb between InDel marker P1 and CAPS marker P3 and co-segregated with the CAPS marker P2 using 6400 F_2 mutant individuals; **c**, The target region contains two predicted candidate genes (*LOC_Os05g34450*, *LOC_Os05g34460*); **d**, A single nucleotide mutation at codon 194(T→G) in the first exon in *LOC_Os05g34460*, consisting of eight exons; **e**, The complemented plants with pCAMBIA1301:CaMV35S:TCM5 (*left*) and transgenic T_0 plants with empty vector pCAMBIA1301(*right*); **f**, Transcript levels of *TCM5* in WT and *tcm5* at the 3-leaf stage grown at 32 and 20 °C, *OsActin*was used as a control for qPCR

To further prove that the lack of *TCM5* was responsible for the mutant phenotype in the *tcm5* mutants, genetic complementation was performed. Resultantly, all of the 11 transgenic plants with pCAMBIA1301:CaMV35S:TCM5 were obtained and displayed wild-type phenotype under high temperatures (Fig. 3e). Meanwhile, 13 independent lines were transformed with empty vector, pCAMBIA1301, all failed to rescue the *tcm5* mutant. These results confirmed that *LOC_Os05g34460* is *TCM5*.

Characterization of TCM5

The *TCM5* encoded a putative Deg protease homologue, also predicated and named as OsDeg10 (Schuhmann et al. 2012). Searching the rice genome database revealed that *TCM5* is a single-copy gene. The *TCM5* gene is comprised of eight exons and encodes a 614-amino acid protein with the molecular mass of approximately 67kD. In addition, TCM5 was predicated to contain not only a chloroplast transit peptide of 94 amino acids at the N terminus, but also two PDZ domains of trypsin-like serine proteases. We found that the *tcm5* mutation site occurred at chloroplast transit peptides.

BLAST searches found that there are 10 Deg protease and 6 Deg-like protease proteins in the whole rice genome as described by Schuhmann et al.(2012) and TCM5 shares the highest sequence similarity to LOC_Os02g50880 (OsDeg9.1) (52.6 % amino acid identity). Additionally, close homology of TCM5 were also indented in *Arabidopsis thaliana*, *Brachypodium distachyon*, *Glycine max*, *Sorghum bicolor*, and *Zea mays*, showing the high conservative property within higher plants. And, Bradi2g25260 from *Brachypodium distachyon*shares the highest homology to TCM5 (88.1 % amino acid identity) (Fig. 4a). However, the functions of these homologs remain unclear. As shown in Fig. 4b, TCM5 homologs can be clearly divided into two groups: monocots and dicotyledons, consistent with the biological taxonomy (Fig. 4b).

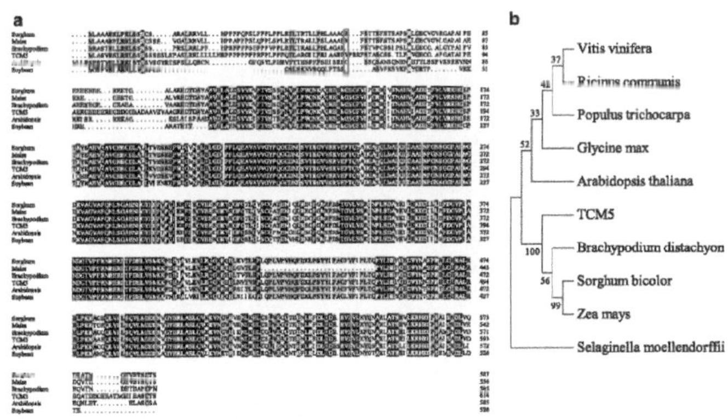

Figure 4: Phylogenic analysis of TCM5; **a**, Amino acid sequence alignment of *TCM5* with the five homologs. Amino acids fully or partially conserved are *shaded black* and *gray*, respectively. In the *red box* is the amino acid substitution in the *tcm5*mutant; **b**, Phylogenic tree of TCM5 and homologs. Protein sequences are *Ricinus communis* (RCOM_1076350),*Populus trichocarpa* (POPTR_0008s07940), *Arabidopsis thaliana* (AT5G36950, AtDeg10), *Glycine max* (100802340),*Vitis vinifera* (100250366), *Brachypodium distachyon* (Bradi2g25260), *Sorghum bicolor* (Sb09g020480), *Zea mays*(GRMZM2G117615), *Selaginella moellendorffii* (SELMO-

DRAFT_62730). The rooted tree is based on a multiple sequence alignment generated with the program Mega6. Scale represents percentage substitution per site. Statistical support for the nodes is indicated

TCM5 Expression

To examine the expression pattern of *TCM5*, we used RT-PCR to investigate the transcript levels of *TCM5* in various tissues (Fig. 5). Resultantly, *TCM5* was highly expressed in all green tissues (leaf and stem) and undetectable in roots and panicles, showing its tissue-specific expression.

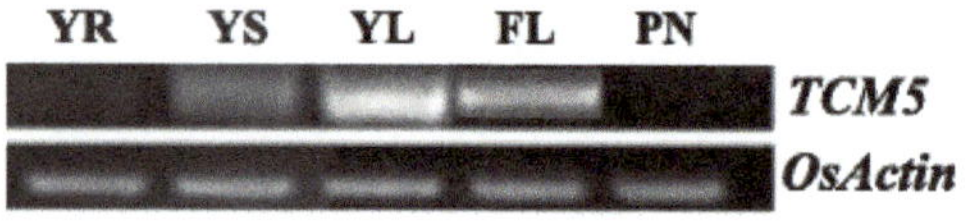

Figure 5: Expression analysis of *TCM5* by RT-PCR analysis; YR, young-seedling roots; YS, young-seedling stem; YL, young-seedling leaf; FL, flag leaf at heading; PN, young panicles. *OsActin* was used as a control (cycle number for *OsActin*was 28, cycle number for *TCM5* was 35)

To further explore whether the *tcm5* mutation and temperature influence *TCM5* expression, we examined *TCM5*transcript levels in 3-leaf *tcm5* and WT seedlings grown at 20 and 32 °C (Fig. 3f). Interestingly, regardless of WT and *tcm5* plants, *TCM5* transcript levels were more strongly expressed at high temperatures than those at low temperatures. Specifically, at 32 °C, *TCM5* transcript levels in the *tcm5* mutants were over 2-fold higher than WT levels. However, at 20 °C, *TCM5* expression levels were relatively low and comparable to WT plants (Fig. 3f). These results suggested that *TCM5* function is more needed for chloroplast development under high temperatures.

Subcellular Localization of TCM5 Protein

The TCM5 protein was predicted to localize to chloroplasts by TargetP (Emanuelsson et al. 2000,http://www.cbs.dtu.dk/services/TargetP/).

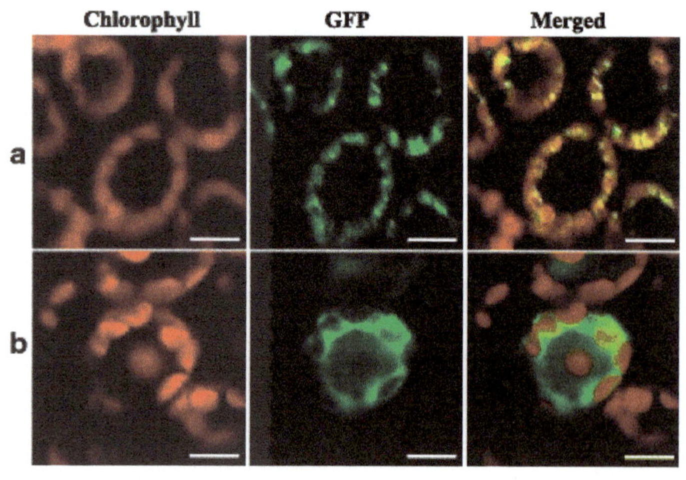

Figure 6: Subcellular localization of TCM5 protein; **a**, TCM5-GFP fusion; **b**, Empty GFP vector without a specific targeting sequence. The scale bar represents 20 μm.

To investigate the actual subcellular localization of TCM5, the cDNA fragment encoding the N-terminal region (amino acids 1–135) of TCM5 was introduced into the N-terminal of the GFP gene in the expression vector pMON530-GFP. Then we transformed the pMON530:TCM5-GFP plasmids into tobacco cells by Agrobacterium-mediated infection, and used empty pMON530-GFP vector without a specific targeting sequence as the positive control. Confocal laser scanning microscopy was used to observe the fluorescent signals after 48 h transformation. The green fluorescent signal of the TCM5-GFP fusion protein perfectly co-localized with the chlorophyll autofluorescence (Fig. 6a) in tobacco cells. By contrast, the cells transformed with the empty GFP vector had green fluorescent signals in both the cytoplasm and the nucleus (Fig. 6b). These findings confirmed that TCM5 was localized to the chloroplast.

The Transcript Expressions of Functionally Related Genes in the *tcm5* Mutants

We examined the transcript levels of genes associated with Chl biosynthesis, photosynthesis and chloroplast development (*PORA, HEMA, CAO1, YGL1, cab1R, rbcL, rbcS, psaA, psbA, LhcpII,V1, V2, V3, OsRpoTp, OsPoLp1, rpoB, 23SrRNA, rps7, rps21, FtsZ*) in the *tcm5* mutants. Under high temperatures, the transcript levels of all tested genes for Chl biosynthesis (Fig. 7a), such as glutamyl tRNA reductase (*HEMA*) and chlorophyllide a oxygenase (*CAO1*), NADPH:protochlorophyllide oxidoreductase (*PORA*) and Chl synthase(*YGL1*), were obviously reduced in*tcm5* mutants, compared with WT plants, consistent with the decreased of Chl content (Fig. 1e) and the albino phenotype (Fig. 1a). We also investigated the photosynthesis-related genes (Fig. 7b) *cab1R*, encoding the Chl a/b binding protein, *rbcL*, encoding the large subunit of ribulose-1,5-bisphosphate carboxylase (Rubisco), *rbcS*, encoding the small subunit of Rubisco, *psaA* and *psbA* for the reaction center polypeptides in photosystems and*LhcpII* encoding light harvesting complex protein in PS II, and found that all these genes were obviously suppressed in *tcm5* mutants especially PSII-associated *psbA* and *LhcpII*, consistent with the low photochemical efficiency and photosynthesis efficiency in *tcm5* mutants under high temperatures.

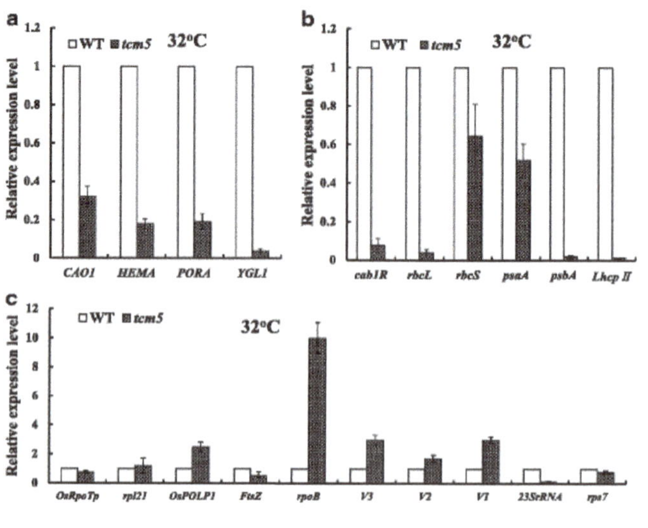

Figure 7: qPCR analysis of those genes associated with Chl biosynthesis, photosynthesis and chloroplast development at 32 °C;**a**, **b**, **c** Expression levels of genes related to Chl biosynthesis (**a**), photosynthesis (**b**) and chloroplast development (**c**) in WT and *tcm5* mutant in the 3rd leaves, respectively. The relative expression level of each gene in WT and mutant was analyzed by qPCR and normalized using the *OsActin* as an internal control. Data are means ± SD ($n=3$)

For chloroplast development-associated transcripts, we investigated the expressions of *V3 (RNRL)* encoding the large subunits of ribonucleotide reductase (Yoo et al. 2009), *V2*, encoding plastid/mitochondrial guanylate kinase (pt/mt GK) (Sugimoto et al. 2007), *V1(NUS1)*, encoding protein NUS1 (Kusumi et al. 2011), *OsRpoTp*, encoding NEP core subunits (Hiratsuka et al. 1989), and *rpoB* encoding one PEP core subunit (Inada et al. 1996; Kusumi et al. 2011) and *OsPoLP1*, encoding one plastidal DNA polymerase (Takeuchi et al. 2007), and *FtsZ*, encoding a component of the plastid division machinery (Vitha et al. 2001). Also we examined the transcript levels of ribosome-associated genes, *23S rRNA* (23S ribosomal RNA), encoding one component of the plastid translation machinery, and *rpl21* and *rps7*, encoding the large and small subunits ribosomal protein, respectively. Resultantly, with the exception of the not significant effects on *FtsZ, OsRpoTp, rps7* and *rpl2* in the *tcm5* mutants grown at 32 °C, other genes (*rpoB, V3, V2, V1, OsPoLP1,* and *23S rRNA*) were significantly affected (Fig. 7c), resulting in abnormal chloroplast (Fig. 2b). Overall, the *tcm5* mutation affected the transcriptional levels of some genes associated with not only Chl biosynthesis, photosynthesis but also the chloroplast development under high temperatures. By contrast, in the *tcm5* mutants grown at 20 °C, all transcripts of the affected genes at 32 °C (Fig. 8) were, at least, partly recovered to WT levels. Accordingly, the differentially expressed levels of so many key genes mentioned above under low or high temperatures might lead to the thermo-sensitivity for leaf-color in the *tcm5* mutant.

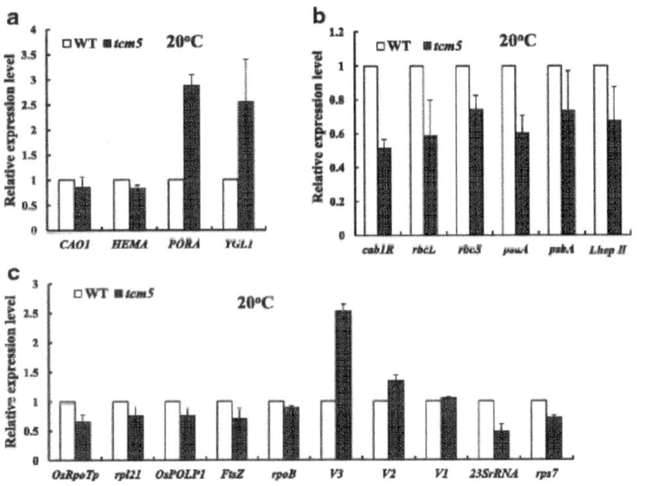

Figure 8: qPCR analysis of those genes associated with Chl biosynthesis, photosynthesis and chloroplast development at 20 °C.**a**, **b**, **c** Expression levels of genes related to Chl biosynthesis (**a**), photosynthesis (**b**) and chloroplast development (**c**) in WT and *tcm5* mutants in the 3rd leaves, respectively. The relative expression level of each gene in WT and mutant was analyzed by qPCR and normalized using the *OsActin* as an internal control. Data are means ± SD ($n = 3$)

DISCUSSION

In this study, we reported a rice nuclear-encoded Deg proteases protein *TCM5*, which is required for chloroplast development under high temperatures. Its disruption led to abnormal chloroplast development and albino phenotype of rice at high temperatures, resulting from abnormally expressed transcripts of those genes associated with chlorophyll biosynthesis, photosynthesis and chloroplast development. Our work demonstrated that *TCM5* plays an important role in chloroplast development under high temperatures.

TCM5 is Essential for Chloroplast Development under High Temperatures

The chloroplast is a semi-autonomous organelle, which contains about 100 genes, more than 3000 proteins function within it (Leister 2003). Thus, tightly coordinated gene expression between the plastid and nuclear genomes is essential for chloroplast development. The mutation of those genes could result in chlorophyll deficient/chloroplast defects in plants. At present, over 70 chlorophyll/chloroplast deficient mutants in rice have been identified (for review, see Kurata et al. 2005). Among them *v1, v2, v3, v5, v7, chs1, chs2, chs3, chs4, and chs5* mutants have been identified to be low temperature sensitive (Kusumi and Iba 2014). Recently, two genes, *TCD9* encoding α subunit of chaperonin protein 60 (Jiang et al. 2014) and *OsV4* encoding a PPR protein (Gong et al. 2014), essential for chloroplast development at low temperature were identified in rice. In contrast, six high-temperature sensitive rice mutants were reported, of which five mutants (Yatou and Cheng 1989) exhibited the chlorotic type over 35 °C, and *cde1(t)* mutant (Liu et al. 2007) displayed the chlorotic type over 26 °C, most likely resulted from the mutation of*OsGluRS*. In this study, the *tcm5* mutants displayed the abnormal chloroplast and even could not survive after the 4–5 leaf stage at 32 °C, but are normal under 20 °C. Judging from these results, there exists different shift sensitive-temperatures and molecular bases for chloroplast development under high temperatures in rice. In addition, the impaired chloroplasts (Fig. 2b) and the abnormal transcripts of genes for chloroplast development (Fig. 7) were observed in *tcm5* mutants at high temperatures, and those affected genes under high temperatures, at least, in part could recover to the comparable levels to WT plants at low temperatures (Fig. 8). These results can explain the changes of chloroplast development and leaf color between 20 and 32 °C (Figs. 1 and 2). Additionally, the highly induced mRNA accumulation (Fig. 3f) of *TCM5* at high temperatures showed that *TCM5* was temperature-regulated gene, consistent with the previous results that Deg proteases are required for the survival of *Escherichia coli* at high temperatures (Lipinska et al. 1988, 1990; Strauch and Beckwith 1988). Our study showed that the *TCM5* was constitutively expressed with high levels of transcripts in all green tissues (Fig. 5) and TCM5 protein distributes throughout the entire chloroplast (Fig. 6a), showing the functions of TCM5 in chloroplast. Also, the lethality of *tcm5*mutants may be due to the block of chloroplast development under high temperatures (Fig. 1b). These results strongly confirm the notion that rice *TCM5* is needed for chloroplast development under high temperatures. However, the reasons why the abnormal chloroplast/albino phenotypes of rice, such as *tcm5* mutant in this study and *cde1(t)* mutant, occur under solely higher temperatures have not been well documented (Yatou and Cheng1989; Liu et al. 2007). A possible explanation in this study is that the *TCM5* function possibly is not prerequisite under lower temperatures, but is essential under higher temperatures for chloroplast development. This explanation was strongly supported by the results from transcriptional analysis (Fig. 3f) that (a): all low expressions and the existence of no discriminated difference at 20 °C regardless of WT and *tcm5* mutant; (b): high expressions at 32 °C regardless of WT and *tcm5* mutant despite of the higher level in the mutant caused by feedback effects.

As stated above, the process of the chloroplast development from proplastids to mature chloroplasts can be divided into three steps (Jarvis and López-Juez 2013; Kusumi et al. 1997; Kusumi and Iba, 2014). The functions of*OsPOLP1*, *FtsZ* and *V3(RNRL)* encoding plastidal DNA polymerase, a component of plastid division machinery and the large subunits of ribonucleotide reductase respectively (Vitha et al. 2001; Takeuchi et al. 2007; Yoo et al. 2009), are known to involve in the first step. In addition, *OsRpoTp*, *V2*, and *rpoB*, encoding NEP, plastidal guanylate kinase, and PEP β core subunit, respectively (Hiratsuka et al. 1989; Kusumi et al. 2004; Sugimoto et al. 2007) are involved in the second step. In view of not significant decreases of transcripts of key genes mentioned above (Fig. 7), *TCM5*could not function in the first and second steps. In addition, the transcripts of all PEP-transcribed plastid genes (*cab1R*, *rbcL*, *psaA*, *psbA*, *LhcpII*) and nuclear gene *rbcS* involved in the third step were severely suppressed under high temperatures in *tcm5* mutants (Fig. 7), showing that *TCM5* functions the third step. Also, the significantly decreased accumulation of *23S rRNA* (Fig. 7c) in *tcm5* mutants under high temperatures indicated that *TCM5*affects the plastid translation machinery in chloroplasts. These results revealed that *TCM5* is necessary in the process of chloroplast development under high temperatures.

TCM5 may be Involved in PSII Function under High Temperatures

The Deg/HtrA family is present in nearly all organisms from bacteria to human and plants. It was known that Deg proteases are essential for the survival of *Escherichia coli* at high temperatures (Lipinska et al. 1988, 1990; Strauch and Beckwith 1988). The chloroplast-targeted Deg proteases were reported to be involved in the degradation of damaged photosynthetic proteins, especially the PSII reaction center D1 protein (for review, see Schuhmann et al.2012, Kato et al. 2012). In Arabidopsis, AtDeg1 acts as a chaperone, assisting in the assembly of PSII dimers and supercomplexes (Sun et al. 2010b). More interestingly, Arabidopsis *deg* mutants *deg5*, *deg7*, and *deg8* all showed decreased growth including chloroplast development, compared with the WT plants under high light, however, have no obvious differences under normal conditions, all AtDeg5, AtDeg7 and AtDeg8 participate in the repair of PSII after photo inhibition, and *AtDeg5*, *AtDeg7*, and *AtDeg8* were all high-light-dependent genes (Sun et al. 2010). Regrettably, any Arabidopsis *deg10* mutant whose AtDeg10 shares the highest homology to TCM5 has not reported yet, still less the function. Judging from the recovered normal chloroplast at low temperatures (<24 °C) in *tcm5* mutants, it could be concluded that *TCM5* is functionally required under high temperatures. The significant decreases of *Fv/Fm*, reflecting the maximum potential capacity of the PSII photochemical reactions, and *PSII*, representing the actual photosynthesis efficiency, in *tcm5* mutants at 32 °C and the similar degrees as WT plants at 20 °C showed that the malfunction of TCM5 barely affected the photosynthesis efficiency under high temperatures. In addition to the similar levels as WT plants at low temperatures (Fig. 8), the transcripts of *psbA* (>50-fold) and *LhcpII* (>80-fold), both encoding the PSII core proteins (D1, Lhcp II), were more significantly suppressed compared to those of *psaA* (<2-fold) encoding the PSI core protein (PsaA) in*tcm5* mutants at high temperatures (Fig. 7b). Thus, it could be reasonably speculated that the TCM5 disruption maybe hinder the PSII function, ultimately leading to the significantly decreased PSII activities (*Fv/Fm* and *PSII*) under high temperatures (32 °C), which was consistent with the previous results of high correlation between thermotolerance and photosystem II activity (Chen et al. 2014). In addition, Arabidopsis Deg7 (AtDeg7, At3g03380), with the low homology to TCM5 (16 % amino acid identity) (data not shown), involving in the primary cleavage of photodamaged D1, D2, CP47, and CP43 in PSII complexes, was reported to repair the PSII function under high-light conditions (Sun et al. 2010a). Considering that *TCM5* is high-temperature-dependent gene, different from high-light-dependent *AtDeg7* (Sun et al. 2010a), we might speculate that, at least, the mechanisms of temperature response are different among them. In addition, TCM5/OsDeg10 was solely located in thylakoid membrane (Fig. 6), not only different from AtDeg1, 5, 8 (on the stromal side) and, AtDeg2, 7(on the lumenal side), but also AtDeg10 (on both chloroplasts and mitochondrion) (Tanz et al. 2014), which clearly showed the different functions even between OsDeg10/TCM5 and AtDeg10.

Taken together, these results show that the *tcm5* mutation affected the third stage of chloroplast development under high temperatures, and the activation of PSII-associated genes is severely weakened, resulting in the suppression of chloroplast development, Chl-biosynthesis and photosynthesis, finally leading to the death under high temperatures. To our knowledge, this is the first experimental evidence that there exists a unique rice Deg/HtrA protease protein essential for chloroplast development under high temperatures. As for *TCM5*, how to participate in chloroplast development and maintain PSII function under high temperatures is needed in the further study.

CONCLUSIONS

The TCM5 encodes chloroplast-targeted Deg protease protein which is important for chloroplast development and the maintenance of PSII function and its disruption would lead to a defective chloroplast and affected expression levels of genes associated with chloroplast development and photosynthesis of rice plants under high temperatures.

METHODS

Plant Materials and Growth Conditions

The thermo-sensitive chlorophyll-deficient mutant, *tcm5*, used in this study, was obtained from *japonica* rice variety, Jiahua 1(WT), and induced by ^{60}Co gamma-radiation in 2006. The F_2 genetic mapping population was generated from a cross between Pei'ai 64S (*indica*) and the *tcm5* mutant. Rice plants were cultivated in rice paddy fields in Shanghai (summer-autumn season, temperate climate), China, following conventional methods. For phenotypic characterization, chlorophyll (Chl) and carotenoid content measurement and RNA extraction, the wild-type (WT) and*tcm5* seedlings were grown in growth chambers. In the F_2 population, thermo-sensitive Chl deficient albino plants were selected for genetic mapping.

Phenotype Characterization, Chlorophyll (Chl) and Carotenoid Measurements

The rice seeds were grown in the growth chambers with 12 h-dark/12 h-light at four temperatures (20, 24, 28, 32 °C) and the light intensity of 120 μmol of photons m^{-2} s^{-1}. The extracts were obtained from 100 mg of the third fresh leaves at the 3-leaf stage seedlings and homogenised in 10 ml of 100 % acetone. Spectrophotometric quantification was carried out in BECKMANCOULTER-DU720. Total Chl, Chla, Chlb and carotenoid contents were determined according to the methods of Arnon (1949) and Alan (1994). In 2010, WT and *tcm5* plants were grown in experimental rice fields of Shanghai Normal University by conventional methods. Leaf chlorophyll SPAD values by means of CHLOROPHYLL METER (SPAD-502, Minolta Co., Ltd, Japan), which can provide a simple, quick portable and non-destructive method for estimating leaf Chl content (Peng et al. 1993; Turner and Jund 1991; Dwyer et al. 1991) were measured every week from transplanting (summer) to heading (autumn). Finally, agronomic traits of rice plants were measured at maturity.

Chlorophyll Fluorescence Analysis and Transmission Electron Microscopy

Rice seedlings were grown at 20 °C or 32 °C in growth chambers. Chlorophyll fluorescence analyses for the third leaves at the 3-leaf stage were performed with a PAM-2000 portable chlorophyll fluorometer (MINI-PAM; Walz;http://walz.com). The variables *Fo*, *Fm*, *Fv*, the *Fv/Fm* ratio and *PSII* were measured and calculated according to Meurer et al. (1996) after seedlings were dark-adapted for 20 min. Meanwhile, transverse sections of top leaves were sampled from the 3-leaf-stage seedlings grown at 20 °C or 32 °C. Samples were fixed in 4 % glutaraldehyde buffer, 2.5 % glutaraldehyde and 1 % osmic acid phosphate buffer at 4 °C for 5 h after vacuum. After staining with uranyl acetate, tissues were further dehydrated in an ethanol series and finally embedded in Spurr's medium prior to ultrathin sectioning. Samples were stained again and examined with a Hitachi-7650 transmission electron microscope.

Positional Cloning of *TCM5*

For genetic analysis, F_2 population from the cross between *tcm5* and Pei'ai 64S was used for fine mapping of the*TCM5* locus. We adopted 53 SSR primers based on data in Gramene (http://www.gramene.org). New SSR, InDel and CAPS markers were developed based on the entire genomic sequences of the*japonica* Nipponbare variety (Goff et al. 2002) and the *indica* variety 9311 (Yu et al. 2002). The candidate genes' function and full length cDNA were acquired using TIGR (http://rice.plantbiology.msu.edu/cgi-bin/gbrowse/rice/) and KOME (http://cdna01.dna.affrc.go.jp/cDNA/index.html) respectively.

RT-PCR and Quantitative Real-time PCR

To understand the tissue expression pattern of *TCM5*, we extracted wild type RNA from young roots, young stems, young leaves, flag-leaves and young panicles with TRIzol Reagent (Invitrogen; http://www.invitrogen.com) and DNase I treated RNA using an RNeasy kit (Qiagen; http://www.qiagen.com) following the manufacturer's instructions. The first-strand cDNA was synthesized with the Revert-Aid first-strand cDNA synthesis kit (Toyobo;http://www.toyobo.co.jp) following the manufacturer's instructions. *OsActin* were internal controls.

For transcriptional expression analysis of *TCM5* and 20 genes (*PORA*, *CAO1*, *YGL1*, *cab1R*, *rbcL*, *rbcS*, *psaA*,*psbA*, *Lhcp II*, *V3 (RNRL)*, *V1*, *V2*, *OsRpoTp*, *OsPoLp1*, *rpoB*, *23SrRNA*, *rps7*, *rps21*, *FtsZ*) associated with Chl-biosynthesis, chloroplast development and photosynthesis, total RNA was obtained from the third fresh leaves of WT and *tcm5* seedlings grown at 20 °C or 32 °C. Quantitative real-time PCR amplifications were carried out with ABI7500 Real-Time PCR System (Applied Biosystems; http://www.appliedbiosystems.com), and the relative quantification of gene expression data was analyzed as described by Livak and Schmittgen (2001). The data set was normalized using *OsActin* as a control.

Complementation Analysis

For complementation analysis, cDNA clone (AK073199) plasmid vector (J0133023K08) (http://cdna01.dna.affrc.go.jp) carrying the full-length *TCM5* (*LOC_Os05g34460*) cDNA sequence, was purchased from National Institute of Agrobiological Sciences of Japan in 2011. Then, the 2426 bp DNA fragment, carrying the full-length CDS (1845 bp) of *TCM5*, was amplified using the primers 5′- CCGGAATTCGTCGACCTCTAT -3′ (*Eco*RI) and 5′-TGGCCGGATCCGGACTGTTT-3′ (*BamHI*). The underlined sequences represent cleavage sites of the enzymes. Then, the resulted DNA fragment was cloned into the pMD18-T vector (TaKaRa). Following sequence verification, the fragments were cloned into the pCAMBIA1301 binary vector (CAMBIA; www.cambia.org.au). Finally, the pCAMBIA1301:CaMV35S:TCM5-cDNA plasmids were transferred into *Agrobacterium tumefaciens* EHA105 and introduced into the *tcm5* mutants by Agrobacterium-mediated transformation (Hiei et al. 1994). In this study, transgenic plants were generated from recovered resistant calli by the procedure of Lu et al. (2001).

Subcellular Localization of GFP Proteins

To investigate the subcellular localization of TCM5 protein, the cDNA fragment encoding the N-terminal region (amino acids 1–135) of TCM5 was amplified from cDNA clone (AK073199) using the primer pairs (pF 5′-GGGGTACCATGCTCGCCTCCGTCC-3′ (*Kpn*I) and pR 5′- CGGCCATGGCGAGTCCAGCGCCAGCT-3′ (*Nco*I) the underlined sequence represents cleavage sites of enzymes) and ligated into the pMON530-GFP vector, in frame with GFP. The resultant pMON530-TCM5-GFP plasmids were introduced into tobacco (*Nicotiana tabacum*) leaves and co-cultured at 25 °C for 2 days. At the same time, the pMON530-GFP empty carrier was used as a control. Then the GFP fluorescences in tobacco cells were observed using a Zeiss confocal laser scanning microscope (LSM 5 PASCAL; http://www.zeiss.com).

Sequence and Phylogenetic Analysis

Gene prediction and structure analysis were performed using TIGR (http://rice.plantbiology.msu.edu/cgi-bin/gbrowse/

rice/).The chloroplast transit peptides (CTP) was predicted by the ChloroP program (http://www.cbs.dtu.dk/services/TargetP/). The full-length amino acid sequences of the TCM5 and homologs identified via BLAST search were aligned with the MUSCLE tool (Edgar 2004) using the default parameters. The phylogenetic tree was constructed and tested by MEGA6 (http://www.megasoftware.net) based on the neighbor-joining method.

ACKNOWLEDGMENTS

We sincerely thank Dr. Youbin Xiang (Stowers Institute for Medical Research, USA) for her critical reading of the manuscript. We thank DNA Bank of National Institute of Agrobiological Sciences, Japan, which kindly offered the rice cDNA clone (AK073199). We are grateful to Dr. Zhongnan Yang from Shanghai Normal University for her skillful technical assistances. The project was financially supported by Natural Science Foundation of China (No. 30971552), Shanghai Municipal Education Commission (No. 09YZ167, No.J50401), Shanghai Municipal Science and Technology Commission (No. 9391912300, 10DZ2271800, 12ZR1422000), and the Scientific Program of Shanghai Normal University (SK20159).

AUTHORS' CONTRIBUTIONS

DL and YD provided the mutant rice seeds. DL, YD, KZ, JC and JZ generated F_2 and F_3 seeds for genotyping and phenotyping. KZ, JZ, DL, HZ and JC performed the experiments of phenotype assays and molecular analysis. KZ, DL, YD, ST and JX designed and discussed the research. DL, YD, KZ and JX wrote the manuscript. All authors approved the manuscript.

REFERENCES

1. Alan RW (1994) The spectral determination of chlorophylls a and b, as well as total carotenoids, using various solvents with spectrophotometers of different resolution. J Plant Physiol 144:307–311
2. Arnon DI (1949) Copper enzymes in isolated chloroplasts. polyphenoloxidase in beta vulgaris. Plant Physiol 24:1–15
3. Chen K, Sun XY, Amombo E, Zhu Q, Zhao ZG, Liang Chen L, Xu QG, Fu JM (2014) High correlation between thermotolerance and photosystem II activity in tall fescue. Photosynth Res 122:305–314
4. Clausen T, Southan C, Ehrmann M (2002) The HtrA family of proteases: implications for protein composition and cell fate. Mol Cell 10:443–455
5. Dwyer LM, Tollenaar M, Houwing L (1991) A nondestructive method to monitor leaf greenness in corn. Can J Plant Sci 71:505–509
6. Edgar RC (2004) MUSCLE: multiple sequence alignment with high accuracy and high throughput. Nucleic Acids Res 32:1792–1797
7. Emanuelsson O, Nielsen H, Brunak S, Heijn GV (2000) Predicting subcellular localization of proteins based on their N-terminal amino acid sequence. J Mol Biol 300:1005–1016
8. Garcia-Lorenzo M, Sjödin A, Jansson S, Funk C (2006) Protease gene families in Populus and Arabidopsis. BMC Plant Biol 6:30
9. Goff SA, Ricke D, Lan TH, Presting G, Wang R, Dunn M, Glazebrook J, Sessions A, Oeller P, Varma H, Hadley D, Hutchison D, Martin C, Katagiri F, Lange BM, Moughamer T, Xia Y, Budworth P, Zhong J, Miguel T, Paszkowski U, Zhang S, Colbert M, Sun WL, Chen L, Cooper B, Park S, Wood TC, Mao L, Quail P, Wing R, Dean R, Yu Y, Zharkikh A, Shen R, Sahasrabudhe S, Thomas A, Cannings R, Gutin A, Pruss D, Reid J, Tavtigian S, Mitchell J, Eldredge G, Scholl T, Miller RM, Bhatnagar S, Adey N, Rubano T, Tusneem N, Robinson R, Feldhaus J, Macalma T, Oliphant A, Briggs S (2002) A draft sequence of the rice genome (*Oryza sativa* L. ssp. *japonica*). Science 296:92–100
10. Gong X, Su Q, Lin D, Jiang Q, Xu J, Zhang J, Teng S, Dong Y (2014) The rice *OsV4* encoding a novel pentatricopeptide repeat protein is required for chloroplast development during the early leaf stage under cold stress. J Integr Plant Biol 56:400–410
11. Hajdukiewicz PT, Allison LA, Maliga P (1997) The two RNA polymerases encoded by the nuclear and the plastid compartments transcribe distinct groups of genes in tobacco plastids. EMBO J 16:4041–4048
12. Helm M, Luck C, Prestele J, Hierl G, Huesgen PF, Fröhlich T, Arnold GJ, Adamska I, Görg A, Lottspeich F (2007) Dual specificities of the glyoxysomal/peroxisomal processing protease Deg15 in higher plants. Proc Natl Acad Sci U S A 104:11501–11506
13. Hiei Y, Ohta S, Komari KT (1994) Efficient transformation of rice (*Oryza sativa* L.) mediated by Agrobacterium and sequence analysis of the boundaries of the T-DNA. Plant J 6:271–282
14. Hiratsuka J, Shimada H, Whittier R, Ishibashi T, Sakamoto M, Mori M, Kondo C, Honji Y, Sun CR, Meng BY (1989) The complete sequence of the rice (*Oryza sativa*) chloroplast genome: intermolecular recombination between distinct tRNA genes accounts for a major plastid DNA inversion during the evolution of the cereals. Mol Gen Gene 217:185–194
15. Huesgen PF, Schuhmann H, Adamska I (2005) The family of Deg proteases in cyanobacteria and chloroplasts of higher plants. Physiol Plant 123:413–420
16. Inada H, Kusumi K, Nishimura M, Iba K (1996) Specific expression of the chloroplast gene for RNA polymerase (*rpoB*) at an early stage of leaf development in rice. Plant Cell Physiol 37:229–232
17. Itzhaki H, Naveh L, Lindahl M, Cook M, Adam Z (1998) Identification and characterization of DegP, a serine protease associated with the luminal side of the thylakoid membrane. J Biol Chem 273:7094–7098
18. Jarvis P, López-Juez E (2013) Biogenesis and homeostasis of chloroplasts and other plastids. Nat Rev Mol Cell Biol 14:787–802
19. Jiang Q, Mei J, Gong XD, Xu JL, Zhang JH, Teng S, Lin D, Dong Y (2014) Importance of the rice *TCD9* encoding alpha subunit of chaperonin protein60 (Cpn60α) for the chloroplast development during the early leaf stage. Plant Sci 215–216:172–179

20. Kato Y, Sakamoto W (2010) New insights into the types and function of proteases in plastids. Int Rev Cell Mol Biol 280:185–218

21. Kato Y, Sun X, Zhang LX, Sakamoto W (2012) Cooperative D1 degradation in the photosystem II repair mediated by chloroplastic proteases in Arabidopsis. Plant Physiol 159:1428–1439

22. Kolmar H, Waller P, Sauer R (1996) The DegP and DegQ periplasmic endoproteases of *Escherichia coli*: specificity for cleavage sites and substrate conformation. J Bacteriol 178:5925–5929

23. Krause GH, Weis E (1991) Chlorophyll fluorescence and photosynthesis: the basics. Annu Rev Plant Physiol Plant Mol Biol 42:313–349

24. Kurata N, Miyoshi K, Nonomura KI, Yamazaki Y, Ito Y (2005) Rice mutants and genes related to organ development, morphogenesis and physiological traits. Plant Cell Physiol 46:48–62

25. Kusumi K, Iba K (2014) Establishment of the chloroplast genetic system in rice during early leaf development and at low temperatures. Front Plant Sci 5:386–391

26. Kusumi K, Mizutani A, Nishimura M, Iba K (1997) A virescent gene *V1* determines the expression timing of plastid genes for transcription/translation apparatus during early leaf development in rice. Plant J 12:1241–1250

27. Kusumi K, Yara A, Mitsui N, Tozawa Y, Iba K (2004) Characterization of a rice nuclear-encoded plastid RNA polymerase gene *OsRpoTp*. Plant Cell Physiol 45:1194–1201

28. Kusumi K, Sakata C, Nakamura T, Kawasaki S, Yoshimura A, Iba K (2011) A plastid protein NUS1 is essential for build-up of the genetic system for early chloroplast development under cold stress conditions. Plant J 68:1039–1050

29. Leister D (2003) Chloroplast research in the genomic age. Trends Genet 19:47–56

30. Lipinska B, Sharma S, Georgopoulos C (1988) Sequence analysis and regulation of the HtrA gene of *Escherichia coli*: a sigma 32-independent mechanism of heat-inducible transcription. Nucleic Acids Res 16:10053–10067

31. Lipinska B, Zylicz M, Georgopoulos C (1990) The HtrA (DegP) protein, essential for *Escherichia coli* survival at high temperatures, is an endopeptidase. J Bacteriol 172:1791–1797

32. Liu W, Fu Y, Hu G, Si H, Zhu L, Wu C, Sun Z (2007) Identification and fine mapping of a thermo-sensitivechlorophyll deficient mutant in rice (*Oryza sativa* L.). Planta 226:785–795

33. Livak KJ, Schmittgen TD (2001) Analysis of relative gene expression data using real time quantitative PCR and the method. Methods 25:402–408

34. Lu HJ, Zhou XR, Gong ZX, Upadhyaya NM (2001) Generation of selectable marker-free transgenic rice using double right-border (DRB) binary vectors. Aust J Plant Physiol 28:241–248

35. Meurer J, Meierhoff K, Westhoff P (1996) Isolation of high-chlorophyll fluorescence mutants of *Arabidopsis thaliana* and their characterisation by spectroscopy, immunoblotting and northern hybridisation. Planta 198:385–396

36. Pendle AF, Clark GP, Boon R, Lewandowska D, Lam YW, Andersen J, Mann M, Lamond AI, Brown JW, Shaw PJ (2005) Proteomic analysis of the Arabidopsis nucleolus suggests novel nucleoli functions. Moll Boil Cell 16:260–269

37. Peng S, Garcia FV, Lasa RC, Classman KG (1993) Adjustment for specific leaf weight improves chlorophyll meter's estimate of rice leaf nitrogen concentration. Agron J 85:987–990

38. Pfalz J, Pfannschmidt T (2013) Essential nucleoid proteins in early chloroplast development. Trends Plant Sci 18:186–194

39. Santis-Maciossek GD, Kofer W, Bock A, Schoch S, Maier RM, Wanner G, Rüdiger W, Koop HU (1999) Targeted disruption of the plastid RNA polymerase genes *rpoA*, B and C1: molecular biology, biochemistry and ultrastructure. Plant J 18:477–489

40. Schuhmann H, Huesgen PF, Gietl C, Adamska I (2008) The DEG15 serine protease cleaves peroxisomal targeting signal 2-containing proteins in *Arabidopsis thaliana*. Plant Physiol 148:1847–1856

41. Schuhmann H, Huesgen PF, Adamska I (2012) The family of Deg/HtrA proteases in plants. BMC Plant Biol 12:52

42. Strauch KL, Beckwith J (1988) An *Escherichia coli* mutation preventing degradation of abnormal periplasmic proteins. Proc Natl Acad Sci U S A 85:1576–1580

43. Sugimoto H, Kusumi K, Noguchi K, Yano M, Yoshimura A, Iba K (2007) The rice nuclear gene, *VIRESCENT2*, is essential for chloroplast development and encodes a novel type of guanylate kinase targeted to plastids and mitochondria. Plant J 52:512–527

44. Sun X, Fu T, Chen N, Guo J, Ma J, Zou M, Lu C, Zhang L (2010a) The stromal chloroplast Deg7 protease participates in the repair of photosystem II after photoinhibition in Arabidopsis. Plant Physiol 152:1263–1273

45. Sun X, Ouyang M, Guo J, Ma J, Lu C, Adam Z, Zhang L (2010b) The thylakoid protease Deg1 is involved in photosystem-II assembly in *Arabidopsis thaliana*. Plant J 62:240–249

46. Takeuchi R, Kimura S, Saotome A, Sakaguchi K (2007) Biochemical properties of a plastidial DNA polymerase of rice. Plant Mol Biol 64:601–611

47. Tanz SK, Castleden I, Hooper CM, Small I, Millar AH (2014) Using the SUBcellular database for Arabidopsis proteins to localize the Deg protease family. Front Plant Sci 5:396

48. Tripathi LP, Sowdhamini R (2006) Cross genome comparisons of serine proteases in Arabidopsis and rice. BMC Genomics 7:200

49. Turner FT, Jund MF (1991) Chlorophyll meter to predict nitrogen topdress requirement for semidwarf rice. Agron J 83:926–928

50. Vitha S, Mcandrew RS, Osteryoung KW (2001) *FtsZ* ring formation at the chloroplast division site in plants. J Cell Biol 153:111–120

51. Walsh NP, Alba BM, Bose B, Gross CA, Sauer RT (2003) OMP peptide signals initiate the envelope-stress response by activating DegS protease via relief of inhibition mediated by its PDZ domain. Cell 113:61–71

52. Yatou O, Cheng XY (1989) Temperature sensitive chlorophyll mutations. Rice Genet Newsletter 6:131

53. Yoo Y, Cho SH, Sugimoto H, Li J, Kusumi K, Koh H, Iba K, Paek N (2009) Rice *Virescent3* and *Stripe1* encoding the largeand small subunits of ribonucleotide reductase are required for chloroplast biogenesis during early leaf development. Plant Physiol 150:388–401

54. Yu J, Hu S, Wang J, Wong GK, Li S, Liu B, Deng Y, Dai L, Zhou Y, Zhang X, Cao M, Liu J, Sun J, Tang J, Chen Y, Huang X, Lin W, Ye C, Tong W, Cong L, Geng J, Han Y, Li L, Li W, Hu G, Huang X, Li W, Li J, Liu Z, Li L et al (2002) A draft sequence of the rice genome (*Oryza sativa* L. ssp. *indica*). Science 296:79–92

CHAPTER 15

METABOLIC PREDICTION OF IMPORTANT AGRONOMIC TRAITS IN HYBRID RICE (ORYZA SATIVA L.)

Zhiwu Dan, Jun Hu, Wei Zhou, Guoxin Yao, Renshan Zhu, Yingguo Zhu, and Wenchao Huang

INTRODUCTION

The continuously growing global population and changing climate have pushed crop breeders to find effective breeding strategies for meeting food and energy demands. Although hybrid crops have made tremendous contributions to yield improvements over the past several decades, breeding new elite hybrid combinations is urgently required to guarantee food security. Quantitative hybridization trials and hybrid performance evaluation, which are time-consuming and laborious, form the foundation of present hybrid breeding programs. Breeding procedures are even tougher for self-pollinated plants such as rice and wheat compared with cross-pollinated plants.

Prediction of hybrid performance based on parental information appears to be a practicable method to increase breeding efficiency. However, parental phenotypes have exhibited low accuracy for the prediction of hybrid performance[1,2]. Additionally, isozymes have proven to be unreliable for the prediction of F_1 yields and heterosis[3]. Then, parental genetic distances based on AFLP (Predictive abilities: 0–0.97, 0.161–0.699 and 0.06–0.8, respectively)[4,5,6], InDel (0.049–0.21)[7], RAPD (0.069–0.785)[8], RFLP (−0.028–0.773)[9], SNP (0.26–0.56 and 0.3–0.46)[1,10], and SSR (0.06–0.82, 0.069–0.785, and −0.028–0.773, respectively)[6,8,9] DNA markers have been widely used to predict hybrid phenotypes, heterosis or combing ability in maize (*Zea Mays* L.)[5,6,10], rice (*Oryza Sativa* L.)[7,8,9], sunflowers (*Helianthus annuus* L.)[4], and *Arabidopsis thaliana*[1]. Some prediction methods based on DNA markers produced no significant correlation or low predictive ability between parental genetic distance and hybrid performance[9,10]. Meanwhile, DNA markers were found to be appropriate for the prediction of specific traits[5,6,9] and had high predictive ability for populations consisting of genetically related parental lines[4,8]. Additionally, parental transcriptome-based predictions of hybrids showed high predictive ability for intra-pool crosses but not inter-pool crosses[11]. Generally speaking, although various types of biomarkers, populations and mathematical models have been tested in various species, problems of low predictive ability or limited application ranges of the prediction approaches remain. And worldwide, numerous studies of hybridization are still performed every year.

The metabolome has manifested potential roles in facilitating crop breeding strategies over the past decades[12,13]. The results from a tomato IL population showed that 73% of the total analysed metabolites were significantly associated with the whole-plant phenotypes[14]. *Arabidopsis thaliana* biomass could be described as a function of metabolic compositions with high predictive power[15,16,17]. Moreover, metabolite profiling of a rice recombinant inbred line population and diverse accessions showed promise in bridging the gap between the genome and phenome[18,19]. Importantly, parental metabolites as biomarkers have been tested for the prediction of traits such as yield in maize[20], biomass and tolerance in *Arabidopsis thaliana*[21,22] and postharvest quality traits in potatoes (*Solanum tuberosum* L.)[23].

In the present study, a complete diallel cross population was built with eighteen rice inbred lines as the parents. Three polygenic traits—yield per plant, heading date and maturation stage plant height—were predicted in the hybrids using 525 metabolites with the partial least square regression method.

Citation: Dan, Z. et al. Metabolic prediction of important agronomic traits in hybrid rice (Oryza sativa L.). Sci. Rep. 6, 21732; doi: 10.1038/srep21732 (2016).

The predictive abilities ranged from 0.858 to 0.977 for the three traits. The population structure and cytoplasmic effects had slight influences on the predictive ability.

RESULTS

Population structure and hybrid performance

Eighteen rice inbred lines were chosen as parental lines for the complete diallel cross design. All the parental lines were divided into equal *indica* (F_i values from 0.39 to 1) and *japonica* (F_i values from 0 to 0.33) groups based on the InDel marker estimating method 24. Principal component analysis (PCA) was also applied to the parental phenotypic and metabolic data to discriminate between the parental lines. PCA of the parental phenotypic performance showed that three *indica* varieties were divided into the *japonica* group and three *japonica* varieties were in the *indica* group (Fig. 1a). At the metabolic level, a total of 525 analytes were detected in all eighteen inbred lines. Unexpectedly, PCA for the parental metabolic parameters produced a similar grouping result as the consequences of the InDel markers and was also highly similar to the dendrogram (Fig. 1b,c).

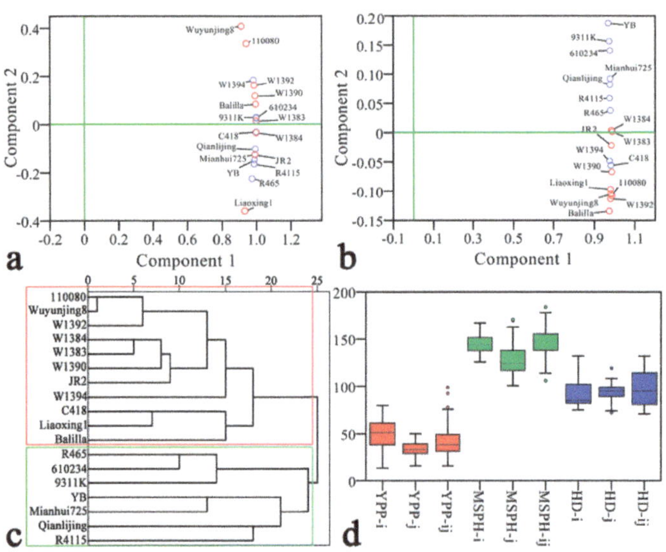

Figure 1: Population structure and hybrid phenotypes in three subgroups. (**a**) Principal component analysis (PCA) of the eighteen rice inbred lines with data for three agronomic traits. Three *indica* varieties were grouped into the *japonica* group, and three *japonica* varieties were grouped into the *indica* group. Red circles indicate *japonica* varieties, and blue circles indicate *indica* varieties. Solid green lines represent the zero values for each component. (**b**) PCA of the eighteen inbred lines with metabolite profiling data. All 525 analytes were used in the PCA. Two *indica* varieties were grouped into the *japonica* group. (**c**) Dendrogram of the eighteen inbred lines. All eighteen inbred lines were divided into two clear groups. Consistent with the grouping result in (**b**), only two *indica* varieties were in the *japonica* group. The red solid box indicates the *japonica* group, and the green solid box indicates the *indica* group. (**d**) Hybrid phenotypes in three subgroups. The yield per plant (YPP), maturation stage plant height (MSPH) and heading date (HD) were evaluated.

Three important agronomic traits were measured: yield per plant (YPP), maturation stage plant height (MSPH), and heading date (HD). Prior to the analysis, the hybrids were divided into an *indica-indica* group (i-group), a *japonica-japonica* group (j-group) and an *indica-japonica*group (ij-group) based on the group to which the corresponding parental lines belonged. For the hybrid phenotypes (trait values *per se*), the highest mean YPP was in the i-group, the lowest was in the j-group, and the ij-group was between the i- and j-group (Fig. 1d). For MSPH and HD, the highest means were observed for the ij-group, which was consistent with the phenomenon that hybridization between inter-subspecific rice has stronger biomass heterosis (increased plant height) and a later heading date. Box plots depicting the relative low-parent heterosis (LPH), mid-parent heterosis (MPH), better-parent heterosis (BPH) and specific combining ability (SCA) for the three traits. For YPP, the hybrid trait values *per se* had the closest relationship with SCA. Trait values of MSPH *per se* had the closest relationship with BPH, and HD was most closely associated with MPH.

Predictions of hybrid performance based on parental traits and genetic distance

First, the sum trait values of the parents (sum), the differences between the parents (difference) and the ratios of the parents (ratio) were used as predictive variables. The highest correlation coefficient (0.714) was obtained between the HD and the sum values of the parents. After performing linear regression analysis of the sum HD values of the parents and the hybrid HD values, the predicted HD values and measured true values displayed dramatic deviances. Therefore, the parental traits were not appropriate for the prediction of hybrid performance.

Second, to evaluate whether genetic distance based on parental metabolic data was suitable for predicting hybrid performance, squared Euclidean distances based on all 525 analytes were calculated to analyse the correlations. The Pearson correlation coefficients between genetic distance and hybrid performance suggested that, although genetic distance was significantly correlated with some traits (such as YPP and MSPH), the predictive abilities were quite low, and some traits were not closely correlated with genetic distance.

Stepwise linear regression and partial least square (PLS) regression with metabolic data and hybrid performance

Next, the sums, differences and ratios of the parents' relative metabolite levels were calculated, and stepwise linear regression was used to identify the appropriate metabolic predictors. Furthermore, the predictive abilities of different traits varied widely, and they were not sufficiently high for most traits.

Then, partial least square regression was used to create a prediction model with metabolic data of the three traits. In Fig. 2a, 17 latent factors were extracted for all the traits using the sums and differences of the parents' relative metabolite levels as variables. Unfortunately, the highest R^2 value obtained was 0.6, and some hybrid traits could not be predicted using these parental variables.

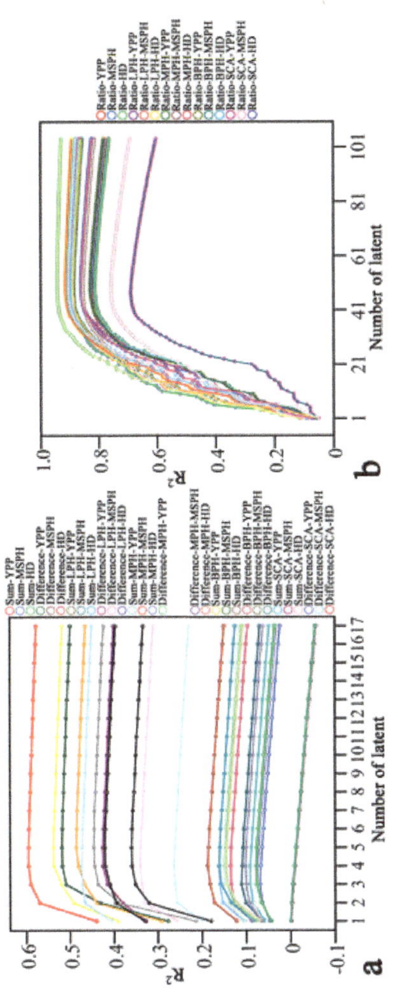

Figure 2: Relationships between PLS latent numbers and R^2 values. (**a**) Sums and differences between parental relative metabolite levels were used as predictive variables for the hybrid phenotypes, relative heterosis and specific combining ability in PLS regression. The R^2 values varied largely between different traits. (**b**) Ratios of parental relative metabolite levels were used for hybrid performance prediction. R^2 values of most traits were highest (above 0.8) at approximately 50 latents. YPP=Yield per plant, MSPH=Maturation stage plant height, HD=Heading date, LPH=Low-parent heterosis, MPH=Mid-parent heterosis, BPH=better-parent heterosis, SCA=Specific combining ability.

Surprisingly, however, high R^2 values were obtained when the ratios of parental metabolic data were used as predictive variables (Fig. 2b). For this analysis, 104 to 107 latent factors were extracted from the metabolic data; most of the traits exhibited the highest R^2 values at approximately 50 latent factors. Then, the number of latent factors at the top R^2 value for each trait was fixed to obtain the values of variable importance in the projection (VIP) of each variable. Because Latent Factor 1 explained the largest proportion of the variance, the VIP values of Latent Factor 1 were reordered to reduce the number of variables. After excluding variables whose VIP values were smaller than 1, the remaining variables were used for PLS regression. Finally, although more than 300 variables were excluded from the regression model, and only slight decreases were observed in R^2 values compared with the values obtained using all 525 metabolites (Table 1). Among the 15 predicted traits, LPH-YPP used the fewest (149) variables and LPH-HD used the most variables (196).

Table 1: **R^2 values of PLS regression with all 525 variables and with variables** with VIP values larger than 1

Traits	Latent number*	R^2	Variable number	Latent number†	R^2
YPP	52	0.824	184	70	0.807
MSPH	51	0.890	161	77	0.871
HD	57	0.944	193	78	0.938
LPH-YPP	49	0.860	149	66	0.828
LPH-MSPH	52	0.893	181	66	0.884
LPH-HD	54	0.916	196	78	0.905
MPH-YPP	51	0.814	162	68	0.784
MPH-MSPH	49	0.855	181	73	0.838
MPH-HD	53	0.896	182	72	0.884
BPH-YPP	52	0.829	166	69	0.810
BPH-MSPH	52	0.887	164	78	0.869
BPH-HD	55	0.905	178	73	0.896
SCA-YPP	48	0.691	183	64	0.661
SCA-MSPH	50	0.758	159	79	0.704
SCA-HD	53	0.861	171	76	0.844

*Latent number for different traits of PLS analysis with all 525 variables. †Latent number for different traits of PLS analysis with variables with VIP values larger than 1. YPP=Yield per plant, MSPH=Maturation stage plant height, HD=Heading date, LPH=Relative low-parent heterosis, MPH=Relative mid-parent heterosis, BPH=Relative better-parent heterosis, SCA=Specific combining ability.

Considering that 149–196 is still a quite large number of variables and might result in overfitting22, the VIP values of each variable were reordered again in the PLS regression model to minimize the amount of variables. Unfortunately, the R^2 values decreased dramatically.

Metabolic prediction of agronomic traits by PLS regression

The coefficient of each variable in the PLS regression results was used to build the equations. Then, a predicted value was assigned for every trait in each hybrid. The relationship between the observed YPP (Fig. 3a), MPH-YPP (Fig. 3b), BPH-YPP (Fig. 3c), and SCA-YPP (Fig. 3d) values and the predicted values demonstrated the high accuracy for this metabolite-based prediction method. The lowest predictive ability was 0.858 with SCA-YPP and the highest was 0.924 with YPP. High predictive abilities were also achieved through the PLS regression approach for the remaining traits (Table 2).

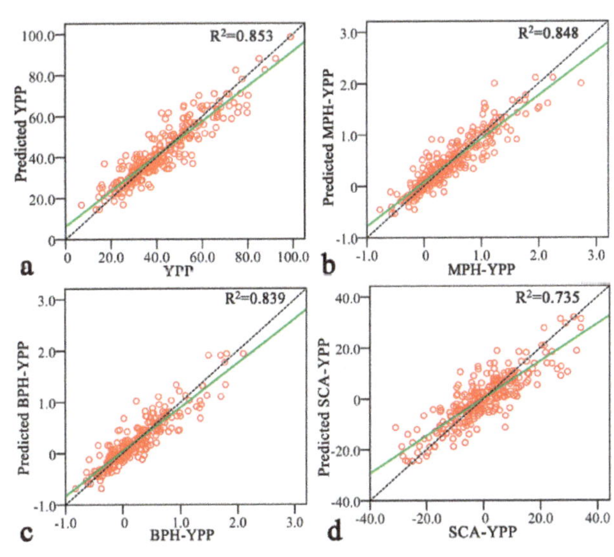

Figure 3: Observed and predicted values of YPP, MPH-YPP, BPH-

YPP, and SCA-YPP. (**a**) Relationships between observed yield per plant and predicted yield per plant. Predicted values were calculated with the equations based on variable coefficients in the PLS regression results. The horizontal axis represents the observed values, and the vertical axis represents the predicted values. The green solid line represents the total fit line, and the black dotted line is $y=x$. (**b–d**) Relationships between the observed relative mid-parent heterosis, better-parent heterosis, specific combining ability of YPP and the corresponding predicted values. YPP=Yield per plant, LPH=Low-parent heterosis, MPH=Mid-parent heterosis, BPH=better-parent heterosis, SCA=Specific combining ability.

Table 2: Predictive abilities of the whole population and three subgroups for the three polygenic traits

Traits	Whole population	i-group	j-group	ij-group
YPP	0.924	0.849	0.806	0.948
MSPH	0.951	0.911	0.901	0.956
HD	0.977	0.987	0.937	0.977
LPH-YPP	0.931	0.927	0.906	0.940
LPH-MSPH	0.954	0.931	0.897	0.965
LPH-HD	0.964	0.963	0.894	0.968
MPH-YPP	0.921	0.883	0.911	0.940
MPH-MSPH	0.938	0.933	0.856	0.948
MPH-HD	0.954	0.963	0.838	0.957
BPH-YPP	0.916	0.848	0.926	0.935
BPH-MSPH	0.949	0.933	0.904	0.949
BPH-HD	0.931	0.953	0.948	0.913
SCA-YPP	0.858	0.783	0.873	0.898
SCA-MSPH	0.885	0.940	0.736	0.868
SCA-HD	0.940	0.963	0.916	0.927

YPP=Yield per plant, MSPH=Maturation stage plant height, HD=Heading date, LPH=Relative low-parent heterosis, MPH=Relative mid-parent heterosis, BPH=Relative better-parent heterosis, SCA=Specific combining ability.

Because the complete diallel cross population consisted of rice inbred lines ranging from typical *indica* to typical *japonica*, the predictive abilities might vary with different hybridization groups. Thus, the predictive abilities for the i-group, j-group and ij-group were calculated to validate the stability of the prediction model (Table 2). As shown inFig. 4a,b, the predictive abilities of YPP in the i-group and j-group were 0.849 and 0.806, respectively. Surprisingly, the predictive ability of the ij-group was as high as 0.948 (Fig. 4c). Moreover, the predictive abilities of relative heterosis and SCA of YPP were all the highest for the ij-group, thereby showing a promising approach for exploiting inter-subspecific heterosis25.

Finally, the cytoplasm from the reciprocal parental lines might have different influences on hybrid performance; these differences might be obstacles to applying the metabolic prediction strategy in screening for potential sterile and restorer lines. Hence, the whole population was divided into two sets of reciprocal hybrids to test whether cytoplasmic effects affected the predictive ability. The results demonstrated that different cytoplasms had weak influences on the predictive ability.

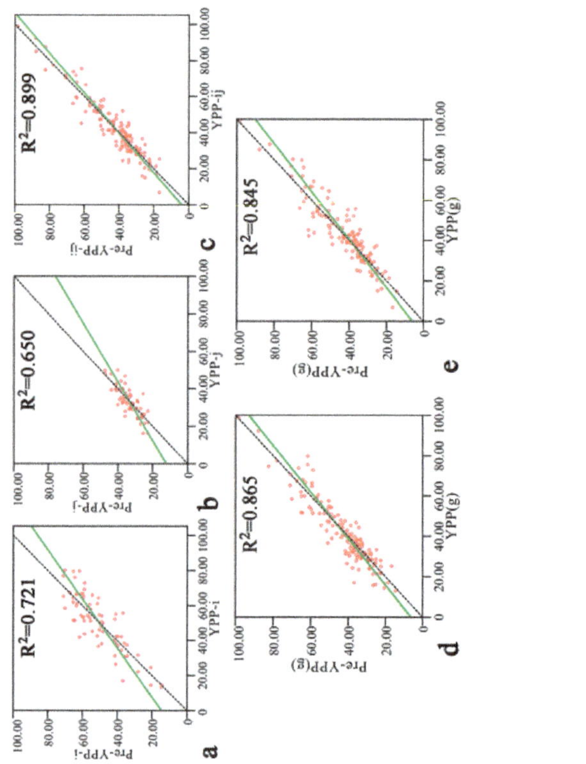

Figure 4: Observed and predicted YPP values in three subgroups and two reciprocal groups. (**a–c**) Relationships between observed yield per plant and predicted values in the i, j and ij three subgroups. The entire hybrid population was divided into three subgroups based on whether a parental line was *indica* or *japonica* (Fig. 1b). The predictive abilities in different population structures showed only slight changes. (**d–e**) Relationships between observed yield per plant and predicted yield per plant in reciprocal hybrids. The whole population was divided into two groups of reciprocal hybrids to test cytoplasmic effects on predictive ability. The results demonstrated that different cytoplasms only weakly influenced predictive ability. YPP=Yield per plant.

DISCUSSION

Accuracy and efficiency are two pivotal indicators for the application of a biotechnology-assisted prediction model in breeding. Because predictive ability might be heavily influenced in populations consisting of genetically distant lines4·11; therefore, assessing whether a prediction model is appropriate is largely determined by its predictive abilities in different genetic structures. To avoid incomplete bias prediction, the predictive model should be built on representative inbred lines. In addition, the richer the information on these inbred lines, the higher the potential accuracy of the model. In this experiment, eighteen rice inbred lines were used to build a prediction model. These lines were collected from a wide range of locations (China, Italy and India) and possessed different degrees of *indica* or *japonica* contents and variable general combining abilities. To some extent, they represented a large range of rice accessions that were suitable for building the prediction model. Therefore, this model may also be applicable to rice inbred lines that were not included in this experiment.

The relationship between parental metabolite profiling and hybrid performance should be nonlinear26 and was treated as a megavariate quantitative structure-activity relationship in this study27. Because of its powerful multi-dimensional information regression ability, partial least squares projections to latent structures28 were applied to build the hybrid trait prediction model. Meanwhile, because cross-validation is built into the PLS algorithms27, the predictive significance is assessed and predictive abilities can be insured for every trait. In addition, taking advantage of the values of variable importance in the projection, the low contribution variables were excluded and the representative variables were reserved for the equations22. The high predictive abilities (0.858–0.977) in Table 2 demonstrate that PLS matched the metabolic prediction model quite well.

Metabolites have close connections with phenotypes14·17·19. Before determining the exact function of a metabolite, correlation analysis between the relative metabolite levels and phenotypes can be used as a type of rough estimating method29. In this study, different groups of metabolites were exploited to predict various traits. Because specific metabolites may have single or multiple functions in different pathways, metabolites such as saccharic acid, p-Cresol, and triacontanoic acid methyl ester were found to be predictive variables for two or more traits. Different correlation levels were detected for the metabolites that were predictive variables for all three traits. Some of these overlapping metabolites manifested significant positive and negative correlations with different traits, which might be the origin of the balance between component traits30 or outcomes of feedback regulation of the biological networks involved in complex traits31. Meanwhile, only a single value was used to represent reciprocal hybrids when calculating the ratios of relative parental metabolite levels. Ignorance of the relationship between the female parent and male parent might decrease the predictive abilities for the reciprocal hybrids. However, almost no predictive ability change was found between the reciprocal hybrids. Therefore, close balances might exist between metabolites and their associations might be controlled by the rules of chemistry26. The influences of the complex associations between metabolites on the hybrid trait prediction model remain unknown.

Nevertheless, predictable metabolites in the prediction model might simply have high predictive abilities for the traits. Therefore, biological functions of the predictive variables in this study were analysed to verify their contributions to the prediction model. Fructose (a predictive variable for yield per plant and maturation stage plant height) contributes strongly to the metabolic efficiency of rosette fresh weight and protein concentration in *Arabidopsis*29. Ferulic acid, a phenolamide32, plays an important role in plant development33 and is a predictive variable for plant height. Spermidine (a predictive variable for heading date and yield per plant) is associated with floral induction and development33·34. Furthermore, significant correlations have been found between alanine and biomass or yield-related traits in *Arabidopsis*35 and tomatoes14. Hence, our results demonstrate that the high predictive abilities for the three traits were not merely coincidental.

However, the metabolic prediction model might not be directly applicable to other breeding programs. In this study, metabolite profiling of 15-day-old seedlings of the parental lines was chosen to provide the predictors for hybrid traits. Thus, the metabolite profiling data were only a snapshot of the whole plant growth procedure. However, plant metabolism is a highly dynamic system that changes substantially over time20·36. Therefore, although high predictive abilities for three agronomic traits were obtained in this study, the metabolite profiling data presented here might be unsuitable for traits such as grain number per panicle or tiller number per plant. Therefore, the appropriate time points must be adapted when forecasting other polygenic traits from assembled metabolite profiling data.

Furthermore, the performance of a plant in nature is the consequence of the combination of genetic information and environmental influences; thus, the effects of environmental inputs on metabolome-assisted breeding strategies might need to be taken into account26. Metabolite profiling data of naturally growing parental seedlings in a specific field environment should be more accurate for the prediction of hybrid performances in corresponding locations.

In summary, a reliable and efficient metabolic prediction strategy was provided by combining parental metabolite profiling with a PLS regression method for rice hybrids. The high predictive abilities for three agronomic traits were implemented with respect to hybrid phenotypes, relative heterosis and specific combining ability. The predictive abilities for the three traits were only slightly influenced by population structures (genetic relatedness) or cytoplasmic effects. Additionally, metabolite-based prediction might be more suitable for traits such as resistance (susceptibility) or tolerance ability21·23·37because small molecules function more directly in these defence processes by confronting biotic/abiotic pressures.

METHODS

Plant materials

The eighteen rice inbred lines were selected based on their proportions of *indica-japonica* content using the InDel marker estimation method24. A complete diallel cross design was used. All the hybrid seeds were produced through manual emasculation at the Hybrid Rice Hainan Experimental Base of Wuhan University in Lingshui (N18° 30′ 22.12″, E110° 2′ 10.72″), Hainan Province, China, in 2012. Seedlings of the eighteen inbred lines and 306 hybrids were planted with a randomized block design in three replicates at the Hybrid Rice Ezhou Experimental Base of Wuhan University in Ezhou (N30° 22′ 19.82″, E114° 44′ 59.17″), Hubei Province, China, in June 2012. Ten plants were planted per row with a spacing of 16.5 × 26.4 cm. Four cytoplasmic male-sterile plants (named YTA) were planted around these experimental plants to decrease marginal effects. The middle five plants of each replicate were chosen for data collection. Three agronomic traits were evaluated: grain yield per plant, heading date and maturation stage plant height. Means among the replications were calculated for each trait and used in the data analysis.

For metabolite profiling analysis, seeds of the eighteen rice inbred lines were first submerged in water for two days at 28 °C. Then, the seeds were placed in an incubator at 28 °C for 24 h to accelerate germination. Next, seedlings at approximately the same stage were transferred to soil containers. Three random replicates were applied for each inbred line; the spacing between seedlings was 2 × 2 cm. Finally, all the seedlings were placed in a phytotron with a temperature of 28 °C, 70% relative humidity and an 8 h light/16 h dark photoperiod. On day 15, 100 mg of the seedlings (root excluded) were collected into

2 ml EP tubes for each replicate and immediately frozen in liquid nitrogen.

Gas chromatography mass spectrometry–based metabolite profiling

For each sample, 0.4 ml of methanol-chloroform ($V_{methanol}$:$V_{chloroform}$ = 3:1) and 20 µl of ribitol (0.2 mg/ml stock in dH_2O, Sigma-Aldrich Co. LLC., USA) were added as internal standards. After vortex mixing for 10 s, steel balls were placed into EP tubes and the samples were homogenized with a ball mill (JXFSTPRP-24, Shanghai Jingxin Experimental Technology, Shanghai, China) for 5 min at 55 Hz. Then, the samples were centrifuged for 15 min at 12,000 rpm at 4 °C. The supernatant (approximately 0.4 ml) was transferred to a new 2 ml GC/MS glass vial. An equal volume of approximately 13 µl (based on the number of samples) from each sample was transferred into a new 2 ml GC/MS glass vial as a mixed sample for quality control. Then, the extracts were dried in a vacuum concentrator without heating at 30 °C for approximately 1.5 h. Next, 80 µl of methoxymethyl amine salt (dissolved in pyridine, final concentration of 20 mg/ml) was added to the dried extracts and incubated at 37 °C for 2 h in an oven after mixing and sealing. The lids were opened, and 100 µl of BSTFA (containing 1% TCMS, v/v, Regis Technologies, Inc., USA) was added to each sample; then, the samples were resealed and incubated at 70 °C for an hour. When the samples had cooled to room temperature, 10 µl of FAMEs (standard mixture of fatty acid methyl esters, 1 mg/ml C8-C16 and 0.5 mg/ml C18-C30 in chloroform) was added to the mixed sample. Then, the sample was mixed well for GC-MS analysis.

GC/TOF MS analysis was performed using an Agilent 7890 (Agilent Technologies, USA) gas chromatograph system coupled with a Pegasus HT time-of-flight mass spectrometer (LECO Corporation, USA). The system utilized a DB-5 MS capillary column coated with 5% diphenyl cross-linked with 95% dimethyl polysiloxane (30 m × 250 µm inner diameter, 0.25 µm film thickness; J&W Scientific, Folsom, CA, USA). A total of 1 µl of the sample was injected in the splitless mode with helium as the carrier gas. The front inlet purge flow was 3 ml min^{-1}, and the gas flow rate through the column was 20 ml min^{-1}. The initial temperature was held constant at 50 °C for 1 min, followed by a 10 °C per min ramp up to 330 °C, and then maintained for 5 min at 330 °C. The injection, transfer line, and ion source temperatures were 280 °C, 280 °C, and 220 °C, respectively. The energy was −70 eV in electron impact mode. Spectra were recorded in full-scan mode with an m/z range of 85–600, at a rate of 20 spectra per second after a solvent delay of 366 s.

The Chroma TOF 4.3X software from LECO Corporation and the LECO-Fiehn Rtx5 database38 were used for raw peak extraction, data baseline filtering and calibration, peak alignment, deconvolution analysis, peak identification and integration of the peak area. The RI (retention time index) method was used for peak identification; the RI tolerance was 5000.

Data analysis

Relative low-parent heterosis (LPH), mid-parent heterosis (MPH), better-parent heterosis (BPH) and specific combining ability (SCA) for each trait were calculated using the following equations: $LPH = (F_1 - P_{Low})/P_{Low}$, $MPH = (F_1 - P_{Mean})/P_{Mean}$, $BPH = (F_1 - P_{High})/P_{High}$, GCA_i(general combining ability) = $P_{i.} - P_{..}$, and $SCA_{ij} = P_{ij} - P_{..} - P_{i.} - P_{j.}$39. For the heading date, earlier heading indicated positive heterosis. F_1 is the trait value of the hybrid; P_{Low}, P_{Mean} and P_{High} are the low value, mean value and high value of the two corresponding parents, respectively. P_{ij} is the trait value of the hybrid from parent i and parent j, $P_{..}$ is the mean of all the hybrids, $P_{i.}$ is the mean of the hybrids from parent i and $P_{j.}$ is the mean of the hybrids from parent j. The relative metabolite levels of each analyte were obtained by calculating the ratio between the areas of the analyte and its corresponding ribitol in each repeat. Means among the replications were used as the raw metabolic data. Metabolic data were log_2-transformed for statistical analysis. Analytes with the same annotations but different mass and similarity (such as 2-hydroxypyridine), the suffixes −1, −2, and −3 were added in retention time order to differentiate between them in the analyses. Microsoft Excel 2010 (Microsoft, USA) was used to calculate the sums (sum = (Female P+Male P)*0.5) of the parents, differences (difference = Female P-Male P) between the parents, and ratios (ratio = Female P/Male P) of the parents' phenotypic data. Female P represents the mean of the female parent values and male P represents the mean of the male parent values. When parental metabolic data were used as predictive variables, only single sum, difference and ratio values were calculated for the reciprocal hybrids. For example, for reciprocal hybrids *YB/Balilla* and *Balilla/YB*, if the relative level of Analyte X is **a1** for *YB*and **a2** for *Balilla*, the sum, difference, and ratio values for Analyte X of these reciprocal hybrids are **(a1+a2)***0.5, **a1−a2**, and **a1/a2**, respectively. The reason for this calculation approach is explained in the Discussion. Correlation analyses and regression analyses were performed with IBM SPSS Statistics 20 (IBM, USA). Average linkage between groups was chosen as the cluster method in the hierarchical cluster analysis. Squared Euclidean distance was calculated as the genetic distance. Principle component analysis was achieved through factor analysis, and eigenvalues greater than 1 were extracted; maximum iterations for convergence was 25; and the non-rotation method was applied for factor analysis. In stepwise linear regression, the probability of F was used as the stepping method criteria. The entry value was 0.05, and the removal value was 0.10. Partial least squares regression was conducted with the PLS extension bundle for SPSS. Main effects were used to specify model effects. Maximum numbers of latent factors were adjusted until no more latent factors could be extracted. Then, in the Proportion of Variance Explained table, the number of latent factor to a trait was determined where the corresponding adjusted R-square value was at the highest. After the number of latent factor was fixed, all the 525 parental metabolic data were applied to PLS again. And variable importance in the projection of Factor 1 was used for reordering. Parameters of independent variables were used to calculate the corresponding values of the dependent variables. Predictive ability was defined based on the Pearson correlation (2-tailed) coefficient of the predicted and observed values2·20.

ACKNOWLEDGMENTS

This study was supported by the National Natural Science Foundation of China (31371236) and the National High Technology Research and Development Program of China (2011AA10A101).

REFERENCES

1. Steinfath M. *et al.* Prediction of hybrid biomass in *Arabidopsis thaliana* by selected parental SNP and metabolic markers. Theor Appl Genet 120, 239–247 (2010).
2. Feher K. *et al.* Deducing hybrid performance from parental metabolic profiles of young primary roots of maize by using a multivariate diallel approach. PLoS ONE 9, e85435 (2014).
3. Yu C. Y., Hu S. W., Zhao H. X., Guo A. G. & Sun G. L. Genetic distances revealed by morphological characters, isozymes, proteins and RAPD markers and their relationships with hybrid performance in oilseed rape (*Brassica napus* L.). Theor Appl Genet 110, 511–518 (2005).
4. Reif J. C., Zhao Y., Würschum T., Gowda M., Hahn V. & Léon J. Genomic prediction of sunflower hybrid performance. Plant Breed 132, 107–114 (2013).
5. Schrag T. A. *et al.* Molecular marker-based prediction of hybrid performance in maize using unbalanced data from multiple experiments with factorial crosses. Theor Appl Genet 118, 741–751 (2009).
6. Schrag T. A. *et al.* Prediction of hybrid performance in maize using molecular markers and joint analyses of hybrids and parental inbreds. Theor Appl Genet 120, 451–461 (2010).
7. Dan Z. *et al.* Balance between a higher degree of heterosis and increased reproductive isolation: a strategic design for breeding inter-subspecific hybrid rice. PLoS ONE 9, e93122 (2014).
8. Xiao J., Li J., Yuan L., McCouch S. & Tanksley S. D. Genetic diversity and its relationship to hybrid performance and heterosis in rice as revealed by PCR-based markers. Theor Appl Genet 92, 637–643 (1996).
9. Zhang Q. F., Zhou Z. Q., Yang G. P., Xu C. G., Liu K. D. & Saghai Maroof M. A. Molecular marker heterozygosity and hybrid performance in indica and japonica rice. Theor Appl Genet 93, 1218–1224 (1996).
10. Windhausen V. S. *et al.* Effectiveness of genomic prediction of maize hybrid performance in different breeding populations and environments. G3: Genes|Genomes|Genetics 2, 1427–1436 (2012).
11. Frisch M., Thiemann A., Fu J., Schrag T. A., Scholten S. & Melchinger A. E. Transcriptome-based distance measures for grouping of germplasm and prediction of hybrid performance in maize. Theor Appl Genet 120, 441–450 (2010).
12. Fernie A. R. & Schauer N. Metabolomics-assisted breeding: a viable option for crop improvement?Trends Genet 25, 39–48 (2009).
13. Stitt M., Sulpice R. & Keurentjes J. Metabolic networks: how to identify key components in the regulation of metabolism and growth. Plant Physiol 152, 428–444 (2010).
14. Schauer N. *et al.* Comprehensive metabolic profiling and phenotyping of interspecific introgression lines for tomato improvement. Nat Biotechnol 24, 447–454 (2006).
15. Meyer R. C. *et al.* The metabolic signature related to high plant growth rate in *Arabidopsis thaliana*.Proc Natl Acad Sci USA 104, 4759–4764 (2007).
16. Lisec J. *et al.* Identification of heterotic metabolite QTL in *Arabidopsis thaliana* RIL and IL populations. Plant J 59, 777–788 (2009).
17. Sulpice R. *et al.* Starch as a major integrator in the regulation of plant growth. Proc Natl Acad Sci USA 106, 10348–10353 (2009).
18. Gong L. *et al.* Genetic analysis of the metabolome exemplified using a rice population. Proc Natl Acad Sci USA 110, 20320–20325 (2013).
19. Chen W. *et al.* Genome-wide association analyses provide genetic and biochemical insights into natural variation in rice metabolism. Nat Genet 46, 714–721 (2014).
20. Riedelsheimer C. *et al.* Genomic and metabolic prediction of complex heterotic traits in hybrid maize. Nat Genet 44, 217–220 (2012).
21. Korn M., Gartner T., Erban A., Kopka J., Selbig J. & Hincha D. K. Predicting *Arabidopsis* freezing tolerance and heterosis in freezing tolerance from metabolite composition. Mol Plant 3, 224–235 (2010).
22. Gärtner T. *et al.* Improved heterosis prediction by combining information on DNA- and metabolic markers. PLoS ONE 4, e5220 (2009).
23. Steinfath M. *et al.* Discovering plant metabolic biomarkers for phenotype prediction using an untargeted approach. Plant Biotechnol J 8, 900–911 (2010).
24. Lu B., Cai X. & Jin X. Efficient *indica* and *japonica* rice identification based on the InDel molecular method:its implication in rice breeding and evolutionary research. *Progress in Nature* Science 19, 1241–1252 (2009).
25. Sun J. *et al.* The contribution of intersubspecific hybridization to the breeding of super-high-yielding*japonica* rice in northeast China. Theor Appl Genet 125, 1149–1157 (2012).
26. Hur M. *et al.* A global approach to analysis and interpretation of metabolic data for plant natural product discovery. Nat Prod Rep 30, 565–583 (2013).
27. Eriksson L., Andersson P. L., Johansson E. & Tysklind M. Megavariate analysis of environmental QSAR data. Part I-a basic framework founded on principal component analysis (PCA), partial least squares (PLS), and statistical molecular design (SMD). Mol Divers 10, 169–186 (2006).
28. Eriksson L., Hermens J. L. M., Johansson E., Verhaar H. J. M. & Wold S. Multivariate analysis of aquatic toxicity data with PLS. Aquat Sci 57, 217–241 (1995).
29. Kleessen S. *et al.* Metabolic efficiency underpins performance trade-offs in growth of *Arabidopsis thaliana*. Nat Commun 5, 3537 (2014).
30. Williams W. Heterosis and the genetics of complex characters. Nature 184, 527–530 (1959).
31. Chen Z. J. Genomic and epigenetic insights into the molecular bases of heterosis. Nat Rev Genet 14, 471–482 (2013).

32. Dong X. *et al.* Spatiotemporal distribution of phenolamides and the genetics of natural variation of hydroxycinnamoyl spermidine in rice. Mol Plant 8, 111–121 (2015).

33. Edreva A. M., Velikova V. B. & Tsonev T. D. Phenylamides in plants. Russ J Plant Physiol 54, 287–301 (2007).

34. Guo D., Sun Y. & Chen Z. Involvement of polyamines in cytoplasmic male sterility of stem mustard(*Brassica juncea* var. *tsatsai*). Plant Growth Regul 41, 33–40 (2003).

35. Sulpice R. *et al.* Network analysis of enzyme activities and metabolite levels and their relationship to biomass in a large panel of *Arabidopsis* accessions. Plant Cell 22, 2872–2893 (2010).

36. Riedelsheimer C. *et al.* Genome-wide association mapping of leaf metabolic profiles for dissecting complex traits in maize. Proc Natl Acad Sci USA 109, 8872–8877 (2012).

37. Kusano M., Yang Z., Okazaki Y., Nakabayashi R., Fukushima A. & Saito K. Using metabolomic approaches to explore chemical diversity in rice. Mol Plant 8, 58–67 (2015).

38. Kind T. *et al.* FiehnLib: mass spectral and retention index libraries for metabolomics based on quadrupole and time-of-flight gas chromatography/mass spectrometry. Anal Chem 81, 10038–10048 (2009).

39. Qu Z. *et al.* QTL mapping of combining ability and heterosis of agronomic traits in rice backcross recombinant inbred lines and hybrid crosses. PLoS ONE 7, e28463 (2012).

CHAPTER 16 SLG CONTROLS GRAIN SIZE AND LEAF ANGLE BY MODULATING BRASSINOSTEROID HOMEOSTASIS IN RICE

Zhiming Feng, Chuanyin Wu, Chunming Wang, Jeehee Roh, Long Zhang, Jun Chen, Shengzhong Zhang, Huan Zhang, Chunyan Yang, Jinlong Hu, Xiaoman You, Xi Liu, Xiaoming Yang, Xiuping Guo, Xin Zhang, Fuqing Wu, William Terzaghi, Seong-Ki Kim, Ling Jiang, and Jianmin Wan

Keywords: BR, grain size, homomer, leaf angle, rice, SLG

INTRODUCTION

Rice (*Oryza sativa*) feeds more than half of the world's population as one of the world's most important cereal crops. Given the rapid increase in the world's population and decrease in cultivated land area, improving rice production remains a great challenge for rice breeding programs. Grain size and leaf angle are two important traits determining rice grain yield and have always been a consideration in breeding programs (Sinclair and Sheehy, 1999; Ikeda *et al.*, 2013).

Grain size, determined by grain length, grain width, and grain thickness, not only contributes to grain yield, but also influences the appearance, processing, cooking, and eating quality of rice. For example, people in Japan, Korea, and Northern China favor medium length round grains, whereas people in the USA, Southeast Asian countries, and Southern China prefer long and slender grains (Unnevehr *et al.*, 1992). The organ size is largely determined by the cell number and cell size during organogenesis (Potter and Xu, 2001; Sugimoto-Shirasu and Roberts, 2003). In recent years, several genes and quantitative trait loci (QTLs) that affect grain size by influencing cell number have been identified in rice, including *GS3*, *GW2*, *GW5*, *GS5*, *GW8*, *qGL3*, *TGW6*, *GW6a*, and *BG1* (Song et al., 2007, 2015; Weng *et al.*, 2008; Mao *et al.*, 2010; Li *et al.*, 2011; Wang*et al.*, 2012; Zhang *et al.*, 2012; Ishimaru *et al.*, 2013; Liu *et al.*, 2015). Some other genes and QTLs that control grain size by influencing cell size have also been isolated in rice, including *PGL1*, *GL7*, and GS2/*GL2* (Heang and Sassa, 2012; Che *et al.*, 2015; Duan *et al.*, 2015; Hu *et al.*, 2015; Wang*et al.*, 2015). It has been documented that at least some of those genes participate in seed size control by regulating biosynthesis and signaling of plant hormones, including brassinosteroids (BRs), cytokinins, gibberellins, and auxin (Ashikari et al., 1999, 2005; Hong *et al.*, 2003; Tanabe *et al.*, 2005; Ishimaru *et al.*, 2013).

Leaf angle, the inclination between the leaf blade and vertical culm, is a key factor determining the plant architecture (Hoshikawa, 1989; Sinclair and Sheehy, 1999). A compact plant type with erect leaves is preferred since it increases photosynthetic efficiency and nitrogen storage for grain filling, and facilitates dense planting (Sinclair and Sheehy, 1999; Sakamoto*et al.*, 2006). A number of genes or QTLs have been reported to have a role in controlling leaf angle, including *Ta1*, *OsDWARF4*, *D2*, *OsBRI1*,*OsBU1*, *ILI1*, *LC2*, and *ILA1* (Li *et al.*, 1998, 1999; Sakamoto *et al.*, 2006;Tanaka *et al.*, 2009; Zhang *et al.*, 2009; Zhao *et al.*, 2010; Ning *et al.*, 2011). The leaf lamina joint that connects the leaf blade and sheath is considered the most important tissue governing the leaf angle. The degree of the leaf angle largely depends on cell division and expansion as well as cell wall composition at the joint (Nakamura *et al.*, 2009; Zhang *et al.*, 2009; Zhao *et al.*, 2010; Ning *et al.*, 2011). Nevertheless, it is well known that BR treatment stimulates leaf inclination in rice (Wada *et al.*, 1981).

BRs are a group of steroidal phytohormones that regulate diverse plant growth and developmental processes, including cell expansion and division, vasculature differentiation, root and leaf development, stem elongation, skotomorphogenesis, and grain filling (Clouse and Sasse, 1998; Fujioka and Yokota, 2003; Wu *et al.*, 2008). In recent decades, researchers have clarified many genes and the main pathway of BR biosynthesis utilizing genetic studies, chemical feeding, and enzymatic analysis. Most of the enzymes known to catalyze

Citation: Jun Chen, Shengzhong Zhang, Huan Zhang, Chunyan Yang, Jinlong Hu, Xiaoman You, Xi Liu, Xiaoming Yang, Xiuping Guo, Xin Zhang, Fuqing Wu, William Terzaghi, Seong-Ki Kim, Ling Jiang, and Jianmin Wan, "SLG controls grain size and leaf angle by modulating brassinosteroid homeostasis in rice," J. Exp. Bot. first published online June 1, 2016 doi:10.1093/jxb/erw204.

BR biosynthesis belong to the cytochrome P450 protein family (Choe, 2006). The BR biosynthesis pathway mainly consists of the early and late C-22 oxidation pathway, and the early and late C-6 oxidation pathway (Choe, 2006). Similarly, research on BR signaling has also developed rapidly, and most of the main participants in BR signaling have been determined in Arabidopsis (Belkhadir and Chory, 2006). BRs are perceived by the receptor kinase BRI1 to transmit signaling (Li and Chory, 1997). In rice, it has been reported that BR plays important roles in the regulation of grain size, leaf angle, and yield potential. For example, most loss-of-function mutants in BR biosynthesis or signaling pathways, such as *d2*, *d11*, and *d61*, display short grains, erect leaves, and dwarf phenotypes (Yamamuro *et al.*, 2000;Hong *et al.*, 2003; Tanabe *et al.*, 2005), while some other mutants or transgenic plants with enhanced BR signaling or increased BR levels, such as *GSK2* knockdown lines, the *DLT* overexpresser, and the *D11* activation mutant *m107*, show longer grains and larger leaf angles (Tanabe *et al.*, 2005; Wan *et al.*, 2009; Tong *et al.*, 2012). More importantly, modulating the expression level of BR-related genes such as *OsDWARF4* and *OsBRI1*has been proven to improve rice grain yield at higher planting densities (Morinaka *et al.*, 2006; Sakamoto *et al.*, 2006).

BR homeostasis is vital for normal growth and development of plants. BRs are synthesized in most plant tissues, and their level is the highest in young developing organs but low in mature organs (Shimada *et al.*, 2003). Unlike the other plant hormones such as auxin that can be transported from the site of synthesis to a distant target site (Berleth and Sachs, 2001), BRs do not undergo long-distance transport and have the same site of synthesis and action (Symons and Reid, 2004). Therefore, there exist mechanisms that cells or tissues use to modulate levels of endogenous BRs precisely to keep cell expansion in balance and ensure normal plant growth and development (Symons and Reid, 2004). Negative feedback regulation is a common mechanism that also regulates BR homeostasis. It is reported that expression of many BR biosynthesis and signaling genes is inhibited by BR treatment, such as *D2*, *D11*, *OsDWARF4*, *BRD1*, *OsBRI1*, and *DLT* in rice (Yamamuro *et al.*, 2000; Hong et al., 2002, 2003; Tanabe*et al.*, 2005; Sakamoto *et al.*, 2006; Tong *et al.*, 2009). However, BR homeostasis is still poorly understood.

In this study, we characterized a rice semi-dominant mutant, *slender grain Dominant* (*slg-D*), with slender grains and enlarged leaf angles, which are caused by enhanced expression of *SLG*, a BAHD acyltransferase-like protein gene. We provide genetic evidence that the BR contents are associated with the expression level of *SLG*. In addition, the plants expressing an RNAi or a truncated version of *SLG* showed a semi-dwarf architecture with smaller leaf angles, which may be useful for rice yield improvement.

MATERIALS AND METHODS

Plant materials

The *slg-D* mutant (3A-10513) was isolated from a collection of activation-tagging T-DNA insertion rice lines (Jeon *et al.*, 2000; Jeong *et al.*, 2006), and kindly provided by Professor Gynheung An. The wild type (WT) of *slg-D* was Dongjin, a *japonica* cultivar. *d61-1*, *d11-2*, and *m107* were kindly provided by Professor Chencai Chu (Tong *et al.*, 2012). The WT of *d61-1*and *d11-2* was a *japonica* cultivar, Zhonghua11, and the WT of *m107* was a *japonica* cultivar, Nipponbare. Rice plants were cultivated in an experimental field under natural long-day conditions in Nanjing, China.

Scanning electron microsocpy (SEM) and light microscopy

For SEM, lemmas were harvested from florets after flowering and fixed in 2.5% (v/v) glutaraldehyde. Fixed samples were soaked in 2% (w/v) OsO_4 for 2h, dehydrated in a graded ethanol series, infiltrated and embedded in butyl methyl methacrylate, treated with critical point drying, and then sputter coated with platinum. The outer and inner epidermal cells of lemmas were observed using a HITACHI S-3400N scanning electron microscope. For light microscopy, lamina joints of the second leaves were harvested 10 d after flowering and fixed with FAA solution, followed by a graded series of dehydration and infiltration steps. Fixed tissues were embedded in paraplast. After sectioning, 10 μm thick sections were dewaxed with xylene, rehydrated, stained with 1% toluidine blue, and observed with a Leica DM5000B microscope. Cell lengths and widths of each organ were measured with IMAGEJ software.

Isolation, cloning, and RNAi suppression of the *SLG* gene

To identify the T-DNA insertion locus in *slg-D*, we searched the flanking sequence database (Jeong *et al.*, 2006; http://orygenesdb.cirad.fr/). The T-DNA loci were confirmed by PCR genotyping, using the primers P1, P2, and P3. To recapitulate the phenotype of *slg-D*, full-length cDNAs of *Loc_Os08g44830* and*Loc_Os08g44840* were amplified by PCR and cloned into the binary vector pCUbi1390 under the control of the maize *Ubi* promoter to create p1390-Ubi-*830*, and p1390-Ubi-*840* constructs, respectively. These constructs were then transformed into the rice variety Dongjin according to a published *Agrobacterium*-mediated method (Hiei *et al.*, 1994).

To obtain *SLG* RNAi plants, the construct pCUbi1390-ΔFAD2 (an FAD2 intron and ubiquitin promoter inserted into pCUbi1390) was used as an RNAi vector (Wu *et al.*, 2007). Both sense and antisense versions of a specific 305bp fragment from the cDNA of *SLG* were amplified with primer pairs SLG-RNAiL and SLG-RNAiR, and cloned into pCUbi1390-ΔFAD2 to create the pUbi-dsRNAiSLG construct, which was then transformed into the rice variety Dongjin by the *Agrobacterium*-mediated method described above.

RNA extraction and quantitative RT-PCR

Total RNA from roots, leaves, leaf sheaths, lamina joints, shoot apices, culms, and different stages of panicles were isolated using the RNAprep Pure Plant Kit (TIANGEN, Beijing, China). First-strand cDNA was reverse transcribed from 1 μg of total RNA using the PrimeScript 1st Strand cDNA Synthesis Kit (TaKaRa). Quantitative RT-PCR was performed using a SYBR Premix Ex TaqTM kit (TaKaRa) on an ABI prism 7500 Real-Time PCR System according to the manufacturer's instructions, and the *ACTIN1* gene was used as an internal control.

GUS staining

To analyze the expression pattern of *SLG*, an ~2.5kb promoter fragment was cloned into the pCAMBIA1381Z vector to create the *PRO SLG :GUS* (β-glucuronidase) reporter gene construct, which was then transformed into the rice variety Dongjin by the *Agrobacterium*-mediated method. GUS staining was performed on *PRO SLG :GUS* T_1 generation transgenic plants according to a method described previously (Jefferson *et al.*, 1987). Images were taken using a Nikon CD5Ri1P camera.

In situ hybridization

RNA *in situ* hybridization was performed as described previously (Bradley*et al.*, 1993). A 305bp gene-specific region of *SLG* amplified with primers SLG-PF and SLG-PR was cloned into the pGEM-T Easy vector (Promega). The linearized templates were amplified from the pGEM-T plasmid containing the gene-specific region of *SLG*using primers Yt7 and Ysp6. Digoxigenin-labeled RNA probes were transcribed in vitro using T7 and SP6 RNA polymerases, respectively, using a DIG Northern Starter Kit (Cat. No. 2039672, Roche) following the manufacturer's instructions. Images were taken using a Leica DM5000B microscope.

Subcellular localization of SLG

To determine the subcellular localization of SLG, green fluorescent protein (GFP) was fused to the C-terminus of SLG under the control of the 35S promoter in the PAN580 vector. In addition, the nuclear marker D53–mCherry was constructed. The SLG–GFP fusion construct was transiently co-transferred into rice protoplasts with the D53–mCherry constructs according to the method described previously (Chen *et al.*, 2006). Next, GFP was fused to the C-terminus of SLG under the control of the*Cauliflower mosiac virus* (CaMV) 35S promoter in the pCAMBIA1305.1 vector. The pCAMBIA1305-SLG-GFP construct was transformed into the rice variety Dongjin by the *Agrobacterium*-mediated method. GFP fluorescence was examined in the young roots of 2-week-old T_1transgenic plants. Fluorescence images were observed using a Zeiss LSM510 confocal laser microscope.

BR and BRZ treatment

The lamina joint bending assay using excised leaf segments was performed as described by Wada *et al.* (1981). Seeds were germinated for 2 d and then grown in the dark for 8 d at 30 °C. Segments of 2cm comprising the second leaf blade, lamina joint, and leaf sheath were floated on distilled water for 24h and then incubated in 2.5mM maleic acid potassium solution containing various concentrations of brassinolide (BL; Sigma, http://www.sigmaaldrich.com/) for 48h in the dark. Lamina joint angles were measured using IMAGEJ software. The coleoptile and root elongation tests were performed using a previously described method (Yamamuro *et al.*, 2000). Seeds were germinated on agar plates containing various concentrations of BL, and then the coleoptile and root lengths were measured 1 d after germination.

To measure the effect of brassinazole (BRZ; TCI) treatment on lamina joint bending, the leaf tips of 8-day-old seedlings of *slg-D* and the WT were spotted with 1 μl of DMSO containing 0 or 10 μM BRZ daily for 3 d, followed by 7 d growth in a controlled growth chamber under long-day conditions (16h light at 28 °C/8h darkness at 24 °C). The angles of the third lamina joints were measured using IMAGEJ software.

Measuring endogenous BRs

BR contents were analyzed using gas chromatography–mass spectrometry (GC-MS) as described previously (Kim *et al.*, 2005). Four-week-old shoots of the WT (88.50g FW) and *slg-D* (65.43g FW) were harvested, lyophilized, and extracted three times with 500ml of 80% methanol. Deuterium-labeled 6-deoxocastasterone (6-deoxoCS), typhasterol (TY), castasterone (CS), and BL were added as internal standards for quantitative analysis of the extracts (200ng each).

Yeast two-hybrid assay

The full-length cDNA of *SLG* was cloned into pGBKT7 (Clontech,http://www.clontech.com). Full-length *SLG* as well as its N- and C-terminal truncated deletions were then subcloned into pGADT7 (Clontech) and all vectors were transformed into yeast strain AH109. A yeast two-hybrid library was constructed from the mRNA of young rice panicles 0.1–5cm long. Yeast transformation and screening procedures were performed according to the manufacturer's instructions (Clontech).

Bimolecular fluorescence complementation (BiFC) assay

The full-length *SLG* cDNA was cloned into the vector pSPYCE(M), and the*SLG* cDNA and its truncated deletions were then subcloned into the vector pSPYNE173. The plasmids were transiently expressed in *Nicotiana benthamiana* leaves as described previously (Waadt and Kudla, 2008). Yellow fluorescent protein (YFP) fluorescent signals were observed under a Zeiss LSM510 confocal laser microscope between 48h and 72h post-transfection.

Pull-down assay

The *SLG* cDNA was cloned into the vectors pMAL-c2x and pGEX4T-2 to generate fusions with maltose-binding protein (MBP) and glutathione *S*-transferase (GST), respectively. Expression of MBP–SLG, GST–SLG, and GST in BL21 Rosetta cells was induced with 0.5mM isopropyl-β-D-thiogalactoside at 16 °C for 20h. The total protein concentration was quantified using the Bio-Rad protein assay reagent. The pull-down assay was performed as reported previously (Miernyk and Thelen, 2008). The proteins were separated on a 10% SDS–PAGE gel and immunoblotted with anti-GST or anti-MBP antibodies (Abmart, http://www.ab-mart.com).

RESULTS

Phenotypic characterization of the semi-dominant mutant *slg-D*

To identify new components involved in regulating rice grain size, we screened a collection of activation-tagging T-DNA insertion rice lines (Jeon *et al.*, 2000; Jeong *et al.*, 2006). As a result, we isolated a mutant (3A-10513) with a slender-grain phenotype, and named it *slender grain Dominant* (*slg-D*).

slg-D showed less compact plant architecture than the WT at both the vegetative and mature stages (Fig. 1A, B). In *slg-D*, the grain length was significantly increased while the grain width decreased, and the 1000-grain weight was slightly decreased (Fig. 1C–F). The lamina joint bending angles of *slg-D* were larger than those of the WT, especially for the flag leaves (Fig. 1G, H). Together, these results indicate that *slg-D*displays slender grain and an enlarged leaf angle.

Figure 1: Phenotype of the *slg-D*mutant. (A, B) Gross morphologies of the WT and*slg-D* at the vegetative (A) and mature (B) stages. (C) Slender grains of *slg*-D compared with the WT. (D–F) Measurements of traits showing longer but narrower grains (D and E), and reduced seed weight (F) in *slg*-D. (G) Comparison of the lamina joint angle of the flag (1), second (2), third (3), and fourth (4) leaves between *slg*-D and the WT (counted from the flag leaf downwards). (H) Quantitation of the flag leaf lamina joint angles. Values are given as means ±SD (n=10 in D–F, H). **P<0.01 compared with the WT by Student's t-test. Scale bars=10cm (A, B), 2mm (C), or 2cm (G).

The F_1 plants from the cross *slg-D*×WT exhibited an intermediate phenotype in grain shape and leaf angle, indicating a semi-dominant nature of the mutation. Genetic analyses of an F_2 population derived from the same cross showed a segregation ratio of 1:2:1 (64 normal:120 intermediate:56 mutant; χ^2=0.09, P>0.05), suggesting that *slg-D* is a single locus mutation. This observation provides a hint that the semi-dominant nature of *slg-D*might be associated with an insertion of the activation-tagging T-DNA.

Cell length change in *slg-D* determines the mutant phenotypes

To investigate the mutant phenotypes in *slg-D* at a cell level, we performed SEM and light microscopy on *slg-D* plants along with the WT. The observations on the outer and inner epidermal cells of lemmas, which determine the shape and size of grains, showed that those cells were stretched longitudinally in *slg-D*, such that the slender-grain phenotype was developed (Fig. 2A–G). The histological analysis on the second leaf lamina joints indicated that there was no significant alteration in cell size in the abaxial sides between *slg-D* and the WT (Fig. 2H–K). In contrast, the cell length of the adaxial sides was increased in *slg-D* (Fig. 2H, L–N). Therefore, it is the asymmetric cell expansion at the opposite sides of the lamina joint that causes a larger leaf bending in *slg-D*. Overall, these results indicate that the changes in cell length are responsible for the phenotype in *slg-D*.

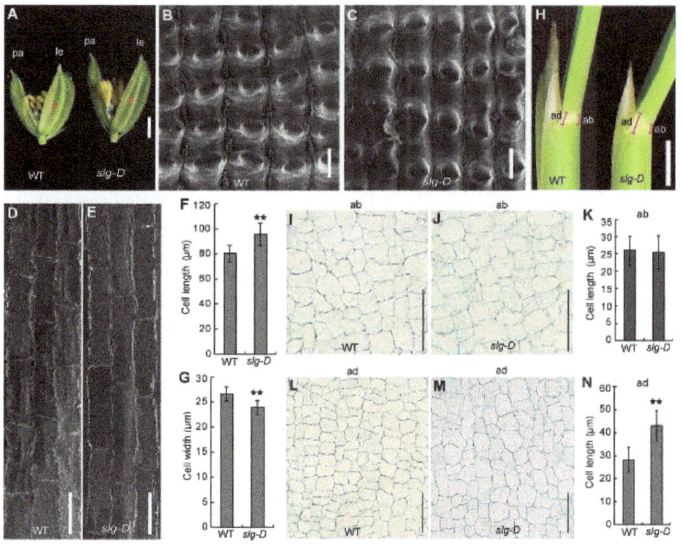

Figure 2: The changes in cell length are responsible for the phenotype in *slg-D*. (A) Spikelets just before heading. The red asterisk indicates the site used for detailed observation in (B–E). pa, palea; le, lemma. (B, C) SEM images of outer epidermal cells of lemmas. (D, E) SEM images of inner epidermal cells of lemmas. (F, G) Average length (F) and average width (G) of inner epidermal cells of lemmas. (H) Comparison of second leaf lamina joints (counted from the flag leaf downwards) between the WT and *slg-D*. Red lines indicate cut sites. ad, adaxial; ab, abaxial. (I, J) Longitudinal sections of the abaxial sides of the second leaf lamina joint shown in (H). (K) Average lengths of cells shown in (I) and (J). (L, M) Longitudinal sections of the adaxial sides of the second leaf lamina joints shown in (H). (N) Average lengths of cells shown in (L) and (M). Values are given as means ±SD (n=3 in F, G, K, N). **P<0.01 compared with the WT by Student's t-test. Scale bars=2mm (A), 50 μm (B–E), 2cm (H), or 100 μm (I, J, L, M).

Enhanced expression of *Loc_Os08g44840* in *slg-D* leads to the mutant phenotypes

To isolate the gene for *slg-D*, we searched the T-DNA insertion database and obtained a genomic flanking sequence (Jeong *et al.*, 2006;http://orygenesdb.cirad.fr/). Based on this information, we designed three PCR primers (P1, P2, and P3) and confirmed the site of the T-DNA insertion in *slg*-D (Fig. 3A, B). To understand whether the mutation in *slg-D* is related to an insertion of the activation-tagging T-DNA, we PCR-genotyped the mutant F_2 plants derived from the cross *slg-D*×WT and found that the mutant phenotypes were always associated with the presence of the T-DNA. A BLAST search (http://www.ncbi.nlm.nih.gov/) showed that the T-DNA was inserted in an intergenic region, with *Loc_Os08g44830* 2000bp upstream and*Loc_Os08g44840* 4500bp downstream (Fig. 3A). Another nearby gene is*Loc_Os08g44820*, upstream of *Loc_Os08g44830* (Fig. 3A). Next we examined if the four enhancer repeats in the T-DNA had an influence on the expression level of the three nearby genes, and found that two of them (*Loc_Os08g44830* and *Loc_Os08g44840*) had elevated expression (Fig. 3C). This result suggests that the changed expression level of the two genes might be responsible for the mutant phenotypes in *slg-D*.

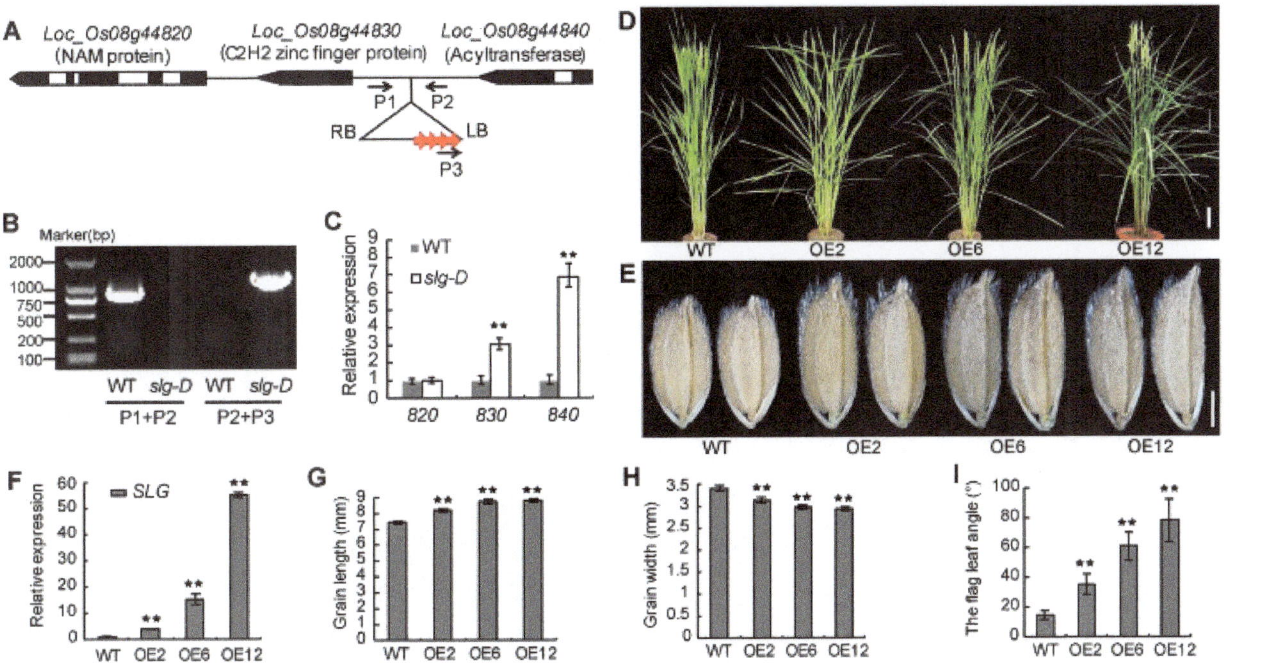

Figure 3: Enhanced expression of*Loc_Os08g44840* in *slg-D*leads to the mutant phenotypes. (A) T-DNA insertion site in *slg-D*. The four copies of the 35S enhancers in the T-DNA are indicated with red arrows, and white boxes indicate introns. P1, P2, and P3 indicate primers used for verification of the insertion site. RB and LB indicate the T-DNA right and left borders, respectively. (B) PCR verification of the insertion site. (C) Quantitative RT-PCR analysis of expression of genes adjacent to the T-DNA insertion site. (D) Gross appearance of*pUbi::Loc_Os08g44840* transgenic plants recapitulating the *slg-D* phenotype in the WT background. OE2, OE6, and OE12 are three independent lines. (E) Grains of the plants shown in (D). (F)*SLG* expression levels in the plants shown in (D). (G, H) Average lengths (G) and widths (H) of grains shown in (E). (I) Quantitation of the flag leaf angles shown in (D). Values are given as means ±SD (n=3 in C, F; n=10 in G–I). **P<0.01 compared with the WT by Student's t-test. Scale bars=10cm (D) or 2mm (E).

We assumed that overexpression of *Loc_Os08g44830*, *Loc_Os08g44840*, or both in the WT might recapitulate the phenotypes observed in *slg-D*. To test this assumption, *Loc_Os08g44840* and *Loc_Os08g44830*, under control of the maize *Ubi* promoter, were individually overexpressed in WT plants. Interestingly, only plants overexpressing *Loc_Os08g44840*, not*Loc_Os08g44830*, showed varying degrees of enlarged leaf angle and slender grain, thus phenocopying *slg-D* . The phenotypic variation in the transgenic plants was well correlated with the expression level of *Loc_Os08g44840* (Fig. 3D–I). We concluded that it is the enhanced expression of *Loc_Os08g44840* that causes the phenotypic changes in *slg-D*. We designated *Loc_Os08g44840* as a*SLENDER GRAIN* (*SLG*) gene.

SLG encodes a putative BAHD acyltransferase-like protein

SLG encodes a protein of 445 amino acids. SLG belongs to the putative BAHD family of acyltransferases, which catalyze formation of a diverse group of plant metabolites using CoA thioesters as substrates (D'Auria, 2006). A phylogenetic analysis revealed that SLG-like proteins largely fall into two groups: dicot and monocot, and SLG is a member of the monocot group. However, to date, none of the genes in this group has been functionally characterized.

The BAHD enzymes typically contain two highly conserved domains: the HXXXDG motif (or HXXXDA) located near the center, and the DFGWG motif located near the C-terminus (D'Auria, 2006). The two conserved motifs in SLG are HAVLDG (167–172) and DFGFG (386–390) with a substitution of tryptophan by phenyalanine.

Expression pattern and subcellular localization of *SLG*

Our quantitative RT-PCR analysis showed that expression of *SLG* is strong in young panicles, relatively high in lamina joints, low in shoot apices, culms and leaves, and very little in roots and leaf sheaths (Fig. 4A). Further, *SLG* has the highest expression level during early panicle development, but drops dramatically as the spikelets reach their final size (Fig. 4B). To investigate further the expression pattern of *SLG*, a genomic sequence ~2.5kb upstream of the translation start site was cloned and introduced into the pCAMBIA1381Z vector, resulting in the *PRO SLG :GUS*reporter construct. Analysis of GUS activity in transgenic lines showed that strong GUS staining was observed in young panicles, lamina joints, and young stem nodes, with faint staining in leaf veins, but not visible in roots and leaf sheaths (Fig. 4C, parts 1–8). Cross-sectioning of the GUS-stained leaf lamina joint and the bottom portion of the stained internode further showed that the GUS signals were mainly restricted to vasculature regions (Fig. 4C, parts 9–12). We also performed RNA *in situ* hybridization to localize *SLG* expression during early panicle development more precisely. Strong *SLG* expression was detected in spikelet meristem primordia, floral meristem primordia, lemma and palea primordia, and vasculature regions (Fig. 4D). The predominant expression of *SLG* in young panicles and lamina joints implies its role in controlling grain shape and leaf angle.

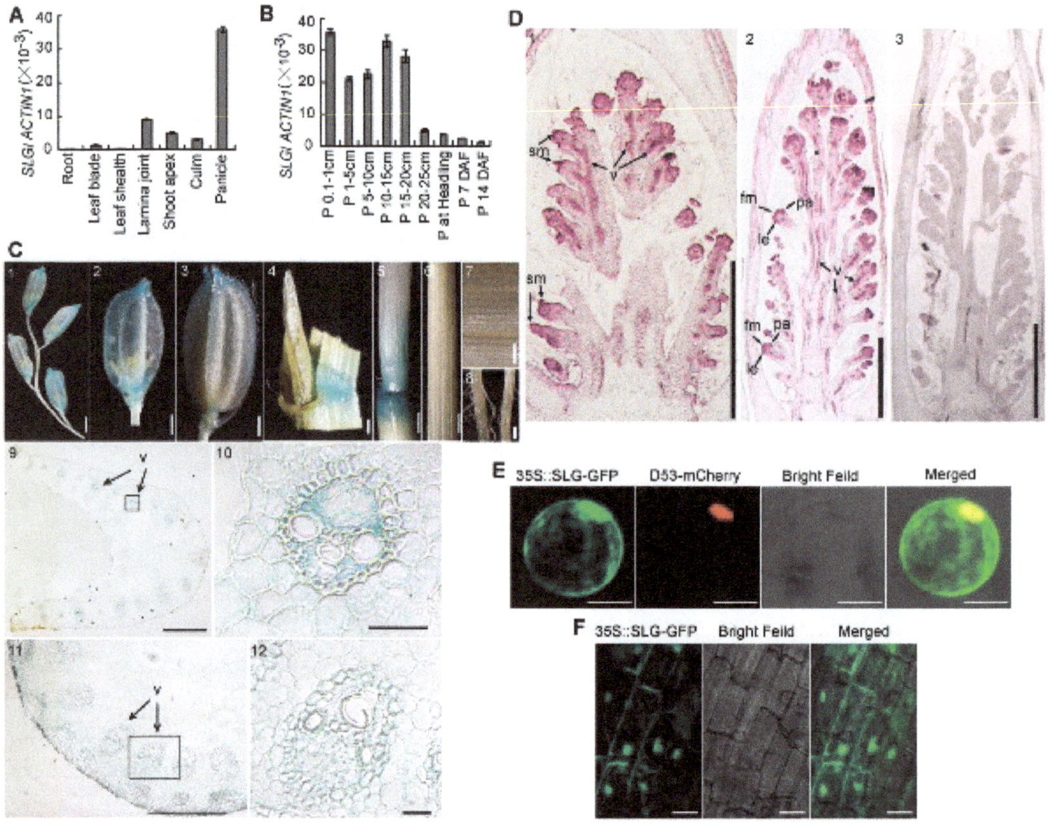

Figure 4: Spatiotemporal expression and subcellular localization of SLG. (A) Quantitative RT-PCR analysis of *SLG*expression in various rice tissues. Roots, leaf blades and sheaths, lamina joints, and culms were harvested from plants at the early booting stage. Shoot apices were collected from 4-week-old seedlings. Panicles were collected when they had reached 1cm length. (B) *SLG* expression in panicle tissues at various stages. P, panicle; DAF, day after flowering. Data (A, B) are ratios of *SLG* to *ACTIN1* signals and are presented as means ±SD (n=3). (C) GUS staining of various *PRO SLG :GUS* transgenic line tissues. (1) Young spikelet. (2) A young floret in (1). (3) A mature floret after flowering. (4) Lamina joint. (5) The bottom portion of the young second internode and stem node. (6) The top portion of the young second internode. (7) Mature leaf blade. (8) Mature root. (9) Cross-section of the lamina joint shown in (4). (10) Magnified image of the region enclosed by the square in (9). (11) Cross-section of the lower part of the tissue shown in (5). (12) Magnified image of the region enclosed by the square in (11). Scale bars=1mm (1–4, 7–9, 11), 1cm (5, 6), or 100 μm (10, 12). (D) *In situ* localization of *SLG* mRNA. (1) An early stage of panicle development. (2) A late stage of panicle development. (3)*SLG in situ* hybridization negative control using a sense probe. Scale bars=1mm (1–3). v, vasculature; sm, spikelet meristem; le, lemma; pa, palea; fm, floral meristem. (E) Transient expression of SLG–GFP and D53–mCherry fusion proteins in rice leaf sheath protoplasts. Scale bar=5 μm. (F) Fluorescent signals in transgenic root cells expressing the SLG–GFP fusion protein. Scale bar=30 μm.

To determine the subcellular localization of SLG, we fused the green fluorescent protein (GFP) to the C-terminus of SLG. Transient expression of this fusion protein in rice protoplasts revealed that the GFP signals were found in both the cytoplasm and the nucleus (Fig. 4E). Transgenic rice plants harboring the same fusion construct also showed the cytoplasmic and nuclear localization pattern of SLG (Fig. 4F).

SLG positively regulates endogenous BR levels

The *slg-D* phenotype resembles that of an activation mutant or transgenic plants with elevated BR accumulation (Wu *et al.*, 2008; Wan *et al.*, 2009), and that of transgenic plants with enhanced BR signaling (Tanaka *et al.*, 2009; Tong *et al.*, 2012), leading us to hypothesize that *SLG* may be involved in regulating the BR pathway. To determine whether *slg-D*responds differently to BR treatment, we first performed lamina joint bending assays using excised leaf segments (Wada *et al.*, 1981). We measured the effects of a range of 24-epibrassinolide (BL; a type of active BR) concentrations on the angle of the lamina joints, and found that lamina joint bending was increased in a dose-dependent manner and the sensitivity to BL treatment was similar in *slg-D* and the WT (Fig. 5A, B). Next we performed another BR response assay involving coleoptile and root elongation using a previously described method (Yamamuro *et al.*, 2000). Comparison of coleoptile and root lengths also showed that *slg-D*had a similar response to BL as the WT. These results indicate that BR signaling is not altered in *slg-D*.

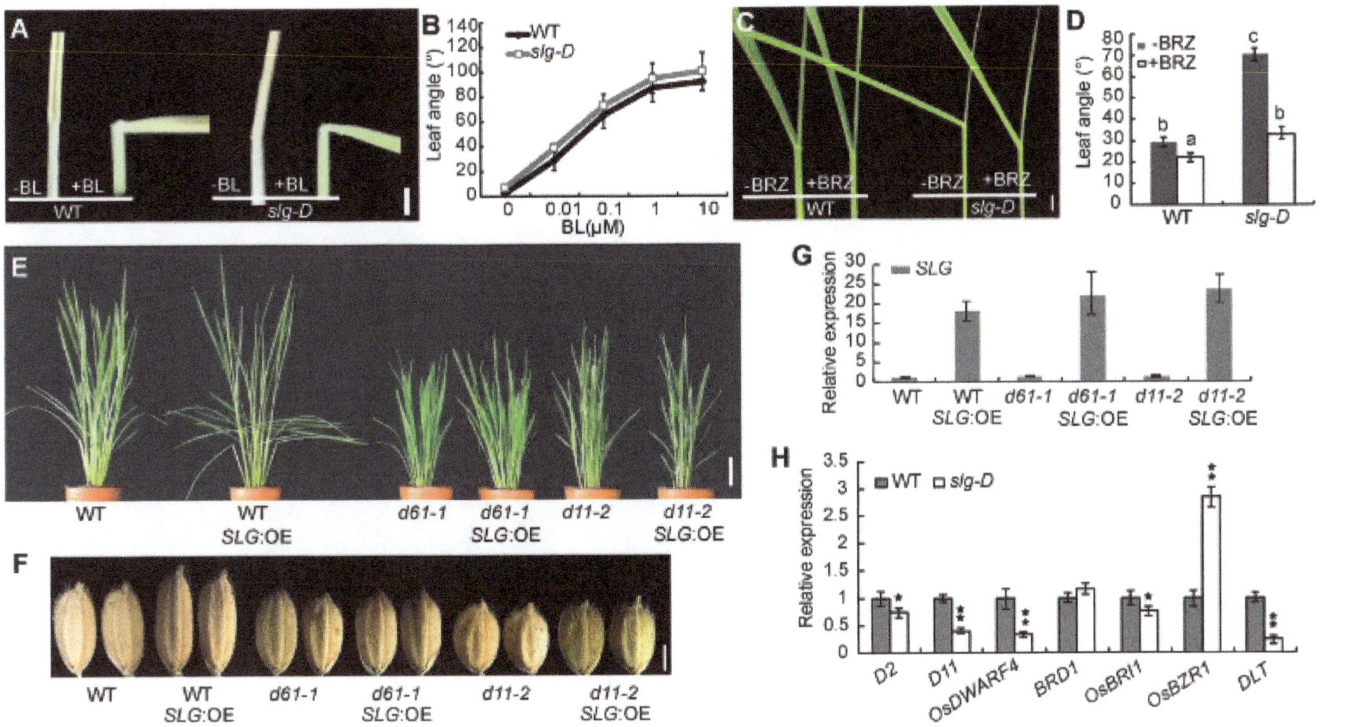

Figure 5: *SLG* is involved in regulating endogenous BR levels but not BR signaling. (A) Response of the second leaf lamina joint to 10 μM BL treatment by the excised leaf segment method. (B) Dose response of the bending angle to various concentrations of BL. (C) Response of the third leaf lamina joint to 10 μM BRZ. (D) Measurement of the lamina joint angles after the 10 μM BRZ treatment shown in (C). (E, F) Gross morphologies (E) and grains (F) of WT, WT *SLG*:OE, *d61-1*, *d61-1 SLG*:OE, *d11-2*, and *d11-2 SLG*:OE. (G) Expression levels of *SLG* in the different lines shown in (E). (H) Quantitative RT-PCR analysis of BR-related genes in young panicles of *slg-D* and the WT. Values are given as means ±SD (n=10 in B, D; n=3 in G, H). Different letters (D) indicate P<0.01 (LSD multiple range tests). *P<0.05; **P<0.01 (H) compared with the WT by Student's t-test. Scale bars=2cm (A, C), 10cm (E), or 2mm (F).

To investigate whether *SLG* functions in regulating endogenous BR levels, we first tested the effect of brassinazole (BRZ; a specific BR biosynthesis inhibitor; Asami *et al.*, 2000) on *slg-D* in a lamina joint bending experiment. The leaf tips of 8-day-old seedlings of *slg-D* and the WT were spotted with 10 μM BRZ daily for 3 d, followed by 7 d growth in a chamber. We observed that the leaf angle of *slg-D* was restored to the WT level by BRZ treatment, whereas the WT seedlings had a milder response to the same treatment (Fig. 5C, D), indicating that *slg-D* is more sensitive to BRZ. A similar result was also seen when the *D11* activation line *m107*, a BR overproduction mutant, was treated with BRZ. Next we introduced the *SLG*-overexpressing construct into *d61-1* (a loss-of-function mutant of the BR receptor gene *OsBRI1*; Yamamuro *et al.*, 2000) and *d11-2* (a mutant deficient in BR biosynthesis; Tanabe *et al.*, 2005), and found that the *d61-1 SLG:OE* and *d11-2 SLG:OE* plants still retained the dwarfism, smaller and round grains, and erect leaves (Fig. 5E–G). Those results suggest a role for *SLG* in regulating BR levels. Consistent with this, our chemical analysis indeed showed a higher content of 6-deoxocastasterone (6-deoxoCS), typhasterol (TY), and castasterone (CS) in *slg-D* than in the WT.

It is known that excessive BRs down-regulate the BR-related genes *D2*,*D11*, *OsDWARF4*, *BRD1*, *OsBRI1*, and *DLT*, but up-regulate *OsBZR1* as a feedback mechanism (Yamamuro *et al.*, 2000; Hong et al., 2002, 2003;Wang *et al.*, 2002; He *et al.*, 2005; Tanabe *et al.*, 2005; Sakamoto *et al.*, 2006; Tong *et al.*, 2009). We analyzed the expression level of those genes in *slg-D* and found that all except *BRD1* had the expected transcription level change as a response to the elevated BR levels (Fig. 5H). As an alternative control, we also measured expression of those BR genes in the mutant *m107*, where *D11* was dramatically enhanced and BR levels increased, and found similar expression changes. Those results further confirm higher BR contents in *slg-D*. However,*SLG* itself did not respond to the exogenous BL treatment.

Taken together, these results suggested that *SLG* positively regulates endogenous BR levels and is a new regulator of BR homeostasis in rice.

Suppression of *SLG* leads to BR-deficient phenotypes

To explore further the function of *SLG*, a *SLG* RNAi vector was constructed and introduced into WT plants. The *SLG* RNAi plants displayed a more compact architecture, reduced plant height, smaller leaf angle, and shorter and rounder grain (Fig. 6A–G). These phenotypes are similar to those of BR-deficient mutants, such as *d61* and *d11* (Hong *et al.*, 2003;Tanabe *et al.*, 2005). In addition, we investigated expression changes of the genes involved in BR synthesis or signaling in R7, a typical *SLG* RNAi line with greatly reduced leaf angle and *SLG* expression (Fig. 6F, G). Six of the genes detected, *D2*, *D11*, *OsDWARF4*, *BRD1*, *OsBRI1*, and *DLT* were up-regulated, but *OsBZR1* was down-regulated in R7 compared with the WT (Fig. 6H). The feedback regulation of those BR-related genes caused by knock-down of *SLG* further supports involvement of *SLG* in regulating the BR level. These results further highlight a role for *SLG* in regulating BR homeostasis. This observation also suggests that an optimized expression level of *SLG* may help create a compact and semi-dwarf ideal plant type.

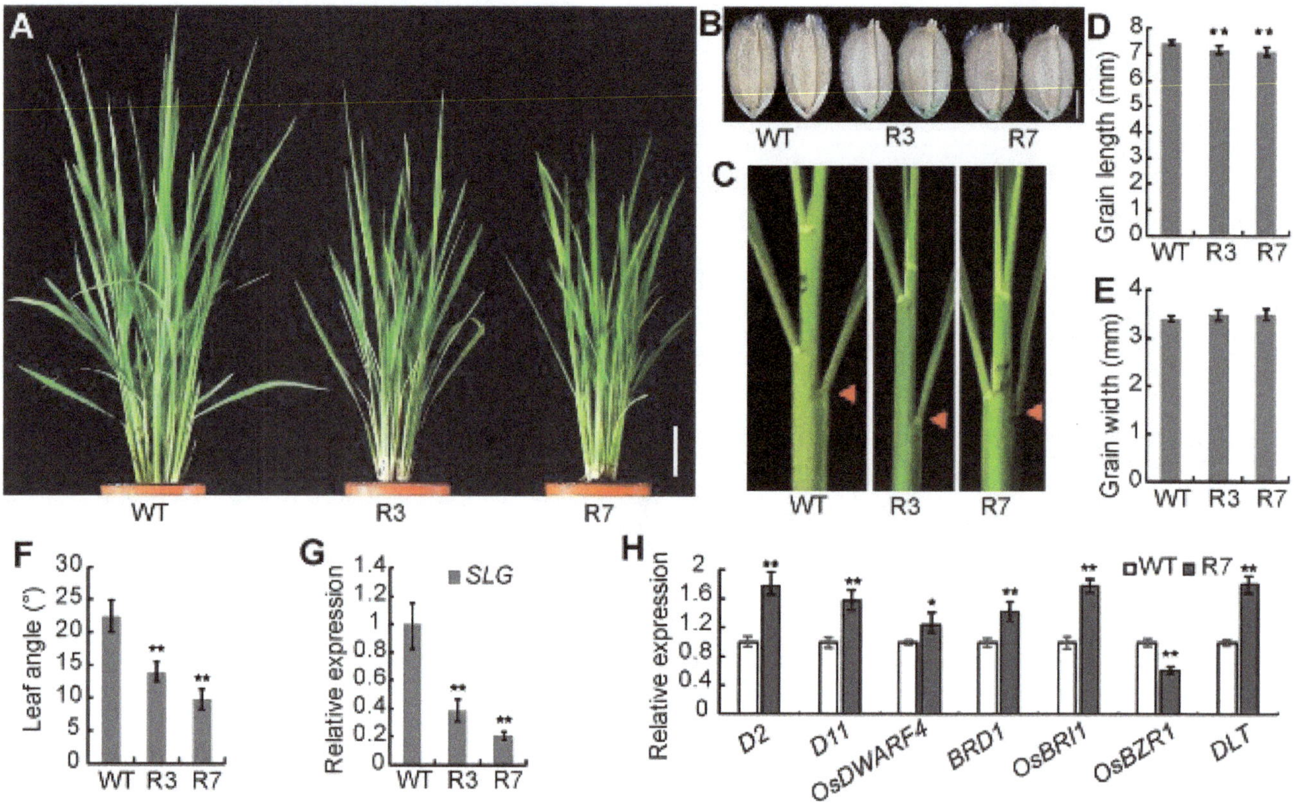

Figure 6: Phenotypes of *SLG* RNAi transgenic plants. (A–C) Gross morphologies (A), grains (B), and leaf angles (C) of *SLG* RNAi transgenic plants in the WT background. R3 and R7 are two independent lines. (D–F) Quantitative comparisons of grain lengths (D), grain widths (E), and leaf angles (F) of the lines shown in (A). Leaf angles were measured at the positions indicated by the arrowheads in (C). (G) *SLG* expression in the lines shown in (A). (H) Quantitative RT-PCR analysis of BR-related genes in R7 and the WT. Values are given as means ±SD (n=10 in D–F; n=3 in G, H). *P<0.05; **P<0.01 compared with the WT by Student's t-test. Scale bars=10cm (A) or 2mm (B).

SLG functions as homomers

It has been reported that enzyme proteins often function as homomers or heteromers (Ali and Imperiali, 2005). To investigate the functional forms of SLG, the full-length SLG protein was used as a bait to screen a yeast two-hybrid library prepared from young rice panicles. We identified four positive clones that contain different *SLG* cDNA fragments from ~1 million yeast transformants. To confirm the self-interaction of SLG, different truncated SLG proteins were used for interaction analysis. As shown in Fig. 7A, a 30 amino acid region in the N-terminus of SLG (SLGΔC4), rather than the two conserved motifs, was required for the self-interaction of SLG. An *in vitro* GST pull-down assay also confirmed the self-interaction (Fig. 7B). In addition, BiFC analysis also showed that SLG physically interacted with itself and this interaction required the N-terminal 30 amino acid region (Fig. 7C).

To study the importance of SLG self-interaction, two truncated *SLG* CDS,*SLGΔC1* with only the 190 N-terminal amino acids, and *SLGΔN3* without the 30 N-terminal amino acids, were individually overexpressed in *slg-D*and the WT. We found that overexpression of *SLGΔC1* but not *SLGΔN3* in both the WT and *slg-D* resulted in shorter and rounder grains, smaller leaf angles, and dwarf phenotypes, similar to those of *SLG* RNAi plants. The truncated SLGΔC1 protein might interfere with formation of functional homomers between the intact SLG proteins, thus leading to a dominant negative mutant phenotype. When the interaction region was removed in SLGΔN3, however, the truncated protein did not exert any effect on SLG, thus providing genetic evidence that SLG indeed functions as homomers *in vivo*.

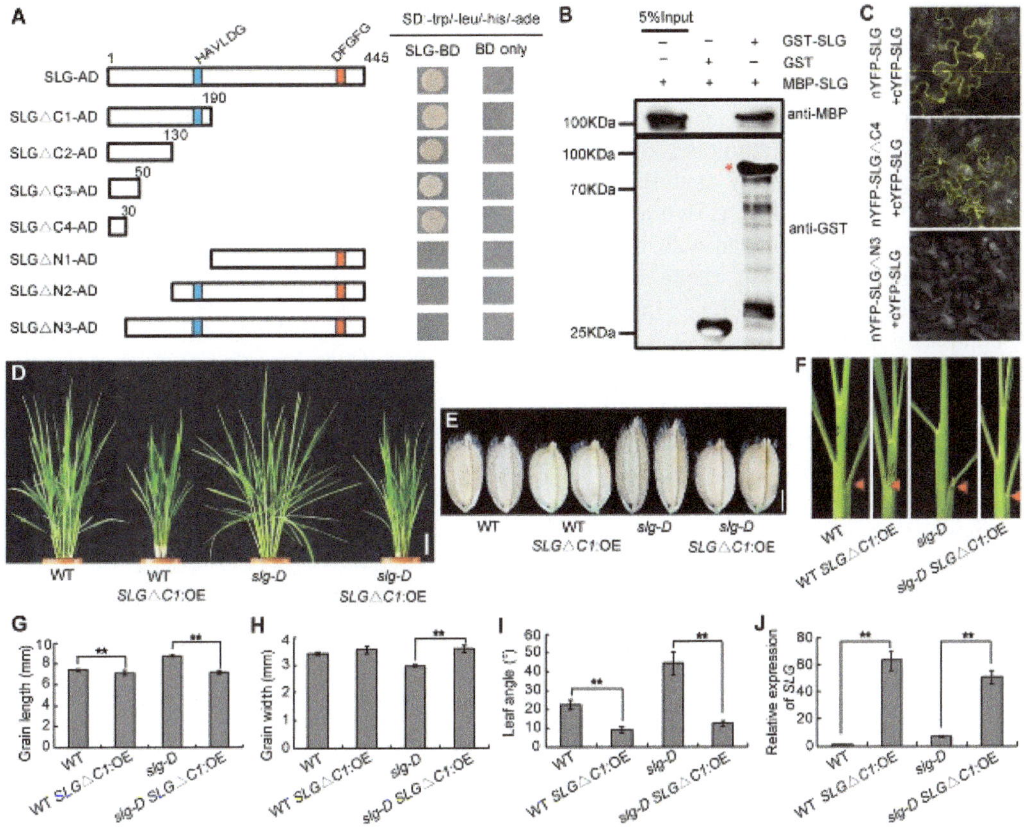

Figure 7: SLG functions as homomers. (A) Yeast two-hybrid assays. Schematic representations of the truncated SLG proteins used for the yeast two-hybrid assays. Cyan and red rectangles represent the two conserved motifs. ΔN and ΔC represent several N- and C-terminally truncated SLG deletions. (B) Pull-down assay showing direct interaction between MBP–SLG and GST–SLG in vitro. (C) BiFC assay showing that cYFP–SLG and nYFP–SLG interacted to form a functional YFP in tobacco leaf cells. (D–F) Gross morphologies (D), grains (E), and leaf angles (F) of WT, WT*SLGΔC1*:OE, *slg-D*, and *slg-D SLGΔC1*:OE plants. (G–I) Quantitative comparisons of grain lengths (G), grain widths (H), and leaf angles (I) of the lines shown in (D). Leaf angles were measured at the positions indicated by the arrowheads in (F). (J) Expression levels of *SLG* in the lines shown in (D). Values are given as means ±SD (n=10 in G–I; n=3 in J). *P<0.05; **P<0.01 by Student's t-test. Scale bars=10cm (D) or 2mm (E).

DISCUSSION

In this study, we have provided evidence that *SLG* is involved in BR homeostasis by positively regulating endogenous BR levels to control grain size and leaf angle in rice. First, the activation-tagging mutant *slg-D*and transgenic plants overexpressing *SLG* displayed longer and narrower grains and larger leaf angles that are similar to the mutants or transgenic plants with enhanced BR signaling or increased BR levels. Secondly, *slg-D*and the WT had similar sensitivity to BL treatment in lamina joint bending, coleoptile elongation, and root elongation assays. Thirdly, the BRZ treatment restored *slg-D* to the WT. Fourthly, overexpression of *SLG* in the BR-related mutants, *d61-1* and *d11-2*, did not lead to slender grains and enlarged leaf angles. Fifthly, the major BRs were increased in *slg-D*. Sixthly, feedback regulation on expression of the known BR genes was seen in *slg-D*. Lastly, knockdown of *SLG* resembled mild BR-deficient mutants.

The size of an organ is determined by cell proliferation and cell expansion (Potter and Xu, 2001; Sugimoto-Shirasu and Roberts, 2003). Our results showed that *SLG* is required for cell expansion in grains. To investigate the possible regulatory relationship between *SLG* and other previously identified genes that control grain size by influencing cell expansion, such as *PGL1*, *GL7*, and *GS2/GL2* (Heang and Sassa, 2012; Che *et al.*, 2015;Duan *et al.*, 2015; Hu *et al.*, 2015; Wang *et al.*, 2015), we examined the transcript level of these genes and found no obvious difference between*slg-D* and the WT. This result suggests that*SLG* may regulate grain size in a pathway independent of *PGL1*, *GL7*, and*GS2/GL2*.

SLG is predicted to encode a BAHD acyltransferase-like protein. Previous studies of BAHD acyltransferase family members have shown that this family is capable of using CoA thioesters and catalyzing the formation of a wide variety of plant metabolites by generating ester or amide bonds (D'Auria, 2006). In Arabidopsis, two BAHD acyltransferases, BIA1 and BAT1, are involved in BR homeostasis, probably by conversion of active BR intermediates into inactive acylated BR conjugates (Roh *et al.*, 2012; Choi*et al.*, 2013). Overexpression of *BIA1* or *BAT1* results in decreased levels of active BRs and typical BR-deficient phenotypes (Roh *et al.*, 2012; Choi *et al.*, 2013). In our study, SLG, as a BAHD acyltransferase, probably converts an as yet unidentified substrate to the corresponding acyl conjugate to affect endogenous BR levels in an opposite way. Overexpression of *SLG*induced increased levels of active BRs and BR-overproduction phenotypes. The difference in Arabidopsis and rice implies that the function of BAHD acyltransferases in BR homeostasis has been differentiated. On the other hand, most of the enzymes known to catalyze BR biosynthesis belong to the cytochrome P450 protein family (Choe, 2006), implying that SLG, as a BAHD acyltransferase, may not work directly on the known BR intermediates, or that it may represent a different class of enzymes mediating BR synthesis. Further studies are needed to clarify how SLG participates in BR homeostasis.

In many cases, homomer formation is an essential biochemical process as it forms the complex quaternary structures of proteins to regulate selectivity against different substrates, enzyme activity, or stability (Dayhoff *et al.*, 2010). Here, we showed that SLG interacted with itself, and its N-terminal 30 amino acid region was required for the interaction. It is likely that the self-interaction of SLG forms a functional enzyme complex with a special quaternary structure that binds the target substrates effectively. A truncated protein SLGΔC1 lacking a 255 amino acid C-terminus is still able to interact with the intact version but may form a complex unable to function properly due to a change in the quaternary structure, thus leading to a dominant negative phenotype. This finding also suggests that the substrate recognition and/or catalyzing domain may be located in the C-terminus. The version without the N-terminal interaction region failed to create a dominant negative phenotype, further confirming that the homomer formation of SLG indeed exists *in vivo*. It will be interesting to investigate further the number of SLG proteins required to form a functional enzyme complex and its structural organization.

The plant architecture determines planting density, and thus yield. In rice, BR-deficient or -insensitive mutants show the erect leaf phenotype, such as *d2*, *d11*, and *d61* (Yamamuro *et al.*, 2000; Hong *et al.*, 2003; Tanabe *et al.*, 2005). More erect leaves that increase light capture and thus enhance photosynthetic efficiency and nitrogen storage for grain filling can be combined with high planting densities to improve grain yield and biomass in rice (Sinclair and Sheehy, 1999; Morinaka *et al.*, 2006; Sakamoto *et al.*, 2006). For example, modulating the expression levels of *OsDWARF4* and*OsBRI1* led to the erect leaf phenotype and efficiently improves rice grain yield and biomass in dense planting conditions (Morinaka *et al.*, 2006;Sakamoto *et al.*, 2006). *SLG*, when knocked-down by RNAi or interfered with by a truncated version, can create a compact semi-dwarf plant type with smaller leaf angles. Therefore, *SLG* can be used as an alternative to manipulate plant height for lodging resistance and leaf angle for planting density by optimizing its expression level, offering the potential for improving rice production.

ACKNOWLEDGEMENTS

We owe special thanks to Professor G. An (Department of Plant Systems Biotech, Kyung Hee University, Korea) for providing *slg-D* lines, and to Professor Chencai Chu (Institute of Genetics and Developmental Biology, Chinese Academy of Sciences, China) for providing *d61-1*, *d11-2*, and*m107* lines. We also thanks the supports by Key Laboratory of Biology, Genetics and Breeding of Japonica Rice in Mid-lower Yangtze River, Ministry of Agriculture, P.R.China, and Jiangsu Collaborative Innovation Center for Modern Crop Production, This research was supported by grants from the National Natural Science Foundation (91535302 and 31571601), 973 National Basic Research Program (2014CB943403), National Transformation Science and Technology Program (2014ZX08001006), 863 National High-tech R&D Program of China (2014AA10A603-15), Jiangsu Science and Technology Development Program (BE2014394), and the National Research Foundation of Korea (NRF-2014R1A1A2056102 to S-KK).

REFERENCES

1. Ali MH Imperiali B . 2005. Protein oligomerization: how and why. Bioorganic and Medicinal Chemistry 13, 5013–5020.
2. Asami T Min YK Nagata N Yamagishi K Takatsuto S Fujioka S Murofushi N Yamaguchi I Yoshida S . 2000. Characterization of brassinazole, a triazole-type brassinosteroid biosynthesis inhibitor. Plant Physiology 123, 93–100.
3. Ashikari M Sakakibara H Lin S Yamamoto T Takashi T Nishimura A Angeles ER Qian Q Kitano H Matsuoka M . 2005. Cytokinin oxidase regulates rice grain production. Science 309, 741–745.
4. Ashikari M Wu J Yano M Sasaki T Yoshimura A . 1999. Rice gibberellin-insensitive dwarf mutant gene Dwarf 1 encodes the alpha-subunit of GTP-binding protein. Proceedings of the National Academy of Sciences, USA 96, 10284–10289.
5. Belkhadir Y Chory J . 2006. Brassinosteroid signaling: a paradigm for steroid hormone signaling from the cell surface. Science 314, 1410–1411.
6. Berleth T Sachs T . 2001. Plant morphogenesis: long-distance coordination and local patterning. Current Opinion in Plant Biology 4, 57–62.
7. Bradley D Carpenter R Sommer H Hartley N Coen E . 1993. Complementary floral homeotic phenotypes result from opposite orientations of a transposon at the plena locus of Antirrhinum. Cell 72, 85–95.
8. Che R Tong H Shi B et al. 2015. Control of grain size and rice yield by GL2-mediated brassinosteroid responses. Nature Plants 2, 15195.
9. Chen S Tao L Zeng L Vega-sanchez ME Umemura K Wang G . 2006. A highly efficient transient protoplast system for analyzing defence gene expression and protein–protein interactions in rice. Molecular Plant Pathology 7, 417–427.
10. Choe S . 2006. Brassinosteroid biosynthesis and inactivation. Physiologia Plantarum 126, 539–548.
11. Choi S Cho YH Kim K Matsui M Son SH Kim SK Fujioka S Hwang I . 2013. BAT1, a putative acyltransferase, modulates brassinosteroid levels in Arabidopsis. The Plant Journal 73, 380–391.
12. Clouse SD Sasse JM . 1998. Brassinosteroids: essential regulators of plant growth and development. Annual Review of Plant Physiology and Plant Molecular Biology 49, 427–451.
13. D'Auria JC . 2006. Acyltransferases in plants: a good time to be BAHD. Current Opinion in Plant Biology 9, 331–340.
14. Dayhoff JE Shoemaker BA Bryant SH Panchenko AR . 2010. Evolution of protein binding modes in homooligomers. Journal of Molecular Biology 395, 860–870.
15. Duan P Ni S Wang J Zhang B Xu R Wang Y Chen H

Zhu X Li Y . 2015. Regulation of OsGRF4 by OsmiR396 controls grain size and yield in rice. Nature Plants 2, 15203.

16. Fujioka S Yokota T . 2003. Biosynthesis and metabolism of brassinosteroids. Annual Review of Plant Biology 54, 137–164.

17. He JX Gendron JM Sun Y Gampala SS Gendron N Sun CQ Wang ZY . 2005. BZR1 is a transcriptional repressor with dual roles in brassinosteroid homeostasis and growth responses. Science 307, 1634–1638.

18. Heang D Sassa H . 2012. Antagonistic actions of HLH/bHLH proteins are involved in grain length and weight in rice. PLoS One 7, e31325.

19. Hiei Y Ohta S Komari T Kumashiro T . 1994. Efficient transformation of rice (Oryza sativa L) mediated by Agrobacterium and sequence analysis of the boundaries of the T-DNA. The Plant Journal 6, 271–282.

20. Hong Z Ueguchi-Tanaka M Shimizu-Sato S et al. 2002. Loss-of-function of a rice brassinosteroid biosynthetic enzyme, C-6 oxidase, prevents the organized arrangement and polar elongation of cells in the leaves and stem. The Plant Journal 32, 495–508.

21. Hong Z Ueguchi-Tanaka M Umemura K Uozu S Fujioka S Takatsuto S Yoshida S Ashikari M Kitano H Matsuoka M . 2003. A rice brassinosteroid-deficient mutant, ebisu dwarf (d2), is caused by a loss of function of a new member of cytochrome P450. The Plant Cell 15, 2900–2910.

22. Hoshikawa K . 1989. Leaf. In: The growing rice plant. Tokyo: Nobunkyo.

23. Hu J Wang Y Fang Y et al. 2015. A rare allele of GS2 enhances grain size and grain yield in rice. Molecular Plant 8, 1455–1465.

24. Ikeda M Miura K Aya K Kitano H Matsuoka M . 2013. Genes offering the potential for designing yield-related traits in rice. Current Opinion in Plant Biology 16, 213–220.

25. Ishimaru K Hirotsu N Madoka Y et al. 2013. Loss of function of the IAA-glucose hydrolase gene TGW6 enhances rice grain weight and increases yield. Nature Genetics 45, 707–711.

26. Jefferson RA Kavanagh TA Bevan MW . 1987. GUS fusions: β-glucuronidase as a sensitive and versatile gene fusion marker in higher plants. EMBO Journal 6, 3901–3907.

27. Medline

28. Jeon JS Lee S Jung KH et al. 2000. T-DNA insertional mutagenesis for functional genomics in rice. The Plant Journal 22, 561–570.

29. Jeong DH An S Park S et al. 2006. Generation of a flanking sequence-tag database for activation-tagging lines in japonica rice. The Plant Journal 45, 123–132.

30. Kim T Hwang JY Joo S Chang S Lee J Takatsuto S Kim S . 2005. Arabidopsis CYP85A2, a cytochrome P450, mediates the Baeyer–Villiger oxidation of castasterone to brassinolide in brassinosteroid biosynthesis. The Plant Cell 17, 2397–2412.

31. Li J Chory J . 1997. A putative leucine-rich repeat receptor kinase involved in brassinosteroid signal transduction. Cell 90, 929–938.

32. Li Y Fan C Xing Y et al. 2011. Natural variation in GS5 plays an important role in regulating grain size and yield in rice. Nature Genetics 43, 1266–1269.

33. Li Z Paterson AH Pinson SRM Stansel JW . 1999. RFLP facilitated analysis of tiller and leaf angles in rice (Oryza sativa L). Euphytica 109, 79–84.

34. Li ZK Paterson AH Pinson SRM Khush G . 1998. A major gene, Tal and QTLs affecting tiller and leaf angles in rice. Rice Genetics Newsletter 15, 79–84.

35. Liu L Tong H Xiao Y Che R Xu F Hu B Liang C Chu J Li J Chu C . 2015. Activation of Big Grain1 significantly improves grain size by regulating auxin transport in rice. Proceedings of the National Academy of Sciences, USA 112, 11102–11107.

36. Mao H Sun S Yao J Wang C Yu S Xu C Li X Zhang Q . 2010. Linking differential domain functions of the GS3 protein to natural variation of grain size in rice. Proceedings of the National Academy of Sciences, USA 107, 19579–19584.

37. Miernyk JA Thelen JJ . 2008. Biochemical approaches for discovering protein–protein interactions. The Plant Journal 53, 597–609.

38. Morinaka Y Sakamoto T Inukai Y Agetsuma M Kitano H Ashikari M Matsuoka M . 2006. Morphological alteration caused by brassinosteroid insensitivity increases the biomass and grain production of rice. Plant Physiology 141, 924–931.

39. Nakamura A Fujioka S Takatsuto S Tsujimoto M Kitano H Yoshida S Asami T Nakano T . 2009. Involvement of C-22-hydroxylated brassinosteroids in auxin-induced lamina joint bending in rice. Plant and Cell Physiology 50, 1627–1635.

40. Ning J Zhang B Wang N Zhou Y Xiong L . 2011. Increased leaf angle1, a Raf-like MAPKKK that interacts with a nuclear protein family, regulates mechanical tissue formation in the lamina joint of rice. The Plant Cell 23, 4334–4347.

41. Potter CJ Xu T . 2001. Mechanisms of size control. Current Opinion in Genetics and Development 11, 279–286.

42. Roh H Jeong CW Fujioka S Kim YK Lee S Ahn JH Choi YD Lee JS . 2012. Genetic evidence for the reduction of brassinosteroid levels by a BAHD acyltransferase-like protein in Arabidopsis. Plant Physiology 159, 696–709.

43. Sakamoto T Morinaka Y Ohnishi T et al. 2006. Erect leaves caused by brassinosteroid deficiency increase biomass production and grain yield in rice. Nature Biotechnology 24, 105–109.

44. Shimada Y Goda H Nakamura A Takatsuto S Fujioka S Yoshida S . 2003. Organ-specific expression of brassinosteroid-biosynthetic genes and distribution of endogenous brassinosteroids in Arabidopsis. Plant Physiology 131, 287–297.

45. Sinclair TR Sheehy JE . 1999. Erect leaves and photosynthesis in rice. Science 283, 1455.

46. Medline

47. Song XJ Huang W Shi M Zhu MZ Lin HX . 2007. A QTL for rice grain width and weight encodes a previously unknown RING-type E3 ubiquitin ligase. Nature Genetics 39, 623–630.

48. Song XJ Kuroha T Ayano M et al. 2015. Rare allele of a previously unidentified histone H4 acetyltransferase enhances grain weight, yield, and plant biomass in rice. Proceedings of the National Academy of Sciences, USA 112, 76–81.

49. Sugimoto-Shirasu K Roberts K . 2003. 'Big it up': endoreduplication and cell-size control in plants. Current Opinion in Plant Biology 6, 544–553.

50. Symons GM Reid JB . 2004. Brassinosteroids do not undergo long-distance transport in pea. Implications for the regulation of endogenous brassinosteroid levels. Plant Physiology 135, 2196–2206.

51. Tanabe S Ashikari M Fujioka S et al. 2005. A novel cytochrome P450 is implicated in brassinosteroid biosynthesis via the characterization of a rice dwarf mutant, dwarf11, with reduced seed length. The Plant Cell 17, 776–790.

52. Tanaka A Nakagawa H Tomita C et al. 2009. BRASSINOSTEROID UPREGULATED1, encoding a helix–loop–helix protein, is a novel gene involved in brassinosteroid signaling and controls bending of the lamina joint in rice. Plant Physiology 151, 669–680.

53. Tong H Jin Y Liu W Li F Fang J Yin Y Qian Q Zhu L Chu C . 2009. DWARF AND LOW-TILLERING, a new member of the GRAS family, plays positive roles in brassinosteroid signaling in rice. The Plant Journal 58, 803–816.

54. Tong H Liu L Jin Y Du L Yin Y Qian Q Zhu L Chu C . 2012. DWARF AND LOW-TILLERING acts as a direct downstream target of a GSK3/SHAGGY-like kinase to mediate brassinosteroid responses in rice. The Plant Cell 24, 2562–2577.

55. Unnevehr LJ Duff B Juliano VO . 1992. Consumer demand for rice grain quality: terminal report of IDRC projects National Grain Quality (Asia) and International Grain Quality Economics (Asia). International Rice Research Institute.

56. Waadt R Kudla J . 2008. In planta visualization of protein interactions using bimolecular fluorescence complementation (BiFC). CSH Protocols 2008, t4995.

57. Wada K Marumo S Ikekawa N Morisaki M Mori K . 1981. Brassinolide and homobrassinolide promotion of lamina inclination of rice seedlings. Plant and Cell Physiology 22, 323–325.

58. Wan S Wu J Zhang Z et al. 2009. Activation tagging, an efficient tool for functional analysis of the rice genome. Plant Molecular Biology 69, 69–80.

59. Wang S Wu K Yuan Q et al. 2012. Control of grain size, shape and quality by OsSPL16 in rice. Nature Genetics 44, 950–954.

60. Wang Y Xiong G Hu J et al. 2015. Copy number variation at the GL7 locus contributes to grain size diversity in rice. Nature Genetics 47, 944–948.

61. Wang ZY Nakano T Gendron J et al. 2002. Nuclear-localized BZR1 mediates brassinosteroid-induced growth and feedback suppression of brassinosteroid biosynthesis. Developmental Cell 2, 505–513.

62. Weng J Gu S Wan X et al. 2008. Isolation and initial characterization of GW5, a major QTL associated with rice grain width and weight. Cell Research 18, 1199–1209.

63. Wu CY Trieu A Radhakrishnan P et al. 2008. Brassinosteroids regulate grain filling in rice. The Plant Cell 20, 2130–2145.

64. Wu Z Zhang X He B et al. 2007. A chlorophyll-deficient rice mutant with impaired chlorophyllide esterification in chlorophyll biosynthesis. Plant Physiology 145, 29–40.

65. Yamamuro C Ihara Y Wu X Noguchi T Fujioka S Takatsuto S Ashikari M Kitano H Matsuoka M . 2000. Loss of function of a rice brassinosteroid insensitive1 homolog prevents internode elongation and bending of the lamina joint. The Plant Cell 12, 1591–1606.

66. Zhang LY Bai MY Wu J et al. 2009. Antagonistic HLH/bHLH transcription factors mediate brassinosteroid regulation of cell elongation and plant development in rice and Arabidopsis. The Plant Cell 21, 3767–3780.

67. Zhang X Wang J Huang J et al. 2012. Rare allele of OsPPKL1 associated with grain length causes extra-large grain and a significant yield increase in rice. Proceedings of the National Academy of Sciences, USA 109, 21534–21539.

68. Zhao SQ Hu J Guo LB Qian Q Xue HW . 2010. Rice leaf inclination 2, a VIN3-like protein, regulates leaf angle through modulating cell division of the collar. Cell Research 20, 935–947.

CHAPTER 17 OVERCOMING INTER-SUBSPECIFIC HYBRID STERILITY IN RICE BY DEVELOPING INDICA-COMPATIBLE JAPONICA LINES.

Guo J, Xu X, Li W, Zhu W, Zhu H, Liu Z, Luan X, Dai Z, Liu G, Zhang Z, Zeng R, Tang G, Fu X, Wang S, and Zhang G

INTRODUCTION

Asian cultivated rice (*Oryza sativa* L.) is the staple food for more than half of the world's population. Continuing production of high-yielding rice is essential for maintaining global food security1·2. Asian cultivated rice is mainly distributed in Asia as well as some other areas in the world. During the course of evolution, the cultivated rice was differentiated into two distinct eco-geographic subspecies, *indica* and*japonica*3. The *indica* subspecies is generally adapted to the humid regions of tropical and subtropical Asia whereas *japonica* rice is mainly distributed in the temperate regions. Since China pioneered *indica*hybrid rice production in the 1970's, a great success has been achieved in the hybrid rice development in China and around the world4·5. The*indica* hybrid rice usually has more about 20% yield increase compared with conventional varieties, and now accounts for more than half of the annual rice planting area in China5·6. Currently used rice varieties with a close genetic relationship have limited heterosis, leading to a yield plateau for *indica* hybrid rice production7. Strong heterosis exists in the inter-subspecific hybrids and the exploitation of this heterosis has long been considered as a promising way to further increase rice yield potential4·5·8. However, the major obstacle of utilizing the heterosis between the subspecies is the strong hybrid sterility3·9·10·11·12.

The hybrid sterility occurs frequently in many remote crosses in rice9. The spikelet fertility of the hybrids varied widely among the crosses from almost completely sterile to fully fertile in a diallel set of 210 crosses involving 21 parents10. Spikelet fertility of F_1 hybrids was found to be negatively related with the genetic divergence index of parental varieties13. The genetic basis of the inter-subspecific hybrid sterility has been extensively investigated in the last several decades. Approximately 50 loci for hybrid fertility have been identified in rice, including loci causing female gamete abortion and those inducing pollen sterility. Among the loci for hybrid fertility, some were identified in the inter-subspecific crosses of *O. sativa* while others were found in the crosses between *O. sativa* and other species of *Oryza* genus11. Among the loci causing female sterility in the inter-subspecific hybrids, *S5* is a major locus14·15. The *S5* locus was mapped onto chromosome 6 and three alleles, namely *indica* allele (*S5*-i), *japonica* allele (*S5*-j) and neutral allele (*S5*-n), were identified at the locus14·16. The *S5* gene was subsequently isolated through map-based cloning approach, which encodes an aspartic protease to manipulate the female fertility of the hybrids17. The*S5*-i and *S5*-j alleles differed only at two nucleotides for the female sterility, while the *S5*-n allele has a loss-function mutation with 136 bp DNA-sequence deletion for the compatibility17·18·19. For F_1 pollen sterility, five loci, i.e. *Sa* (*S-E3*), *Sb* (*S-E2*), *Sc* (*S-E5*), *Sd* and *Se*, were identified through a series of allelic test-crosses20·21·22·23·24. Among five identified F_1 pollen sterility loci, the *Sa* locus was mapped to a region of 30 kb on chromosome 125·26, and was subsequently cloned27; the *Sb*locus was delimited to a region of 27 kb on chromosome 528·29; the *Sc*locus was narrowed down to a region of 46 kb on chromosome 330·31; the *Sd* locus was delimited to a region of 67 kb on chromosome 132; and the *Se* locus was mapped to an interval of 5 cM on chromosome 1233. Interestingly, some loci for hybrid sterility, such as *S24* and *S31*34·35,*S35*36, *S25* and *S36*37·38 were also identified in the same regions as the*Sb*, *Sd* and *Se* loci,

Citation: Guo, J. et al. Overcoming inter-subspecific hybrid sterility in rice by developing indica-compatible japonica lines. Sci. Rep. 6, 26878; doi: 10.1038/srep26878 (2016).

respectively, by different research groups. To overcome the inter-specific hybrid sterility, a strategy of developing*indica*-compatible *japonica* lines (ICJLs) which carry *S*-i alleles at the loci for hybrid sterility in *japonica* genetic background, was proposed23,39. The pyramiding effect of the genes for hybrid sterility was evaluated by developing pyramiding lines with *S*-i alleles at multiple loci of *Sa*, *Sb* and *Sc*24. Here, we report that the inter-subspecific hybrid sterility can be overcome by using ICJLs. The ICJLs carrying the *S*-i alleles at *Sb*, *Sc*, *Sd* and *Se* loci for F_1 pollen fertility and the *S*-n allele at*S5* locus for F_1 female fertility in *japonica* genetic background are compatible with *indica* while incompatible with *japonica*, reversing the compatibility of general *japonica* rice. Our results showed that the ICJLs would be an important genetic stock for overcoming the inter-specific hybrid sterility in rice.

RESULTS

F_1 pollen sterility caused by allelic interaction at the *Sb, Sc, Sd* and *Se*loci

To assess F_1 pollen sterility caused by the loci for F_1 pollen sterility, a set of Taichung65 (T65) isogenic F_1-sterile lines (TISLs) with the *indica*allele at the locus for F_1 pollen sterility in the genetic background of T65 (a *japonica* variety) were developed by marker-facilitated backcrossing. Each TISL carried only one chromosomal substituted segment from a donor *indica* cultivar detected by a survey of the genome with molecular markers. Of the 11 TISLs, 3, 4, 3 and 1 TISLs carried *S*-i at the loci *Sb*, *Sc*, *Sd* and *Se*, respectively (Fig. 1). All the TISLs were test-crossed with T65 which has *S*-j allele at the all loci to generate heterozygote *S*-i/*S*-j at the corresponding locus. The F_1 hybrids in the crosses showed partial pollen fertility, ranging from 36.59% to 79.00%. In all the crosses, distorted segregation from the Mendelian segregation ratio was observed in the F_2 populations, in which the ratio of genotype JJ (*S*-j/*S*-j) was significantly lower than that of genotype II (*S*-i/*S*-i) (Fig. 1). The results indicated that the interaction between the *S*-i allele from the TISLs and the *S*-j allele from T65 at the loci caused pollen with the *S*-j allele aborted in varying degrees in the crosses. Similar results were obtained from the test-crosses of TISLs with other *japonica* testers.

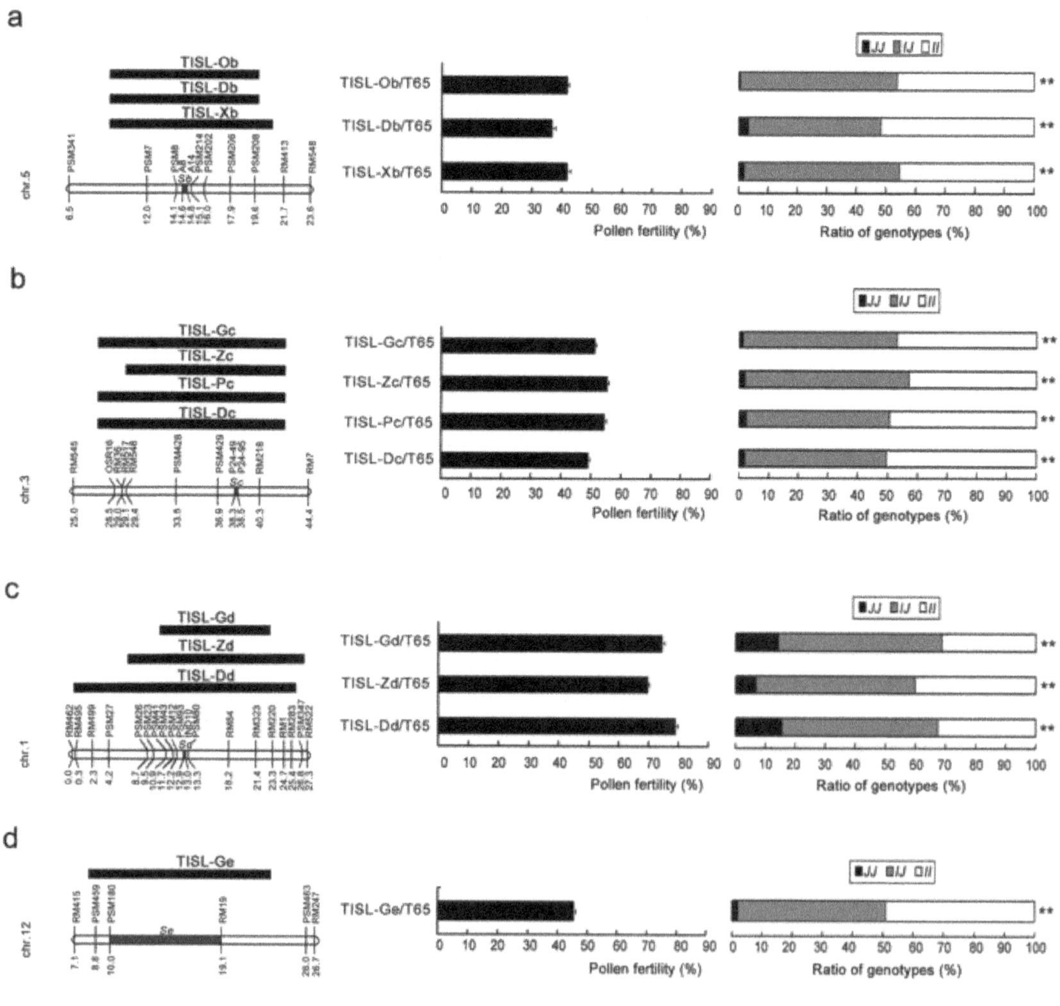

Figure 1: Location and genetic effect of the genes at the *Sb*, *Sc*, *Sd*and *Se* loci on the chromosomal substituted segments in the TISLs. (**a**) The *Sb* locus in TISL-Ob, TISL-Db and TISL-Xb. (**b**) The *Sc* locus in TISL-Gc, TISL-Zc, TISL-Pc and TISL-Dc. (**c**) The *Sd* locus in TISL-Gd, TISL-Zd and TISL-Dd. (**d**) The *Se* locus in TISL-Ge.*Left*: Chromosome locations of the *Sb*, *Sc*, *Sd* and *Se* loci on the substituted segments in the TISLs. The black horizontal bars represent substituted segments in the TISLs. The positions of the *Sb*, *Sc*, *Sd* and *Se* loci are indicated in the chromosome maps. *Middle*: Pollen fertility of F_1 hybrids from the crosses of TISLs with T65. Error bars represent the SD. *Right*: Ratios of genotypes at the *Sb*, *Sc*, *Sd* and *Se* loci in the F_2 populations from the crosses of TISLs with T65. *JJ*, Genotype of T65 (*S*-j/*S*-j). *IJ*, Heterozygous genotypes (*S*-i/*S*-j). *II*, Genotypes of TISLs (*S*-i/*S*-i). **Significantly different at the 0.01 probability level.

Pyramiding effect of the genes for F_1 pollen sterility at the *Sb, Sc, Sd*and *Se* loci

To analyse the pyramiding effect of the genes for F_1 pollen sterility at the *Sb*, *Sc*, *Sd* and *Se* loci, a set of 7 TISLs with different numbers of *S*-i allele at the *Sb*, *Sc* and *Sd* loci were developed using the *indica* variety Dijiaowujian as the *S*-i allele donor. In the test-crosses of the TISLs with T65, the F_1 pollen fertility was 43.5%, 81.3% and 62.0% in the genotypes of *S*-i/*S*-j at the single locus of *Sb*, *Sc* and *Sd*, respectively. The pollen fertility was 42.4%, 19.7% and 51.7% in the double heterozygotes at *Sb*and *Sc*, *Sb* and *Sd*, and *Sc* and *Sd*, respectively, and the fertility was only 14.0% in the trihybrid at the *Sb*, *Sc* and *Sd* loci (Fig. 2a). These results indicated that the pollen fertility of heterozygotes decreased with the increase of the number of heterozygous loci for the F_1 pollen sterility, and the pollen fertility of heterozygotes at multiple loci was approximate to the product of pollen fertility in heterozygotes at each of the loci. Similar results were observed in a BC_4F_2 population derived from the backcross between Guangluai4 (a donor of the *S*-i allele at the*Sb*, *Sd* and *Se* loci) and T65. The F_1 pollen fertility was 33.1%, 85.4% and 46.8% for the single heterozygous locus, respectively, whereas it was 21.3%, 15.0% and 31.3% for the two heterozygous loci of *Sb* and *Sd*, *Sb*and *Se*, and *Sd* and *Se*, respectively, and the fertility was only 7.0% for the line with three heterozygous alleles at *Sb*, *Sd* and *Se* loci (Fig. 2b).

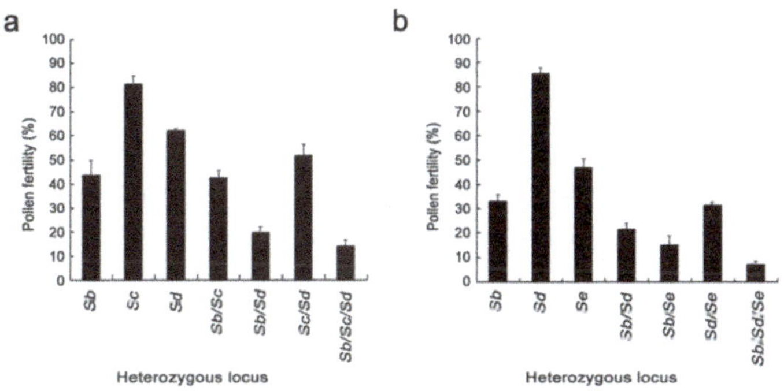

Figure 2: F_1 pollen sterility caused by allelic interactions at different combinations of the *Sb*, *Sc*, *Sd* and *Se* loci. (**a**) F_1 pollen fertility in the crosses of TISLs developed from the Dijiaowujian donor with T65 in the first cropping season in 2002. (**b**) F_1 pollen fertility in the crosses of TISLs developed from the Guangluai4 donor with T65 in the second cropping season in 2005. Error bars represent the SD.

Overcoming the pollen sterility of the inter-subspecific hybrids by developing of the TISLs with *S*-i alleles at the *Sb*, *Sc*, *Sd* and *Se* loci

TISL-Dbcd with *S*-i at the *Sb*, *Sc* and *Sd* loci from Dijiaowujian and TISL-Gde with *S*-i at the *Sd* and *Se* loci from Guangluai4 were selected to develop pyramiding line with *S*-i at the *Sb*, *Sc*, *Sd* and *Se* loci. The pyramiding line TISL-Dbc-Gde carried the *S*-i allele at the *Sb* and *Sc* loci from Dijiaowujian and the *S*-i allele at the *Sd* and *Se* loci from Guangluai4. TISL-Dbc-Gde carried 62.1 cM of substituted segments with the *S*-i allele from the *indica*donors that accounted for only 4.06% of the rice genome (Fig. 3c). TISL-Dbc-Gde was similar to T65 in terms of plant type, panicle type and grain type (Fig. 3a,b). No significant difference between TISL-Dbc-Gde and T65 was detected on the traits of days from sowing to heading, plant height, thousand grain weight, grain length, grain width, and panicle number per plant. These results indicated that TISL-Dbc-Gde was a *japonica* line with a genotype and phenotype similar to T65. To evaluate the compatibility, TISL-Dbc-Gde was test-crossed with a set of testers consisting of typical *indica* and *japonica* varieties for two years. In the F_1hybrids from the crosses of TISL-Dbc-Gde with the *japonica* testers, the pollen fertility was 0.97% and 2.40% and the spikelet fertility was close to complete sterility in the two experiments. In the F_1 hybrids from the crosses of TISL-Dbc-Gde with the *indica* testers, the pollen fertility was 88.22% and 89.17% and the spikelet fertility was 61.25% and 57.15%. As a control, the pollen fertility was 91.40% and 17.83% and the spikelet fertility was 94.60% and 6.96% in the F_1 hybrids from the crosses of T65 with the *japonica* and *indica* testers, respectively. These results indicated that TISL-Dbc-Gde showed reverse compatibility with T65. In other words, TISL-Dbc-Gde was compatible with the *indica* testers but incompatible with the *japonica* testers in terms of pollen fertility. Thus, the pollen sterility of the inter-subspecific hybrids was effectively overcome in the crosses of TISL-Dbc-Gde with the *indica* testers.

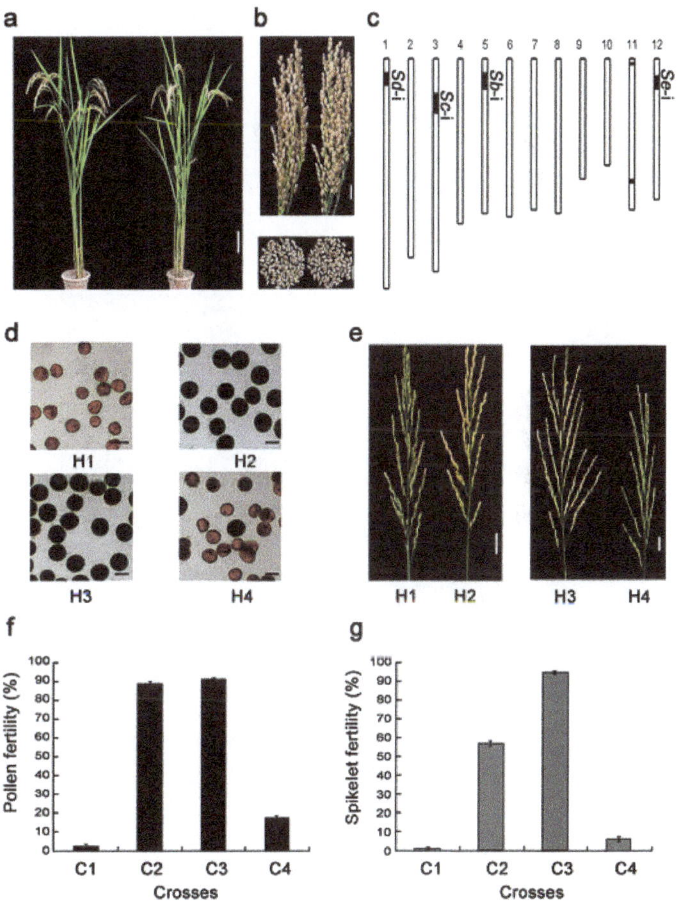

Figure 3: Phenotypes and genotypes of TISL-Dbc-Gde with *S*-i at the *Sb*, *Sc*, *Sd* and *Se* loci. (**a**) Plant morphology of TISL-Dbc-Gde (left) and T65 (right). Scale bar, 10 cm. (**b**) Panicle-type and grain-type of TISL-Dbc-Gde (left) and T65 (right). Scale bar, 2 cm. (**c**) Positions of the substituted segments in the genome of TISL-Dbc-Gde. *Vertical* bars are a graphical representation of chromosomes. *Deep* parts are substitution segments from donors, and*Light* parts are the genetic background from T65. (**d**) Pollen grains stained by I_2-KI solution in the F_1 hybrids of four crosses. Scale bar, 30 μm. (**e**) Spikelet fertility in the panicles of the F_1 hybrids of four crosses. Scale bar, 2 cm. (**f**) Pollen fertility of the F_1 hybrids from the crosses between TISL-Dbc-Gde and the testers in the second cropping season of 2012 (n=15).

(**g**) Spikelet fertility of the F_1 hybrids from the crosses between TISL-Dbc-Gde and the testers in the second cropping season of 2012 (n=15). H1, TISL-Dbc-Gde/Ballila; H2, T65/Ballila; H3, TISL-Dbc-Gde/Aijiaonante; H4, T65/Aijiaonante; C1, TISL-Dbc-Gde/*japonica*-testers; C2, TISL-Dbc-Gde/*indica* testers; C3, T65/*japonica* testers; C4, T65/*indica* testers. Error bars represent the SD.

Overcoming the spikelet sterility of the inter-subspecific hybrids by developing ICJLs

Although the pollen fertility was almost normal, the spikelet fertility was still low in the F_1 hybrids from the crosses of TISL-Dbc-Gde with the *indica* testers. To survey the problem, the genetic behaviour of six loci for hybrid sterility was investigated in all available crosses of TISL-Dbc-Gde with the *indica*testers. No distorted segregation was found at the *Sa, Sb, Sc, Sd* and *Se*loci for F_1 pollen sterility in eighteen F_2 populations, except for two populations at the *Sa* locus and one population at the *Sd* locus. However, distorted segregation was found in all five F_2 populations at the *S5* locus for F_1 embryo sac sterility.

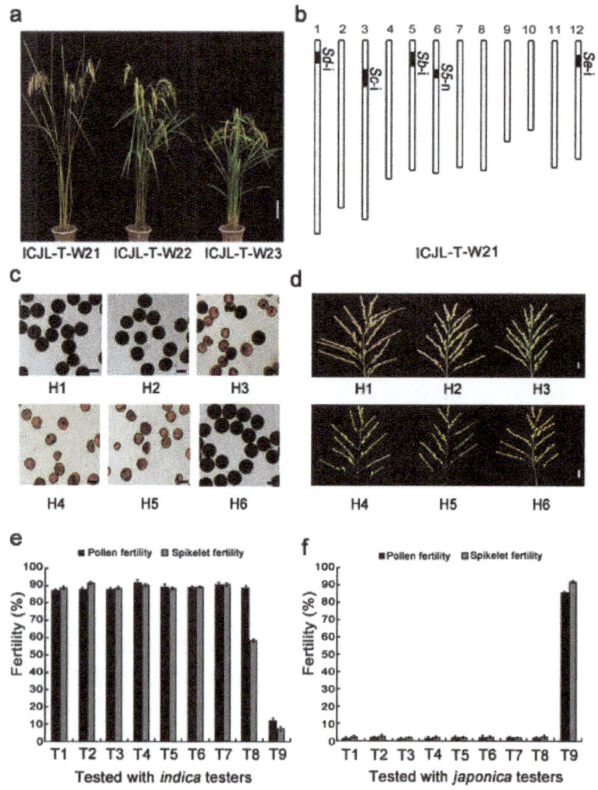

Figure 4: Phenotypes and genotypes of ICJLs with *S*-i at the *Sb, Sc,Sd* and *Se* loci and *S*-n at the *S5* locus. (**a**) Plant morphologies of ICJL-T-W21, ICJL-T-W22 and ICJL-T-W23. Scale bar, 10 cm. (**b**) Positions of the substituted segments in the ICJL-T-W21 genome. *Vertical* bars are a graphical representation of the chromosomes. *Deep* parts are substitute segments from donors, and *light* parts are the genetic background from *japonica* donors. (**c**) Pollen grains of the F_1 hybrids from six crosses. Pollen grains were stained with the I_2-KI solution. Scale bar, 30 µm. (**d**) Spikelet fertility in the panicles of the F_1 hybrids from six crosses. Scale bar, 2 cm. (**e**) Pollen fertility and spikelet fertility of the F_1 hybrids from the crosses of ICJLs (TISL-Dbc-Gde and T65 as controls) with the *indica* testers in the second cropping season in 2014 (n=10). (**f**) Pollen fertility and spikelet fertility of the F_1 hybrids from the crosses of ICJLs (TISL-Dbc-Gde and T65 as controls) with the *japonica* testers in the second cropping season in 2014 (n=10). H1, ICJL-T-W21/9311; H2, TISL-Dbc-Gde/9311; H3, T65/9311; H4, ICJL-T-W21/T65; H5, TISL-Dbc-Gde/T65; H6, T65. T1, ICJL-T-W6; T2, ICJL-T-W19; T3, ICJL-T-W21; T4, ICJL-T-W22; T5, ICJL-T-W23; T6, ICJL-T-W24; T7, ICJL-T-W27; T8, TISL-Dbc-Gde; T9, T65. Error bars represent the SD.

These results implied that the partial spikelet sterility in the crosses of TISL-Dbc-Gde with the *indica* testers was mainly controlled by the allelic interaction of *S*-i/*S*-j at the *S5* locus. To identify the *S5*-n gene, 171 accessions of *O. sativa* collected throughout the world were used to identify the allele types at the *S5* locus using functional molecular markers of the *S5* gene. Seventeen of the accessions were identified as carrying the *S5*-n allele. To develop the ICJLs, seven *japonica* accessions with the *S5*-n allele were selected as donors and crossed with TISL-Dbc-Gde. Totally seven ICJLs from the seven crosses were developed through marker-assisted backcrossing. The ICJLs carried homozygous *S*-i allele at the *Sb, Sc, Sd* and *Se* loci from TISL-Dbc-Gde and the *S5*-n allele from the *S5*-n donors. To evaluate the compatibility, the ICJLs were test-crossed with a set of testers consisting of typical *indica* and *japonica* varieties. The F_1 pollen fertility of the ICJLs was similar to that of TISL-Dbc-Gde when test-crossed with the *indica* or *japonica* testers. All of the ICJLs were compatible with the *indica* testers but incompatible with the *japonica* testers in terms of pollen fertility. The F_1 spikelet fertility of the ICJLs was different from that of TISL-Dbc-Gde when test-crossed with the *indica* testers. All of the ICJLs were compatible with the *indica* testers and had high spikelet fertility in the crosses while they were incompatible with the *japonica* testers and had high spikelet sterility in the crosses. These results indicated that the ICJLs showed reverse compatibility with T65 and other *japonica* varieties since they were compatible with *indica* while incompatible with *japonica* rice. Thus, the spikelet sterility of the inter-subspecific hybrids was effectively overcome.

DISCUSSION

Hybrid sterility is the most common form of postzygotic reproductive isolation in plants. The best-known example is perhaps the hybrid sterility between *indica* and *japonica* subspecies of Asian cultivated rice40. The hybrid sterility between *indica* and *japonica* subspecies of Asian cultivated rice is a complex trait. It consists of male sterility or pollen sterility and female sterility or embryo-sac sterility, and varies in different crosses and in different environments15·24·40·41. In some studies, the hybrid sterility was only detected by spikelet fertility. Thus, it is not known to what extent male and female gamete abortions influence the spikelet fertility15. Among approximate 50 loci for the hybrid sterility identified in rice, some loci were detected in inter-subspecific hybrids of *O. sativa* while others were found in inter-specific hybrids between *O. sativa* and other species of *Oryza* genus11. Thus, it is not clear how many loci involving in the inter-specific hybrid sterility. In this study, the ICJLs, carrying the *S*-i allele at the *Sb, Sc, Sd*and *Se* loci and the *S*-n allele at the *S5* locus, were almost completely compatible with *indica* testers and incompatible with *japonica* testers in pollen fertility and in spikelet fertility. These results indicated that the hybrid sterility in inter-subspecific hybrids was mainly controlled by the genes at the *Sb, Sc, Sd, Se* and *S5* loci. Understanding the genetic basis of hybrid sterility in rice lays the foundation for overcoming the hybrid sterility in inter-subspecific hybrids.

At least three strategies have been proposed for overcoming the hybrid sterility in *indica-japonica* crosses. First, the neutral alleles, or wide-compatibility genes (WCGs) can be introgressed from the wide-compatibility varieties (WCVs) into the parents whose hybrids exhibit strong yield heterosis. The second strategy is to breed 'ICJLs' by introgressing *indica* alleles of several hybrid sterility loci into *japonica*lines by backcrossing. The third is to produce artificial neutral alleles by suppressing expression of the genes causing hybrid sterility with RNAi or microRNA technology, if such gene silencing does not affect the plant growth or development40. The *S5*-n gene was identified as a WCG and the varieties carrying *S5*-n, named WCVs, were believed to be compatible with both *indica* and *japonica*14. However, it has been frequently found that the *S5*-n gene alone is not sufficient for producing *indica-japonica* hybrids with normal fertility. *S5*-n can only overcome the sterility caused by embryo sac abortion15. We proposed the strategy of development of 'ICJLs' which carry the *S*-i alleles at the loci for hybrid sterility in the *japonica* genetic background by backcrossing and MAS23·39. Following this strategy, we developed a set of the ICJLs which carry the *S*-i allele at the *Sb*, *Sc*, *Sd* and *Se* loci and the *S*-n allele at the *S5* locus, as an outcome of two decades of continuing pursuit. The ICJLs are compatible with *indica* while incompatible to *japonica* rice. Thus, the hybrid sterility of *indica-japonica* rice was effectively overcome in the crosses of the ICJLs with the *indica* rice. It provides the first example of overcoming hybrid sterility in plants.

The breeding of *indica-japonica* rice has been practiced since the last decades. On the one hand, the utilization of inter-subspecific crosses has been considered as an efficient approach to develop traditional varieties with high yield potential1·2. On the other hand, the hybrids between *indica* and *japonica* varieties often show strong heterosis compared with intra-subspecific hybrids. For example, in China's "super" rice breeding, the two-line or three-line method was used to develop F_1 hybrid combinations by crossing an intermediate type between *indica* and *japonica* with an *indica* parent in order to use inter-subspecific heterosis1·5. The ICJLs will be an important genetic stock for breeding of *indica-japonica* rice. First, the ICJLs will be used to develop traditional varieties with higher yield potential. In the crosses of ICJLs with *indica* varieties, recombination of the genes from *indica* and*japonica* (ICJLs) will be no longer limited by hybrid sterility. Second, the ICJLs will be used to develop inter-subspecific hybrid rice. In the two-line or three-line rice hybrid system, ICJLs can be used to develop sterile line (A line), or restorer line (R line). Recently, a platform for breeding by design of CMS sterile lines based on an SSSL library in rice was developed42. By the combination of the strategy of the ICJLs with the platform for breeding by design, the three-lines of ICJLs for hybrid rice will be developed effectively. It will lead to open a new horizon in utilization of heterosis in the *indica-japonica* hybrid rice.

METHODS

Plant materials and growth conditions

All of the TISLs, ICJLs, *indica* testers, *japonica* testers and other plant materials in this study were planted in an experimental station at South China Agricultural University, Guangzhou (23°07'N, 113°15'E). The materials were planted in two cropping seasons each year. The first cropping season was from late February to middle July, and the second cropping season was from late July to middle November. The seeds were sowed on seed beds, and the seedlings were transplanted to the fields. Field management, including irrigation, fertilizer application and pest control, followed essentially normal agricultural practices.

Marker development and assay

DNA was extracted from fresh young leaves using the CTAB method43. The PCR profile used for amplification followed a previously described protocol44. The SSR markers used in this study were selected on rice microsatellite maps45·46. To identify the genotypes at the *S5* locus, we selected five functional markers and designed several markers based on the DNA sequences of *S5* locus12·19·47. To identify the genotypes at the *Sa*locus, we selected marker G02-14827. To conduct marker-assisted selection at other loci, SSR markers PSM8, PSM12 and PSM180 and the insertion/deletion (InDel) markers IND19 and ID5 were developed in our lab.

Examination of pollen and spikelet fertility

To examine pollen fertility, 6–9 mature flowers were collected from the upper one-third of the panicles of plants during the flowering time and fixed in FAA solution (ethanol, formaldehyde and acetic acid at a ratio of 89:6:5). The pollen was stained with a 1% I_2-KI solution containing 0.1% (w/v) iodine and 1% (w/v) potassium iodide. More than 300 pollen grains were randomly scanned per plant. The pollen was divided into three types: normal pollen (normal size and fully stained), stained abortive pollen (small size and lightly stained) and empty abortive pollen (small size and empty)20. To examine spikelet fertility, three panicles per plant were harvested during the maturation time. Ten to twenty plants were recorded for each variation.

Statistical analysis

The statistical model $y_{ij}=\mu+G_i+{}_{ij}$ was used to analyse the variance (ANOVA) of all data obtained from the experimental materials in one environment, where y_{ij} was the *j*th observed value of the target trait for the *i*th genotype and μ, *G* and were the population mean value, genotypic effects and residual error, respectively. For the estimation of genotypic effects on target traits, the significance of the difference between one genotype and the control selected at the probability level was tested by the least significant difference (LSD) method. Data were given as the mean ± standard deviation (SD) and transferred by $\arcsin^{-1}$ prior to analysis if they were provided as a percentage. Statistical analysis and graphing were performed using SPSS version 1848 and SigmaPlot49.

REFERENCES

1. Peng, S., Khush, G. S., Virk, P., Tang, Q. & Zou, Y. Progress in ideotype breeding to increase rice yield potential. *Field Crops Res*108, 32–38 (2008).
2. Khush, G. S. Strategies for increasing the yield potential of cereals: case of rice as an example. *Plant Breeding* 132, 433–436 (2013).

3. Kato, S., Kosaka, H. & Hara, S. On the affinity of rice varieties as shown by fertility of hybrid plants. *Bull. Sci. Fac. Agric. Kyushu. Univ.* 3, 132–147 (1928).

4. Yuan, L. & Virmani, S. Status of hybrid rice research and development. In *Hybrid rice* (Internationa Rice Research Institute, Manila, Philippines), 7–24 (1988).

5. Cheng, S., Zhuang, J., Fan, Y., Du, J. & Cao, L. Progress in research and development on hybrid rice: a super-domesticate in China.*Ann. Bot.* 100, 959–966 (2007).

6. Huang, J. E. Z., Zhang, H. & Shu, Q. Workable male sterility systems for hybrid rice: Genetics, biochemistry, molecular biology, and utilization. *Rice* 7, 13 (2014).

7. Peng, S., Laza, R. C., Visperas, R. M., Khush, G. S., Virk, P. & Zhu, D.Rice: progress in breaking the yield ceiling "New directions for a diverse planet". In: Proceedings of the 4th international crop science congress. Brisbane, Australia. 26 (2004).

8. Xiong, Z. *et al.* Latitudinal distribution and differentiation of rice germplasm: its implications in breeding. *Crop Sci.* 51, 1050–1058 (2011).

9. Oka, H. Genic analysis for the sterility of hybrids between distantly related varieties of cultivated rice. *J. Genet.* 55, 397–409 (1957).

10. Liu, K., Zhou, Z., Xu, C., Zhang, Q. & Maroof, M. An analysis of hybrid sterility in rice using a diallel cross of 21 parents involving*indica*, *japonica* and wide compatibility varieties. *Euphytica* 90, 275–280 (1996).

11. Ouyang, Y., Chen, J., Ding, J. & Zhang, Q. Advances in the understanding of inter-subspecific hybrid sterility and wide-compatibility in rice. *Chin. Sci. Bull.* 54, 2332–2341 (2009).

12. Yang, J. *et al.* A killer-protector system regulates both hybrid sterility and segregation distortion in rice. *Science* 337, 1336–1340 (2012).

13. Liu, P. *et al.* Predicting hybrid fertility from maker-based genetic divergence index of parental varieties: implications for utilizing inter-subspecies heterosis in hybrid rice breeding. *Euphytica* 203, 47–57 (2015).

14. Ikehashi, H. & Araki, H. Genetics of F_1 sterility in remote crosses in rice. In *Rice Genetics: Proceedings of the First Rice Genetics Symposium* (International Rice Research Institute), 119–130 (1986).

15. Song, X., Qiu, S., Xu, C., Li, X. & Zhang, Q. Genetic dissection of embryo sac fertility, pollen fertility, and their contributions to spikelet fertility of intersubspecific hybrids in rice. *Theor. Appl. Genet.* 110, 205–211 (2005).

16. Yanagihara, S. *et al.* Molecular analysis of the inheritance of the *S*-5 locus, conferring wide compatibility in *indica/japonica* hybrids of rice (*Oryza sativa* L.). *Theor. Appl. Genet.* 90, 182–188 (1995).

17. Chen, J. *et al.* A triallelic system of *S5* is a major regulator of the reproductive barrier and compatibility of *indica-japonica* hybrids in rice. *Proc. Natl Acad. Sci. USA* 105, 11436–11441 (2008).

18. Ji, Q. *et al.* Two sequence alterations, a 136 bp InDel and an A/C polymorphic site, in the *S5* locus are associated with spikelet fertility of *indica-japonica* hybrid in rice. *J. Genet. Genom.* 37, 57–68 (2010).

19. Du, H., Ouyang, Y., Zhang, C. & Zhang, Q. Complex evolution of *S5*, a major reproductive barrier regulator, in the cultivated rice*Oryza sativa* and its wild relatives. *New Phytol.* 191, 275–287 (2011).

20. Zhang, G. & Lu, Y. Genetic studies of the hybrid sterility in cultivated rice (*Oryza sativa*). I. Diallel analysis of the hybrid sterility among isogenic F_1 sterile lines (in Chinese with English abstract). *Chinese J. Rice Sci.* 3, 97–101 (1989).

21. Zhang, G. & Lu, Y. Genetic studies of the hybrid sterility in cultivated rice (*Oryza sativa*). II. A genic model for F_1 pollen sterility (in Chinese with English abstract). *Acta Genet. Sin.* 20, 222–228 (1993).

22. Zhang, G., Lu, Y., Liu, G., Yang, J. & Zhang, H. Genetic studies on the hybrid sterility in cultivated rice (*Oryza sativa*). III. Allele differentiation of F_1 pollen sterility in different types of varieties (in Chinese with English abstract). *Acta Genet. Sin.* 20, 541–551 (1993).

23. Zhang, G., Lu, Y., Zhang, H., Yang, J. & Liu, G. Genetic studies on the hybrid sterility in cultivated rice (*Oryza sativa*). IV. Genotypes for F_1 pollen sterility (in Chinese with English abstract). *Acta Genet. Sin.* 21, 34–41 (1994).

24. Zhang, G. & Lu, Y. Genetics of F_1 pollen sterility in *Oryza sativa*. In*Rice Genetics III* (International Rice Research Institute, Manila, Philippines), 418–422 (1996).

25. Zhuang, C., Zhang, G., Mei, M. & Lu, Y. Molecular mapping of the*S-a* locus for F_1 pollen sterility in cultivated rice (*Oryza sativa* L.).*Acta Genet. Sin.* 26, 213–218 (1999).

26. Su, J. & Liu, Y. Fine mapping and cloning of the gene *S-a* for F_1pollen sterility in cultivated rice (*Oryza sativa* L.). *Mol Plant Breed.*1, 757–758 (2003).

27. Long, Y. *et al.* Hybrid male sterility in rice controlled by interaction between divergent alleles of two adjacent genes. *Proc. Natl Acad. Sci. USA* 105, 18871–18876 (2008).

28. Zhuang, C., Mei, M., Zhang, G. & Lu, Y. Chromosome mapping of the *S-b* locus for F_1 pollen sterility in cultivated rice (*Oryza sativa*L.) with RAPD markers (in Chinese with English abstract). *Acta Genet. Sin.* 29, 700–705 (2002).

29. Li, W., Zeng, R., Zhang, Z., Ding, X. & Zhang, G. Fine mapping of locus *S-b* for F_1 pollen sterility in rice (*Oryza sativa* L.). *Chin. Sci. Bull.* 51, 675–680 (2006).

30. Zhang, Z. & Zhang, G. Fine mapping of the *S-c* locus and marker-assisted selection using PCR markers in rice (in Chinese with English abstract). *Acta Agronomica Sinica* 27, 704–709 (2001).

31. Yang, C., Chen, Z., Zhuang, C., Mei, M. & Liu, Y. Genetic and physical fine-mapping of the *Sc* locus conferring *indica-japonica*hybrid sterility in rice (*Oryza sativa* L.). *Chin. Sci. Bull.* 49, 1718–1721 (2004).

32. Li, W., Zeng, R., Zhang, Z., Ding, X. & Zhang, G. Identification and fine mapping of *S-d*, a new locus conferring the partial pollen sterility of intersubspecific F_1 hybrids in rice (*Oryza sativa* L.).*Theor. Appl. Genet.* 116, 915–922 (2008).

33. Zhu, W. *et al.* Preliminary identification of F_1 pollen sterility gene*S-e* in *Oryza sativa* (in Chinese with English abstract). *J. South. China Agricult. Univ.* 29, 1–5 (2008).

34. Zhao, Z. *et al.* Molecular analysis of an additional case of hybrid sterility in rice (*Oryza sativa* L.). *Planta* 233, 485–494 (2011).

35. Zhao, Z. *et al.* Fine mapping of *S31*, a gene responsible for hybrid embryo-sac abortion in rice (*Oryza sativa* L.). *Planta* 226, 1087–1096 (2007).

36. Kubo, T., Yamagata, Y., Eguchi, M. & Yoshimura, A. A novel epistatic interaction at two loci causing hybrid male sterility in an inter-subspecific cross of rice (*Oryza sativa* L.). *Genes Genet. Syst.*83, 443–453 (2008).

37. Kubo, T. & Yoshimura, A. Linkage analysis of an F_1 sterility gene in*japonica/indica* cross of rice. *Rice Genet Newslett* 18, 52–54 (2001).

38. Wen, J., Zhang, W., Jiang, L., Chen, L., Zhai, H. & Wan, J. Two novel loci for pollen sterility in hybrids between the weedy strain Ludao and the*japonica* variety Akihikari of rice (*Oryza sativa* L.). *Theor. Appl. Genet.* 114, 915–925 (2007).

39. Zhang, G. & Lu, Y. Breeding of the *indica*-compatible *japonica*lines and their use in the breeding of super-high-yield hybrid rice (in Chinese with English abstract). *Hybrid Rice* 14, 3–5 (1999).

40. Ouyang, Y., Liu, Y. & Zhang, Q. Hybrid sterility in plant: stories from rice. *Curr. Opin. Plant Biol.* 13, 186–192 (2010).

41. Oka, H. I. Analysis of genes controlling F_1 sterility in rice by the use of isogenic lines. *Genet.* 77, 521–534 (1974).

42. Dai, Z. *et al.* Development of a platform for breeding by design of CMS lines based on an SSSL library in rice (*Oryza sativa* L.).*Euphytica* 205, 63–72 (2015).

43. Murray, M. G. & Thompson, W. F. Rapid isolation of high molecular weight plant DNA. *Nucleic Acids Res.* 8, 4321–4325 (1980).

44. Panaud, O., Chen, X. & McCouch, S. R. Development of microsatellite markers and characterization of simple sequence length polymorphism (SSLP) in rice (*Oryza sativa* L.). *Mol. Gen. Genet.* 252, 597–607 (1996).

45. McCouch, S. R. Development and mapping of 2240 new SSR markers for rice (*Oryza sativa* L.). *DNA Res.* 9, 199–207 (2002).

46. Zhang, Z. *et al.* A genome-wide microsatellite polymorphism database for the *indica* and *japonica* Rice. *DNA Res.* 14, 37–45 (2007).

47. Sundaram, R. M. *et al.* Development and validation of a PCR-based functional marker system for the major wide-compatible gene locus *S5* in rice. *Mol. Breeding* 26, 719–727 (2010).

48. Coakes, S. J. & Ong, C. SPSS Version 18.0 for Windows *Analysis without anguish* (John Wiley & Sons) (2011).

49. Charland, M. B. *SigmaPlot for Scientists* (William C Brown Press) (1995).

INDEX

A

Abscisic acid (ABA) 111
Analyse the variance (ANOVA) 193

B

Brown rice rate (BRR) 91

C

Central imaging and flow cytometry facility (CIFF) 82
Coefficients of variation (CV) 138
crop growth rate (CGR) 138, 143

D

Days to flowering (DF) 52
Dicer-like (DCL) 111
discriminate function analysis (DFA) 137
Domesticated rice 25, 27, 30, 39

F

Filled grains per panicle (FG/P) 52
Flag leaf area (FLA) 52

G

Gene ontology (GO) 82
General linear model (GLM) 124, 138
Genome-wide associations (GWAS) 26
Genome-wide association studies (GWAS) 1, 15
Genotype-phenotype 31
Global food 79
Glutathione S-transferase (GST) 179
Grain appearance quality 91, 99
Grain length (GL) 52, 91
Grain length to width ratio (GLWR) 91
Grain thickness (GT) 91
Grain width (GW) 91
Green fluorescent protein (GFP) 179

H

Haplotype diversity (HD) 71
Hardy weinberg equilibrium (HWE) 143
Harvest index (HI) 47, 52
High-density rice array (HDRA) 1
Hours after imbibition (HAI) 111
Hybrid sterility 192, 195

I

Indian council of agricultural research 77
International rice research institute (IRRI) 15, 109
Inter-quartile range (IQR) 138

L

Large veins (LVs) 150
Least significant difference (LSD) 193
Linkage disequilibrium (LD) 1, 77
Low quality 82

M

Maltose-binding protein (MBP) 179
Million tonnes (MT) 69
Minimal folding free energy index (MFEI) 112

N

Next-generation sequencing (NGS) 15, 109
Nuclear-encoded plastid RNA polymerase (NEP) 159

P

Peptidyltransferase centre (PTC) 148
Percentage of grains with chalkiness (PGWC) 91
Plant height 46, 47, 48, 49, 52
Plastid-encoded RNA polymerase (PEP) 159
Polyvinylidene difluoride (PVDF) 149
Principal component analysis (PCA) 124
Principal-components analysis (PCA) 58
Principal components (PCs) 2

R

Rapid visco analyzer (RVA) 57
Recombinant inbred lines (RILs) 124

Ribosomal proteins (RPs) 147, 153
RNA-induced silencing complex (RISC) 111
Root architecture 27, 34, 35
Rural development administration (RDA) 15

S

Secondary branch length (SBL) 52
Single-nucleotide polymorphism (SNP) 57
Small veins (SVs) 150
Standard deviation (SD) 193
Standard evaluation system (SES) 70, 72

V

Vigor index (VI) 136, 138, 143

W

Wide-compatibility genes (WCGs) 193